Protein Kinase Functions

Frontiers in Molecular Biology

SERIES EDITORS

B. D. Hames

*Department of Biochemistry
and Molecular Biology
University of Leeds, Leeds LS2 9JT, UK*

D. M. Glover

*Cancer Research Laboratories,
Department of Anatomy and Physiology,
University of Dundee, Dundee DD1 4HN, UK*

TITLES IN THE SERIES

1. Human Retroviruses
Bryan R. Cullen

2. Steroid Hormone Action
Malcolm G. Parker

3. Mechanisms of Protein Folding
Roger H. Pain

4. Molecular Glycobiology
Minoru Fukuda and Ole Hindsgaul

5. Protein Kinases
Jim Woodgett

6. RNA–Protein Interactions
Kyoshi Nagai and Iain W. Mattaj

7. DNA–Protein: Structural Interactions
David M. J. Lilley

8. Mobile Genetic Elements
David J. Sherratt

9. Chromatin Structure and Gene Expression
Sarah C. R. Elgin

10. Cell Cycle Control
Chris Hutchinson and D. M. Glover

11. Molecular Immunology (Second Edition)
B. David Hames and David M. Glover

12. Eukaryotic Gene Transcription
Stephen Goodbourn

13. Molecular Biology of Parasitic Protozoa
Deborah F. Smith and Marilyn Parsons

14. Molecular Genetics of Photosynthesis
Bertil Andersson, A. Hugh Salter, and James Barber

15. Eukaryotic DNA Replication
J. Julian Blow

16. Protein Targeting
Stella M. Hurtley

17. Eukaryotic mRNA Processing
Adrian Krainer

18. Genomic Imprinting
Wolf Reik and Azim Surani

19. Oncogenes and Tumour Suppressors
Gordon Peters and Karen Vousden

20. Dynamics of Cell Division
Sharyn A. Endow and David M. Glover

21. Prokaryotic Gene Expression
Simon Baumberg

22. DNA Recombination and Repair
Paul J. Smith and Christopher J. Jones

23. Neurotransmitter Release
Hugo J. Bellen

24. GTPases
Alan Hall

25. The Cytokine Network
Fran Balkwill

26. DNA Virus Replication
Alan J. Cann

27. Biology of Phosphoinositides
Shamshad Cockcroft

28. Cell Polarity
David G. Drubin

29. Protein Kinase Functions
Jim Woodgett

Protein Kinase Functions

EDITED BY

Jim Woodgett

Ontario Cancer Institute
Princess Margaret Hospital
Toronto Canada

OXFORD

UNIVERSITY PRESS

OXFORD
UNIVERSITY PRESS

Great Clarendon Street, Oxford OX2 6DP

Oxford University Press is a department of the University of Oxford
It furthers the University's objective of excellence in research, scholarship,
and education by publishing worldwide in

Oxford New York

Athens Auckland Bangkok Bogotá Buenos Aires Calcutta
Cape Town Chennai Dar es Salaam Delhi Florence Hong Kong Istanbul
Karachi Kuala Lumpur Madrid Melbourne Mexico City Mumbai
Nairobi Paris São Paulo Singapore Taipei Tokyo Toronto Warsaw
with associated companies in
Berlin Ibadan

Oxford is a registered trade mark of Oxford University Press
in the UK and in certain other countries

Published in the United States
by Oxford University Press Inc., New York

British Library Cataloguing in Publication Data
Data available

Library of Congress Cataloging in Publication Data

Protein kinase functions / Jim Woodgett [editor].—[2nd ed.].
p. cm.—(Frontiers in molecular biology; 29)
Includes bibliographical references.
1. Protein kinases. I. Woodgett, James Robert. II. Series.
QP606.P76 P378 2000   572'.792—dc21   00-032391

1 3 5 7 9 10 8 6 4 2

ISBN 0 19 963771 7 (Hbk)
ISBN 0 19 963770 9 (Pbk)

Typeset by Footnote Graphics, Warminster, Wilts

Printed in Great Britain on acid free paper by
The Bath Press, Avon

Preface

The first edition of *Protein Kinases* was published in the Frontiers in Molecular Biology series in 1994. Since that time there have been dramatic advances in the field of signal transduction. Of course, more protein kinases have been discovered and characterized, but that was to be expected. Indeed Tony Hunter predicted 1001 protein kinases almost a decade ago and has since upgraded the estimate. The most significant progress has been in our understanding of what this remarkably diverse class of enzymes actually do within cells. Whilst contemplating a significant modernization and revision of the original book, it became clear that simply rejuvenating the topic would be a missed opportunity to reflect the recent advances in kinase biology. In addition, the original *Protein kinases* book is still timely (a testament to the forward thinking and writing style of the contributors). The trend over the past 5 years has been towards understanding the context and function of signalling molecules. As a consequence, this new book is intended largely to complement the first edition rather than act as its replacement.

The contributors to *Protein Kinase Functions* are a mix of new and previously invited authors. New topics include in-depth descriptions of the phosphatidylinositol 3' kinase, Jak-Stat, and TGFβ/activin pathways, the protein kinases involved in the response to DNA damage, and an exciting new class of kinase regulators defined by the SOCS proteins. In the first edition, discussion of the biological contexts of protein kinases was largely confined to the chapter on Genetic Approaches, which has been refreshed here. For the current volume, all authors were asked to emphasize the functional aspects of their areas resulting in 'pathway-centric' analyses. In this age of proteomic science, it's not only relevant but necessary to think about the complexity of interactions between proteins and signalling systems. Indeed, protein characterization has 'come of age', moving beyond introspective analysis into study of the relationships between processes and molecules. Our aim for this book is to assist readers to assimilate, compare, and integrate the molecular machinery used by cells to coordinate and respond to their environments. This may be a big challenge, but the penalty for not embracing the complexity of proteome-wide systems' regulation includes exclusion from the fastest growing area of biological research. Hopefully, resources such as this volume will ease the transition.

J.W.

Toronto
May 2000

Contents

List of contributors   *xv*

Abbreviations   *xvii*

**1 Phosphatidylinositol 3-kinase signalling: a tale of two
   kinase activities**   1

PAUL J. COFFER

  1. **Introduction**   1

      1.1 Identification and function of D3-phosphatidylinositol lipids   1

      1.2 Structure of PI3Ks   3

      1.3 A protein serine-kinase activity for PI3K   6

      1.4 Tools for analysing PI3K function   7

  2. **Mechanisms regulating PI3K activation**   9

      2.1 Phosphotyrosine-mediated PI3K activation   10

      2.2 Activation of PI3K by p21Ras   10

      2.3 Serpentine receptor-mediated PI3K activation   12

      2.4 SHIP and PTEN: modulating the action of PI3K lipid products   13

  3. **Downstream effectors of PI3K activation**   15

      3.1 Protein kinase B (c-akt)   15

      3.2 PtdIns(3,4,5)P$_3$-dependent kinases (PDKs)   16

      3.3 The p70 S6 kinase   18

      3.4 Protein kinase C isoforms   18

      3.5 Guanine nucleotide exchange factors   20

  4. **Functions of PI3K activation**   21

      4.1 Proliferation and oncogenesis   21

      4.2 Apoptosis   24

      4.3 Cytoskeletal reorganization   26

      4.4 Membrane trafficking   27

      4.5 Protein synthesis and regulation of metabolism   29

      4.6 Differentiation   30

  5. **Conclusions**   32

      References   33

2 Mammalian MAP kinase pathways   40

JOHN M. KYRIAKIS

1. **General considerations**   40
   1.1 The MAP3K → MEK → MAPK core signalling module, an emerging paradigm   40
2. **The Ras MAPK pathway**   42
   2.1 The Ras → MAPK pathway, a MAPK pathway in mammalian cells that is activated by insulin and mitogens—general considerations   42
   2.2 Substrates of ERK1 and -2: other protein kinases and transcription factors   44
   2.3 The regulation of ERK1 and -2 by MAPK/ERK-kinases-1 and -2   52
   2.4 MEK1 and -2 are substrates of the Raf-1 and Mos proto-oncoproteins   55
   2.5 Ras, a molecular switch that couples tyrosine kinases to Raf-1 and the ERK pathway   56
   2.6 Differential regulation of Raf-1 and B-Raf by Rap1   61
   2.7 Regulation of the ERK pathway by the scaffold protein MEK partner-1 (MP1)   63
   2.8 Regulation of ERK function by nucleocytoplasmic shuttling   63
3. **The stress-activated protein kinase (SAPK)/c-Jun-NH$_2$-terminal kinase (JNK), p38 and ERK5/big MAPK1 (BMK1) pathways: mammalian MAPK signalling pathways activated by environmental stress and inflammatory cytokines**   65
   3.1 General considerations   65
   3.2 The stress-activated protein kinases/c-Jun NH$_2$-terminal kinases (SAPKs), a family of MAPKs activated by environmental stresses and inflammatory cytokines   65
   3.3 The p38 MAPKs, a second stress-activated MAPK family   66
   3.4 ERK5/big MAP kinase-1 (BMK1), a third class of stress-activated MAPK   67
   3.5 SAPK, p38, and ERK5 substrates   68
   3.6 MEKs upstream of the SAPKs and p38s   78
   3.7 The SAPKs and p38s are activated by several divergent families of MAP3Ks   80
   3.8 Regulation of the SAPKs and p38s by Rho family GTPases   90
   3.9 Regulation of the SAPKs and p38s by heterotrimeric G proteins   95
   3.10 Regulation of the SAPKs by scaffold proteins   97
   3.11 Regulation of the SAPKs by the germinal centre kinases, a family of putative regulated scaffold kinases   100
   3.12 Regulation of the SAPKs and p38s by adapter proteins that couple to TNF family receptors   104

3.13 Coupling stress-activated MAP3Ks to upstream signals   109

3.14 Biological functions of the SAPKs and p38s pathways   125

4. **Concluding remarks**   133

4.1 Oligomerization, adapter/G protein binding and translocation as themes in MAP3K regulation   134

4.2 MAPK pathway biology   134

References   135

3. Genetic approaches to protein kinase function in lower eukaryotes   157

SIMON E. PLYTE

1. **Introduction**   157

2. **Genetic manipulation in *Drosophila***   158

2.1 Signal transduction from the Sevenless receptor   158

3. **Vulval development in *Caenorhabditis elegans***   164

3.1 Vulval induction   164

3.2 Lateral induction/inhibition   168

4. **Signal transduction in slime mould**   169

4.1 Identification of protein kinases   171

4.2 The role of cAMP-dependent protein kinase   171

4.3 Glycogen synthase kinase-3   173

4.4 Restriction enzyme mediated integration (REMI)   174

5. **Serine/threonine kinases in yeast**   174

5.1 The yeast mating response   176

5.2 Signalling cross-talk   179

5.3 Yeast interaction trap assays   180

6. **Concluding remarks**   183

References   183

4 Specificity in JAK-STAT signalling   194

ADRIENNE HALUPA and DWAYNE L. BARBER

1. **Introduction**   194

2. **Jak-Stat signalling—a general paradigm**   195

3. **Gene-targeting studies**   197

3.1 JAK1   197

3.2   JAK2   200

3.3   JAK3   203

3.4   Tyk2   206

3.5   Stat1   206

3.6   Stat2   208

3.7   Stat3   209

3.8   Stat4   211

3.9   Stat6   212

3.10 Stat4/Stat6 (−/−)   214

3.11 Stat5   215

4.   The Jak–Stat pathway and development   219

4.1   *Drosophila*   220

4.2   *C. elegans*   221

4.3   *Dictyostelium*   221

5.   Involvement of the Jak–Stat pathway in disease   222

5.1   Hopscotch   222

5.2   v-src   223

5.3   v-abl   223

5.4   HTLV-1   223

5.5   STAT activation in leukaemia   224

5.6   BCR–ABL   224

5.7   TEL–JAK2   225

5.8   JAK3 and SCIDS   227

5.9   Models of constitutive activation   229

6.   Future directions   230

References   232

5   On the road to destruction: suppression of protein tyrosine kinase signalling by SOCS family proteins   246

ROBERT ROTTAPEL, SUBBURAJ ILANGUMARAN, and PAULO DE SEPULVEDA

1.   Introduction   246

2.   Discovery   247

3.   Nomenclature   247

4.   Genomic organization and domain topography of SOCS family members   248

5.   Tissue expression and induction of SOCS family members   249

6. **Structural aspects of SOCS-1 required for signal suppression**   251
    6.1   SOCS-1 is a pseudosubstrate inhibitor of JAK kinases   251
    6.2   Receptor tyrosine kinase signalling is modulated by SOCS-1   254

7. **SOCS-1 binds to multiple signal transduction molecules**   255
    7.1   SOCS-1 binds to SH3-containing proteins   255
    7.2   VAV is a SOCS-1 binding partner   255
    7.3   SOCS-1 suppresses VAV function and targets VAV to the ubiquitin proteasome pathway   256

8. **The SOCS box couples SOCS family proteins to the Elongin B/C ubiquitin ligase complex**   256
    8.1   Elongin B/C is a putative ubiquitin ligase   257
    8.2   Molecular structure of the VCB complex   259
    8.3   Problems with the SOCS–Elongin B/C hypothesis   260

9. **Biology**   261
    9.1   CIS   261
    9.2   SOCS-1 deficient mice   262
    9.3   SOCS-2   264
    9.4   SOCS-3   265

10. **Conclusions and perspectives**   267
References   268

6   Cyclin-dependent protein kinases   277

KATHLEEN L. GOULD

1. **Introduction**   277
2. **Discovery of Cdks**   277
    2.1   Nomenclature   278
    2.2   A brief history of Cdc2   278
    2.3   A family of Cdks   279

3. **Cyclins**   280
    3.1   A proliferation of cyclins   281
    3.2   Specificities of interactions   281

4. **Regulation**   282
    4.1   Transcriptional   282
    4.2   Phosphorylation   283
    4.3   Inhibitors   286
    4.4   Regulated cyclin proteolysis   287

5. **Interaction with Suc1p/Csk1p**   289

6. Structure   289

7. Localization   290

8. Substrates   291

9. Conclusions   294

References   294

7. Mechanisms and biology of signalling by serine/threonine kinase receptors form the TGFβ superfamily   303

MICHAEL KLÜPPEL, PAMELA A. HOODLESS, JEFFREY L. WRANA, and LILIANA ATTISANO

1. Overview   303

2. The ligands   303

    2.1 The TGFβ superfamily of ligands   303

    2.2 Structure of the TGFβ superfamily members   304

    2.3 Interaction of the TGFβ superfamily ligands with soluble proteins   305

3. The receptors   307

    3.1 The TGFβ superfamily signals through a heteromeric serine/threonine kinase receptor complex   307

    3.2 Structural requirements in type-II receptors   307

    3.3 Structural requirements in type-I receptors   309

    3.4 Betaglycan and endoglin   310

4. The signalling pathway   310

    4.1 Receptor-interacting proteins   310

    4.2 The Smad family   313

5. The biology of TGFβ signalling   323

    5.1 Mutations in TGFβ signalling components in mice   324

    5.2 Gastrulation   325

    5.3 Somite patterning   336

6. TGFβ signalling and disease   336

    6.1 Disrupted TGFβ signalling in cancer   337

    6.2 Hereditary diseases   338

7. Conclusions and future perspectives   339

Acknowledgements   340

References   340

8. The ATM and DNA-PK proteins: the sensing, signalling, and repair of DNA damage 357

ROBERT BRISTOW, HANS BLUYSSEN, ANNELIES DE KLEIN, DIK VAN GENT

1. **Overview: DNA damage responses** 357
   1.1 DNA repair 357
   1.2 Cell-cycle checkpoints and genomic stability 358
   1.3 Evolutionary conservation and biochemical function within the PI3K-related kinase family 358
2. **DNA-dependent protein kinase (DNA-PK)** 360
   2.1 Biochemistry of DNA binding and phosphorylation by DNA-PK 360
   2.2 Protein–protein interactions 362
   2.3 Other functions of DNA-PK 363
3. **The cell-cycle checkpoint gene *ATM*** 364
   3.1 Basic biochemistry and phosphorylation substrates of ATM 364
   3.2 ATM in DNA repair and cell-cycle checkpoint control 365
4. **Other PI3K family members** 366
   4.1 ATR 366
   4.1 TRAPP and mTOR 366
5. **DNA-PK- and ATM-deficient mice as models for human disease** 367
   5.1 Mutations in DNA-PK 367
   5.2 Mutations in ATM 369
   5.3 ATM and DNA-PK protein function in cancer prognosis and therapy 370
6. **Conclusions and outstanding questions** 370
   References 371

Index 378

Contributors

LILIANA ATTISANO
Department of Anatomy and Cell Biology, University of Toronto, Toronto, Ontario, Canada M5S 1A8.

DWAYNE L. BARBER
Departments of Medical Biophysics and Laboratory Medicine and Pathobiology, University of Toronto; Department of Laboratory Medicine and Pathobiology, Toronto General Hospital; Division of Cellular and Molecular Biology, Ontario Cancer Institute, 610 University Avenue, Toronto, Ontario, Canada M5G 2M9.

HANS BLUYSSEN
Department of Cell Biology and Genetics, Faculty of Medicine, Erasmus University Rotterdam, Dr. Molewaterplein 50, 3015 GE Rotterdam, The Netherlands.

ROBERT BRISTOW
Department of Radiation Oncology and Division of Experimental Therapeutics, Ontario Cancer Institute/Princess Margaret Hospital, 610 University Avenue, Toronto, Ontario, Canada M5G 2M9.

PAUL J. COFFER
Department of Pulmonary Diseases, G03.550, University Hospital Utrecht, Heidelberglaan 100, 3584 CX Utrecht, The Netherlands.

ANNELIES DE KLEIN
Department of Cell Biology and Genetics, Faculty of Medicine, Erasmus University Rotterdam, Dr. Molewaterplein 50, 3015 GE Rotterdam, The Netherlands.

PAULO DE SEPULVEDA
Departments of Medicine, Immunology, and Medical Biophysics, University of Toronto, Ontario Cancer Institute/Princess Margaret Hospital 610 University Avenue, Toronto, Ontario, Canada M5G 2M9.

KATHLEEN L. GOULD
HHMI and Department of Cell Biology, Vanderbilt University School of Medicine, Nashville, Tennessee 37232, USA.

ADRIENNE HALUPA
Department of Medical Biophysics, Division of Cellular and Molecular Biology, Ontario Cancer Institute, 610 University Avenue, Toronto, Ontario, Canada M5G 2M9.

PAMELA A. HOODLESS
The Jacob and Wolf Lebovich Center for Molecular Medicine at the Samuel Lunenfeld Research Institute, Mount Sinai Hospital, 600 University Avenue, Toronto, Ontario, Canada M5G 1X5.

SUBBURAJ ILANGUMARAN
Departments of Medicine, Immunology, and Medical Biophysics, University of Toronto, Ontario Cancer Institute/Princess Margaret Hospital, 610 University Avenue, Toronto, Ontario, Canada M5G 2M9.

MICHAEL KLÜPPEL
The Jacob and Wolf Lebovich Center for Molecular Medicine at the Samuel Lunenfeld Research Institute, Mount Sinai Hospital, 600 University Avenue, Toronto, Ontario, Canada M5G 1X5.

JOHN M. KYRIAKIS
Diabetes Research Laboratory, Massachusetts General Hospital East, 149 13th Street, Charlestown, MA 02129, USA.

SIMON E. PLYTE
Pharmacia and Upjohn, Via per Pogliano, 20014, Nerviano, Italy.

ROBERT ROTTAPEL
Departments of Medicine, Immunology, and Medical Biophysics, University of Toronto, Ontario Cancer Institute/Princess Margaret Hospital 610 University Avenue, Toronto, Ontario, Canada M5G 2M9.

DIK VAN GENT
Department of Cell Biology and Genetics, Faculty of Medicine, Erasmus University Rotterdam, Dr. Molewaterplein 50, 3015 GE Rotterdam, The Netherlands.

JEFFREY L. WRANA
The Jacob and Wolf Lebovich Center for Molecular Medicine at the Samuel Lunenfeld Research Institute, Mount Sinai Hospital, 600 University Avenue, Toronto, Ontario, Canada M5G 1X5.

Abbreviations

20αSDH	20α-hydroxysteroid dehydrogenase
4E-BP	translational repressor protein, 4E binding protein (see PHAS-I)
ABP	actin binding protein
ALL	acute lymphoblastic leukaemia
AML	acute myeloid leukaemia
ANF	atrial natriuretic factor
ASB	ankyrin SOCS box
AP	activator protein
APC/C	anaphase-promoting complex/cyclosome
APCs	antigen-presenting cells
APRF	acute-phase reactive factor
AraC	cytosine arabinoside
ARE	activin response element
ARF	activin response factor
ARK	adrenergic receptor kinase
ASK	apoptosis signalling kinase
ASV	avian sarcoma virus
ATF	activating transcription factor
ATM	mutated in ataxia telangiectasia
ATR	ATM and Rad3-related protein
Bcr	breakpoint cluster region
BFU	blast-forming unit
BH	Bcr-homology
BIR	baculoviral IAP repeat
BMK	big MAPK
BMP	bone morphogenic protein
BMT	bone-marrow transplant
bp	base pair
BP-1	binding protein-1
BrdU	bromodeoxyuridine
Btk	Bruton's Tyr kinase
C/EBP	CCAAT/enhancer binding protein
CAK	Cdk activating kinase
CaN	calcineurin/PP2B
cAR	cAMP receptor
CDC	cell division cycle
CDDP	*cis*-platinum
Cdk	cyclin-dependent kinase

CDMP cartilage-derived morphogenetic protein
CEF chicken embryo fibroblast
CFU colony-forming unit
CFU-E erythroid colony-forming unit
CH CIS homology
CHOP CREB homologous protein
CIITA MHC class-II *trans* activating protein
CIS cytokine-induced SH2 domain
CKI CDK inhibitor
CML chronic myelogenous leukaemia
CMML chronic monomyelocytic leukaemia
CNTF ciliary neurotrophic factor
ConA concanavalin A
CR conserved region
CRD cysteine-rich, zinc-finger domain
CRE cyclic AMP response element
CREB cyclic AMP response-element binding protein
CRM cross-reacting material
CRIB Cdc42/Rac interaction and binding
cs catalytic subunit
CSAID cytokine-suppressive anti-inflammatory drug
CSBP CSAID binding protein
CSF colony-stimulating factor
CTD C-terminal domain
Cul Cullin
DAD daughters against dpp
DAG diacylglycerol
DBD DNA binding domain
DBP DNA binding protein
DEP Dsh/Egl-10/pleckstrin domain
DER *Drosophila* EGF receptor
DH Dbl homology (domain)
DIF differentiation inducing factor
DIX dishevelled/axin domain
DLK dual leucine zipper kinase (aka MUK or ZPK)
DNA-PK DNA-dependent protein kinase
DPP decapentaplegic
DR death receptor
dsb double-strand break
DTH delayed-type hypersensitivity
E embryonic day
EBV Epstein–Barr virus
ECB early cell-cycle box
EGF epidermal growth factor

EKLF	erythroid Kruppel-like factor
EMSA	electrophoretic mobility shift assay
EPO	erythropoietin
EPOR	erythropoietin receptor
ERK	extracellular signal-regulated protein kinase
ES	embryonic stem (cell)
EST	expressed sequence tag
FADD	Fas receptor-associated death domain
FAK	focal adhesion kinase
FAST	forkhead activin signal transducer
FGD	factigenital dysplasia
FGF	fibroblast growth factor
FKBP	FK506 binding protein
fMLP	formylMet-Leu-Phe
FT	farnesyl transferase
G-CSF	granulocyte colony-stimulating factor
GADD	growth arrest and DNA damage
GAP	GTPase activating protein
GAS	interferon γ-activated sequence
GBP	guanylate binding protein
GCK	germinal centre kinase
GCKR	germinal kinase-related kinase (see KHS)
GEF	GTPase exchange factor
GH	growth hormone
GLK	GC-like kinase
GM-CSF	granulocyte/macrophage colony-stimulating factor
GPCR	G protein-coupled receptor
GS	glycogen synthase (or glycine–serine)
GSK	glycogen synthase kinase
HDPK	homeodomain-interacting protein kinase
HECT	homologous to E6-AP C-terminus
HHT	hereditary haemorrhagic telangiectasia
HIF	hypoxia-inducible factor
HLH	helix–loop–helix
HNPCC	hereditary non-polyposis colon cancer
HPK	haemopoietic precursor (or progenitor) kinase
HSP	heat-shock protein
IAP	inhibitor of apoptosis protein
ICAM	intercellular adhesion molecule
ICE	interleukin-1 converting enzyme
IF	initiation factor
IFN	interferon
IFNAR	interferon-α receptor
IFNGR	interferon-γ receptor

IGF	insulin-like growth factor
ILK	integrin-linked kinase
INOS	inducible nitric oxide synthase
INSR	insulin receptor
IPTG	isopropyl β-D-thiogalactoside
IRAK	IL-1 receptor-associated kinase
IRF	interferon regulated factor
IRS	insulin receptor substrate
ISGF	interferon-stimulated gene factor
ISRE	interferon-stimulated response element
JAK	Janus kinase (just another kinase)
JH	JAK homology
JIP	JNK-interacting protein
JNK	Jun N-terminal kinase
KHS	kinase homologous to Ste20 (see GCKR)
Ksr	kinase suppressor of ras
LAP	latency associated protein
LF	lethal factor (anthrax)
LIF	leukaemia inhibitory factor
LMP	latent membrane protein
LOH	loss of heterozygosity
LPA	lysophosphatidic acid
LPS	lipopolysaccharide
LT	lymphotoxin
LTBP	latent TGFβ binding protein
M-CSF	macrophage colony-stimulating factor
MAD	mothers against *dpp*
MADS	MCM1-agamous and deficiens-SRF
MAP2	microtubule-associated protein 2
MAPK	mitogen-activated protein kinase
MAPKAP	mitogen-activated protein kinase-activating kinase
MBP	myelin basic protein
MEF	myocyte enhancing factor
MEK	MAPK and ERK kinase
MEKK	MEK kinase
MGF	mammary gland factor
MH	MAD homology
MHC	major histocompatibility complex
MHV	mouse hepatitis virus
MIS	Müllerian inhibiting substance
MKK	MAPK kinase
MLK	mixed lineage kinase
MMS	methyl methanesulfonate
MNK	MAPK interacting kinase

MP	MEK partner
MPF	maturation promotion factor
MPSV	myeloproliferative sarcoma virus
MSK	mitogen- and stress-activated kinases
MST	MKN28 cell-derived Ser/Thr kinase (aka MLK2)
MTK1	MAP three kinase-1 (aka MEKK4)
mTOR	mammalian target of ropamycin
MUK	MAPK upstream kinase
MUP	major urinary protein
NckIK	Nck interacting kinase (NIK)
NES	nuclear export signal
NF-κB	nuclear factor κB
NFAT	nuclear factor of activated T cells
NGF	nerve growth factor
NIK	NF-κB-inducing kinase
NK	natural killer (cell)
NLK	nemo-like kinase
NMR	nuclear magnetic resonance
NO	nitric oxide
NTR	neurotrophin receptor
OPG	osteoprotegerin
OSM	oncostatin M
PAI	plasminogen activator inhibitor
PAK	p21 activated kinase
PCR	polymerase chain reaction
PBL	peripheral blood leucocytes
PDGF	platelet-derived growth factor
PDK	3'-phosphoinositide-dependent kinase
PDZ	PSD95/disks large/Z0–1 domain
PEST	proline/glutamic acid/serine/threonine
PH	pleckstrin homology
PHAS-I	phosphorylated, heat- and acid-stable protein regulated by insulin (see 4E-BP)
PI	phosphoinositide
PI3K	phosphatidylinositol 3-kinase
PI4K	phosphatidylinositol 4-kinase
PIAS	protein inhibitor of activated STAT
PIK	phosphoinositide kinase
PIX	PAK interacting exchange factor
PKA	cyclic AMP-dependent protein kinase
PKB	protein kinase B
PKC	protein kinase C
PLC	phospholipase-C
PMA	phorbol myristate acetate (see TPA)
PMDS	persistent Müllerian duct syndrome

POMC	pro-opiomelanocortin
POSH	plenty of SH3 domains
PP1-G	G-subunit of phosphatase-1
PRAK	p38-regulated/activated kinase
PRS	proline-rich sequence
PtdIns	phosphatidylinositol
PTK	protein tyrosine kinase
R	receptor
RANK	receptor activator of NF-κB (aka OPG or TRANCE)
Rb	retinoblastoma
RBD	ras binding domain
Rbx	Ring box
REMI	restriction-enzyme mediated integration
RGD	arginine–glycine–aspartic acid
RGS	regulators of G-protein signalling
RIP	receptor interacting protein
Roc	regulator of cullin
ROI	reactive oxygen intermediates
RSS	recombination signal sequences
RTK	receptor tyrosine kinase
SAGA	histone acetyltransferase complex consisting of Spt, Ada and Gen5 proteins
SAPK	stress-activated protein kinase
SARA	SMAD anchor for receptor activation
SBE	SMAD binding element
SCF	Skp1–Cdc53/CUL1-F box
SCF	stem-cell factor
SCID	severe combined immune deficiency
SEK	SAPK and ERK kinase
Ser	serine
SH	src homology
Shh	sonic hedgehog
SHIP	SH2-domain-containing inositol 5-phosphatase
sIgM	surface immunoglobulin M
SLC	small latent complex
SMAD	similar to mothers against decapentaplegic
SOCS	suppressor of cytokine signalling
SPB	spindle pole body
SPC	subtilisin-like proprotein convertase
SPRK	SH3 domain-containing proline-rich kinase (aka MLK3 or PTK1)
SSB	SPRY SOCS box
SRE	serum response element
SRF	serum response factor
SSI	STAT-induced STAT inhibitor
STAT	signal transducers and activators of transcription

STRAP	serine–threonine kinase receptor-associated protein
SWiP	SOCS box and WD-repeats in protein
SYM	symphalangism
SYNS	multiple synostoses syndrome
TAB	TAK1 binding protein
TAD	transactivation domain
TAK	TGFβ-activated kinase
TAO	Thousand and one
TARE	TGFβ/activin-responsive region
TβRI/II	TGFβ type-I/-II receptor
TCF	ternary complex factor
TCR	T-cell receptor
TGF	transforming growth factor
T_H	T-helper (cell)
Thr	threonine
TNF	tumour necrosis factor
TOR	target of rapamycin
TPA	tetradecanoyl phorbol myristate acetate (see PMA)
Tpl	tumour progression locus
TPO	thrombopoietin
TRADD	TNF receptor-associated death domain
TRAF	TNF receptor-associated factor
TRAILR	TNF-related, apoptosis-inducing ligand receptor
TRANCE	TNF-related, activation-induced cytokine
TRAP	TGFβ receptor-associated protein
TRAPP	transformation/transcription domain-associated protein
TRCP	transducing repeat-containing protein
TRE	TPA response element
TRF	TGFβ/activin response factor
TRIP	TGFβ-receptor interacting protein
Trx	thioredoxin
TUNEL	TdT-mediated dUTP-X nick end labelling
Tyr	tyrosine
UBC	ubiquitin conjugating (enzyme)
UTR	Untranslated region
VCB	VHL-Elongin C/Elongin B
VDR	vitamin D receptor
VHL	von Hippel-Lindau
VPC	vulval precursor cell
VSV	vesicular stomatitis virus
WAP	whey acidic protein
WSB	WD-40 SOCS box
XIAP	X-chromosome-linked inhibitor of apoptosis protein
XNR	*Xenopus* nodal-related factor

ZPA   zone of polarizing activity
ZPK   zipper-containing kinase

1 | Phosphatidylinositol 3-kinase signalling: a tale of two kinase activities

PAUL J. COFFER

1. Introduction

Over the last 25 years considerable emphasis has been placed on the phosphorylation of proteins on tyrosine and serine/threonine residues as a mechanism of regulating cellular function. The importance of protein phosphorylation is highlighted in the other chapters of this book. However, there is another class of kinases that has recently taken centre stage, those kinases that phosphorylate phosphoinositide (PI) lipids. In the original concept of membrane structure, it was thought that the membrane proteins were solely responsible for regulating membrane functions. More recently however, membrane lipids have been found to play not only a structural role, by isolating the intracellular environment from the extracellular, but also a critical role in the generation of second messengers. For example, the hydrolysis of phosphatidyl-inositol (4,5)-bisphosphate (PtdIns(4,5)P_2) to diacylglycerol (DAG) and inositol(1,4,5)P_3 results in the release of Ca^{2+} from intracellular stores and the activation of protein kinase C (PKC). While the significance of PtdIns(4,5)P_2 hydrolysis has long been appreciated, the role of PI-lipid phosphorylation is only now beginning to be fully understood. The reason for the relatively slow progress in this field of research is that, unlike protein kinases that were first cloned in the 1980s, the first PI-lipid kinase, phosphatidylinositol 3-kinase (PI3K), was only cloned in 1992. This chapter will focus on recent discoveries concerning the structure and regulation of PI3K and its role in cellular functioning.

1.1 Identification and function of D3-phosphatidylinositol lipids

Phosphatidylinositol is unique among the lipids in that it can undergo reversible phosphorylation at multiple sites to generate five different phosphoinositides (PIs).

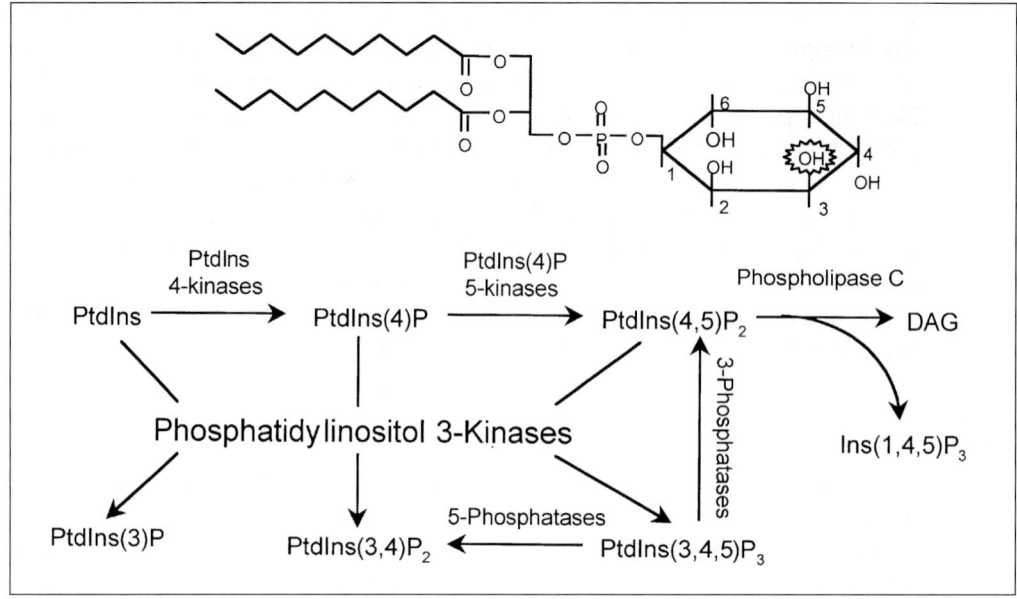

Fig. 1 Pathways for phosphoinositide synthesis. Above, the structure of phosphatidylinositol is shown with the D3-inositol position highlighted. Below, the sequence of phosphorylation reactions that generate the different phosphorylated isomers of PtdIns.

The pathways for phosphoinositide synthesis are schematically represented in Fig. 1. Identification of an activity capable of phosphorylating the D3 position of phosphatidylinositol to produce phosphatidylinositol 3-phosphate (PtdIns(3)P) was first detected in complex with viral Src and the middle-T antigen of the polyoma virus around a decade ago. This interesting observation was quickly followed by the identification of a novel tetrakisphosphate-containing phospholipid, PtdIns(3,4,5)P₃, in activated human neutrophils. Subsequent work identified a number of 3-phosphorylated lipids in the cell, and their role as second messengers was implied since they rapidly accumulated after cellular stimulation with a wide variety of agonists.

While early studies demonstrated the presence of several D3-phospholipids in the cell, it was subsequently shown that cellular responses can be associated with the selective production of individual PIs. During mitogenic stimulation, for example, there is a rapid accumulation of PtdIns(3,4,5)P₃ followed by a delayed increase in PtdIns(3,4)P₂ (1). The rise in PtdIns(3,4,5)P₃ is transient and the duration is both stimulus- and cell type-specific, ranging from between seconds and hours. PtdIns (3,4,5)P₃ unlike other PI-lipids is not degraded by phospholipases but by the action of specific regulated phosphatases. It is the regulation of these phosphatases that determines the duration of PtdIns(3,4,5)P₃ signalling and this is discussed below (Section 2.4).

PtdIns(3,4,5)P₃ can directly modulate downstream signalling events as shown by the addition of membrane-permeable forms to cells. While it has been postulated that PtdIns(3,4,5)P₃ could directly modulate vesicle budding and fusion by altering membrane curvature due to its high charge density (2), it is interaction with down-stream signalling molecules that is thought to modulate function. Phosphoinositides have been demonstrated to mediate protein–lipid interactions through the recently identified pleckstrin-homology or PH-domain (3). These structural modules with high sequence variability were first described in 1993 and there are now over 100 proteins that have been shown to contain this domain. The general model that has emerged from many studies is that PH-domains function as signal-dependent membrane adapters, allowing the translocation of cytosolic proteins to the membrane after cellular stimulation. This binding may result in alterations of the target proteins' catalytic activity, conformational change resulting in secondary phosphorylation or simply co-localization of signalling complexes.

1.2 Structure of PI3Ks

A 110-kDa kinase with a D3-phosphorylating activity was initially cloned and characterized in 1992 (4). Many isoforms of PI3K have now been identified and all have highly homologous catalytic domains (Fig. 2). Analysis of catalytic domain structure reveals some identity with phosphatidylinositol 4-kinases (PI4Ks), suggesting a structural integrity necessary for lipid phosphorylation in general. Interestingly, the catalytic domains of PI3K have also been shown to have homology with a series

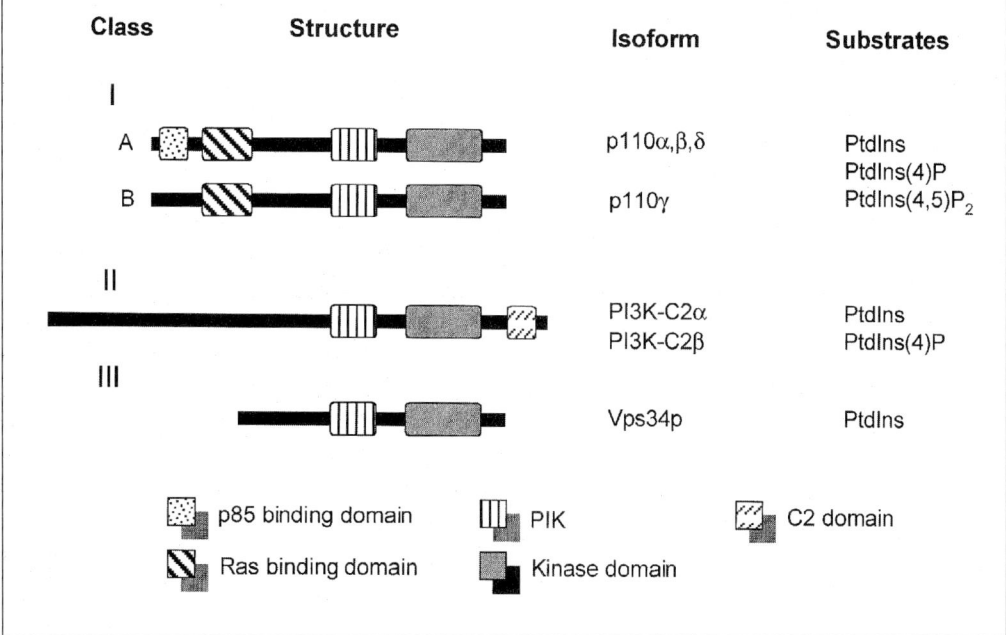

Fig. 2 Classification of phosphatidylinositol 3-kinase family members. A schematic representation of PI3K catalytic subunits showing key structural motifs. The specific lipid substrates of these kinases is also listed (far right). PIK, phosphoinositide kinase homology domain.

of protein kinases, with functional consequences that will be discussed later (Section 1.3).

Diversity in function between isoforms can occur as a result of variation in PI3K substrate specificity, which determines the range of lipid products generated and results in selectivity in downstream signalling events. Additionally, differential control of PI3K activation can occur through the presence of an intrinsic regulatory domain or extrinsic adapter subunit. Indeed, both these mechanisms have been found to occur. While having homologous catalytic domains, PI3K isoforms have recently been found to have both heterogeneous specificity's for lipid substrates as well as diverse regulatory subunits. This was first observed following the characterization of a bovine PI3K that was a heterodimer of an 85-kDa (p85) and a 110-kDa (p110) protein (5). The cloning, expression, and biochemical characterization of p85 showed that it serves as an adapter, mediating the selective association of this enzyme with phosphorylated tyrosine residues via its two Src-homology (SH2) domains (6).

The cloning of several catalytic and regulatory subunits has resulted in the classification of three major classes of PI3K as shown in Fig. 2. The class I PI3Ks include p110α, p110β, p110δ, and p110γ, which are by far the best characterized and widely expressed. They all phosphorylate PtdIns, PtdIns(4)P, and PtdIns(4,5)P$_2$ *in vitro*, although the preferred substrate *in vivo* is thought to be PtdIns(4,5)P$_2$. All class I PI3Ks are able to associate with active, GTP-loaded p21ras, a small GTPase that can regulate many cellular functions. Class I PI3Ks can be further subdivided into two groups according to the associated subunits. The 'classical' PI3Ks are p110α/β. These exist as heterodimers that consist of a regulatory adapter-subunit which constitutively associates with the catalytic subunit (4, 6, 7). The associated adapter molecules can also be subdivided into various classes containing homologous isoforms (Fig. 3). All contain two SH2-domains and an inter-SH2 p110-binding region. The class IA adapter subunits are encoded by at least three genes generating homologous products. Two isoforms generate products of 85 kDa termed p85α and p85β. Both contain domains N-terminal to the SH2-domains: an SH3-domain, a Bcr-homology (BH) domain, and two proline-rich regions. SH3-domains interact with specific proline-rich regions, and intramolecular interactions between the SH3-domain and proline-rich regions of p85α may occur that function in an autoregulatory manner. Similarly, the proline-rich regions of p85 have also been shown to interact with other proteins such as the SH3-domains of Src kinases (8). These interactions may result in the generation of specific tertiary signalling complexes, depending on cell-type and stimulus, resulting in the activation of specific downstream signalling events. While the specific role of SH3 and proline-rich domains *in vivo* is still unclear, the function of the BH domain is almost completely uncharacterized. Two truncated isoforms of p85α have been identified termed p55α and p50α. These lack the SH3-, proline-rich-, and BH-domains and are preferentially expressed in brain and muscle (9). Furthermore, a third adapter subunit termed p55γ has also been described, although, so far, the functional significance of these truncated p85-isoforms remains unclear.

In 1995, a novel p110-related PI3K termed p110γ was cloned that was found to be activated by the G-protein subunits of activated serpentine receptors (10). The p110γ

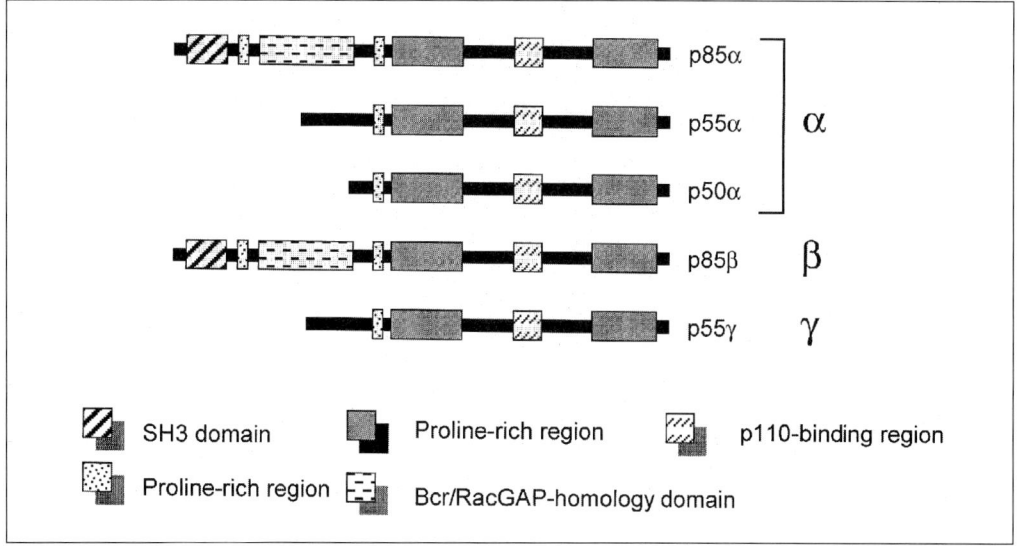

Fig. 3 Adapter subunits for class Ia phosphatidylinositol 3-kinases. The p85-related adapter subunits demonstrate significant structural homology. p50α and p55α are splice variants of p85, while p85β and p55γ are encoded by unique genes.

catalytic subunit is similar to p110α/p110β in the C-terminal kinase domain, but does not have any homology in the N-terminal p85-binding domain. Indeed, while the catalytic subunits of class IA share homology with those of the G protein-regulated class IB, their mechanisms of activation are distinct as discussed below (Section 2). The p110γ subunit is tightly associated with a novel adapter molecule termed p101 that greatly increases the ability of the Gα-subunits of heterotrimeric G-proteins to stimulate PtdIns(3,4)P$_2$ and PtdIns(3,4,5)P$_3$ accumulation (11). This regulatory subunit has no homology with the classical adapter molecules represented in Fig. 3 and appears to modulate the Gβγ sensitivity of p110γ (see below). The recently cloned p110δ is even less well characterized and is predominantly expressed in leukocytes (12). p110δ harbours a unique, potential leucine-zipper protein–protein interaction domain, suggesting a potentially novel method of regulation.

The class II PI3Ks have both N- and C-terminal extensions compared with the class I enzymes, resulting in a larger molecular weight (> 200 kDa). Little is known concerning the regulatory mechanisms activating this class of PI3Ks, although several isoforms have now been cloned. All class II enzymes contain a C-terminal C2-domain, which may allow specific regulation. C2-domains comprise approximately 130 residues and were first identified in protein kinase C. Around 100 C2-domain-containing proteins have now been described, most of which appear to function in signal transduction or membrane trafficking. These protein modules have been implicated in Ca^{2+}- and phospholipid-binding and, uniquely, Ca^{2+} can itself regulate phospholipid-binding to many C2-domains. The lipid specificity of this class appears to be restricted to PtdIns and PtdIns(4)P *in vitro* (13). So far it is unclear whether this group of PI3Ks can be regulated by extracellular stimuli, although the presence of a C2-like domain hints at a distinct regulatory mechanism and function.

The third class of PI3Ks are closely related to the product of the *S. cerevisiae VPS34p* gene originally implicated in vesicle-mediated membrane trafficking in yeast. Yeast strains deleted in the *VPS34* gene or carrying Vps34p point mutations were found to lack detectable PI3K activity and to exhibit severe defects in the trafficking of newly formed proteins from the Golgi apparatus to the vacuole, the equivalent of the mammalian lysosome (14). These PI3Ks only phosphorylate PtdIns, a phospholipid whose amounts do not vary after cellular stimulation, therefore suggesting some sort of housekeeping role. Members of this class, similarly to class I, also occur as heterodimers. Yeast Vsp34p is found in complex with Vps15p, a 170-kDa serine/threonine kinase which both activates and recruits Vsp34p to membranes. Similarly, human Vps34p associates with a 150-kDa Vps15p homologue, and it is thought that class III PI3Ks fulfil a role in constitutive membrane trafficking (15, 16). Again, similar to the class II PI3Ks, there is still very little known concerning the regulation of this class of PI3Ks in mammalian cells.

1.3 A protein serine-kinase activity for PI3K

PI3K is, by virtue of its name, a lipid kinase and the generation of D3-phosphorylated lipids by this kinase family is critical in cellular functioning. However, many laboratories have observed a protein-serine kinase activity associated with PI3K isolated from a variety of sources. Shortly after the cloning and characterization of the p110 catalytic subunit a surprising observation was made independently by two groups (17, 18). An *intrinsic* protein serine-kinase activity of p110α was identified that phosphorylates the p85 regulatory adapter molecule. This protein-kinase activity was detectable upon binding of p110 to the p85 subunit. Phosphopeptide mapping also revealed this serine, serine-608 in p85, to be phosphorylated *in vivo* and that phosphorylation resulted in an 80% decrease in PI3K activity. Subsequently, the yeast Vps34 was found to autophosphorylate on serine, threonine, and tyrosine *in vitro* in the absence of lipid substrate (19). To further complicate matters, p110δ does not phosphorylate p85 but does have an intrinsic autophosphorylation activity whose function is so far unclear (12).

PI3K γ (p110γ) has also been demonstrated to have an intrinsic protein kinase activity (10). Although a functional role for this activity in phosphorylating exogenous substrates has not been clearly defined, a recent report suggests that it is relevant. Ingeniously, a p110γ mutant was engineered so that it retained protein kinase activity but lost lipid kinase activity (20). However, it was found that transiently expressed p110γ mutants were still able to activate a specific subset of downstream signalling events, thus suggesting a critical role for this protein kinase activity.

These data suggest that PI3K is a dual-specificity enzyme, which can autoregulate its own activity. Analysis of the PI3K structure reveals that it does indeed contain sequence motifs characteristic of classical protein kinases. Mutations in these regions result in reductions of both lipid and protein kinase activities (18). Conceptually it is surprising that the catalytic site of p110 can accommodate and phosphorylate both a hydroxyl on an inositol ring linked to a phospholipid as well as a hydroxyl on a serine residue in a peptide backbone. It may simply be through evolutionary diversification that the protein kinase domain has been modified to accommodate inositol lipid products, and that the protein kinase activity observed is simply a residual artefact of this evolutionary process. This will not be resolved until distinct targets of PI3K protein kinase activity are unmasked.

1.4 Tools for analysing PI3K function

The ability to link the activation of PI3K to specific cellular functions has been greatly aided by the use of two distinct pharmacological inhibitors. The most widely used inhibitor is wortmannin, a fungal metabolite derived from *Penicillium wortmannii*, whose use in the inhibition of diverse cell functions preceded its discovery as an inhibitor of PI3Ks (Table 1; Fig. 4). Wortmannin inhibits PI3K at nanomolar concentrations by covalent modification of lysine-802 of the catalytic subunit (21). The highly efficient inhibition observed is because wortmannin not only blocks substrate-binding but also alkylates the lysine residue, whose nucleophilic nature is critical for the catalytic process. Recent evidence has underlined the problems of interpreting results obtained using wortmannin as an inhibitor (22). It appears that wortmannin can also potently inhibit two other kinases, PtdIns 4-kinase and phospholipase A_2. Wortmannin inhibition of PtdIns 4-kinase may prevent the formation of PtdIns$(4,5)P_2$, a substrate for phospholipase C, and thus effect signalling independently of PI3K (see Fig. 1). Similarly, inhibition of phospholipase A_2 will have consequences on processes requiring arachidonic acid production. Thus a certain amount of caution must be observed when interpreting results based solely on the use of this inhibitor. Furthermore, while many PI3K isoforms are inhibited by wortmannin, recently the class II PI3K C2α has been shown to exhibit a greatly reduced sensitivity to this inhibitor (13).

Importantly, a structurally unrelated PI3K inhibitor has been developed, LY-294002 related to quercetin, also known as the 'Lilly compound' (see Fig. 4). This inhibitor also appears to potently inhibit most PI3K isoforms at micromolar concentrations. Class II kinases are again somewhat insensitive to inhibition by LY294002 with p110α

Table 1 Functional roles for PI3K based on inhibitor studies

Response	Receptor	Cell
Respiratory burst	fMLP	Neutrophil
Apoptotic rescue	NGF	PC12
Chemotaxis	RANTES	T cell
Actin rearrangement	PDGF	Fibroblast
Membrane ruffling	PDGF	Endothelial
Glucose transport	Insulin	Fibroblast
IL-2 production	CD28	T cell
Germinal vesicle breakdown	Insulin	*Xenopus* oocyte

Wortmannin

MW = 428.4
Fungal metabolite
Irreversible, covalent binding
Highly cell permeable
IC_{50} = 5 nM

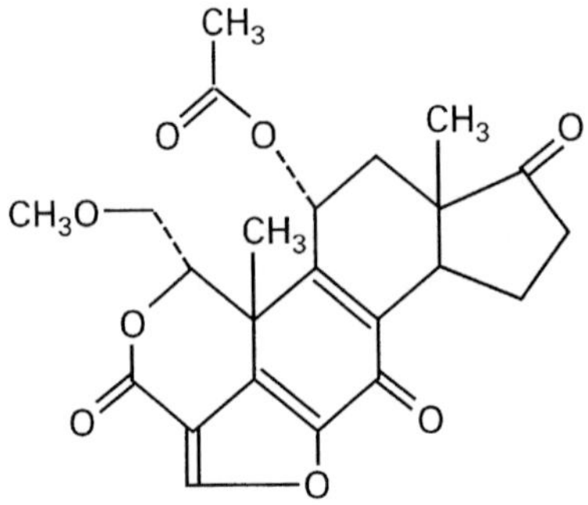

LY294002

MW = 307.4
2-(4-Morpholinyl)-8-phenyl-4H-1-benzopyran-4-one
Reversible binding
Cell permeable
IC_{50} = 1.4 μM

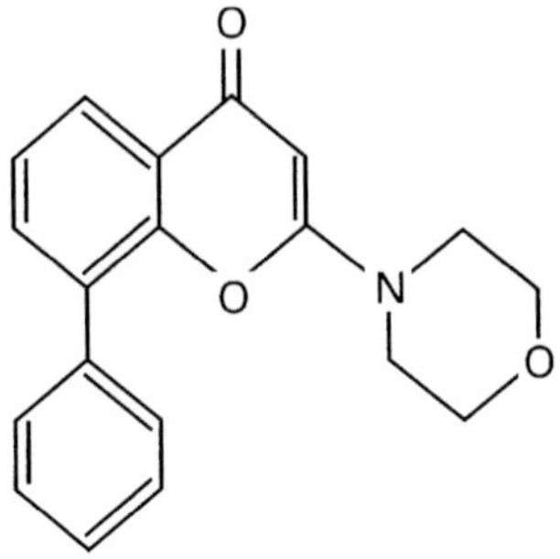

Fig. 4 Structural formulae and properties of pharmacological inhibitors of PI3K.

IC_{50} approximately 1 μM, while PI3K C2α IC_{50} is approximately 20 μM (13). LY294002 is mechanistically different from wortmannin as it directly competes for the ATP-binding site of the catalytic subunit in a reversible manner. Utilizing both these PI3K inhibitors in a careful dose-response analysis of PI3K inhibition compared with inhibition of the cellular event of interest helps to validate the role of PI3K in a particular process.

Aside from pharmacological inhibition, cDNAs engineered to express either active or dominant-negative forms of PI3K have been widely utilized. Altered forms of p85 that lack the p110-binding site have been shown to act as dominant-negative proteins blocking the activation of class I PI3K. A form of p110, termed p110*, engineered to contain an intrinsic activating domain based on the p110-binding domain of p85, also appears to be active independently of p85 (23). Surprisingly, a single point mutant of a conserved lysine residue (K227E) preventing the interaction of PI3K by the small GTPase p21ras (see below), demonstrates a more than fourfold increase in *basal* PI3K activity allowing its use as a dominant active protein (24). Its mechanism of action is unknown, but it is suggested that p110 undergoes a conformational change during the activation process that is mimicked by the K227E mutation. The most common method to constitutively activate PI3K is to selectively target the p110 catalytic subunit to the membrane using either N-terminal myristoylation (myr_p110) or C-terminal farnesylation (p110_CAAX). Generation of these fusion constructs in combination with the p110* mutation have provided valuable tools for evaluating the role of active PI3K in transfected cell systems (25).

It is the use of both pharmacological inhibition combined with mutants of PI3K that thus provides the best means of evaluating the role of this kinase in a particular physiological process.

2. Mechanisms regulating PI3K activation

Initiation of signalling events by growth factor and cytokine receptors involves the activation of an intrinsic or associated protein tyrosine kinase (PTK) activity. PTKs then phosphorylate the associated receptor, or exogenous substrate proteins, resulting in the recruitment of SH2 domain-containing proteins to the receptor. The structure of class I PI3Ks described in Section 1.2 immediately suggests that PI3K is an example of a protein recruited in this way. However, simple recruitment to phosphotyrosine residues is only one facet of the activation process. Activation of PI3K by G protein-coupled receptors (GPCRs) does not require this classical phosphotyrosine SH2-domain interaction but utilizes an unrelated mechanism. Furthermore, as previously discussed, class II PI3Ks may be regulated by Ca^{2+}- and/or phospholipid-binding, while the mechanisms regulating class III enzymes remain unknown: possibly they are constitutively active since the levels of PtdIns(3)P do not vary after cellular stimulation.

Activation of PI3K by PTKs and GPCRs results in the initiation of many downstream signalling events (see Fig. 5). The regulation of specific signalling events are

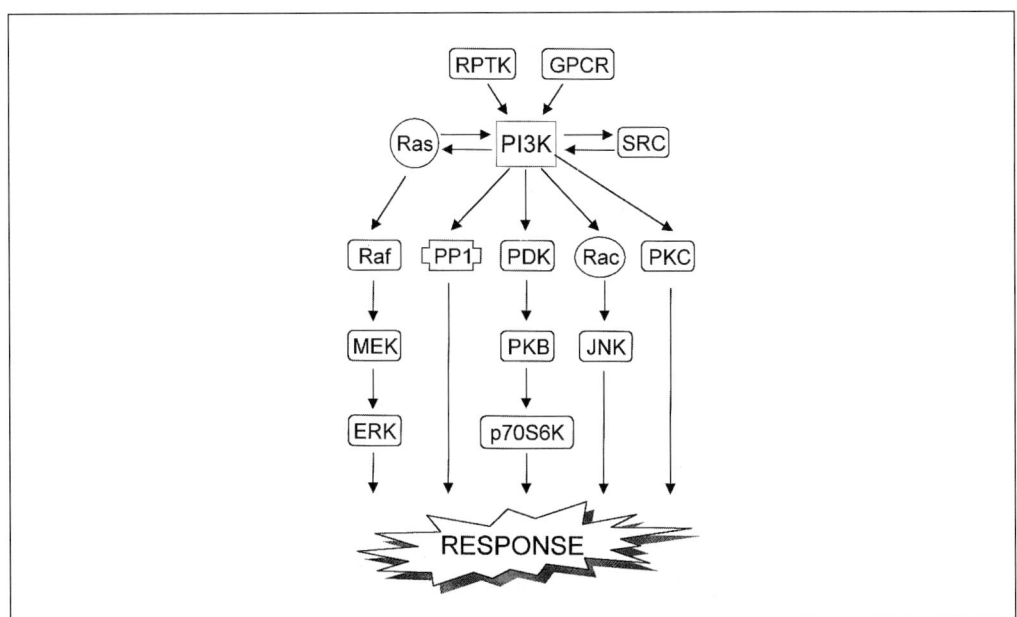

Fig. 5 Signal transduction pathways activated by PI3K. Activation of PI3K by receptor protein tyrosine kinases (RPTK) or G protein-coupled receptors (GPCR) results in activation of PI3K. This results in the subsequent activation of many downstream signal transduction pathways, some of which are highlighted, followed by the initiation of various cellular effector responses. The double arrows between Ras/PI3K and SRC/PI3K represent both upstream and downstream effects of their interaction. See text for detailed explanation and abbreviations.

detailed in Section 3 and the effector responses subsequently activated are discussed in Section 4. This section will highlight some of the mechanisms regulating PI3K activity so far described, and, furthermore, will review recent findings concerning how this system of second-messenger generation is turned off.

2.1 Phosphotyrosine-mediated PI3K activation

Since the discovery of PI-kinase activity in antiphosphotyrosine immunoprecipitates from platelet-derived growth factor (PDGF)-stimulated cells, there has been detailed analysis of the mechanisms by which growth factors and cytokines activate lipid kinase activity. Initial studies demonstrated the association of an 85-kDa phospho-protein with the activated PDGF receptor and, after the cloning of the regulatory and catalytic subunits of PI3K, it was demonstrated that p85 specifically interacts with phosphotyrosine residues present on several growth-factor receptors (26).

The major role of the SH2-domains of p85 thus appears to be the relocation of the constitutively bound, cytoplasmic p110 catalytic subunit to the membrane where its lipid substrate is located (see Fig. 6A). Analysis of the amino acid sequences surrounding phosphotyrosine residues known to bind p85 has revealed a consensus p85-binding site to be tyrosine–X–X–methionine (YXXM), a fairly specific sequence for the p85 SH2-domain (27). All p85-related adapter molecules are thought to bind to this consensus motif, suggesting functional redundancy.

As well as recruitment of the p110 catalytic subunit to tyrosine phosphorylated receptors via p85, it has also been demonstrated that activated receptors can induce tyrosine phosphorylation of both the p85 and p110 subunits (28, 29). Although levels of p85 phosphorylation *in vivo* are low and the site of phosphorylation varies depending on the stimulus, it has been proposed that phosphorylation increases PI3K activity (30, 31). There are many stimuli, however, that simply do not induce p85 tyrosine phosphorylation so this does not appear to be a prerequisite for activation *in vivo*.

2.2 Activation of PI3K by p21Ras

The *RAS* proto-oncogene product is a 21-kDa GTPase (p21Ras), a member of a much larger family of small GTPases. This protein has been implicated in cellular growth and transformation, and activating mutations in p21Ras are found in a large percentage of human tumours. All members of this superfamily are GTP-binding proteins that possess an intrinsic GTPase activity. They act as molecular switches being active in a GTP-bound 'on' state, generated by guanine-nucleotide exchange factors (GEFs) and inactivated by stimulation of their intrinsic GTPase activity by GTPase-activating proteins (GAPs) therefore generating a GDP-bound 'off' state. Ligand binding to receptors activates p21Ras by stimulating GEF-mediated nucleotide exchange. When GTP-bound, p21Ras can interact and activate several downstream effector molecules thus transducing a signal.

How may the activation of p21Ras influence PI3K or vice versa? It was initially found that PI3K co-immunoprecipitates with p21Ras, suggesting that this protein could be acting as either an effector or regulator of p21Ras (32). Subsequently, p21Ras was found to interact directly with the p110 catalytic subunit of PI3K in a GTP-dependent manner (Fig. 6B). Interfering mutants of p21Ras that lock the molecule in a GDP- or GTP-bound state can be utilized as dominant-negative or active proteins, respectively. Expression of a dominant-negative p21Ras (Asn17) was found to inhibit the production of 3'-phosphorylated phosphoinositides, while overexpression of an active p21Ras (Val12) resulted in a large elevation in the level of these lipids (33). Furthermore, utilizing an *in vitro* purified liposome-reconstitution system, p21Ras–

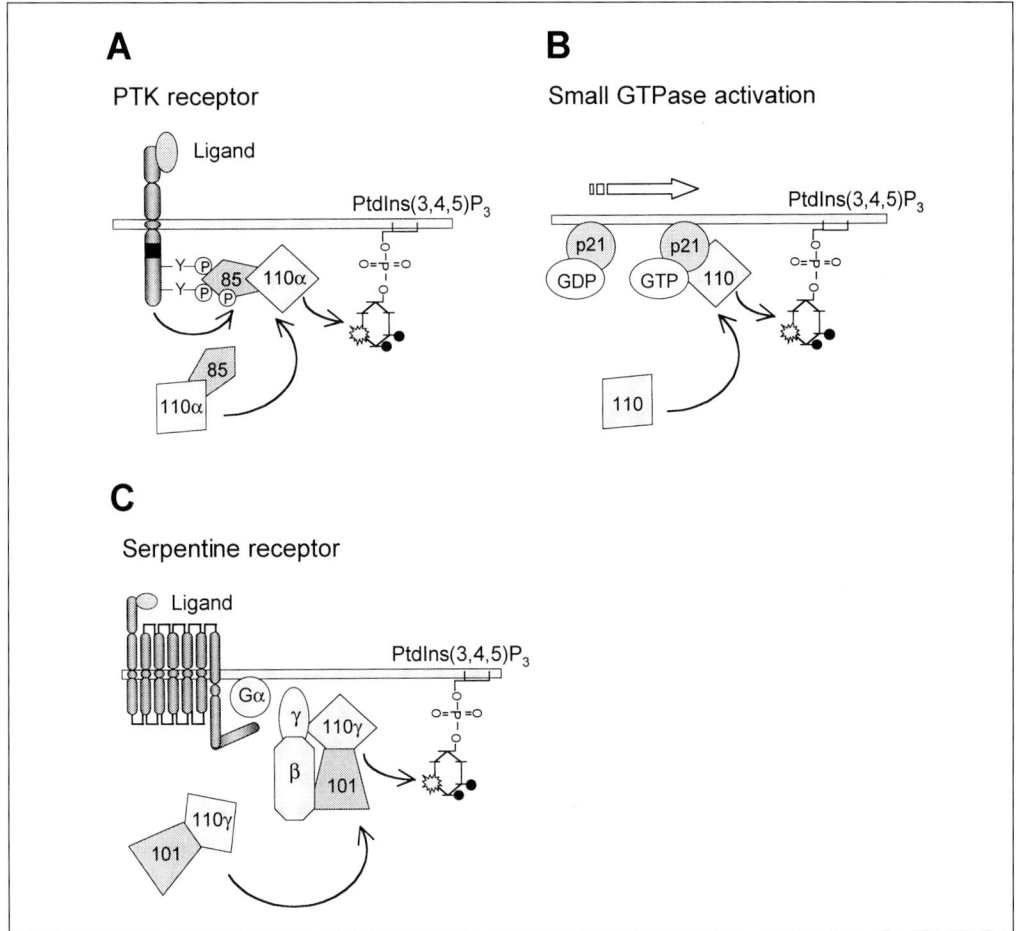

Fig. 6 Mechanisms of PI3K activation. Various methods are utilized by cellular stimuli to activate PI3K and translocate the enzyme into proximity with its lipid substrates. (A) Protein tyrosine kinase (PTK) receptors autophosphorylate on tyrosine residues after agonist stimulation. These tyrosines provide binding sites for the p85 adapter subunit of PI3K, recruiting it to the membrane. Here it may be phosphorylated and further activated. (B) Small GTPases, such as p21Ras, are activated upon GTP-for–GDP exchange after receptor stimulation. GTP-bound p21-proteins can then associate with the PI3K p110-catalytic subunit resulting in translocation. (C) Upon activation of serpentine receptors, heterotrimeric G-protein subunits dissociate from each other. The βγ complex is then able to associate with PI3K through a novel adapter protein, p101.

GTP was found to activate the lipid kinase activity of PI3K directly (24). These data suggest a potential role for p21Ras in the activation of PI3K by growth factors. It is conceivable that an interaction of PI3K with p21Ras targets p110 to the membrane in a similar way to the interaction of p85 with phosphotyrosine residues. While the site of p21Ras binding on p110 has been mapped to the N-terminus (see Fig. 2), recent studies have demonstrated that the level of PI3K activity obtained by direct p21Ras stimulation varies depending on the PI3K isoform—p110α is potently activated by p21Ras, while p110γ binds to, but is only very modestly activated in the same cells (34).

The function of the PI3K and small GTPase interactions is complicated by the finding that PI3K can also *stimulate* activation of p21Ras. This suggests that PI3K can be both upstream and downstream of p21Ras-activation (23). Experiments utilizing an active PI3K mutant (p110*) revealed that dominant-negative p21Ras blocked p110*-induced transcription and, when expressed in *Xenopus* oocytes, p110* increased the levels of GTP-bound p21ras inducing oocyte maturation (23).

These data suggest that the interaction between activated small GTPases and PI3K results in a complex regulation of downstream signalling events. The functional consequences of this interaction are discussed further in Section 4.

2.3   Serpentine receptor-mediated PI3K activation

In 1988 a PI3K activity was described in fMLP-stimulated neutrophils. This PI3K was stimulated by agonists of G protein-coupled receptors (GPCRs) and its activation was inhibited by pertussis toxin but not inhibitors of PTKs. While the partially purified kinase could be stimulated by the Gβγ subunits of GPCRs, unlike p85-p110, it was not activated by phosphotyrosine peptides (35). Furthermore, the kinetics of activation of this myeloid kinase was extremely rapid compared with the slower activation of PI3K by growth factor and cytokine receptors.

Together, these facts suggested that the PI3K activity measured was distinct from that of the classical p85-p110 kinase. However, it was several years before a cDNA was cloned encoding a novel p110-related enzyme termed p110γ (10). This p110γ was activated *in vitro* by the βγ subunits of heterotrimeric G-proteins and did not interact with p85. However, the activity obtained by the addition of Gβγ was much less than that obtained in the partially purified preparations, suggesting the existence of another subunit *in vivo* regulating p110γ activation (35).

The missing link in p110γ activation was recently identified in complex with Gβγ-activated p110γ purified from pig neutrophil cytosol (36). A protein of 101 kDa with no homology to any previously identified peptides was subsequently cloned and found to constitutively associate with p110γ. It appears that the p101–p110γ complex is the counterpart of the p85–p110α complex previously described. Although p101 confers Gβγ-sensitivity on the p101–p110γ heterodimer, it is unclear whether one or both subunits bind Gβγ *in vivo*. At this moment little more is known concerning the precise role of p101. Since the activity of p110γ is lower when complexed to p101 in the absence of Gβγ, it appears that this subunit also acts to repress its activity in unstimulated cells (36). A model for the activation of PI3K by G protein-coupled receptors is presented in Fig. 6C. The translocation of p101–p110γ to membrane-bound Gβγ parallels the translocation of p85–p110α to tyrosine-phosphorylated receptors (Fig. 6A).

2.4 SHIP and PTEN: modulating the action of PI3K lipid products

As discussed in Section 1.1, activation of PI3K results in the production of several D3-phosphatidylinositol lipid variants (see Fig. 1), one of which is PtdIns(3,4,5)P_3. Recently, much attention has been paid to an inositol 5-phosphatase responsible for mediating the removal of the 5-phosphate from this lipid. Activation of this phosphatase does not simply result in the removal of the lipid second messenger, turning off the signal, but also acts to re-route these messengers to alternative signalling pathways.

While the inositol 5-phosphatase family comprises several members, it is the SHIP-subgroup that regulates PI3K signalling (SHIP, *SH*2 domain-containing *i*nositol 5-*p*hosphatases). SHIP only dephosphorylates 3-phosphorylated substrates, which places it as a direct regulator of PI3K products (37, 38). As defines most signalling molecules, SHIP contains several protein domains likely to be critical for functioning, including the N-terminal SH2-domain that gives it its name and a proline-rich region (see Fig. 7). SHIP has been found to accumulate in phosphotyrosine complexes in

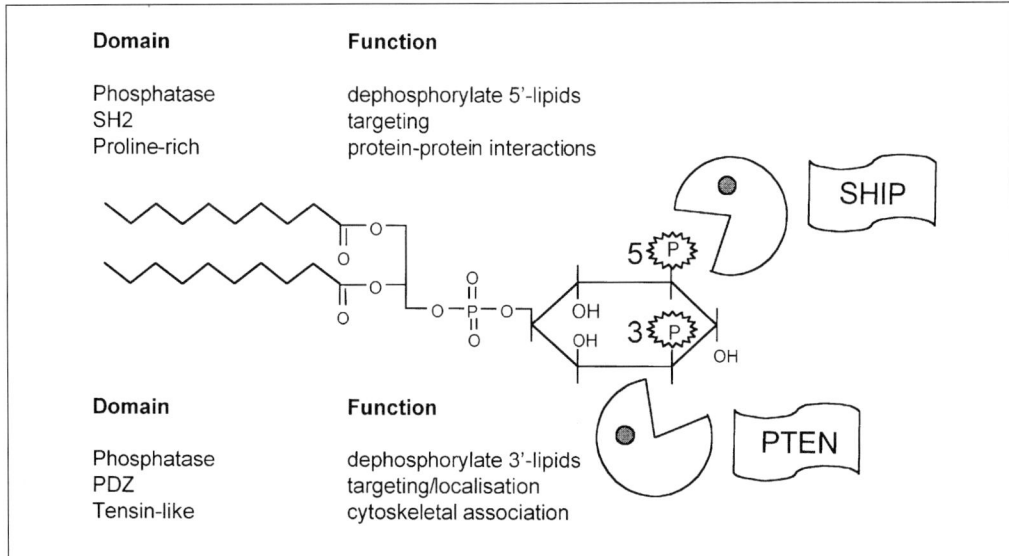

Domain	Function
Phosphatase	dephosphorylate 5'-lipids
SH2	targeting
Proline-rich	protein-protein interactions

Domain	Function
Phosphatase	dephosphorylate 3'-lipids
PDZ	targeting/localisation
Tensin-like	cytoskeletal association

Fig. 7 Specific phosphatases modulate the action of PI3K lipid products. Two lipid phosphatases have been demonstrated to modulate PI3K activity. The 5-phosphatase, SHIP (above), acts to re-route 3-phosphorylated PI-lipid products. However, the 3-phosphatase, PTEN (below), removes PI3K-generated lipids providing substrates for phospholipase C (see Fig. 1).

response to a number of agonists and stimulation results in tyrosine phosphorylation of SHIP, although this in itself does not affect 5-phosphatase activity (39). Thus it is likely that, similar to the mechanisms previously discussed for the activation of PI3K, the targeting of SHIP to the membrane via complex formation is likely to be critical in its activation, placing it in close proximity to lipid substrates.

How does the dephosphorylation of the PtdIns products of PI3K regulate signalling events? Several studies have revealed a role for SHIP in specific signalling processes. First, overexpression of SHIP in haemopoietic cells inhibits cytokine-induced proliferation, suggesting a critical role for 5-phosphorylated lipids (40). Furthermore, overexpression of SHIP in *Xenopus* oocytes blocks the induction of germinal vesicle breakdown induced by p110*, suggesting that $PtdIns(3,4,5)P_3$ is critical for this process (41). Most recently, it has been demonstrated in B-cells that $PtdIns(3,4,5)P_3$ regulates receptor-mediated Ca^{2+}-release from internal stores, through membrane targeting of the Tec kinase via its PH-domain (42). This is blocked by engagement of B-cell 'inhibitory receptors' that activate SHIP, thereby providing a system for regulating Ca^{2+}-mediated signalling.

Apart from this simple 'off' signal generated by the removal of $PtdIns(3,4,5)P_3$, SHIP concomitantly increases the levels of $PtdIns(3,4)P_2$. As mentioned previously, PH domain-containing proteins bind to specific phospholipids. One of the downstream targets of PI3K is protein kinase B, discussed in Section 3.1, which preferentially binds $PtdIns(3,4)P_2$ (43). This suggests that SHIP may actually play a positive role in dynamically directing PI3K-induced signals towards PH domain-containing proteins with a $PtdIns(3,4)P_2$ preference.

Recently, the cloning of another novel phosphatase has led to a much greater understanding of the mechanisms by which 3-phosphorylated inositol lipids can be regulated. Mutations in the *PTEN* gene (phosphatase and tensin homologue deleted on chromosome ten) have been implicated in the development of a variety of neoplasias including glioblastomas, melanomas, and prostate cancer. Cloning of PTEN initially revealed it to be a protein tyrosine phosphatase (PTPase) with activity against highly negatively charged substrates (44). However, recent work has demonstrated that PTEN may not act as a PTPase, but as a specific PI-lipid phosphatase, dephosphorylating $PtdIns(3,4,5)P_3$ on the D3-position (45). Overexpression of PTEN was found to inhibit insulin-induced $PtdIns(3,4,5)P_3$ production without affecting the activity of PI3K. Furthermore, overexpression of a catalytically inactive PTEN mutant (C124S) resulted in the accumulation of $PtdIns(3,4,5)P_3$ in the absence of cellular stimulation. More convincingly, purified PTEN was found to dephosphorylate $PtdIns(3,4,5)P_3$, $PtdIns(3,4)P_2$, and $PtdIns(3)P$ *in vitro*, although the effect on the latter two PI-lipids was much less potent. In cells derived from knockout mice lacking PTEN, the downstream PI3K target PKB is constitutively active (Fig. 5; ref. 46) and cells exhibit decreased sensitivity to cell death in response to apoptotic stimuli (see below). Together, these observations suggest that PTEN is indeed a PI-phosphatase playing a critical role in directly dephosphorylating products of PI3K. Unlike SHIP, PTEN will not re-route PI3K products but will simply remove these lipids, effectively turning the signal off. The fact that this phosphatase is a tumour suppresser demonstrates the importance of regulating PI3K activity for the control of cell proliferation (see Section 4.1). What is still unclear are the precise roles of the various PTEN domains in the regulation of its function. For example, PTEN contains a putative PDZ-domain, a protein–protein association module, often linked to the targeting of signalling molecules to membranes (see Fig. 7). Whether this plays a critical role in the regulation of PTEN *in vivo* remains to be seen.

3. Downstream effectors of PI3K activation

The previous sections discussed the mechanisms by which PI3K can be activated by cellular stimuli resulting in the generation of specific PI-lipid products. These models are the result of a large amount of work carried out in the late 1980s and early 1990s and, together with the use of pharmacological inhibition, PI3K was implicated in the regulation of a multitude of cellular processes. It was known that 3-phosphorylated lipid products were responsible for mediating the receptor-induced signals transduced by PI3K, but how the generation of these lipids could activate downstream signalling events was unclear. Particularly relevant was the question of how phosphorylated inositol lipids were able to activate downstream protein-kinase cascades known to mediate these signalling events. In the last 5 years there has been an enormous advance in the understanding of the mechanisms by which PI3K activates downstream signalling components resulting in the regulation of cellular processes. In this section, the major targets of PI3K will be discussed in detail.

3.1 Protein kinase B (c-akt)

The first component to be put into place was the identification of a serine/threonine kinase named protein kinase B (PKB) or alternatively c-akt (for a detailed review see ref. 43). This protein kinase was independently cloned by three groups in 1991, but its relevance as a direct target of PI3K-mediated signalling was not appreciated until some 4 years later. PKB, the product of the transforming AKT8 retrovirus (47), was subsequently found to be activated by growth factors in a PI3K-dependent manner as demonstrated by inhibition studies utilizing (a) wortmannin/LY294002, (b) dominant-negative PI3K constructs, and (c) receptor mutants incapable of activating PI3K (48, 49). In the 2 years proceeding these findings more than 200 papers have been published concerning the mechanism and function of PKB activation.

PKB was found to have an N-terminal PH-domain of around 100 residues, a catalytic domain, and a C-terminal tail of approximately 60 amino acids. It appears that PtdIns(3,4)P$_2$-binding to the PH-domain facilitates the dimerization of PKB (50) and targets PKB to the membrane. That this was important in activation was supported by the finding that mutation of the PH-domain resulted in a drastic reduction in kinase activity in some, although not all, cell systems (49). Supporting a role for this translocation, studies have indeed shown that PKB does relocate to the membrane following growth-factor stimulation and that membrane-targeted PKB is constitutively active (48, 49).

This model was, however, found to be an oversimplification following the discovery that PKB is also phosphorylated after receptor-mediated PI3K activation (48, 49, 51). Dephosphorylation of PKB, similarly to PH-domain mutagenesis, results in kinase inactivation. This implied a dual mechanism of PI-lipid-mediated recruitment and phosphorylation in the activation of PKB (see Fig. 8). Two sites were found to be phosphorylated on PKB after stimulation, threonine-308, which lies within the kinase-domain, and serine-473 at the C-terminus. The surrounding residues of these phosphorylation sites are very different, suggesting the participation of two different kinases. Mutagenesis of each of these residues to alanine revealed that both are required for full activation (51). The kinases responsible for these phosphorylation events and their regulation are discussed in detail below (Section 3.2).

Activation of PKB appears to be critical for mediating many of the effects of PI3K, such as the regulation of proliferation, survival, protein synthesis, and metabolism (see Section 4). Indeed, at present PKB is thought to be a master controller of PI3K-mediated events—removal of this gene from the *Drosophila* fruit fly is embryonic-lethal—supporting a critical role for PKB in cellular homeostasis (52).

3.2 PtdIns(3,4,5)P$_3$-dependent kinases (PDKs)

As mentioned above, while the membrane localization of PKB by binding to PtdIns(4,5)P$_2$ is an important step in its regulation, subsequent phosphorylation is critical to complete this process. To identify the kinases responsible for PKB phosphorylation, several groups embarked on a purification of a PKB-phosphorylating activity. The result was the identification of a PH domain-containing kinase termed PDK1 (for PtdIns(3,4,5)P$_3$-dependent kinase) (53, 54). As suggested by its name, PDK1 activity towards PKB was completely dependent on PtdIns(3,4,5)P$_3$, making it a direct target of PI3K-generated phospholipids. Analysis of the residue phosphorylated on PKB revealed it to be threonine-308 and phosphorylation of PKB *in vitro* by purified PDK1 resulted in activation. The subsequent cloning of PDK1 revealed a kinase domain with approximately 35% homology to PKA, PKB, and PKC (54, 55). More interestingly, PDK1 exhibited homology to the *Drosophila DSTPK61* gene which has been implicated in the regulation of sex differentiation, oogenesis, and spermatogenesis.

While overexpression of PDK1 clearly resulted in the phosphorylation and activation of PKB, it was unclear whether the agonists shown to stimulate PI3K also stimulated PDK1 activity. An initial study suggested this not to be the case since insulin treatment of target cells did not actually increase the activity or phosphorylation of PDK1 (53). The concentration of 3-phosphorylated inositol lipid required to translocate PDK1 to lipid vesicles is very small, suggesting that maybe in unstimulated cells there is already membrane-bound PDK1 (55). The idea presented was that the translocation of PKB to the membrane through phospholipid-binding results in a conformational change exposing threonine-308 for phosphorylation by PDK1. Indeed, removal of the PH-domain of PKB allows lipid-independent phosphorylation by PDK1 (55). However, recent work has shown that PDGF does in fact also stimulate

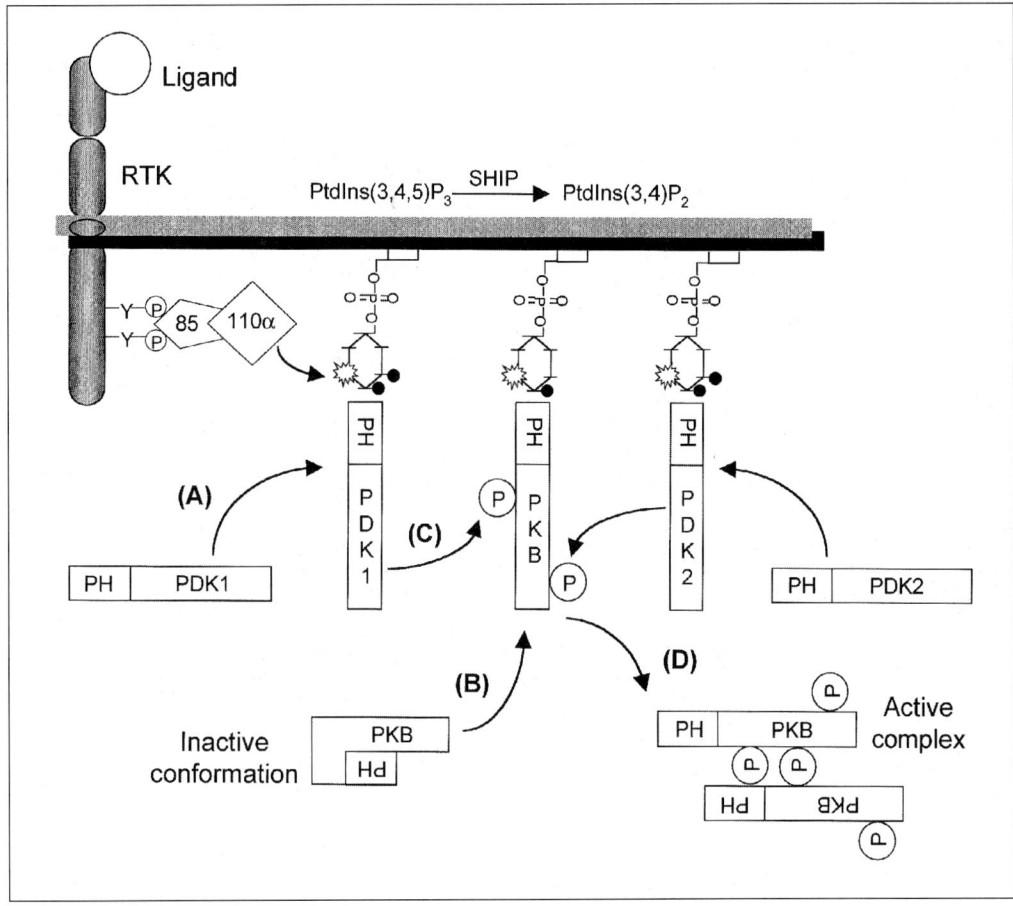

Fig. 8 Mechanism of protein kinase B activation. Activation of PI3K by receptor stimulation results in the production of PtdIns(3,4,5)P$_3$ at the plasma membrane. (A) The protein kinase PDK1 translocates to the membrane, binding to PtdIns(3,4,5)P$_3$. (B) PKB also subsequently translocates to the membrane and binds to PtdIns(3,4)P$_2$. (C) This binding results in a conformational change exposing sites for phosphorylation by PDKs. (D) Phosphorylated PKB then dimerizes and translocates to the cytosol where it is able to phosphorylate substrates.

the translocation of PDK1 to the plasma membrane in a PI3K-dependent manner, although this lipid-binding does not affect PDK1 activity *per se* (56). A schematic model of this activation process is shown in Fig. 8.

Having described the cloning and characterization of PDK1, the PKB threonine-308 kinase, a question still remains concerning the identification of the PDK2, the PKB serine-473 kinase, necessary for optimal PKB activation. At this time there are few clues hinting to the identity of this protein kinase. Integrin-linked kinase (ILK) has, however, been demonstrated to phosphorylate PKB on serine-473 and a dominant-negative ILK impairs PKB phosphorylation (57). It is possible that this kinase or a related kinase may indeed correspond to the missing PDK2.

3.3   The p70 S6 kinase

The regulation of protein synthesis is a complex event, but one of the key players in this process is the p70 S6 protein kinase (p70S6K; see ref. 58). This kinase is also implicated in the progression of cells from G_1 to S phase of the cell cycle by mitogenic stimuli. As suggested by its name, it phosphorylates the S6 protein of the 40S ribosomal subunit, which appears to increase the translation of specific mRNAs containing a polypyrimidine tract in their 5′-untranslated regions. This family of mRNAs encodes components of the translational machinery and is thought to be important for cell-cycle progression.

Much evidence has pointed to PI3K playing a role in the activation of p70S6K (59). For example, the activity of p70S6K is increased by the expression of constitutively active PI3K (25), while both dominant-negative forms of PI3K (48) and wortmannin/LY294002 block the activation of p70S6K. Indeed, these data implicate PI3K directly in the regulation of protein synthesis by various stimuli such as insulin (60). Since p70S6K is regulated by protein phosphorylation on multiple sites, it was clear that there must be a downstream kinase mediating the activation of p70S6K by PI3K. With the discovery of PKB, one potential intermediate was uncovered. Expression of a constitutively active form of PKB resulted in a potent activation of p70S6K, although this does not appear to be due to direct phosphorylation (48).

The mechanism by which PI3K activates p70S6K has been taken to a further level of complexity by the discovery that PDK1 can directly phosphorylate and activate p70S6K *in vitro* and *in vivo* (61). This was a surprising and at first puzzling result since PDK1 is an upstream activator of PKB (see Fig. 8). PDK1 phosphorylated p70S6K within the kinase domain at a position homologous to the threonine-308 site of PKB. Unlike the phosphorylation of PKB, p70S6K phosphorylation was independent of the presence of PtdIns(3,4,5)P_3, leading to the question of whether PDK1 is actually the mitogen-stimulated p70S6K kinase *in vivo*. While p70S6K has no lipid-binding activity it is regulated through multi-site phosphorylation. This raises the possibility that the initial phosphorylation of p70S6K regulates the subsequent availability of the PDK1 target site (threonine-229) allowing the secondary phosphorylation to occur. Indeed it was demonstrated that prior phosphorylation of threonine-389 was critical for the subsequent PDK1-mediated phosphorylation step (61). Perhaps PKB is involved in the phosphorylation of threonine-389, setting the stage for PDK1 to complete the activation process. A possible model incorporating the results described above is shown in Fig. 9. But what is the mechanism for bringing cytosolic p70S6K into the vicinity of membrane-associated PDK1? The observation that p70S6K forms complexes with the small GTPases p21Rac and Cdc42 suggest a mechanism for targeting p70S6K to the membrane allowing its subsequent phosphorylation by PDK1.

3.4   Protein kinase C isoforms

The role of membrane lipids in the activation of protein kinase C (PKC) isoforms has been studied for many years. The classical pathway involves phospholipase C-

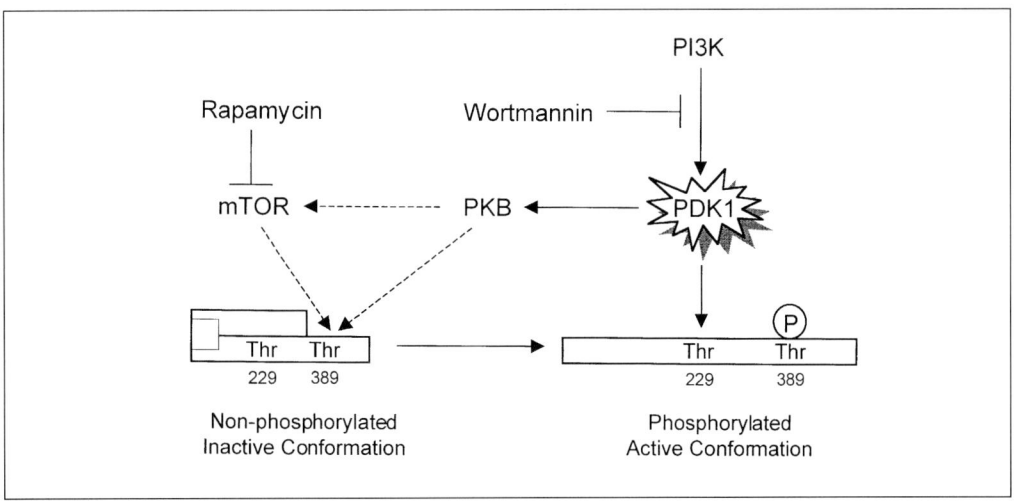

Fig. 9 p70 S6 kinase is activated by PI3K-mediated phosphorylation. Activation of PKB and/or mTOR results in phosphorylation of C-terminal threonine-389 in p70S6K. This phosphorylation results in a conformational change exposing a threonine-229. This residue is then phosphorylated by PDK1 resulting in a fully active enzyme.

mediated breakdown of PtdIns(4,5)P$_2$ to InsP$_3$ and diacylglycerol (DAG), resulting in PKC activation and Ca^{2+}-release from internal stores (see Fig. 1). While this pathway is utilized for the activation of the conventional PKC isoforms such as PKCα, there are several novel PKC isoforms whose activation requires only DAG and atypical PKCs that require neither Ca^{2+} nor DAG. A direct role for the lipid products of PI3K in the activation of PKCs was implied by the finding some years ago that both the atypical PKC, PKCζ, and the Ca^{2+}-independent PKCs, PKC$\delta/\varepsilon/\eta$, could be activated by PtdIns(3,4,5)P$_3$ (62).

Several studies have begun to suggest that there is indeed a role for PI3K in the regulation of PKC isoforms. For example, phosphorylation of the major PKC substrate, pleckstrin, in platelets is inhibited by wortmannin and stimulated by the addition of PtdIns(3,4,5)P$_3$ (62). A more recent study in adipocytes has demonstrated that insulin provokes a rapid increase in PKCζ phosphorylation and activity in a wortmannin-sensitive fashion (63). These data suggest that PI3K can indeed function upstream of PKC.

The most compelling evidence for a PI3K-mediated activation of PKC isoforms comes from studies investigating the activation of PKCλ and PKCε by growth-factor receptors (64, 65). PKCλ was found to be phosphorylated in response to growth-factor treatment and this resulted in a translocation from the nucleus to the cytosol. Both these effects were inhibited by wortmannin and, furthermore, overexpression of PI3K resulted in PKCλ activation (64). A similar study with PKCε also revealed growth factor-mediated activation of this enzyme resulting in translocation from the cytosol to the membrane (65). These effects were similarly sensitive to wortmannin treatment.

Table 2 Comparison of PDK-activated effector kinases

Kinase	Phosphorylated site	PtdIns(3,4,5)P$_3$-dependence	Membrane targeting
PKB	GATMK$\underline{\text{T}}$FCGTP	Necessary	Constitutively active
p70S6K	GTVTH$\underline{\text{T}}$FCGTI	ND	No effect
PKCζ	GDTTS$\underline{\text{T}}$FCGTP	Modest effect	Constitutively active

The threonine residues phosphorylated within the kinase domain by PDK1 are underlined. Membrane targeting refers to the construction of constitutively membrane-bound versions of the kinases. ND, not determined.

The question of how PI3K could be activating these atypical PKCs if not through direct PI-lipid binding was recently answered and, similar to PKB and p70S6K, appears to involve the PDKs. As mentioned above, PKCλ is phosphorylated upon growth-factor treatment of cells, a phenomenon that appears to be critical in the regulation of PKC isoforms. One of the phosphorylated residues is within the kinase domain (threonine-401 for PKCζ), in a similar position to that phosphorylated by PDK1 in PKB (threonine-308). The region surrounding this phosphorylation site in PKCζ indeed conforms to a potential PDK consensus target sequence (see Table 2). Recent studies have demonstrated that PDK1 is likely to be the *in vivo* kinase mediating PI3K activation of PKCs (66, 67). Overexpression of PDK1 resulted in the constitutive activation of PKCζ, while a dominant-negative PDK1 reduced PKCζ phosphorylation of threonine-401 *in vivo*. Furthermore, PDK1 was found to directly phosphorylate PKCζ *in vitro* and both were found to associate when overexpressed in cells. These studies demonstrate that similarly to PKB and p70S6K, PKC isoforms are another direct target of PDKs. In all three cases, release from an autoinhibitory state results in the subsequent PDK-mediated threonine phosphorylation and activation of the kinases in question. Since the phosphorylation of PKCζ by PDK1 was modestly enhanced by PtdIns(3,4,5)P$_3$, it is suggested that binding of this PI-lipid to PKCζ may allow a conformational change to occur exposing threonine-401 for subsequent phosphorylation.

While there are thus parallels in these PDK-mediated activation mechanisms there are also differences. For example, while PKCζ constitutively associates with PDK1, the association of PKB with PDK1 is PDGF-dependent. The relationship between these activation mechanisms is summarized in Table 2 and suggests considerable cross-talk between effector molecules in the PI3K signalling pathways.

3.5 Guanine nucleotide exchange factors

As already discussed, small GTPases act as binary switches, cycling between an active GTP-bound state and an inactive, GDP-bound state. Guanine nucleotide exchange factors (GEFs) act to enhance the transition from the inactive to active state by promoting the exchange of GDP for GTP on the requisite GTPase. Several GEFs have been shown to contain PH-domains and the functions of these proteins are thus thought to be regulated, in part, by their ability to bind PI-lipids. A prime example of this is found in the exchange factor Vav which mediates the activation of the small GTPase p21Rac in haemopoietic cells. Vav, named after the sixth letter in the Hebrew alphabet, appears to play a critical role in lymphocytes since Vav-deficient mice display various defects in B- and T-cell development. Several years ago it was reported that PI3K was necessary for the activation of p21Rac (68). The PH-domain of Vav was subsequently found to bind *in vitro* to PtdIns(3,4,5)P_3, and this promoted the phosphorylation of Vav by tyrosine kinases resulting in enhanced nucleotide exchange (69). PtdIns(4,5)P_2, on the other hand, inhibited the exchange activity and tyrosine phosphorylation of Vav. These data suggests a model whereby PtdIns(4,5)P_2-bound Vav is held in an inactive state that is then released upon activation of PI3K and binding of PtdIns(3,4,5)P_3 resulting in activation of exchange-activity. This was supported by the finding that PtdIns(3,4,5)P_3 has a higher affinity for Vav than PtdIns(4,5)P_2. The activation of Vav by first lipid-binding and then phosphorylation is analogous to the mechanism of PKB activation (Section 3.1). Interestingly in the case of Vav, PI3K activation also acts to remove an inhibitor and at the same time generate an activator of GEF function.

4. Functions of PI3K activation

It is apparent that the 3-phosphorylated lipid products of PI3K play critical roles in the activation of various intracellular signalling molecules. The themes of translocation followed by phosphorylation are recurrent in the activation of these downstream effectors. In this section, the role of PI3K signalling in the regulation of diverse physiological processes will be discussed in detail. The main focus will be on the role of this lipid kinase in regulating proliferation and survival, but roles in other diverse cellular processes will also be discussed.

4.1 Proliferation and oncogenesis

Much work has implicated PI3K in the regulation of cellular proliferation by a plethora of mitogenic stimuli. The use of pharmacological inhibitors of PI3K in cell lines has defined a critical role in cellular survival and these events are discussed below. However, since pharmacological inhibition of PI3K can lead to cell death via the induction of an apoptotic programme, it has been through selective activation of PI3K, rather than inhibition, that its role in cellular proliferation has been studied. Early studies linked PI3K to PDGF-stimulated proliferative signals utilizing a system of PDGF receptor-mutants. Removal of tyrosine residues responsible for activating PI3K perturbed the mitogenic signal. Furthermore, cells expressing receptors containing these tyrosine residues alone were still viable, suggesting that PI3K is both necessary and sufficient for cell-cycle progression. Studies utilizing the microinjection of p110α antibodies demonstrated that while PDGF- and epidermal growth factor (EGF)-induced DNA synthesis was inhibited, colony-stimulating factor (CSF)-1, bombesin, and lysophosphatidic acid (LPA) were not (70). This demonstrates that p110α is critical for a subset of mitogens to induce a proliferative response. Of course, since these antibodies do not cross-react with other PI3K isoforms, such experiments do not address the possibility that other catalytic subunits may be utilized by bombesin and LPA.

Recently, an elegant study into the role of PI3K in cell-cycle progression utilized a novel, inducibly active, PI3K-oestrogen receptor fusion protein (ER-p110*). The addition of an oestrogen derivative to transfected cells results in the rapid activation of PI3K independently of the other signalling pathways normally activated by mitogenic receptors (71). Utilizing this system, several important observations have been made concerning the role of PI3K in cell-cycle progression. First, activation of PI3K in the absence of external factors was sufficient to promote entry into the S-phase of the cycle and induce DNA synthesis. However, activation of PI3K alone was not sufficient to produce progression through the entire cell cycle in the absence of serum. Indeed, prolonged activation of PI3K in the absence of serum resulted in apoptosis. Cells appeared to have a problem in re-entering the cell cycle and this may be the cause of the induced apoptotic programme. Finally, however, when PI3K activity was induced in the presence of serum, cellular changes occurred that resembled those of oncogenic transformation. Thus there is likely to be a subtle interplay between PI3K and other mitogen-induced signalling components to efficiently regulate proliferation and cell survival.

Various lower organisms have been used to further address the role of PI3K in cellular functioning. PI3K isoforms have been cloned from organisms as diverse as fruit flies, slime moulds, yeasts, worms, and soybeans. All have a similar primary structure and *in vitro* substrate-specificity. The fruit fly, *Drosophila melanogaster*, contains a class IA PI3K termed Dp110 (72). Flies overexpressing a constitutively active, membrane-bound form of Dp110 exhibited both increased wing size and bulging, roughened eyes. These structures are increased in size, and the reverse is seen upon ectopic expression of a mutated, catalytically inactive Dp110. These effects are due to both increased cell size and number, thus implicating Dp110 in the control of cell growth. Interestingly, the overexpression of class II or class III *Drosophila* PI3K isoforms did not produce the same phenotype (72). This suggests that it is the PtdIns(3,4,5)P_3 product of PI3K that is the mediator of proliferative function rather than other 3-phosphorylated PI-lipids.

The identification of a PI3K gene in the nematode worm, *Caenorhabditis elegans*, has provided another interesting insight into its function. Normal development of the worm involves four larval stages, with a lifespan of 2–3 weeks. When food is scarce and population density high, a pheromone-induced neurosecretory pathway causes the worms to form what is known as 'dauer larvae' that are developmentally arrested and which can endure stressful environmental conditions. This effectively results in an increase in longevity of up to 6 months. Mutants of a gene termed *age-1* exist only as dauer larvae, while worms with maternal but not zygotic *age-1* develop as non-dauers and live two to three times longer than normal. Identification and cloning of *age-1* revealed it to be a class IA PI3K (73). These data are complex to interpret but suggest that PI3K normally acts to repress dauer formation under normal conditions, repressing genes responsible for initiating this developmental arrest programme. However, a reduced level of PI-lipids results in the suppression of only a subset of these genes, allowing further development, which is a selective advantage against environmental stress and results in a longer lifespan. Most interestingly, a gene termed *daf-16* has been cloned from *C. elegans* and found to encode a transcription factor. It appears that the role of *age-1* is to antagonize *daf-16* thus inhibiting the transcription of target genes responsible for longevity. DAF-16 has a human homologue, AFX, and these data suggest a possible mechanism for PI3K to mediate transcription in mammalian systems. Further analysis of genetic interactions within this intriguing system may provide answers to the mechanisms by which PI3K may influence growth and proliferation in higher organisms.

Oncogenesis results from the dysfunctional regulation of signal transduction pathways, often involving mutation in an upstream component resulting in constitutive activation. While such mutations are found in a wide variety of proteins including growth-factor receptors, protein kinases, and small GTPases, surprisingly no PI3K transforming oncogene had been identified until very recently. One possible reason for this is suggested by the fact that prolonged activation of PI3K in the absence of serum results in the induction of an apoptotic programme (71), which would lead to the removal of cells harbouring activating PI3K mutations when growth conditions were suboptimal. However, more recently there have been two cases of transforming mutations in PI3K.

The avian sarcoma virus (ASV16) is a retrovirus that induces haemangiosarcomas in chickens. Analysis of the ASV16 genome revealed that it encoded an oncogene derived from the cellular gene for PI3K (p110) referred to as v-*p3k* (74). Interestingly, the c-*p3k* gene was fused to the viral *gag* gene resulting in the generation of a membrane-targeted PI3K. This is similar to the *gag* fusion observed in the transforming oncogene of PKB, v-*akt* (47). Chicken embryo fibroblasts (CEFs) transformed with v-*p3k* showed elevated levels of PtdIns(3,4)P_2 and PtdIns(3,4,5)P_3 and activation of PKB.

In an analysis of murine lymphoid tumours, transformed thymic cell lines were systematically analysed for potential defects in signalling molecules. One such cell line was found to express a 65-kDa fusion protein encoding the N-terminal 571 amino acids of p85 linked to a heterologous 24-amino acid, cysteine-rich domain (75). This protein when it was both associated with the catalytic p110 subunit and overexpressed, in contrast to p85, was found to increase the levels of 3-phosphoinositides and to activate PKB. At first, it was puzzling as to why this truncated form of p85 should activate p110 while overexpression of the full-length regulatory subunit actually decreases PI3K activity in transfected cells. Fractionation assays suggested that p65 localizes to the plasma membrane more stably than p85, thereby increasing the opportunities of the associated p110 to phosphorylate lipid substrates independently of cellular stimulation. The fact that PI3K can indeed act as an oncogene is conformation of its important role in cellular signalling processes regulating growth and proliferation. Only time will tell us whether such mutations also play a role in the progression of human neoplasias.

4.2 Apoptosis

'*Apoptosis*', the Greek word for the 'shedding of leaves from a tree', has been coined to delineate a structurally distinct mode of cell death responsible for cell loss within living tissues. In apoptosis, the cell is an active participant in its own demise and performs cellular 'suicide'. This occurs during physiological processes as diverse as embryogenesis and immune tolerance. Cells undergoing apoptosis exhibit character-istic morphological changes that include membrane blebbing, chromatin aggrega-tion, and DNA fragmentation. Since apoptotic cells are rapidly recognized by phago-cytes and disposed of, no inflammatory response is elicited *in vivo*.

An initial study implicating PI3K in the regulation of apoptosis involved trans-fected PC12 cells. In these cells, nerve growth factor (NGF) prevents apoptosis in-duced by serum withdrawal, an effect that is blocked by the pharmacological inhibition of PI3K (76). Transfection of these cells with the PDGF receptor results in the same protective effect as NGF, and by introducing receptor mutants only capable of activating specific signalling pathways an analysis of those pathways regulating apoptosis was possible. In PC12 cells expressing a mutant receptor, unable to activate PI3K, the induction of apoptosis occurred even in the presence of PDGF. This ability of PI3K to prevent apoptosis was also observed in other cell systems. For example, in the absence of serum, overexpression of c-myc in fibroblasts also induces apoptosis and this is blocked by overexpression of constitutively active p110 mutants (77).

The mechanism by which PI3K could regulate apoptotic events has been the subject of intensive study (Fig. 10). While several potential targets of PI3K activation have been implicated in the regulation of apoptosis, it is the downstream kinase, PKB, which has received the most attention (43). Initial studies on the effect of insulin-like growth factor-1 (IGF-1) on cerebellar neurons demonstrated an essential role for PI3K in their survival. Transfection of these cells with dominant-negative versions of PKB also enhanced apoptosis in both the presence and absence of survival factors (78). Furthermore, the overexpression of wild-type PKB enhanced survival even in the presence of pharmacological inhibitors of PI3K, demonstrating that PKB plays a critical role in the protective function of survival factors. How PI3K-mediated PKB activation acts to prevent apoptosis is beyond the scope of this review. However, a multitude of potential mechanisms have now been described including direct phosphorylation and inactivation of caspases, phosphorylation and inactivation of the proapoptotic factor Bad, and induction of the antiapoptotic *Bcl-2* gene (Fig. 10; ref. 43). The antiapoptotic effect of PI3K signalling induced by survival factors can itself be regulated by cellular stress. In response to activation of the Fas receptor (CD95), γ-irradiation, UV radiation, and hyperosmolar stress, the intracellular ceramide levels increase. These factors have also been found to inhibit PI3K activity and, more im-portantly, to induce apoptosis (79). This induction of apoptosis was again reduced by overexpression of active PI3K mutants. Although at this moment the precise mech-anism by which ceramide can inhibit PI3K is unknown, it does provide a model for explaining the mechanisms by which environmental stress can activate a suicide programme within target cells. Furthermore, during tumorigenesis when ceramide

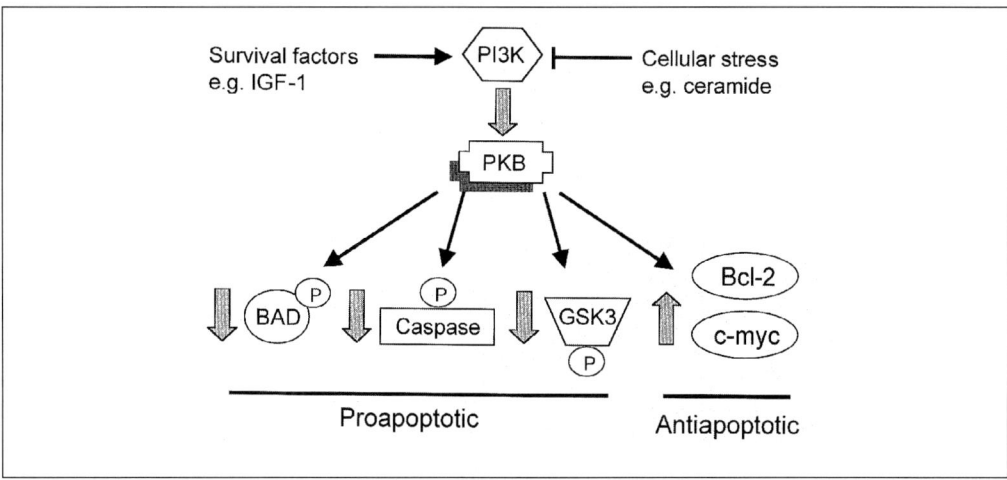

Fig. 10 Antiapoptotic signalling mediated by PI3K. The regulation of programmed cell death is critical in maintaining cellular homeostasis. Cellular stress can activate apoptosis through the generation of ceramide that inhibits PI3K. Survival factors, however, activate PI3K resulting in the phosphorylation and inhibition of various proapoptotic factors. Furthermore, there is an increase in the expression of several antiapoptotic components.

levels are limiting, PI3K is left unchecked and thus cells have an increased survival advantage. However, in degenerative diseases, where ceramide levels are elevated, cells will tend towards a destructive pathway due to a general inhibition of PI3K.

The regulation of epithelial- and endothelial-cell apoptosis upon detachment from the matrix is termed '*anoikis*', the ancient Greek word for 'homelessness'. This mechanism ensures that cells do not survive outside the context of their normal environment, thus preventing cells colonizing elsewhere if they become detached. This system has also recently been demonstrated to be regulated by PI3K (80). In canine kidney cells, detachment from the extracellular matrix leads to the induction of an apoptotic programme resulting in cell death. This apoptosis is concomitant with a rapid decrease in PI3K activity, while attachment conversely stimulates the generation of PI-lipid products—probably through focal adhesion kinase (FAK)-mediated activation of PI3K. If these cells are transfected with active mutants of p21Ras, PI3K, or PKB, detachment of the cells no longer results in this decrease of PI3K activity and the cells no longer become apoptotic. This suggests that the engagement of integrins by extracellular-matrix interactions results in the activation of PI3K and thus PKB, resulting in cellular survival. Detachment, however, results in a rapid inhibition of PI3K, induction of apoptosis, and thus the prevention of cell survival outside its normal microenvironment. This provides one explanation of how transformed cells are able to survive: activation of PI3K in non-adhered cells overrides the normal death signal.

The regulation of apoptosis by PI3K is critical in the maintenance of normal tissue homeostasis. It appears that PKB is one of the key mediators of the antiapoptotic activity of PI3K, although there are also PKB-independent systems. The mechanisms by which PI3K prevents apoptosis are likely to be critical in maintaining the trans-formed phenotype of many cells. In particular, tumour cells containing activating p21Ras mutations, which also serve to activate PI3K, will have a selective survival advantage based on repression of the normal apoptotic programme.

4.3 Cytoskeletal reorganization

Actin filaments, microtubules, and intermediate filaments comprise the major cytoskeletal protein networks regulating cell shape, movement, and various cellular processes. Local actin polymerization, underlying the plasma membrane, results in the generation of membrane protrusions critical in the processes of cellular migra-tion. The cytoskeleton can thus be considered as a dynamic structure that can be reshaped in response to diverse cellular stimuli. Stimulation of endothelial cells, for example, results in a rearrangement of cortical actin filaments in a criss-cross net-work, resulting in curtain-like extensions, lamellipodia, and so-called membrane 'ruffling'. These ruffles are lamellipodia that have lifted up from the substratum at the leading edge of the cells. As discussed above, mutants of the PDGF receptor have been utilized to define the role of specific signalling pathways in cellular functions. Mutant PDGF receptors lacking their major PI3K-binding site no longer stimulate this endothelial cell-membrane ruffling (81). The pharmacological inhibition of PI3K by the addition of wortmannin or the overexpression of p85, also both result in a block of PDGF-stimulated membrane ruffling. Furthermore, a transient overexpression of an activated PI3K mutant, p110*, is sufficient for the formation of actin membrane ruffles. Together this provides a clear role for PI3K in cytoskeletal reorganization leading to lamellipodia formation and membrane ruffling. Genetic analyses have also supported a role for PI3K in this process. Deletion of PI3K genes in the slime mould, *Dictyostelium discoideum*, is accompanied by dramatic defects in a subset of F-actin enriched structures such as ruffles and pseudopodia (82).

Cytoskeletal reorganization and the formation of ruffles requires the small GTPase p21Rac. Since PI3K is known to stimulate p21Rac through the activation of exchange factors (see Section 3.5) and dominant-negative p21Rac(N17) inhibits PI3K-induced ruffling (83), it is likely that p21Rac is indeed the target of PI3K activation, mediating cytoskeletal changes. How precisely PI3K-mediated p21Rac activation results in local actin polymerization is, as yet, unknown, although p21Rac is known to activate a wide variety of downstream signalling events.

One consequence of the role of PI3K in cytoskeletal reorganization is that it also appears to play a critical role in various aspects of the process regulating cell migra-tion. For example, transformation of mammary epithelial cells results in a switch to an invasive, more motile cell phenotype. Direct PI3K activation through the use of constitutively active mutants is sufficient to induce this motility, resulting in an invasive cell phenotype (84). Furthermore, inhibition of PI3K results in the disruption of actin structures and prevents the actin polymerization that normally results in cell spreading.

4.4 Membrane trafficking

Despite the continuous transport of membrane components throughout the cell, the distinct composition of the various intracellular compartments is maintained. This involves the targeting and fusion of specific vesicles to the appropriate target organelle. This process is highly regulated and complex, involving the docking of proteins, termed SNARES, on the vesicle (v-SNARES) with those on the target membrane (t-SNARES). Growing evidence has now implicated PI-lipids such as PtdIns(4,5)P$_2$ as well as products of PI3K in these processes.

As discussed earlier (see Section 1.2), the yeast *VPS34* gene was found to encode a PI3K. Yeast has proved to be a valuable system for studying the role of PI3K in trafficking since there is a single gene and it is an organism readily susceptible to genetic manipulation. The *vps* genes encode proteins whose functions are required for sorting and delivering soluble vacuole hydrolases from the late Golgi to the vacuole—the yeast equivalent of the lysosome. Yeast strains deleted for the *VPS34* gene or carrying the Vps34 protein (Vps34p) point mutations lack detectable PI3K activity and exhibit severe defects in vacuolar protein sorting (2). Further analysis revealed that Vps34p is associated with a serine/threonine kinase termed Vps15p, which both activates and recruits it to the membrane. It appears that Vps15p regulates the sorting of proteins to the vacuole from the Golgi by selectively recruiting Vps34p to the appropriate membrane site where vacuolar hydrolases are packaged into vesicles. Perhaps the local production of PtdIns(3)P by Vps34p may allow the recruitment of effector molecules, to catalyse the transport reaction (Fig. 11). Other possibilities include a change in membrane curvature due to incorporation of PtdIns(3)P stimulating vesicle budding. A similar system has now been identified in mammalian cells (15) and pharmacological inhibition of PI3K has been demonstrated to disrupt normal sorting of mammalian lysosomal hydrolases.

Again a link has been made between PI3K activation and small GTPases. One large family of GTPases, the Rabs, is thought to facilitate the docking of vesicles with target membranes. The Rab family contains a member, Rab5, which is thought to participate in the early stages of the endocytotic pathway. In its GTP-bound form, Rab5 activates fusion between coated vesicles and early endosomes, and stimulates horseradish peroxidase uptake and transferrin endocytosis. Inhibition of PI3K by the addition of wortmannin to cells blocks these functions, linking PI-lipid production to these endocytotic processes. However, PI3K inhibition did not affect peroxidase uptake in cells expressing a constitutively active Rab5(Q79L), suggesting that PI3K mediates its effects upstream of Rab5 possibly by activation of an exchange factor (85). It has been postulated that localized PI-lipid production might target proteins critical for cellular trafficking to the appropriate location. Indeed, it has been recently demonstrated in yeast that several proteins of the FYVE subfamily, previously shown to be required for vacuolar/lysosomal trafficking, bind specifically to PtdIns(3)P. These findings provide insight into the role of PI3K in the regulation of protein sorting.

PI3K has also been linked to other membrane trafficking events, in particular with respect to the stimulation of glucose uptake. One of the major metabolic responses of

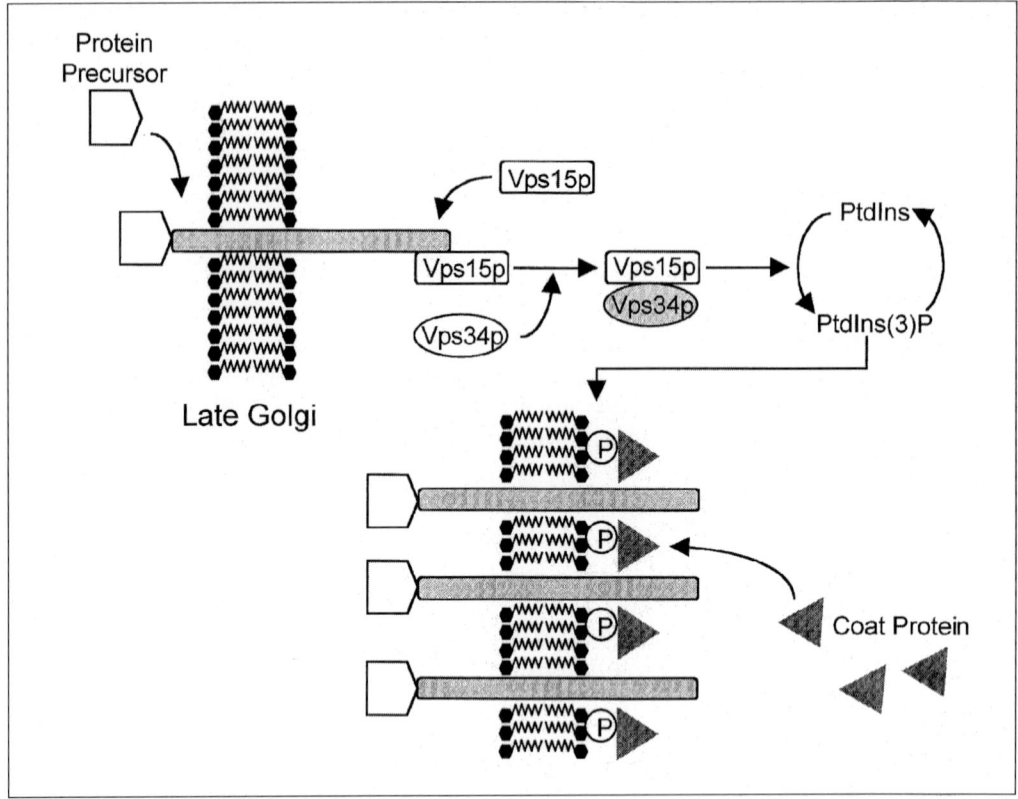

Fig. 11 Regulation of protein sorting by PI3K. The binding of vacuolar protein precursors to receptor molecules in the Golgi results in the activation of Vps15p. This provides a binding site for the Vps34p PI3K which then translocates to the membrane. Phosphorylation of PI-lipids by Vps34p and production of PtdIns(3)P results in the local recruitment of coat proteins required for vesicle formation and transport.

insulin is the stimulation of facilitated glucose transport in muscle and adipose tissues, primarily due to the translocation of the glucose transporter GLUT4 from an intracellular pool to the plasma membrane. The pharmacological inhibition of PI3K or introduction of dominant-negative mutants abolishes the recruitment of GLUT4 to the cell surface in a variety of adipose and muscle cells. Furthermore, constitutively active p110 variants act to stimulate GLUT4 translocation, thus demonstrating a role for PI3K in this process. The precise mechanism of this regulation is unclear, although several studies have implicated PKB and PKCζ as being components of this exocytotic pathway. Again, the Rab-family of small GTPases is thought to play a regulatory role. The Rab4 small GTPase is thought to be involved in the insulin-stimulated release of GLUT4-containing vesicles from the microsomal membrane fraction, thereby allowing their translocation to the plasma membrane. Pharmacological inhibition of PI3K acts to inhibit Rab4 activation placing PI3K directly upstream of Rab4-mediated vesicle release, analogous to PI3K-mediated activation of p21Rac.

The studies described above seem to suggest a general mechanism for the regulation of membrane trafficking either in endocytotic or exocytotic pathways. The localized generation of PI-lipids by PI3K acts to relocate exchange factors stimulating activating members of the Rab-family of small GTPases. This process results in the release, or docking, of specific vesicles from, or to, specific target membranes directing the flow of membrane traffic.

4.5   Protein synthesis and regulation of metabolism

Insulin stimulation of muscle, liver, and adipose tissue results not only in an increase in the rate of protein translation, but also the uptake of glucose from the blood, enhancing its conversion to glycogen and triacylglycerol. Major advances have been made in the understanding of these processes and, through the use of pharmacological inhibitors, a critical role for PI3K has been revealed.

Translation of mRNA into protein can be divided into the stages of initiation, elongation, and termination (60). The first two stages have been shown to be regulated directly by PI3K. Initiation involves the phosphorylation and association of the preinitiation ribosomal complex. As discussed in Section 3.3, phosphorylation of the S6 ribosomal subunit by p70S6K, results in the enhanced translation of certain mRNAs containing a 5'-polypyrimidine tract, many of which themselves are components of the translational machinery. This is one mechanism by which mitogen stimulation, mediated by PI3K, can enhance the rate of protein translation.

Enhanced initiation of translational can also occur by the regulation of initiation factors such as the initiation factor 4E (eIF-4E). Normally eIF-4E is complexed with a second protein termed 4E-BP1. The binding of BP-1 inhibits eIF-4E, preventing it from binding to the mRNA 5'-cap structure, which ensures the correct positioning of the ribosome at the start of translation. In response to insulin, BP-1 becomes phosphorylated and dissociates from eIF-4E, thus stimulating the initiation of transcription (Fig. 12). Phosphorylation of BP-1 is completely abolished by wortmannin and partially blocked by rapamycin, suggesting a dual activation mechanism involving at least two components downstream of PI3K (86). Overexpression of constitutively active PKB results in increased 4E-BP1 phosphorylation, even in the presence of wortmannin, suggesting that PKB may be one of the critical components downstream of PI3K.

Another initiation factor, eIF-2B, mediates the recycling of eIF-2, which is itself responsible for recruiting the initiator Met-tRNA to the ribosome (60). eIF-2 is active in a GTP-bound state where it forms an eIF2–GTP/Met-tRNA complex which subsequently binds to the 40S ribosomal subunit. eIF-2B itself, acts as a GEF, mediating the release of GDP from eIF2–GDP and allowing GTP-binding. Regulation of eIF-2B activity provides a general mechanism for controlling the rate of peptide-chain initiation and is thought to be critical in translational responses to viral infection and cellular stress. So how is this process regulated by insulin and what role does PI3K play in this response? Insulin dramatically activates eIF-2B and this activation is accompanied by dephosphorylation of the protein. Glycogen synthase kinase-3

(GSK-3) was subsequently found to phosphorylate and inactivate eIF-2B and was itself inactivated in response to insulin. Through the use of inhibitors, the insulin-mediated activation of GSK-3 has been demonstrated to be PI3K-dependent. Furthermore, recent work has demonstrated that PI3K effects are mediated by direct GSK-3 phosphorylation by PKB, inhibiting GSK-3 activity (87). Thus insulin stimulation results in PI3K activation, PKB activation, and phosphorylation of GSK-3 resulting in its inactivation. eIF-2B can be dephosphorylated, leading to activation, and mediates GTP for GDP exchange on eIF-2. In this way the rate of translational initiation can be carefully regulated (Fig. 13).

Along with a role in the fine control of protein synthesis, insulin-stimulated PI3K activation and the subsequent inactivation of GSK-3 by PKB, is also responsible for the regulation of metabolic pathways. Increased deposition of glycogen, for example, results as a consequence of increased glucose availability due to enhanced glucose transport, as well as increases in the activity of glycogen synthase. In fat and muscle cells, this is due to a dephosphorylation of key, inhibitory serine residues in the C-terminus of glycogen synthase, the key rate-limiting step in glycogen synthesis. The role of PI3K in this process is similar to that described above for the regulation of eIF-2B. Again, GSK-3 is responsible for phosphorylating and inhibiting glycogen synthase, and insulin stimulation results in GSK-3 inhibition through phosphorylation by PKB.

Lipogenesis, the control of degradation and synthesis of triacylglycerols by insulin, also appears to be regulated by a PI3K-dependent pathway. For example, activation of acetyl-CoA carboxylase by insulin stimulates *de novo* fatty acid synthesis. This correlates with increased phosphorylation and, again, PI3K inhibitors block this mechanism.

In summary, the activation of relatively few key-signalling molecules allows insulin to regulate diverse effector functions with concomitant changes in protein synthesis and metabolism. This paradigm is depicted in Fig. 14.

4.6 Differentiation

The differentiation of both muscle and adipose tissue is regulated by insulin-like growth factors (IGFs), although the molecular mechanisms of this are ill-defined. In muscle cells this differentiation is accompanied by cell fusion, myogenin, and GLUT4 gene expression. In adipocytes the expression of adipogenic-related genes such as fatty acid synthase, glycerol 3-phosphate dehydrogenase, and acetylcoenzyme A are markers of a terminally differentiated phenotype. The use of pharmacological inhibitors and dominant-negative mutants has again implicated PI3K as being a critical component of both the skeletal muscle and adipocyte differentiation programmes, as based on the expression of differentiation markers (88). This work is supported by the finding that the introduction of oncogenic, constitutively active v-p3k strongly enhances the myogenic differentiation of chicken myoblasts (89). The mechanisms by which PI3K can regulate gene expression on the transcriptional level remain unknown. There is, however, a clue suggested by the analysis of transcriptional regulation by the transcription factor C/EBPα in adipocytes (90). C/EBPα plays a critical

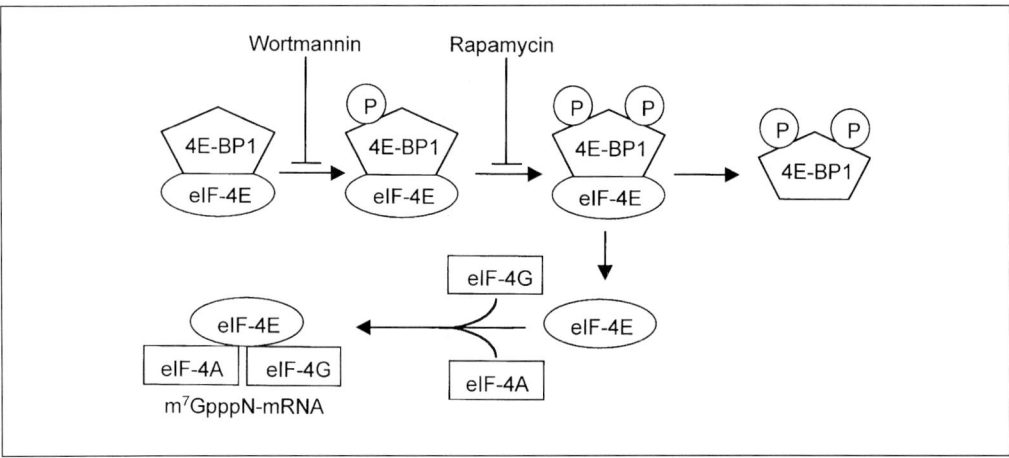

Fig. 12 PI3K can regulate protein synthesis through 4E-BP1 phosphorylation. Phosphorylation of 4E-BP1 prevents its association with eIF-4E allowing the formation of a complex with capped mRNA molecules (m^{7}GpppN–mRNA). A role for PI3K has been implicated in this phosphorylation through the use of pharmacological inhibitors (see text for details).

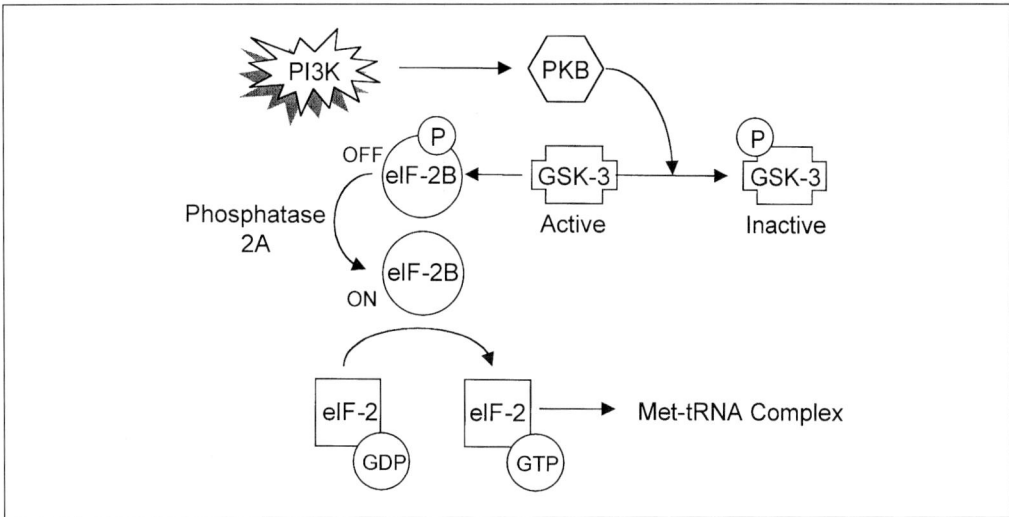

Fig. 13 Regulation of protein synthesis through PI3K-mediated GSK3 inactivation. Activation of PKB by PI3K results in the phosphorylation and inactivation of GSK3. This allows dephosphorylation of eIF-2B and subsequent exchange of GDP for GTP on eIF-2. Active eIF-2 is then able to associate with the initiator methionine–tRNA complex, recruiting it to the ribosome, and protein translation can proceed (see text for details).

role in adipocyte differentiation and is thought to continue to regulate transcription in mature cells by regulating, for example, GLUT4 expression. Insulin triggers C/EBPα dephosphorylation in a signalling pathway involving PI3K. Thus it is possible that through PI3K, regulating C/EBPα phosphorylation, cellular agonists can directly influence the differentiation programme.

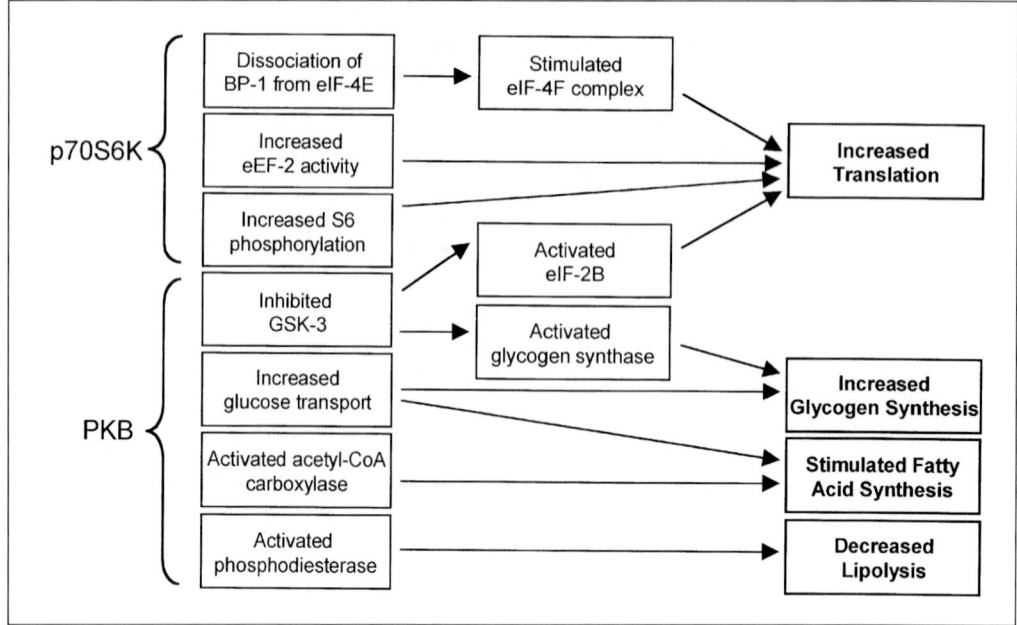

Fig. 14 Summary of PI3K-mediated signalling pathways regulating metabolism.

In contrast to the above studies, analysis of the signal transduction pathways mediating endocrine differentiation of fetal-derived pancreatic epithelial cells has revealed another picture. In this model system, inhibition of PI3K resulted in a dramatic *increase* in the number of differentiated cells, as measured by the marker of hormone production (91). This inhibitor-induced differentiation is also accompanied by a decrease in proliferation, as measured by DNA synthesis, thus coordinating division and differentiation. These unexpected results indicate that PI3K can also play a role as a negative regulator of cellular differentiation during fetal neogenesis of the endocrine system.

Taken together, the limited data so far suggest that PI3K can play both a positive and negative role in the regulation of differentiation in mammalian cells. It is difficult to analyse the role of PI3K in the development of an organism *per se*, since, as observed in the fruit fly, PI3K knockouts will probably be lethal due to the subsequent dysfunctional regulation of apoptosis that will occur (see Section 4.2).

5. Conclusions

This chapter has dealt in detail with the generation of phosphatidylinositol 3-lipids and their activation of downstream signalling pathways and cell-effector functions. It should be apparent to the reader that, while the activation and function of protein kinases has been the major focus work over the last 20 years, the cloning of PI3Ks has led to an invaluable understanding of the integration of lipid signalling into these protein kinase signalling routes. The reversible phosphorylation of phosphatidyl-inositol on multiple sites and the discovery of PH-domains provide a clue to signal-specificity and the realization of how signalling complexes can be generated within a cell after agonist stimulation. These protein relocalization events have homology with phosphotyrosine-SH2 interactions discussed elsewhere in this publication.

While much work has focused on the class I PI3Ks in terms of regulation and function, it is the C2 domain-containing class II enzymes whose function requires detailed investigation. The mechanisms of activation of this subfamily of PI3Ks, as well as downstream targets, will surely be the subject of much future work. Indeed, the general question of specificity within PI3K isoforms in terms of the activation of cellular functions still needs to be answered. Preliminary studies utilizing micro-injected antibodies against particular PI3K isoforms have thus far provided some limited evidence for specific roles for these enzymes in events such as proliferation. Furthermore, one facet of PI3K activation that still remains somewhat of a mystery is its protein kinase activity. The relevance of this in the autoregulation of PI3K or in downstream functioning requires further analysis. Is this second kinase activity simply the remnant of an ancestral protein kinase domain, whose ability to phosphorylate proteins is irrelevant? Recent studies suggest this not to be the case and further work will shed light on this interesting phenomenon.

There appears to be a complex network of events regulating mitogenesis, survival, metabolism, and differentiation, all regulated by PI3K. Indeed, this means that pharmacological modulation of PI3K activity is unlikely to provide a suitable target for the development of novel therapeutic agents. The generation of 3-phosphorylated PI-lipids is simply too important for cellular homeostasis, inhibition can result in cell death through apoptosis, while unregulated PI3K activity can lead to uncontrolled mitogenesis and neoplasia. Perhaps the modulation of downstream effectors of PI3K, such as PKB or PKCζ, will allow the development of novel therapeutic strategies utilizing pharmacological agents that prove to be less toxic.

References

1. Auger, K. R., Serunian, L. A., Soltoff, S. P., Libby, P., and Cantley, L. C. (1989). PDGF-dependent tyrosine phosphorylation stimulates production of novel polyphosphoinositides in intact cells. *Cell*, **57**, 167.
2. Schu, P. V., Takegawa, K., Fry, M. J., Stack, J. H., Waterfield, M. D., and Emr, S. D. (1993). Phosphatidylinositol 3-kinase encoded by yeast VPS34 gene essential for protein sorting. *Science*, **260**, 88.
3. Shaw, G. (1996). The pleckstrin homology domain: an intriguing multifunctional protein module. *Bioessays*, **18**, 35.
4. Hiles, I. D., Otsu, M., Volinia, S., Fry, M. J., Gout, I., Dhand, R., Panayotou, G., Ruiz-Larrea, F., Thompson, A., Totty, N. F., *et al.* (1992). Phosphatidylinositol 3-kinase: structure and expression of the 110 kd catalytic subunit. *Cell*, **70**, 419–29.
5. Morgan, S. J., Smith, A. D., and Parker, P. J. (1990). Purification and characterization of bovine brain type I phosphatidylinositol kinase. *Eur. J. Biochem.*, **191**, 761.
6. Otsu, M., Hiles, I., Gout, I., Fry, M. J., Ruiz-Larrea, F., Panayotou, G., Thompson, A.,

Dhand, R., Hsuan, J., Totty, N., *et al.* (1991). Characterization of two 85 kd proteins that associate with receptor tyrosine kinases, middle-T/pp60c-src complexes and PI3-kinase. *Cell*, **65**, 91.

7. Dhand, R., Hara, K., Hiles, I., Bax, B., Gout, I., Panayotou, G., Fry, M. J., Yonezawa, K., Kasuga, M., and Waterfield, M. D. (1994). PI 3-kinase: structural and functional analysis of intersubunit interactions. *EMBO J.*, **13**, 511.

8. Pleiman, C. M., Hertz, W. M., and Cambier, J. C. (1994). Activation of phosphatidyl-inositol-3 kinase by Src-family kinase SH3-binding to the p85 subunit. *Science*, **263**, 1609.

9. Antonetti, D. A., Algenstaedt, P., and Kahn, C. R. (1996). Insulin receptor substrate 1 binds two novel splice variants of the regulatory subunit of phosphatidylinositol 3-kinase in muscle and brain. *Mol. Cell. Biol.*, **16**, 2195.

10. Stoyanov, B., Volinia, S., Hanck, T., Rubio, I., Loubtchenkov, M., Malek, D., Stoyanova, S., Vanhaesebroeck, B., Dhand, R., *et al.* (1995). Cloning and characterization of a G protein-activated human phosphoinositide-3 kinase. *Science*, **269**, 690.

11. Stephens, L. R., Eguinoa, A., Erdjument-Bromage, H., Lui, M., Cooke, F., Coadwell, J., Smrcka, A. S., Thelen, M., Cadwallader, K., Tempst, P., and Hawkins, P. T. (1997). The G βγ sensitivity of a PI3K is dependent upon a tightly associated adaptor, p101. *Cell*, **89**, 105.

12. Vanhaesebroeck, B., Welham, M. J., Kotani, K., Stein, R., Warne, P. H., Zvelebil, M. J., Higashi, K., Volinia, S., Downward, J., and Waterfield, M. D. (1997). P110delta, a novel phosphoinositide 3-kinase in leukocytes. *Proc. Natl Acad. Sci. USA*, **94**, 4330.

13. Domin, J., Pages, F., Volinia, S., Rittenhouse, S. E., Zvelebil, M. J., Stein, R. C., and Waterfield, M. D. (1997). Cloning of a human phosphoinositide 3-kinase with a C2 domain that displays reduced sensitivity to the inhibitor wortmannin. *Biochem. J.*, **326**, 139.

14. Stack, J. H., DeWald, D. B., Takegawa, K., and Emr, S. D. (1995). Vesicle-mediated protein transport: regulatory interactions between the Vps15 protein kinase and the Vps34 PtdIns 3-kinase essential for protein sorting to the vacuole in yeast. *J. Cell. Biol.*, **129**, 321.

15. Volinia, S., Dhand, R., Vanhaesebroeck, B., MacDougall, L. K., Stein, R., Zvelebil, M. J., Domin, J., Panayatou, G., and Waterfield, M. D. (1995). A human phosphatidylinositol 3-kinase complex related to the yeast Vps34p-15p protein sorting system. *EMBO J.*, **14**, 3339.

16. Stephens, L., Cooke, F. T., Walters, R., Jackson, T., Volinia, S., Gout, I., Waterfield, M. D., and Hawkins, P. T. (1994). Characterization of a phosphatidylinositol-specific phospho-inositide 3-kinase from mammalian cells. *Curr. Biol.*, **4**, 203.

17. Carpenter, C. L., Auger, K. R., Duckworth, B. C., Hou, W. M., Schaffhausen, B., and Cantley, L. C. (1993). A tightly associated serine/threonine protein kinase regulates phospho-inositide 3-kinase activity. *Mol. Cell. Biol.*, **13**, 1657.

18. Dhand, R., Hiles, I., Panayotou, G., Roche, S., Fry, M. J., Gout, I., Totty, N. F., Truong, O., Vicendo, P., Yonezawa, K., *et al.* (1994). PI 3-kinase is a dual specificity enzyme: autoregulation by an intrinsic protein-serine kinase activity. *EMBO J.*, **13**, 522.

19. Stack, J. H. and Emr, S. D. (1994). Vps34p required for yeast vacuolar protein sorting is a multiple specificity kinase that exhibits both protein kinase and phosphatidylinositol-specific PI 3-kinase activities. *J. Biol. Chem.*, **16**, 31552.

20. Bondeva, T., Pirola, L., Bulgarelli-Leva, G., Rubio, I., Wetzker, R., and Wymann, M. P. (1998). Bifurcation of lipid and protein kinase signals of PI3Kgamma to the protein kinases PKB and MAPK. *Science*, **282**, 293.

21. Wymann, M. P., Bulgarelli-Leva, G., Zvelebil, M. J., Pirola, L., Vanhaesebroeck, B., Waterfield, M. D., and Panayotou, G. (1996). Wortmannin inactivates phosphoinositide 3-kinase by covalent modification of Lys-802, a residue involved in the phosphate transfer reaction. *Mol. Cell. Biol.*, **16**, 1722.

22. Cross, M. J., Stewart, A., Hodgkin, M. N., Kerr, D. J., and Wakelam, M. J. (1995). Wortmannin and its structural analogue demethoxyviridin inhibit stimulated phospholipase A2 activity in Swiss 3T3 cells. Wortmannin is not a specific inhibitor of phosphatidylinositol 3-kinase. *J. Biol. Chem.*, **270**, 25352.

23. Hu, Q., Klippel, A., Muslin, A. J., Fantl, W. J., and Williams, L. T. (1995). Ras-dependent induction of cellular responses by constitutively active phosphatidylinositol-3 kinase. *Science*, **268**, 100.

24. Rodriguez-Viciana, P., Warne, P. H., Vanhaesebroeck, B., Waterfield, M. D., and Downward, J. (1996). Activation of phosphoinositide 3-kinase by interaction with Ras and by point mutation. *EMBO J.*, **15**, 2442.

25. Klippel, A., Reinhard, C., Kavanaugh, W. M., Apell, G., Escobedo, M. A., and Williams, L. T. (1996). Membrane localization of phosphatidylinositol 3-kinase is sufficient to activate multiple signal-transducing kinase pathways. *Mol. Cell. Biol.* **16**, 4117.

26. McGlade, C. J., Ellis, C., Reedijk, M., Anderson, D., Mbamalu, G., Reith, A. D., Panayotou, G., End, P., Bernstein, A., Kazlauskas, A., *et al.* (1992). SH2 domains of the p85 alpha subunit of phosphatidylinositol 3-kinase regulate binding to growth factor receptors. *Mol. Cell. Biol.* **12**, 991.

27. Songyang, Z., Shoelson, S. E., Chaudhuri, M., Gish, G., Pawson, T., Haser, W. G., King, F., Roberts, T., Ratnofsky, S., Lechleider, R. J., *et al.* (1993). SH2 domains recognize specific phosphopeptide sequences. *Cell*, **72**, 767.

28. Kavanaugh, W. M., Klippel, A., Escobedo, J. A., and Williams, L. T. (1992). Modification of the 85-kilodalton subunit of phosphatidylinositol-3 kinase in platelet-derived growth factor-stimulated cells. *Mol. Cell. Biol.* **12**, 3415

29. Roche, S., Dhand, R., Waterfield, M. D., and Courtneidge, S. A. (1994). The catalytic subunit of phosphatidylinositol 3-kinase is a substrate for the activated platelet-derived growth factor receptor, but not for middle-T antigen-pp60c-src complexes. *Biochem. J.*, **301**, 703.

30. Ruiz-Larrea, F., Vicendo, P., Yaish, P., End, P., Panayotou, G., Fry, M. J., Morgan, S. J., Thompson, A., Parker, P. J., and Waterfield, M. D. (1993). Characterization of the bovine brain cytosolic phosphatidylinositol 3-kinase complex. *Biochem. J.*, **290**, 609.

31. Hayashi, H., Kamohara, S., Nishioka, Y., Kanai, F., Miyake, N., Fukui, Y., Shibasaki, F., Takenawa, T., and Ebina, Y. (1992). Insulin treatment stimulates the tyrosine phosphorylation of the alpha-type 85-kDa subunit of phosphatidylinositol 3-kinase *in vivo*. *J. Biol. Chem.*, **267**, 22575.

32. Sjolander, A., Yamamoto, K., Huber, B. E., and Lapetina, E. G. (1991). Association of p21ras with phosphatidylinositol 3-kinase. *Proc. Natl Acad. Sci. USA*, **88**, 7908.

33. Rodriguez-Viciana, P., Warne, P. H., Dhand, R., Vanhaesebroeck, B., Gout, I., Fry, M. J., Waterfield, M. D., and Downward, J. (1994). Phosphatidylinositol-3-OH kinase as a direct target of Ras. *Nature*, **370**, 527.

34. Rubio, I., Rodriguez-Viciana, P., Downward, J., and Wetzker, R. (1997). Interaction of Ras with phosphoinositide 3-kinase gamma. *Biochem. J.*, **326**, 891.

35. Stephens, L., Smrcka, A., Cooke, F. T., Jackson, T. R., Sternweis, P. C., and Hawkins, P. T. (1994). A novel phosphoinositide 3 kinase activity in myeloid-derived cells is activated by G protein beta gamma subunits. *Cell*, **77**, 83.

36. Stephens, L. R., Eguinoa, A., Erdjument-Bromage, H., Lui, M., Cooke, F., Coadwell, J., Smrcka A. S., Thelen, M., Cadwallader, K., Tempst, P., and Hawkins P. T. (1997). The G beta gamma sensitivity of a PI3K is dependent upon a tightly associated adaptor, p101. *Cell*, **89**, 105.

37. Kavanaugh, W. M., Pot, D. A., Chin, S. M., Deuter-Reinhard, M., Jefferson, A. B., Norris, F. A., Masiarz, F. R., Cousens, L. S., Majerus, P. W., and Williams, L. T. (1996). Multiple forms of an inositol polyphosphate 5-phosphatase form signaling complexes with Shc and Grb2. *Curr. Biol.*, **6**, 438.

38. Damen, J. E., Liu, L., Rosten, P., Humphries, R. K., Jefferson, A. B., Majerus, P. W., and Krystal, G. (1996). The 145-kDa protein induced to associate with Shc by multiple cytokines is an inositol tetra-phosphate and phosphatidylinositol 3,4,5-triphosphate 5-phosphatase. *Proc. Natl Acad. Sci. USA*, **93**, 1689.

39. Habib, T., Hejna, J. A., Moses, R. E., and Decker, S. J. (1998). Growth factors and insulin stimulate tyrosine phosphorylation of the 51C/SHIP2 protein. *Biochem. J.*, **273**, 18605.

40. Lioubin, M. N., Algate, P. A., Tsai, S., Carlberg, K., Aebersold, A., and Rohrschneider, L. R. (1996). p150Ship, a signal transduction molecule with inositol polyphosphate-5-phosphatase activity. *Genes Dev.*, **10**, 1084.

41. Deuter-Reinhard, M., Apell, G., Pot, D., Klippel, A., Williams, L. T., and Kavanaugh, W. M. (1997). SIP/SHIP inhibits *Xenopus* oocyte maturation induced by insulin and phosphatidylinositol 3-kinase. *Mol. Cell. Biol.*, **17**, 2559.

42. Scharenberg, A. M. and Kinet, J. P. (1998). PtdIns-3,4,5-P3: a regulatory nexus between tyrosine kinases and sustained calcium signals. *Cell*, **94**, 5.

43. Coffer, P. J., Jin, J., and Woodgett, J. R. (1998). Protein kinase B (c-Akt): a multifunctional mediator of phosphatidylinositol 3-kinase activation. *Biochem. J.*, **335**, 1.

44. Li, J., Yen, C., Liaw, D., Podsypanina, K., Bose, S., Wang, S. I., Puc, J., Miliaresis, C., Rodgers, L., McCombie, R., Bigner, S. H., Giovanella, B. C., Ittmann, M., Tycko, B., Hibshoosh, H., Wigler, M. H., and Parsons, R. (1997). PTEN, a putative protein tyrosine phosphatase gene mutated in human brain, breast and prostate cancer. *Science*, **275**, 1943.

45. Maehama, T. and Dixon, J. E. (1998). The tumor suppressor, PTEN/MMAC1, dephosphorylates the lipid second messenger, phosphatidylinositol 3,4,5-trisphosphate. *J. Biol. Chem.*, **273**, 13375.

46. Stambolic, V., Suzuki, A., de la Pompa, J. L., Brothers, G. M., Mirtsos, C., Sasaki, T., Ruland, J., Penninger, J. M., Siderovski, D. P., and Mak, T. W. (1998). Negative regulation of PKB/Akt-dependent cell survival by the tumor suppressor PTEN. *Cell*, **95**, 29.

47. Bellacosa, A., Testa, J. R., Staal, S. P., and Tsichlis, P. N. (1991). A retroviral oncogene, akt, encoding a serine-threonine kinase containing an SH2-like region. *Science*, **254**, 274.

48. Burgering, B. M. and Coffer, P. J. (1995). Protein kinase B (c-Akt) in phosphatidylinositol-3-OH kinase signal transduction. *Nature*, **376**, 599.

49. Franke, T. F., Yang, S. I., Chan, T. O., Datta, K., Kazlauskas, A., Morrison, D. K., Kaplan, D. R., and Tsichlis, P. N. (1995). The protein kinase encoded by the Akt proto-oncogene is a target of the PDGF-activated phosphatidylinositol 3-kinase. *Cell*, **81**, 727.

50. Franke, T. F., Kaplan, D. R., Cantley, L. C., and Toker, A. (1997). Direct regulation of the Akt proto-oncogene product by phosphatidylinositol-3,4-bisphosphate. *Science*, **275**, 665.

51. Alessi, D. R. Andjelkovic, M., Caudwell, B., Cron, P., Morrice, N., Cohen, P., and Hemmings, B. A. (1996). Mechanism of activation of protein kinase B by insulin and IGF-1. *EMBO J.*, **15**, 6541.

52. Staveley, B. E., Ruel, L., Jin, J., Stambolic, V., Mastronardi, F. G., Heitzler, P., Woodgett, J. R., and Manoukian, A. S. (1998). Genetic analysis of protein kinase B (AKT) in *Drosophila*. *Curr. Biol.*, **8**, 599.

53. Alessi, D. R., James, S. R., Downes, C. P., Holmes, A. B., Gaffney, P. R., Reese, C. B., and Cohen, P. (1997). Characterization of a 3-phosphoinositide-dependent protein kinase which phosphorylates and activates protein kinase B alpha. *Curr. Biol.*, **7**, 261.

54. Stephens, L., Anderson, K., Stokoe, D., Erdjument-Bromage, H., Painter, G. F., Holmes, A. B., Gaffney, P. R., Reese, C. B., McCormick, F., Tempst, P., Coadwell, J., and Hawkins, P. T. (1998). Protein kinase B kinases that mediate phosphatidylinositol 3,4,5-trisphosphate-dependent activation of protein kinase B. *Science*, **279**, 710.

55. Alessi, D. R., Deak, M., Casamayor, A., Caudwell, F. B., Morrice, N., Norman, D. G., Gaffney, P., Reese, C. B., MacDougall, C. N., Harbison, D., Ashworth, A., and Bownes, M. (1997). 3-Phosphoinositide-dependent protein kinase-1 (PDK1): structural and functional homology with the *Drosophila* DSTPK61 kinase. *Curr. Biol.*, **7**, 776–89.

56. Anderson, K. E., Coadwell, J., Stephens, L. R., and Hawkins, P. T. (1998). Translocation of PDK-1 to the plasma membrane is important in allowing PDK-1 to activate protein kinase B. *Curr. Biol.*, **8**, 684.

57. Delcommenne, M., Tan, C., Gray, V., Rue, L., Woodgett, J., and Dedhar, S. (1998). Phosphoinositide-3-OH kinase-dependent regulation of glycogen synthase kinase 3 and protein kinase B/AKT by the integrin-linked kinase. *Proc. Natl Acad. Sci. USA*, **95**, 11211.

58. Proud, C. G. (1996). p70S6 kinase: an enigma with variations. *Trends Biochem. Sci.*, **21**, 181.

59. Chung, J., Grammer, T. C., Lemon, K. P., Kazlauskas, A., and Blenis, J. (1994). PDGF- and insulin-dependent pp70S6k activation mediated by phosphatidylinositol-3-OH kinase. *Nature*, **370**, 71.

60. Proud, C. G. and Denton, R. M. (1997). Molecular mechanisms for the control of translation by insulin. *Biochem. J.*, **328**, 329.

61. Alessi, D. R., Kozlowski, M. T., Weng, Q. P., Morrice, N., and Avruch, J. (1998). 3-Phosphoinositide-dependent protein kinase 1 (PDK1) phosphorylates and activates the p70 S6 kinase *in vivo* and *in vitro*. *Curr. Biol.*, **8**, 69.

62. Toker, A., Meyer, M., Reddy, K. K., Falck, J. R., Aneja, R., Aneja, S., Parra, A., Burns, D. J., Ballas, L. M., and Cantley, L. C. (1994). Activation of protein kinase C family members by the novel polyphosphoinositides PtdIns-3,4-P2 and PtdIns-3,4,5-P3. *J. Biol. Chem.*, **269**, 32358.

63. Standaert, M. L., Galloway, L., Karnam, P., Bandyopadhyay, G., Moscat, J., and Farese, R. V. (1997). Protein kinase C-zeta as a downstream effector of phosphatidylinositol 3-kinase during insulin stimulation in rat adipocytes. Potential role in glucose transport. *J. Biol. Chem.*, **272**, 30075.

64. Akimoto, K., Takahashi, R., Moriya, S., Nishioka, N., Takayanagi, J., Kimura, K., Fukui, Y., Osada, S., Mizuno, K., Hirai, S., Kazlauskas, A., and Ohno, S. (1996). EGF or PDGF receptors activate atypical PKClambda through phosphatidylinositol 3-kinase. *EMBO J.*, **15**, 788.

65. Moriya, S., Kazlauskas, A., Akimoto, K., Hirai, S., Mizuno, K., Takenawa, T., Fukui, Y., Watanabe, Y., Ozaki, S., and Ohno, S. (1996). Platelet-derived growth factor activates protein kinase C epsilon through redundant and independent signaling pathways involving phospholipase C gamma or phosphatidylinositol 3-kinase. *Proc. Natl Acad. Sci. USA*, **93**, 151.

66. Le Good, J. A., Ziegler, W. H., Parekh, D. B., Alessi, D. R., Cohen, P., and Parker, P. J. (1998). Protein kinase C isotypes controlled by phosphoinositide 3-kinase through the protein kinase PDK1. *Science*, **281**, 2042.

67. Chou, M. M., Hou, W., Johnson, J., Graham, L. K., Lee, M. H., Chen, C. S., Newton, A. C., Schaffhausen, B. S., and Toker, A. (1998). Regulation of protein kinase C zeta by PI 3-kinase and PDK-1. *Curr. Biol.*, **8**, 1069.

68. Hawkins, P. T., Eguinoa, A., Qiu, R. G., Stokoe, D., Cooke, F. T., Walters, R., Wennstrom, S., Claesson-Welsh, L., Evans, T., Symons, M., *et al.* (1995). PDGF stimulates an increase in GTP-Rac via activation of phosphoinositide 3-kinase. *Curr. Biol.*, **5**, 393.

69. Han, J., Luby-Phelps, K., Das, B., Shu, X., Xia, Y., Mosteller, R. D., Krishna, U. M., Falck, J. R., White, M. A., and Broek, D. (1998). Role of substrates and products of PI 3-kinase in regulating activation of Rac-related guanosine triphosphatases by Vav. *Science*, **279**, 558.

70. Roche, S., Koegl, M., and Courtneidge, S. A. (1994). The phosphatidylinositol 3-kinase alpha is required for DNA synthesis induced by some, but not all, growth factors. *Proc. Natl Acad. Sci. USA*, **91**, 9185.

71. Klippel, A., Escobedo, M. A., Wachowicz, M. S., Apell, G., Brown, T. W., Giedlin, M. A., Kavanaugh, W. M., and Williams, L. T. (1998). Activation of phosphatidylinositol 3-kinase is sufficient for cell cycle entry and promotes cellular changes characteristic of oncogenic transformation. *Mol. Cell. Biol.* **18**, 5699.

72. Leevers, S. J., Weinkove, D., MacDougall, L. K., Hafen, E., and Waterfield, M. D. (1996). The *Drosophila* phosphoinositide 3-kinase Dp110 promotes cell growth. *EMBO J.*, **15**, 6584.

73. Morris, J. Z., Tissenbaum, H. A., and Ruvkun, G. (1996). A phosphatidylinositol-3-OH kinase family member regulating longevity and diapause in *Caenorhabditis elegans*. *Nature*, **382**, 536.

74. Chang, H. W., Aoki, M., Fruman, D., Auger, K. R., Bellacosa, A., Tsichlis, P. N., Cantley, L. C., Roberts, T. M., and Vogt, P. K. (1997). Transformation of chicken cells by the gene encoding the catalytic subunit of PI 3-kinase. *Science*, **276**, 1848.

75. Jimenez, C., Jones, D. R., Rodriguez-Viciana, P., Gonzalez–Garcia, A., Leonardo, E., Wennstrom, S., von Kobbe, C., Toran, J. L., R-Borlado, L., Calvo, V., Copin, S. G., Albar, J. P., Gaspar, M. L., Diez, E., Marcos, M. A., Downward, J., Martinez, A. C., Merida, I., and Carrera, A. C. (1998). Identification and characterization of a new oncogene derived from the regulatory subunit of phosphoinositide 3-kinase. *EMBO J.*, **17**, 743.

76. Yao, R. and Cooper, G. M. (1995). Requirement for phosphatidylinositol-3 kinase in the prevention of apoptosis by nerve growth factor. *Science*, **267**, 2003.

77. Kauffmann-Zeh, A., Rodriguez-Viciana, P., Ulrich, E., Gilbert, C., Coffer, P., Downward, J., and Evan, G. (1997). Suppression of c-Myc-induced apoptosis by Ras signalling through PI(3)K and PKB. *Nature*, **385**, 544.

78. Dudek, H., Datta, S. R., Franke, T. F., Birnbaum, M. J., Yao, R., Cooper, G. M., Segal, R. A., Kaplan, D. R., and Greenberg, M. E. (1997). Regulation of neuronal survival by the serine-threonine protein kinase Akt. *Science*, **275**, 661.

79. Zundel, W. and Giaccia, A. (1998). Inhibition of the anti-apoptotic PI(3)K/Akt/Bad pathway by stress. *Genes Dev.*, **12**, 1941–6

80. Khwaja, A., Rodriguez-Viciana, P., Wennstrom, S., Warne, P. H., and Downward, J. (1997). Matrix adhesion and Ras transformation both activate a phosphoinositide 3-OH kinase and protein kinase B/Akt cellular survival pathway. *EMBO J.*, **16**, 2783.

81. Wennstrom, S., Hawkins, P., Cooke, F., Hara, K., Yonezawa, K., Kasuga, M., Jackson, T., Claesson-Welsh, L., and Stephens L. (1994). Activation of phosphoinositide 3-kinase is required for PDGF-stimulated membrane ruffling. *Curr. Biol.*, **4**, 385.

82. Zhou, K., Pandol, S., Bokoch, G., and Traynor-Kaplan, A. E. (1998). Disruption of *Dictyostelium* PI3K genes reduces [^{32}P]phosphatidylinositol 3,4 bisphosphate and [^{32}P]phosphatidylinositol trisphosphate levels, alters F-actin distribution and impairs pinocytosis. *J. Cell. Sci.*, **111**, 283.

83. Rodriguez-Viciana, P., Warne, P. H., Khwaja, A., Marte, B. M., Pappin, D., Das, P., Waterfield, M. D., Ridley, A., and Downward, J. (1997). Role of phosphoinositide 3-OH kinase in cell transformation and control of the actin cytoskeleton by Ras. *Cell*, **89**, 457.

84. Keely, P. J., Westwick, J. K., Whitehead, I. P., Der, C. J., and Parise, L. V. (1997). Cdc42 and Rac1 induce integrin-mediated cell motility and invasiveness through PI(3)K. *Nature*, **390**, 632.

85. Li, G., D'Souza-Schorey, C., Barbieri, M. A., Roberts, R. L., Klippel, A., Williams, L. T., and Stahl, P. D. (1995). Evidence for phosphatidylinositol 3-kinase as a regulator of endocytosis via activation of Rab5. *Proc. Natl Acad. Sci. USA*, **24**, 10207.

86. Diggle, T. A., Moule, S. K., Avison, M. B., Flynn, A., Foulstone, E. J., Proud, C. G., and Denton, R. M. (1996). Both rapamycin-sensitive and -insensitive pathways are involved in the phosphorylation of the initiation factor-4E-binding protein (4E-BP1) in response to insulin in rat epididymal fat-cells. *Biochem. J.*, **316**, 447.

87. Cross, D. A., Alessi, D. R., Cohen, P., Andjelkovich, M., and Hemmings, B. A. (1995). Inhibition of glycogen synthase kinase-3 by insulin mediated by protein kinase B. *Nature*, **378**, 785.

88. Kaliman, P., Vinals, F., Testar, X., Palacin, M., and Zorzano, A. (1996). Phosphatidylinositol 3-kinase inhibitors block differentiation of skeletal muscle cells. *J. Biol. Chem.*, **271**, 19146.

89. Jiang, B. H., Zheng, J. Z., and Vogt, P. K. (1998). An essential role of phosphatidylinositol 3-kinase in myogenic differentiation. *Proc. Natl Acad. Sci. USA*, **95**, 14179.

90. Hemati, N., Ross, S. E., Erickson, R. L., Groblewski, G. E., and MacDougald, O. A. (1997). Signaling pathways through which insulin regulates CCAAT/enhancer binding protein alpha (C/EBPalpha). phosphorylation and gene expression in 3T3-L1 adipocytes. Correlation with GLUT4 gene expression. *J. Biol. Chem.*, **272**, 25913.

91. Ptasznik, A., Beattie, G. M., Mally, M. I., Cirulli, V., Lopez, A., and Hayek, A. (1997). Phosphatidylinositol 3-kinase is a negative regulator of cellular differentiation. *J. Cell. Biol.*, **137**, 1127.

2 | Mammalian MAP kinase pathways

JOHN M. KYRIAKIS

1. General considerations

1.1 The MAP3K → MEK → MAPK core signalling module, an emerging paradigm

Mitogen-activated protein kinase (MAPK) signal transduction pathways are among the most widespread mechanisms of cellular regulation. All eukaryotic cells possess multiple MAPK pathways, each of which is preferentially recruited by distinct sets of stimuli, thereby allowing the cell to respond in parallel to multiple divergent inputs. Mammalian MAPK pathways can be recruited by a wide variety of different stimuli ranging from hormones, such as insulin and growth hormone, to mitogens (i.e. epidermal growth factor, EGF; platelet-derived growth factor, PDGF; fibroblast growth factor, FGF), vasoactive peptides (angiotensin-II, endothelin), inflammatory cytokines of the tumour necrosis factor (TNF) family, and environmental stresses such as osmotic shock, ionizing radiation, and ischaemic injury.

All MAPK pathways include central, three-tiered 'core signalling modules' (Fig. 1) wherein MAPKs are activated by concomitant Thr and Tyr phosphorylation catalysed by a family of dual-specificity kinases referred to as MAPK/extracellular signal-regulated kinase (ERK)-kinases (MEKs or MKKs). MEKs, in turn, are regulated by Ser/Thr phosphorylation catalysed by several protein kinase families collectively referred to as MAPK-kinase-kinases (MAP3Ks). The core signalling modules are themselves regulated by a divergent variety of upstream activators and inhibitors including GTPases of the Ras superfamily and adapter proteins coupled to cytokine receptors (1–3).

The notion of multiple, parallel, MAPK signalling cascades was first appreciated from studies of simple eukaryotes such as the budding yeast *Saccharomyces cerevisiae*. To date, six *S. cerevisiae* MAPK signalling pathways have been identified (3). These are also discussed in Chapter 3. The features of yeast MAPK pathways, as well as early biochemical studies, have revealed several common principles shared by all MAPK pathways:

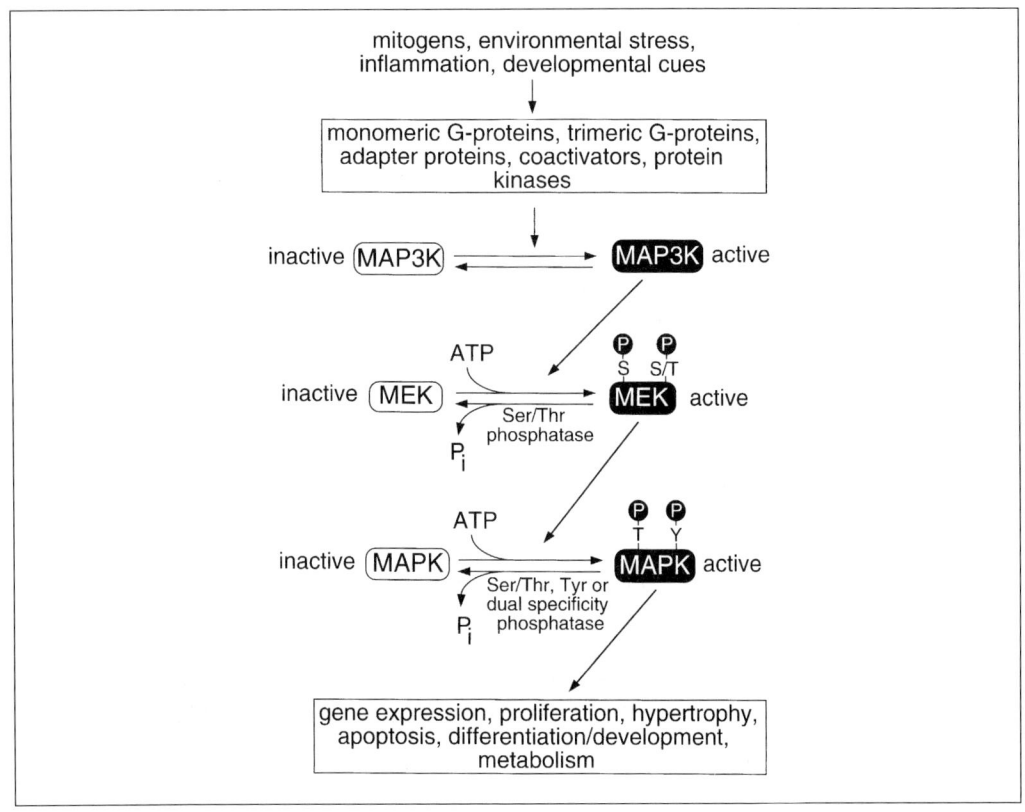

Fig. 1 The MAPK core signalling module. Divergent inputs feed into MAP3K → MEK → MAPK core pathways which then recruit appropriate responses.

1. *MAPKs are proline-directed kinases; however, substrate selectivity is often conferred by specific MAPK docking sites present on physiological substrates*: The proline-directed substrate specificity of the MAP kinases was established using peptide substrates corresponding to the sequences surrounding Thr 669 of the EGF receptor (Glu–Leu–Val–Glu–Pro–Leu–<u>Thr</u>–Pro–Ser–Gly–Glu–Ala–Pro–Asn–Gln–Ala–Leu–Leu–Arg) and the site on myelin basic protein (Ala–Pro–Arg–<u>Thr</u>–Pro–Gly–Gly–Arg) phosphorylated *in vitro* by the 42-kDa insulin- and mitogen-stimulated MAP kinase ERK2. Systematic variation in the sequences surrounding the single Thr phosphorylation site in the EGFR peptide and the single Thr in the MBP peptide, verified the essential role of the proline at +1 carboxyl-terminal to the site of phosphorylation (4–6). However, this proline directedness is not sufficient to account for the high degree of substrate selectivity manifested by different MAPK subgroups. It has become apparent that many of the physiological substrates of the MAPKs possess specific binding sites that allow for a strong interaction with select MAPK subfamilies to the exclusion of others (7–9). This confers a striking substrate specificity on a family of protein kinases with an otherwise apparently broad *in vitro* substrate profile.

2. *Signalling components with more than one biological function/signalling components under multiple forms of regulation*: Within MAPK core signalling modules, there are instances wherein individual elements can function in more than one pathway. For example, the *S. cerevisiae* MAP3K Ste11p functions as part of the mating-pheromone response pathway and the osmosensing pathway (3, 10), while the mammalian MAP3K tumour progression locus-2 (*Tpl-2*) can activate both mitogenic (ERK)- and stress-activated protein kinase pathways (11). Conversely, the yeast osmosensing MEK Pbs2p can be regulated by three MAP3Ks (Ssk2p, Ssk22p, and Ste11p), while the mammalian stress-regulated, MEK stress-activated, protein kinase/ERK-kinase-1 (SEK1)/MAPK-kinase-4 (MKK4) is a putative substrate for at least 10 known MAP3Ks (1, 3, 10, 12).

3. *Pathway segregation by scaffolding proteins*: Given the extraordinary complexity and diversity of MAPK regulation and function, it is critical that the speed and selectivity of MAPK pathways be preserved. Scaffold proteins, by binding and sequestering select MAPK pathway components, help to maintain pathway integrity and to permit the coordinated and efficient activation and function of MAPK components in response to specific types of stimuli (13). Some yeast signalling pathways include distinct scaffolding proteins that bind and segregate groups of signalling components. Alternatively, in some MAPK pathways, the signalling components themselves possess intrinsic scaffolding properties (Fig. 2). Thus, Ste5p of the yeast mating-pheromone pathway is a scaffolding protein that selectively binds a MAP3K (Ste11p), a MEK (Ste7p), and a MAPK (Fus3p) and couples them to upstream activators. Therefore, although Ste11p can function in both the mating and osmosensing pathways, it selectively activates different MEKs in each pathway: Ste7p for the mating pathway and Pbs2p for the osmosensing pathway. This selectivity is due in part to the fact that Ste5p maintains signalling pathway specificity by selectively binding Ste7p and not Pbs2p (3, 10).

Conversely, Pbs2p, in addition to serving as a MEK, acts as a scaffold protein, selectively binding Ste11p and the osmosensing MAPK Hog1p. Pbs2p does not bind Fus3p; and thus Pbs2p maintains signalling pathway integrity by interacting specifically with Hog1p and not with Fus3p or Kss1p (3, 10). Likewise, the mammalian scaffold proteins c-Jun-N-terminal kinase (JNK) interacting protein-1 (JIP-1), and MEK partner-1 (MP1), like Ste5p, couple, respectively, elements of mammalian stress- and mitogen-activated MAPK core signalling modules; while mammalian SEK1, and germinal centre kinase (GCK), like Pbs2p, possess both the properties of a protein kinase and a scaffold protein (14–18).

2. The Ras MAPK pathway

2.1 The Ras → MAPK pathway, a MAPK pathway in mammalian cells that is activated by insulin and mitogens—general considerations

The first mammalian MAPK was detected as an insulin-stimulated 40–44-kDa Ser/Thr kinase that could phosphorylate microtubule-associated protein-2 (MAP2);

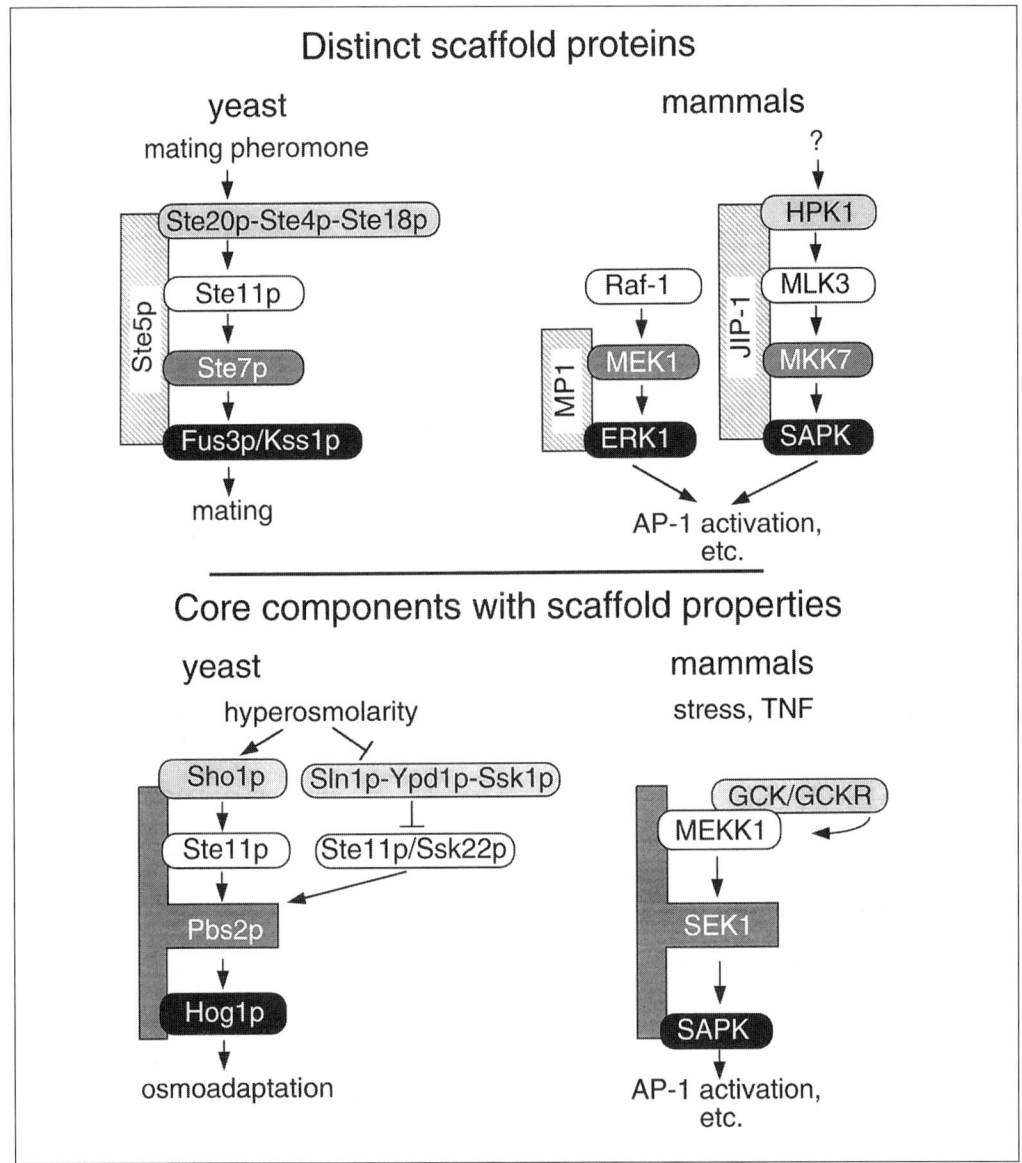

Fig. 2 Scaffold proteins and the regulation of MAPK pathway specificity and selectivity. Scaffold proteins can be distinct polypeptides (top) or elements of core pathways (bottom).

accordingly, this kinase was designated microtubule-associated protein kinase (MAPK). Activation of the insulin-stimulated MAPK was rapid—preceding the activation of other known mitogen-activated Ser/Thr kinases—suggesting a proximal role in signal transduction. MAPK was rechristened mitogen-activated protein kinase when it was observed that this kinase was activated not only by insulin, but by a wide variety of mitogens that couple to Tyr kinases (19, 20).

Table 1 MAPK nomenclature

Name	Alternate names	Substrates
ERK1	**p44-MAPK**	MAPKAP-K1, MNKs, MSKs, Elk1
ERK2	**p42-MAPK**	MAPKAP-K1, MNKs, MSKs, Elk1
SAPK-α	**JNK2, SAPK1a**	c-Jun, JunD, ATF2, Elk1
SAPK-p54α1	JNK2β2	
SAPK-p54α2	JNK2α2	
SAPK-p46α1	JNK2β1	
SAPK-p46α2	JNK2α1	
SAPK-β	**JNK3, SAPK1b**	c-Jun, JunD, ATF2, Elk1
SAPK-p54β1	JNK3β2	
SAPK-p54β2	JNK3α2	
SAPK-p46β1	JNK3β1	
SAPK-p46β2	JNK3α1	
SAPK-γ	**JNK1, SAPK1c**	c-Jun, Jun D, ATF2, Elk1
SAPK-p54γ1	JNK1β2	
SAPK-p54γ2	JNK1α2	
SAPK-p46γ1	JNK1β1	
SAPK-p46γ2	JNK1α1	
p38α	**SAPK2a, CSBP1**	MAPKAP-K2/3, MSKs, ATF2, Elk1, MEF2C
p38β	**SAPK2b, p38-2**	MAPKAP-K2/3, MSKs, ATF2
p38γ	**SAPK3, ERK6**	ATF2
p38δ	**SAPK4**	ATF2

Tables 1–4: nomenclature for mammalian MAPK pathway components. Included are commonly accepted names found in the primary literature. Not all these names are included in the text.

Molecular cloning of MAPK revealed a small family of kinases comprising 44- and 42-kDa protein Ser/Thr kinases. These cDNAs were designated, respectively, extra-cellular signal-regulated kinases (ERKs)-1 and -2 (Table 1). ERK1 and ERK2 were shown to be remarkably homologous to *S. cerevisiae* Fus3p and Kss1p, kinases of the yeast mating-pheromone pathway (Chapter 3) (3, 21, 22). This was the first indication of the conservation of MAPK pathways.

2.2 Substrates of ERK1 and -2: other protein kinases and transcription factors

ERK1 and -2 phosphorylate and activate both transcription factors and other protein kinases, thereby influencing a large variety of cellular processes.

2.2.1 MAPK-activated protein kinase-1/ribosomal S6 kinase (MAPKAP-K1/Rsk)

MAPKAP-K1/Rsk was the first insulin-stimulated protein kinase with S6 phos-phorylating activity to be purified and cloned (23). However, MAPKAP-K1/Rsk does not represent the physiological S6 kinase activated by insulin and mitogens in somatic cells. This function is performed by the p70 S6 kinase. At least three MAPKAP-

K1/Rsk isoforms have been cloned (MAPKAP-K1a/Rsk1, MAPKAP-K1b/Rsk2, and MAPKAP-K1c/Rsk3). The MAPKAP-K1s/Rsks have a distinct molecular structure in that each possesses two complete protein kinase domains. Both domains are necessary for MAPKAP-K1/Rsk regulation and function (23, 24).

MAPKAP-K1/Rsk is thought to be important in insulin regulation of glycogen metabolism. Among the first polypeptides shown to be regulated by insulin in a phosphorylation-dependent manner was glycogen synthase (GS); and this regulation involved dephosphorylation. Inactive GS is phosphorylated at up to seven sites: 1, 2, 3a, 3b, 3c, 4, and 5. Sites 3a–c and 4 reside near the GS carboxyl terminus. Insulin stimulates the glucose-6-phosphate-independent activation of GS by fostering GS dephosphorylation—primarily of sites 2, 3a-c and, to a lesser extent, 4. GS is phosphorylated and inactivated by protein kinases that are active in the resting cell such as casein kinase-II and glycogen synthase kinase-3 (GSK3, Chapter 3), which phosphorylate sites 2, 3a–c, and 4 (25, 26).

GS is also inactivated, albeit to a slightly lesser extent, by agonists that elevate cAMP and activate the cAMP-dependent protein kinase (PKA) (25, 26). PKA activation culminates in the phosphorylation of sites 1 and 2. The PKA pathway can be antagonized by insulin; and the opposing effects of insulin and cAMP agonists on GS phosphorylation led to the view that insulin's primary effect on Ser/Thr phosphorylation would be to promote dephosphorylation. This view came into question with the identification of several polypeptides that undergo rapid insulin-stimulated Ser/Thr phosphorylation *in vivo*. Most prominent among these was ribosomal S6 (26). The identification of insulin-stimulated Ser/Thr phosphorylation led to a reassessment of the effect of insulin and other receptor Tyr kinases on protein phosphorylation, and it is now accepted that the major mode of insulin action is to promote protein phosphorylation.

Insulin can trigger the phosphorylation and inactivation of GSK3 (Chapter 1) (27). However, GSK3 can only phosphorylate GS at sites 3a–c and 4; site 2 is not phosphorylated by GSK3, and insulin-stimulated dephosphorylation of GS involves dephosphorylation of sites 2 and 3a–c, and, to a lesser extent, site 4 (25, 26). Thus, insulin-mediated inhibition of GSK3 cannot account for all insulin's action on GS. MAPKAP-K1/Rsk phosphorylation of the G-subunit of phosphatase-1 (PP1-G) represents a mechanism by which MAPK pathways, in conjunction with the inactivation of GSK3, can regulate skeletal muscle GS activation (28). PP1-G is expressed selectively in skeletal muscle where it binds phosphatase-1 (PP1) in a reversible fashion and interacts constitutively with the glycogen granule. Of note, glycogen synthase is constitutively associated with skeletal muscle glycogen granules (25, 26). PP1-G can be phosphorylated at two sites referred to, somewhat confusingly, as sites 1 and 2. PP1-G site 2 is a substrate for cAMP-dependent protein kinase. Dephosphorylation of PP1-G site 2, which occurs under conditions wherein PKA is inactive, promotes the binding of PP1 to PP1-G (25, 26). Phosphorylation of PP1-G site 1 by MAPKAP-K1/Rsk significantly increases the rate at which PP1 dephosphorylates GS (28). By these processes, PP1 is targeted, in an insulin-stimulated manner, to GS and can efficiently dephosphorylate (at sites 2 and 3a–c, but not site 4)

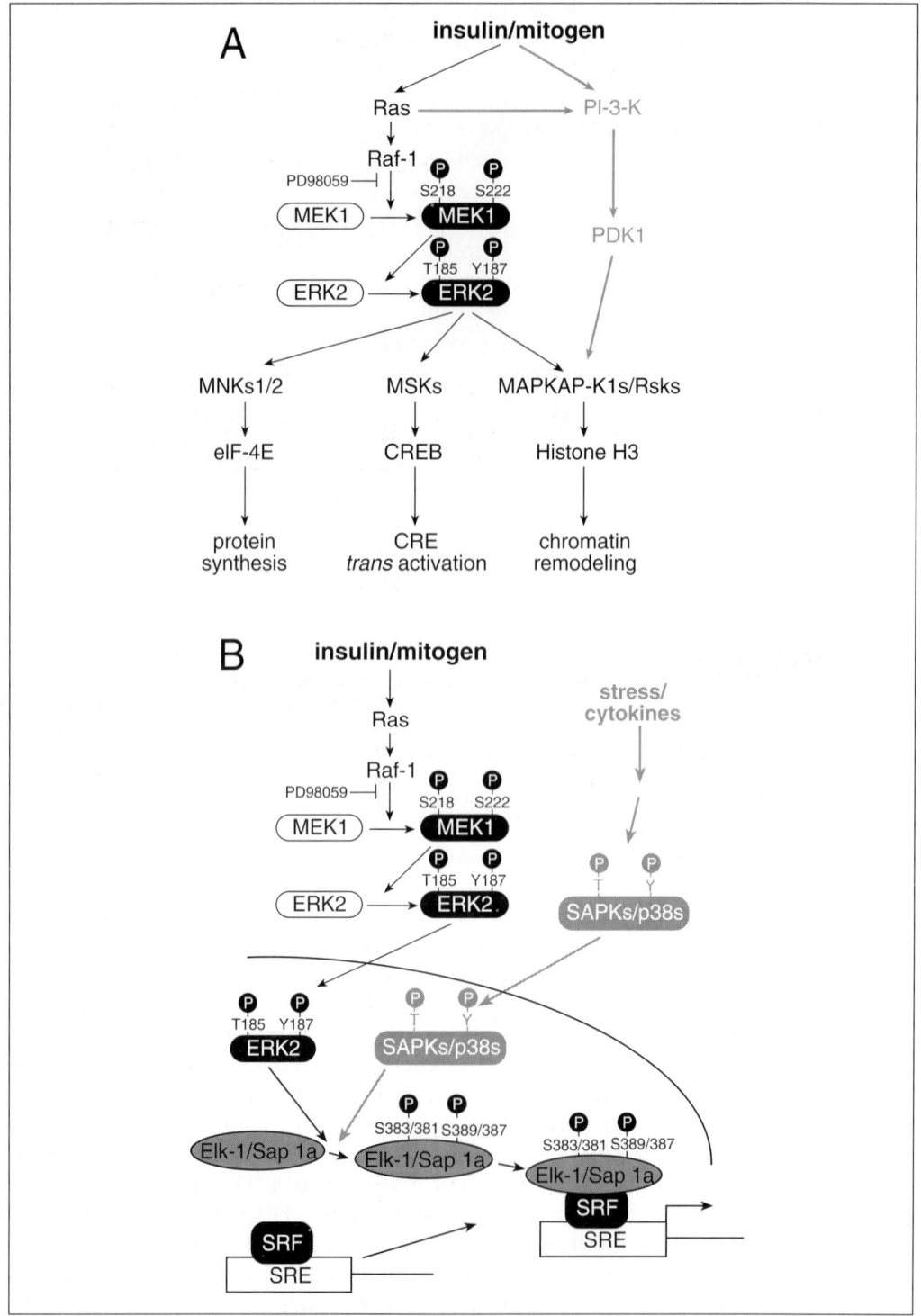

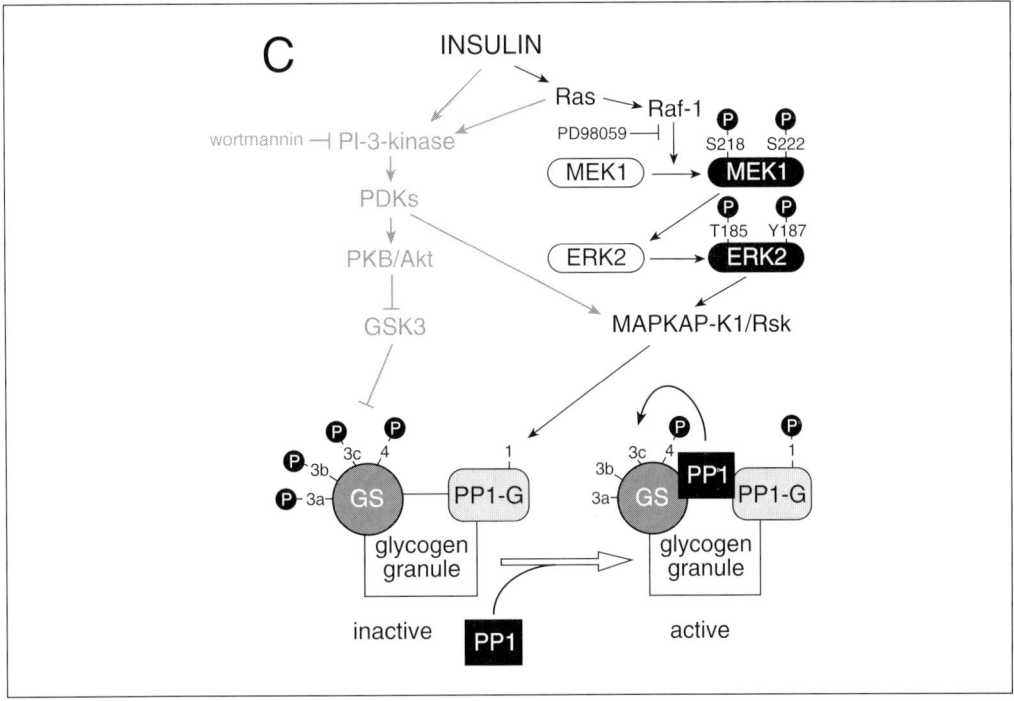

Fig. 3 Important mechanisms regulated by the Ras → ERK pathway. (A) Protein kinases activated by ERKs in response to mitogen and insulin. Regulation by ERK2 is shown. (B) Regulation of TCF family transcription factors by ERKs. Regulation by SAPKs and p38s is also shown. TCF phosphorylation leads to binding to SRF and *trans* activation of the SRE. SAPKs and ERKs phosphorylate Elk-1 (at Ser383 and Ser389) while the p38s can phosphorylate Sap1a at Ser381 and Ser387. (C) Regulation of skeletal-muscle glycogen synthase (GS) by insulin. While the ERK/Rsk mechanism is probably relevant, the PI-3-kinase pathway is probably the dominant mechanism by which GS is regulated. Black circles with a white P indicate phosphorylation. GS, glycogen synthase; PP1, protein phosphatase-1 catalytic subunit; PP1-G, PP1-GS targeting subunit.

and contribute to the activation of GS (Fig. 3C) (25, 26, 28). It should be noted that this mechanism is specific for skeletal muscle inasmuch as PP1-G is a muscle-specific protein. Whether or not analogous mechanisms exist in other tissues is unclear.

The gene encoding MAPKAP-K1b/Rsk2 activity is ablated in patients with Coffin–Lowry syndrome, a disease marked by mental retardation, facies, and skeletal anomalies as well as cardiac dysfunction. Allis and colleagues have identified at least one molecular mechanism that defines Coffin–Lowry disease, at least in part, as a defect in mitogen-induced gene expression (29). Histone H3 is rapidly phosphorylated at Ser10 in response to mitogen, and this phosphorylation is thought to underlie the loosening of chromatin structure associated with mitogen-stimulated gene expression. Indeed, the use of antiphosphoantibodies to immunoprecipitate chromatin containing phospho-Ser10 H3 from mitogen-treated or oncogenically transformed cells results in the co-precipitation of genomic DNA encoding the immediate-early genes c-*fos* and c-*myc* (30). Fibroblasts from patients with Coffin–Lowry disease display a marked reduction in EGF-stimulated H3 phosphorylation which is restored upon reintroduction of *mapkap-k1b/rsk2*. Moreover, biochemically purified MAPKAP-K1b/Rsk2 can phosphorylate H3 *in vitro* at Ser10, suggesting that it is a direct mitogen-stimulated H3 kinase *in vivo* (29).

In addition to phosphorylation of PP1-G, and H3, MAPKAP-K1/Rsk can phosphorylate several other proteins that are important to cell survival and gene expression; however, the physiological significance of this phosphorylation is unclear. The transcription factor cAMP response-element binding protein (CREB) is activated upon phosphorylation of Ser133. MAPKAP-K1/Rsk can phosphorylate CREB at Ser133 (31); however, it is unclear if MAPKAP-K1/Rsk is a physiological CREB kinase inasmuch as MSK1 (32) appears to phosphorylate CREB far more efficiently (see Section 2.2.2). MAPKAP-K1/Rsk can also phosphorylate IκB and stimulate its nuclear translocation (33); however, the significance of this phosphorylation is unclear inasmuch as the IKKs are more efficient IκB kinases.

MAPKAP-K1/Rsk, once activated by insulin or mitogens, can be completely inactivated with protein Ser/Thr phosphatases. Both ERK1 and ERK2 can phosphorylate and partially reactivate phosphatase-inactivated MAPKAP-K1/Rsk (34). That ERK1 and -2 are physiologically relevant MAPKAP-K1/Rsk kinases is evidenced by the observation that the ERKs are activated *in vivo* prior to MAPKAP-K1/Rsk activation.

The mechanism by which ERK1 and -2 activate MAPKAP-K1/Rsk is complex. ERKs activate the carboxyl-terminal catalytic domain by phosphorylating Thr573 (the numbering system used here is for the rat Rsk1 isoform) in the carboxyl-terminal kinase domain activation loop, and participate in the activation of the amino-terminal catalytic domain by phosphorylating Ser363 which resides just outside the amino-terminal kinase domain (35). The activated carboxyl-terminal catalytic domain then phosphorylates Ser380 (*trans* autophosphorylation), which also resides in between the two kinase domains (35) (Fig. 4).

Optimal MAPKAP-K1/Rsk activation also requires phosphorylation of the amino-terminal kinase domain at Ser221 which resides within the amino-terminal kinase domain, activation loop. This phosphorylation can be catalysed by 3'-phosphoinositide-dependent kinase-1 (PDK1), a Ser/Thr kinase implicated in the inositol lipid-dependent phosphorylation and activation of protein kinase B (PKB/Akt) and p70 S6 kinase (36) (Fig. 4, see also Chapter 1). PDK1 phosphorylation of MAPKAP-K1/Rsk appears to occur constitutively. This is not surprising, inasmuch as PDK1 is, in most instances, constitutively active, and regulation of PDK1 in large part involves inositol lipid-dependent gating of access to phosphoacceptor sites on some, but not all, PDK1 substrate proteins (37, 38). MAPKAP-K1/Rsk Ser221 is apparently one such site (36, see also Chapter 1). Ser221 phosphorylation may also be the result of autophosphorylation (*cis* autophosphorylation), but the relative contribution of these two events to MAPKAP-K1/Rsk regulation is unclear (35).

Once MAPKAP-K1/Rsk is fully phosphorylated and activated, it is the amino-terminal kinase domain that is probably responsible for the phosphorylation of MAPKAP-K1/Rsk substrates.

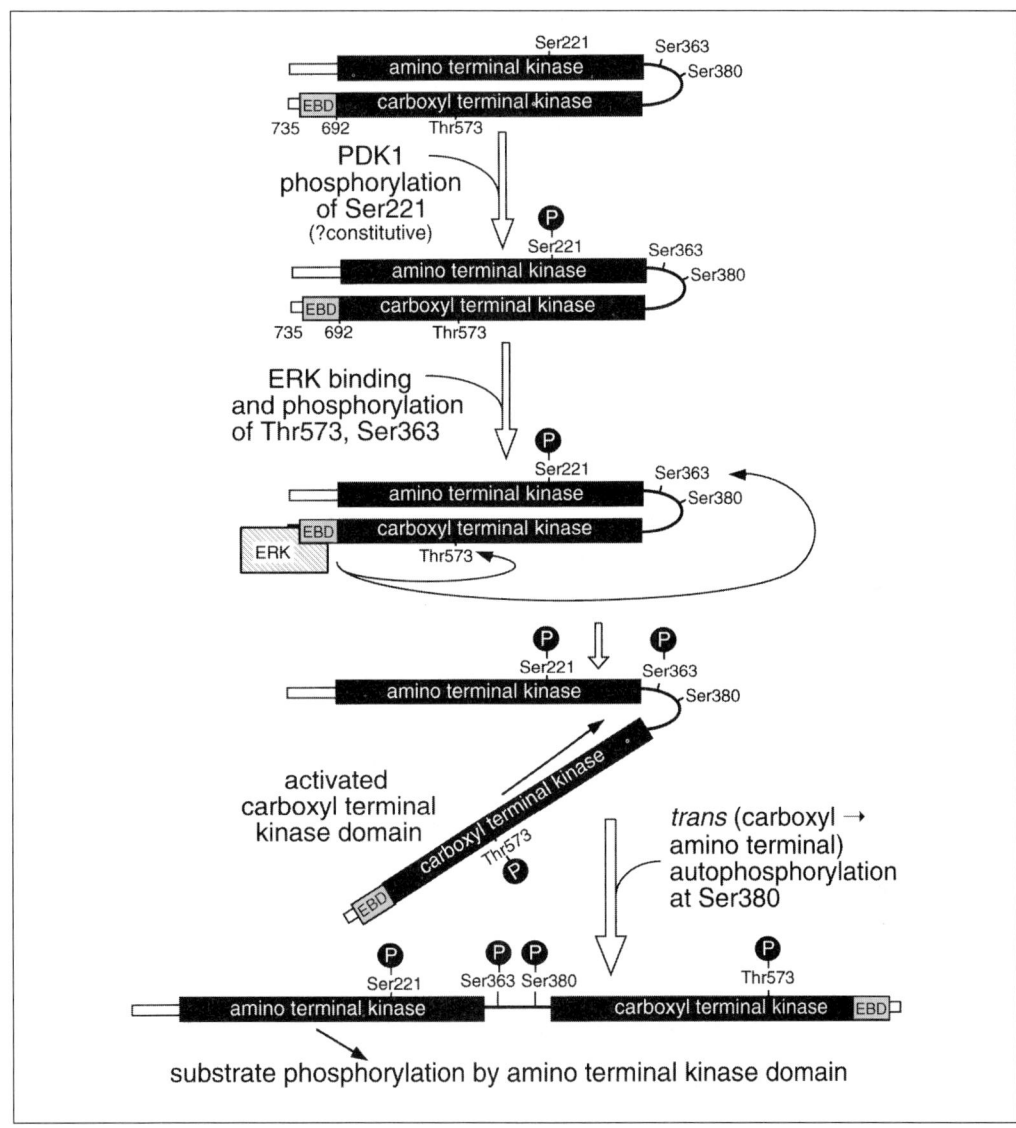

Fig. 4 Regulation of Rsk by ERKs and cotranslational phosphorylation by PDK1. Ser221 phosphorylation by PDK1, while required for activation, is not sufficient. Rsks are actually activated *in vivo*, in response to mitogens and insulin, by ERKs. Note the docking site for ERK (EBD) at the Rsk carboxyl terminus. Black circles with a white P indicate phosphorylation.

The considerable efficiency with which the ERKs (and not other MAPKs) phosphorylate and trigger the activation of MAPKAP-K1/Rsk can be attributed, in large part, to the presence of a specific ERK docking site at the extreme carboxyl terminus of the MAPKAP-K1/Rsk polypeptide (39) (Fig. 4). This binding site as well as the SAPK binding site on c-Jun are examples of mammalian MAPK docking sites that are responsible for conferring MAPK substrate specificity (Section 1.1). Gavin and Nebreda showed that *Xenopus* ERKs could interact with the carboxyl-terminal 43 amino acids of *Xenopus* MAPKAP-K1/Rsk. Interestingly, if these 43 amino acids were fused to the carboxyl terminus of MAPKAP-kinase-2 (MAPKAP-K2, a stress-regulated kinase that is not normally an ERK substrate, see Section 3.5.1), the MAPKAP-K2 could then bind ERK and undergo ERK-catalysed phosphorylation. Likewise, deletion of the carboxyl-terminal 43 amino acids of Rsk drastically reduced the efficiency of ERK-catalysed phosphorylation and activation of MAPKAP-K1/Rsk. More refined mapping has narrowed the ERK binding site on MAPKAP-K1/Rsk to the carboxyl-terminal 25 amino acids of MAPKAP-K1/Rsk. This sequence is conserved and is present in several other ERK substrates including the MNKs and MSK1 (Sections 2.2.2, 2.2.3, and 3.5.1) (39). However, it has not been demonstrated if these conserved sequences are also bona fide ERK binding sites.

2.2.2 MSK1

Mitogen- and stress-activated protein kinases (MSKs)-1 and -2 are a newly identified family of Ser/Thr protein kinases with an overall structure similar to that of the MAPKAP-K1s/Rsks (Section 2.2.1)—notably, the MSKs possess two tandem protein kinase domains. MSKs1/2 were identified in a search of databases for DNA sequences homologous to p70 S6 kinase (Chapter 1). *In vitro* MSK1 will phosphory-late the synthetic peptide Crosstide (Gly–Arg–Pro–Arg–Thr–Ser–Ser–Phe–Ala–Glu–Gly); however, MSK1 purified from unstimulated cells possesses low Crosstide kinase activity. By contrast, the Crosstide kinase activity of MSK1 is activated more than 102-fold upon incubation *in vitro* with ERK2 (see Figs 3A and 8). Consistent with the activation of MSK1 by ERKs, MSK1 is activated *in vivo* by mitogens in a manner inhibitable by the MAPK pathway inhibitor PD98059 (Section 2.4) (32). MSK1 is also a substrate for the p38 MAPKs (Section 3.5.1), and is activated by environmental stresses and inflammatory cytokines in a manner inhibitable by the p38 inhibitory drug SB203580 (Sections 3.3 and 3.5.1) (32) (see Figs 3A and 8).

cAMP response-element binding protein (CREB) is a bZIP transcription factor that binds and *trans* activates genes containing the cAMP response element (CRE—consensus sequence: TGACGTCA). CREB *trans* activating activity can be activated by a number of divergent classes of stimuli including mitogens, stresses, and, of course, agonists that elevate cAMP. Activation of CREB *trans* activating activity correlates with phosphorylation at Ser133. PKA probably represents the major CREB kinase recruited by cAMP (40). By contrast, several protein kinases including MAPKAP-K1s/Rsks (31) (Section 2.2.1) and the stress-regulated kinase MAPKAP-K2 (41) (Section 3.5.1) have been shown to possess mitogen- or stress-activated CREB Ser133 kinase activity. MSK1 is a potent CREB kinase; and a substantial amount of evidence indicates that, in fact, MSK1 (and, possibly, by extension the related, but less well characterized, MSK2), and not MAPKAP-K1/Rsk or MAPKAP-K2/3, is the physio-logically relevant stress- and mitogen-activated CREB kinase.

First, the K_{cat} for the MSK1-catalysed phosphorylation of CREB is nearly two orders of magnitude greater than that for MAPKAP-K1/Rsk-catalysed CREB phosphorylation. The K_m for MSK1 phosphorylation of CREB is about 20-fold lower than that for MAPKAP-K1, while the V_{max} is 4–5-fold greater. Second, the MSK1 polypeptide has a bipartite nuclear localization signal, and is localized exclusively in the nucleus, where CREB resides. MAPKAP-K1s/Rsks are largely cytosolic. Third, the inhibition of stress- and mitogen-activated CREB phosphorylation *in vivo* closely mirrors that of MSK1 activation. Thus, activation of CREB by stress and mitogens is blocked by the broad-specificity kinase inhibitor Ro318220. MSK1 is strongly inhibited *in vitro* by Ro318220, whereas MAPKAP-K2 and MAPKAP-K3 are not. Finally, CREB activation *in vivo* by TNF can be blocked with SB203580 (Section 3.3) and PD98059 (Section 2.4), with kinetics that parallel those described above for the inhibition of MSK1 (32).

2.2.3   MNKs

Cellular mRNAs contain a 5′ cap structure, the N^7-methylguanosine cap. Efficient translation of proteins is dependent on the binding of multimeric translational initiation factor eIF-4F to the methylguanosine cap. The binding of eIF-4F to mRNA is thought to result in a relaxation of mRNA secondary structure, thereby facilitating the binding of the 40S small ribosomal subunit. The eIF-4F complex consists of eIF-4A, -B, -G, and -E. eIF-4A is an RNA helicase that acts with eIF-4B, an RNA binding protein, to unwind mRNA thereby allowing for ribosome binding (42). eIF-4G is a multi-functional scaffolding protein, that gathers the eIF-4F subunits into a complex and delivers them to the 5′ cap. The ability of the eIF-4F complex to bind to the 5′-cap is conferred through an association between the eIF-4E and eIF-4G subunits. This association is thought to recruit the remaining eIF-4F subunits and foster the formation of a complete eIF-4F complex capable of contributing to translational initiation (42).

The eIF-4E/eIF-4G interaction is negatively regulated by the translational re-pressor protein 4E-binding protein-1 (4E-BP1 is also called phosphorylated heat- and acid-stable protein regulated by insulin, PHAS-I). 4E-BP1 is rapidly phosphorylated at Ser64 in response to insulin and mitogens; and this phosphorylation results in the dissociation of 4E-BP1 from eIF-4E. *In vivo*, both the phosphorylation of 4E-BP1 and translational initiation are completely inhibited by rapamycin, consistent with the fact that dissociation of 4E-BP1 is regulated by mTOR, and that this dissociation is probably an important step in the formation of a functional eIF-4F complex (42, see also Chapter 1). However, eIF-4E itself also undergoes a regulatory phosphorylation at Ser209 in response to both insulin/mitogen and environmental stress. This phosphorylation is thought to increase the affinity of eIF-4E for the 5′ cap; and both crystallographic and biochemical data indicate that phosphorylated eIF-4E is preferentially associated with the 5′ cap (42).

MAPK interacting kinases (MNKs)-1 and -2 are two closely related kinases that are probably the physiologically relevant eIF-4E Ser209 kinases. As the name implies, MNKs associate *in vivo* with MAPKs and are *in vitro* and *in vivo* MAPK substrates. MNKs are phosphorylated and activated both by ERK-1 and -2 (in response to insulin and mitogens) and by the p38 MAPKs (in response to stress). The site(s) on MNKs phosphorylated by MAPKs have not been determined (43, 44) (see Figs 3A and 8).

2.2.4 Elk-1

One of the earliest transcriptional events known to occur in response to mitogen is the induction of c-*fos* expression. c-Fos is a bZIP transcription factor that, together with c-Jun, comprises one form of the AP-1 transcription factor (Section 3.5). AP-1 regulation is quite complex; and serum- or growth-factor induction of c-*fos* is one mechanism for AP-1 activation. Elevated levels of c-Fos polypeptide correlate well with elevations in AP-1 *trans* activating activity (45, 46).

The *fos* promoter contains a *cis* acting element, the serum response element (SRE), that participates in the recruitment of transcription factors which induce *fos* expression. The SRE binds a heterodimeric transcription factor containing two poly-peptides, the serum response factor (SRF) and the ternary complex factor (TCF) (45, 46).

The TCFs are a family of Ets-domain transcription factors that includes Elk-1 and Sap-1a. The regulation of Elk-1 by MAPKs has been extensively studied. Activation of ERK1 and -2 (and, indeed, all MAPKs) coincides with the translocation of a portion of the ERK1/2 pool to the nucleus, thereby enabling phosphorylation of nuclear substrates (Section 2.8). ERK1 and -2, as well as SAPKs (Section 3.5.2), can phosphorylate two critical residues in the Elk-1 C-terminus (Ser383, Ser389; the p38s can also phosphorylate Sap-1a, Section 3.5.2). This enhances the binding of TCFs to the SRF and thereby elevates *trans* activation at the SRE. By this process, both stress- and mitogen-regulated MAPKs contribute to c-*fos* induction (9, 47–50) (Fig. 3B see also Section 3.5.5).

2.3 The regulation of ERK1 and -2 by MAPK/ERK-kinases-1 and -2

Purification of ERK2 from ^{32}P-labelled, mitogen-stimulated cells revealed that, upon insulin or phorbol ester stimulation, the 42- and 43-kDa MAPKs (ERK2 and -1) underwent concomitant Tyr and Thr phosphorylation (20). It was subsequently demonstrated that ERK2 and -1 could be inactivated with Tyr-specific or Ser/Thr-specific protein phosphatases (51). These results indicated that ERK2 required both Tyr and Thr phosphorylation for activity. The sites of phosphorylation were mapped to Thr185 and Tyr187 for ERK2 and Thr203 and Tyr 207 for ERK1 in the activation loop of subdomain VIII of the catalytic domain (52).

The existence of a MAPK 'activator' was first demonstrated by fractionating cytosolic extracts of EGF-treated cells on ion exchange or gel filtration columns and assaying the fractions for an activity that could activate ERK1 and/or ERK2 *in vitro*. It was observed that a single peak of activity could catalyse both the Tyr and Thr phosphorylation and activation of ERK1 and -2 (53, 54). Purification and molecular cloning of this activity revealed a family of novel dual-specificity (Thr/Tyr) protein kinases designated either MAPK-kinases-1 and -2 (MAPKKs-1/2) or MAPK/ERK kinases-1 or -2 (MEKs-1/2) (54–57) (Table 2).

As with ERK1 and -2, MEK1 and -2 share a remarkable homology with kinases

Table 2 MEK nomenclature

Name	Alternate names	Substrate(s)
MEK1	MAPKK1, MKK1	ERK1, ERK2
MEK2	MAPKK2, MKK2	ERK1, ERK2
SEK1	MKK4, JNK kinase (JNKK)-1, MEK4, SAPK-kinase (SKK)-1	SAPKs
MKK7	JNKK2, MEK7, SKK4	SAPKs
MKK3	MEK3, SKK2	p38s
MKK6	MEK6, SKK3	p38s

from lower eukaryotes. Thus, Fus3p and Kss1p, MAPKs of the *S. cerevisiae* mating-pheromone response pathway, are regulated *in vivo* by a MEK homologue, Ste7p. Likewise, Hog1p, a MAPK of the *S. cerevisiae* osmosensing pathway, is activated by the MEK Pbs2p (3, 54–57).

2.3.1 Structural features of MEK1 reveal domains important for function

Recent analysis of the structure of MEKs has revealed important features that convey biological and biochemical functions (Fig. 5A). MEK1 is phosphorylated at Ser218 and Ser222 (see Section 2.4, below) resulting in activation (58, 59). Marshall and co-workers demonstrated that mutation of both these residues to acidic residues (Asp) mimics the charge of phosphorylation and results in a 52–82-fold activation of the mutant MEK construct. Expression of these mutants is sufficient to transform 3T3 fibroblasts or to stimulate neurite outgrowth in PC12 phaeochromocytoma cells (60).

Ahn and colleagues demonstrated that further activation of MEK1 can be achieved upon deletion of an amino-terminal domain (amino acids 32–51, Fig. 5A) which is rich in hydrophilic residues and suggestive of an alpha helix (61). A similar alpha-helical region is present in the cAMP-dependent protein kinase (amino acids 11–30) as well as several other protein Ser/Thr and Tyr kinases. By contrast, deletion of amino acids 1–32 or 1–52 abrogates MEK1 activity. Consistent with this, it has been shown that amino acids 1–32 of MEK1 constitute an ERK binding site implicated in the regulation of ERK subcellular localization (Fig. 5A and Section 2.8) (62). Combining dual Ser118Asp/Ser221Asp with Δ32–52 results in a mutant MEK construct with a nearly 400-fold greater activity than the inactive, wild-type enzyme (61). This mutant is a potent oncogene which can transform NIH3T3 cells (61). Interestingly, residues 34–53 overlap with the nuclear export signal of MEK1 (Fig. 5A and Section 2.8) (62). Thus, deletion of this domain may not only activate MEK1, but may foster the dysregulated subcellular localization of MEK1 and associated ERK1 and ERK2 contributing to enhanced ERK activation.

Finally, MEK1 (but not MEK2) contains a proline-rich domain (amino acids 261–307, Fig. 5A). As is discussed in Section 2.7, this domain has been implicated in binding the scaffold protein MEK interacting partner-1 (MP1). Deletion of this domain substantially reduces the efficiency of *in vivo* ERK activation catalysed by

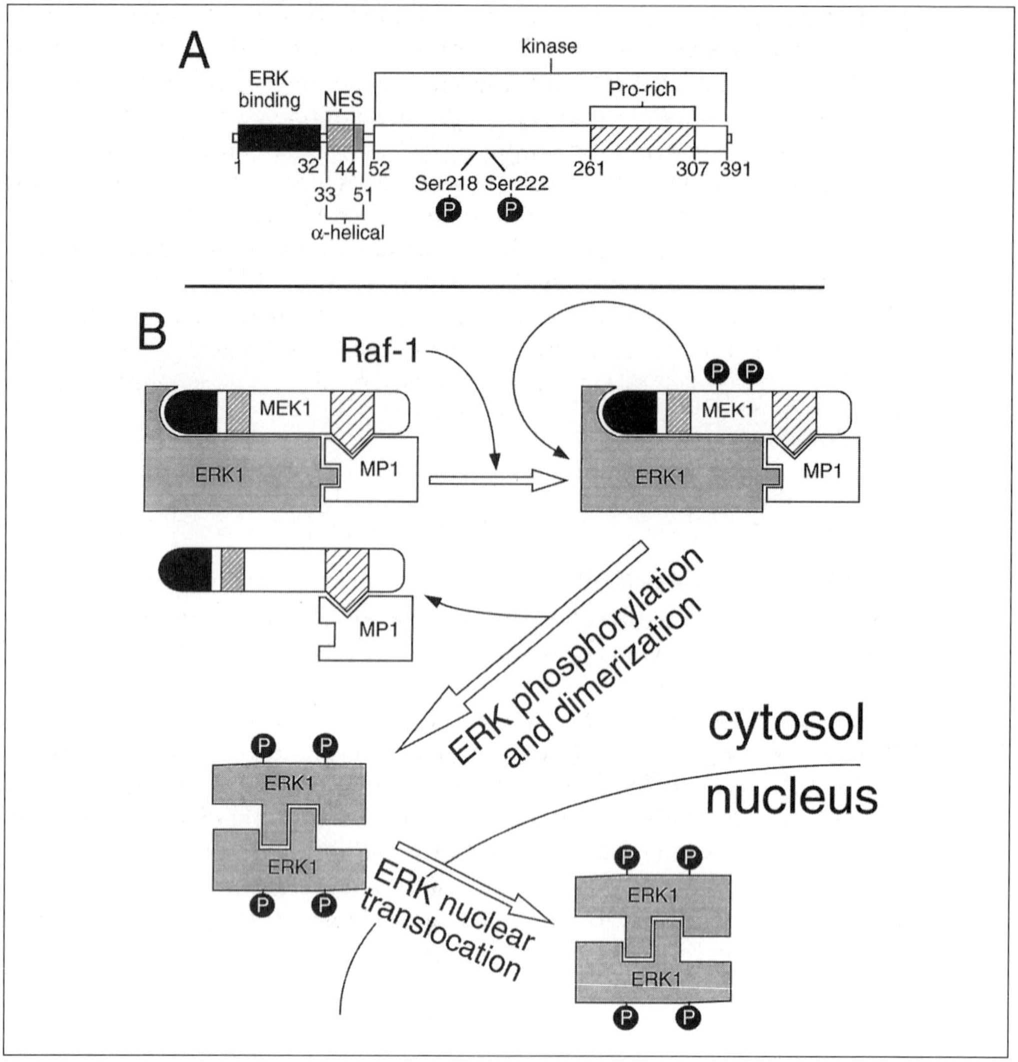

Fig. 5 MEK1 structure and function, and the role of ERK phosphorylation and dimerization in nucleocytoplasmic shuttling. (A) MEK1 structural features. NES, nuclear export signal. Black circles with a white P indicate phosphorylation. (B) Control of ERK localization. ERK1, MEK1, and MP1 probably form a cytosolic complex. Phosphorylation of ERK by MEK fosters its dissociation from MEK, ERK dimerization, and translocation. Black circles with a white P indicate phosphorylation.

MEK1, suggesting that MP1 binding is necessary for optimal MEK function *in vivo* (16, 63).

Both MEK1 and MEK2 are substrates for the anthrax lethal factor (LF) (64). LF, produced by *Bacillus anthracis*, is the major cause of death in animals infected with anthrax; and with the possibility that anthrax could be employed as a biological weapon, interest in abrogating the effects of LF is considerable. LF is a 716-amino acid protein containing a zinc binding site (amino acids 686–690) reminiscent of metallo-proteases. A screen of the activity profile of LF revealed that its *in vivo* effects were similar to those of PD98059, an inhibitor of the Raf → MEK step in the Ras MAPK pathway (Section 2.4) (64). LF can block progesterone activation of the Mos → MAPK *Xenopus* oocyte maturation pathway (Section 2.4) coincident with an observed cleavage of both MEK1 and MEK2 that removes a conserved sequence in the extreme amino terminus (PKKKPTP, amino acids 1–7 for MEK1 or LARRKPVLP, amino acids 1–9 for MEK2). In the case of MEK1 this sequence resides in the ERK binding site and is essential for ERK function (62); thus, the cleaved MEK polypeptides are no longer able to activate ERK, in spite of being largely intact (64). MKK3, a MEK upstream of the p38 MAPKs (Section 3.6.2)—but not SEK1, a MEK upstream of the SAPKs (Section 3.6.1)—contains a similar putative LF cleavage sequence; accordingly additional MEKs may be LF targets (64).

2.4   MEK1 and -2 are substrates of the Raf-1 and Mos proto-oncoproteins

MEKs can be deactivated by Ser/Thr-specific phosphatases but not by Tyr phosphatases (54), indicating that at least one Ser/Thr kinase lies between MEK1 and -2 and receptor Tyr kinases.

Raf-1 is the normal cellular homologue of v-*raf*, an acutely transforming oncogene that encodes a Ser/Thr protein kinase. Raf-1 is one of a small family of related Ser/Thr kinases, A-Raf, B-Raf, and Raf-1, all of which share similar properties (26, 65). Several observations pointed to the possibility that Raf-1 was upstream of ERK1 and -2. First, some v-*raf*-transformed cells manifest constitutively active ERK1 and -2 (66). Second, expression of dominant inhibitory constructs of Raf-1 could block the induction of AP-1 in response to mitogens (67). As was noted above, AP-1 can be regulated in part by the ERK pathway via Elk-1/SRF induction of c-*fos*. However, there remained the possibility that Raf-1 was not upstream of the ERKs; a hypothesis based on the observation that not all *raf*-transformed cells displayed constitutive MAPK activation (67). Moreover, yeast studies had identified protein kinases homologous to STE11 as MEK activators (Fig. 2, see also Chapter 3). Although mammalian kinases homologous to STE11 have been identified (e.g. MEK-kinase-1, MEKK1, see below), the sequence of Raf-1 has no significant homology to STE11 outside the regions shared by all protein kinases.

The placement of Raf-1 as a direct upstream activator of MEK1 and -2 was established with the observation that upon phosphatase inactivation, purified MEK could be phosphorylated and reactivated *in vitro* with purified, oncogenic Raf-1 (68–70). It was subsequently shown that endogenous Raf-1 activity toward MEK1 was stably activated by insulin and mitogens (71). Detailed analysis of Raf-1 phosphorylation of MEK1 indicated that MEK1 was phosphorylated at Ser218 and Ser222 (58, 59). Again, these residues lie within the activation loop of subdomain VIII of the catalytic domain (55).

The phosphorylation of MEK1 (but not MEK2) by Raf-1 can be inhibited by the Parke-Davis compound PD98059 (72, 73). This inhibition is remarkably specific, and enables the use of this inhibitor to identify cellular and biochemical processes regulated by the Raf-1 → ERK pathway. PD98059 was discovered in a biochemical screen for inhibitors of the ERK pathway. It is a potent inhibitor of the biological processes for which the ERK pathway is a dominant player. Thus, NGF stimulation of PC12 cell neurite outgrowth is blocked with PD98059 (72). By contrast, insulin activation of glycogen synthase (GS) is not significantly inhibited with PD98059, in spite of the observed role of ERK MAPKAP-K1/Rsk in the regulation of PP1-G (Fig. 3C) (74). Thus MEK2 may be important in skeletal muscle GS regulation. Alternatively, inasmuch as the PI-3-kinase pathway is also important in insulin activation of GS (74, see also Chapter 1), the ERK pathway's role in GS regulation by insulin may not be prominent.

MEK1 and -2 are also substrates for the product of the c-*mos* proto-oncogene (75, 76). c-*mos* encodes a protein Ser/Thr kinase that is an important component in oocyte maturation. Thus, progesterone-induced maturation of *Xenopus laevis* oocytes requires ongoing protein synthesis. Specifically, several important mRNAs are translated at the onset of maturation including cyclins A and B1, Cdk2 (Chapter 6), and c-*mos*. The progesterone-induced signalling pathway culminates in the completion of meiosis-I and the progression to metaphase of meiosis–II. Entry into meiosis-I correlates with germinal vesicle breakdown, activation of maturation-promoting factor (MPF)/cdc2, and activation of MAP kinase (ERK2). Antisense inhibition of c-*mos* expression blocks oocyte maturation—an indication that pathways triggered by c-Mos coincide with the activation and stabilization of MPF and are critical to oocyte maturation (75). However, c-Mos is not sufficient in and of itself to activate MPF. Thus, the pathway c-Mos → MEK → ERK is critical to oocyte maturation via activation of those mechanisms parallel to MPF which perhaps are necessary for the ultimate activation of MPF but are not sufficient to activate MPF *per se* (75).

2.5 Ras, a molecular switch that couples tyrosine kinases to Raf-1 and the ERK pathway

At least 30% of human cancers can be attributed in part to mutations in one or more of the Ras subfamily proto-oncogenes. Accordingly, the role of Ras subfamily proteins in mitogenic signalling has attracted intense interest. The Ras subfamily consists of Ha-Ras, Ki-Ras, and N-Ras and is part of a large superfamily of small monomeric GTPases referred to as the Ras superfamily (2, 77–79).

Ras superfamily proteins are molecular switches that relay signals from receptor complexes to downstream effectors. All Ras proteins bind the guanine nucleotides GTP and GDP and possess a slow, intrinsic GTPase activity. When in the GTP-bound state, Ras proteins are competent to signal downstream. GDP-bound Ras proteins are inactive (77–79).

The members of the Ras superfamily share a common structural configuration. At the N-terminus is a GTP binding domain, this is followed by the effector loop which binds downstream effector proteins. There are also two switch domains (switch-I and II) which undergo conformational changes upon exchanging GDP for GTP (77–79). The switch-I domain overlaps considerably with the effector loop, while the switch-II domain contains residues important to Ras GTPase activity. Crystallographic studies indicate that GTP binding causes conformational changes that render switch-I accessible to effector proteins (77–79). At the extreme C-terminus of mature Ras family proteins is the CAAX domain (C = cysteine, A = alanine, X is any amino acid). The CAAX domain undergoes post-translational modification: prenylation (farnesylation in the case of the Ras subfamily, and geranylgeranylation in the case of the Rho subfamily) of the cysteine followed by proteolytic removal of the AAX residues. These lipid modifications render Ras proteins hydrophobic at the carboxyl terminus and serve to localize Ras proteins at the inner leaflet of the plasma membrane. Plasma membrane localization is essential for Ras protein function—one of the major roles of Ras superfamily proteins is to recruit effector proteins to the plasma membrane (77–79).

Activation of Ras superfamily proteins is catalysed by guanine nucleotide exchange factors (GEFs). These proteins act to accelerate the rate of dissociation of GDP from their target Ras proteins. Inasmuch as GTP is in excess in the cytosol, this GDP dissociation is quickly followed by GTP binding and acquisition of signalling capacity (2, 65, 77–79). Inactivation of Ras superfamily proteins is catalysed by GTPase activating proteins (GAPs). These act to accelerate the rate of Ras-protein GTPase activity. By extension, then, GTPase-deficient mutants of Ras proteins, such as Val12–Ras, are constitutively active and oncogenic. Conversely, Ras superfamily proteins, such as Asn17–Ras, which cannot exchange GDP for GTP if overexpressed, can titre-out or sequester GEFs and prevent the activation of endogenous, wild-type Ras proteins (2, 77–79).

That proteins of the Ras subfamily of the Ras superfamily were involved in insulin and mitogen signalling, and, in particular, activation of the ERKs, became clear as a consequence of several independent lines of investigation. First, treatment of cells with insulin or mitogen was shown to stimulate the GTP loading of Ras (65, 80, 81). Moreover, either the ectopic expression or scrape-loading of cells with Val12–Ha-Ras resulted in activation of the ERKs; and the addition of active, GTP-loaded Val12–Ha-Ras plus an ATP regenerating system to crude cell extracts could trigger ERK activation (65–67, 82–84). In addition, several genetic models of tyrosine kinase signalling, in particular the *sevenless* and *torso* Tyr kinase pathways of *Drosophila* photoreceptor development and the vulval induction pathway in the nematode worm *Caenorhabditis elegans*, implicated Ras as a downstream effector of receptor Tyr kinases (85–87) (Chapter 3). Finally, ectopic overexpression of dominant inhibitory Asn17-Ha-Ras could effectively block the activation of ERKs by mitogens and insulin, suggesting that Ras was downstream of mitogen receptors in mammalian cells (88, 89).

GTP–Ras activates the ERK pathway by directly binding and promoting the activation of Raf family kinases (65). The identification of Raf-1 as a direct effector for GTP-loaded Ras came from biochemical and genetic studies demonstrating that the two polypeptides could interact *in vivo* and *in vitro* (65, 90–94). The Raf-1 polypeptide

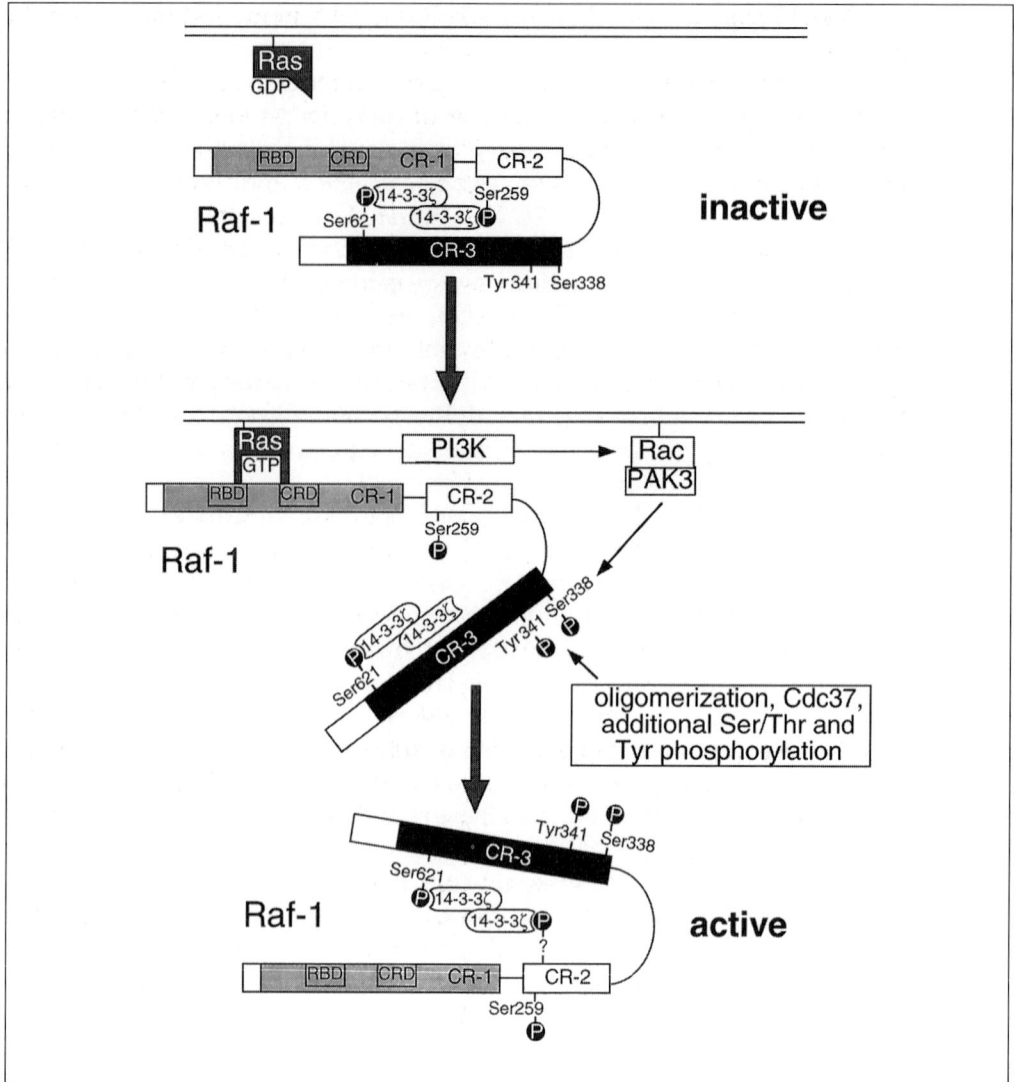

Fig. 6 Model for the complex regulation of Raf-1. Ras binding to Raf-1 changes the interaction between Raf-1 and 14–3–3, and exposes phosphoacceptor sites. In addition, membrane translocation makes Raf-1 available to membrane-localized activators. Phosphorylation at Tyr and Ser/Thr residues results in Raf-1 activation. Structural integrity of Raf-1 during its activation may be mediated by Cdc37 and Hsp90. See text for details. CR, conserved region; RBD, Ras binding domain; CRD, Cys-rich domain. Black circles with a white P indicate phosphorylation.

consists of three domains conserved among Raf family members: a carboxyl-terminal kinase domain (conserved region-3, CR-3), CR-2, a Ser/Thr-rich hinge domain; CR-1, an amino-terminal regulatory domain that contains a canonical Ras binding domain (RBD); and a cysteine-rich, Zn-finger motif (the CRD) (65) (Fig. 6). In large part, the interaction of Ras and Raf-1 is mediated by interactions between the Raf-1 RBD and the Ras-effector loop and between the Ras CAAX motif and the CRD of Raf-1 CR-1 (95, 96).

The binding of Ras to Raf-1 is a critical step in the activation of Raf-1 and is required for Ras oncogenicity. Thus, a number of Ras-effector loop mutants with a demonstrated lack of transforming capability, were subsequently shown to be unable to interact with Raf-1 in yeast two-hybrid assays (92). Moreover, Raf-1 constructs that cannot bind Ras are inactive (67). The primary function of the Ras–Raf-1 interaction is to recruit Raf-1 to the plasma membrane and to provide access to Raf-1 for membrane-localized Raf-1 activators (see below). This translocation function can be replaced by insertion of a Ras CAAX motif at the Raf-1 carboxyl terminus. The resulting Raf-1 construct is constitutively localized at the plasma membrane inner leaflet and is partially activated (97). Interestingly, however, this Raf-CAAX can be further activated with mitogen, implying that Raf-1 is activated by additional inputs whose regulation is not necessarily Ras-dependent, but whose access to Raf-1 requires Raf-1 localization at the membrane (97).

While required for Raf-1 activation, the binding of Raf-1 to Ras is insufficient for Raf-1 activation; and the exact mechanism of Raf-1 activation is still somewhat of an enigma. It is apparent that Ras triggers Raf-1 oligomerization, which appears to be necessary for Raf-1 activation (98, 99). Indeed, when Raf-1 is expressed from a recombinant baculovirus in SF9 cells, it migrates on gel filtration columns largely as a monomer, with a small but significant portion appearing in a high molecular-weight form. Coexpression with v-Src and v-Ras leads to a quantitative shift to a high molecular-weight pool that exceeds the predicted M_r of a 1:1 dimer of Ras and Raf-1, or even a 1:1:1:1 complex of Ras, Raf-1, and the heat-shock proteins Cdc37 and Hsp90 (see below) (100). Moreover, coexpression with Ras or treatment of cells with mitogen permits the coimmunoprecipitation of heterologously-tagged Raf-1 constructs, indicating that Raf-1 undergoes stimulus-dependent oligomerization (98).

FK506 is a macrolide immunosuppressant that binds to a receptor-protein FK506 binding protein-12 (FKBP12). This complex, in turn, can bind to and inhibit the calcium-sensitive phosphatase calcineurin. By this process, activation of NF-ATs during T-cell activation is inhibited, thus accounting for the immunosuppressive activity of FK506 (101). Fusion proteins consisting of the FK506 binding domain of FKBP12 linked to Raf-1 were constructed. These constructs, when expressed in transfected cells, were forced to oligomerize upon addition of FK1012, a non-immunosuppressive dimeric analogue of FK506. In these cells, FK1012 could foster the activation of coexpressed ERK1 (98). Similar results were obtained using a fusion of DNA-gyrase and Raf-1 induced to dimerize with coumermycin. In the case of FKBP12–Raf-1, but, curiously, not coumermycin–Raf-1, optimum drug-dependent dimerization apparently requires Ras, inasmuch as Asn17–Ras significantly blunted FK1012-dependent ERK activation in cells expressing FKBP12–Raf-1 (98, 99). The reason for this discrepancy is unclear; and the mechanism by which dimerization contributes to Raf-1 activation is unknown.

In addition to oligomerization, Raf-1 undergoes several additional Ras-dependent modifications during activation. Current evidence suggests that the activation state of

Raf-1 is dependent in part upon differential interactions with the dimerized proteins 14–3–3 family (102–105). 14–3–3s are a large class of abundant cytosolic polypeptides that can simultaneously homodimerize and heterodimerize with several signalling proteins in a regulatory manner. 14–3–3 proteins require prior phosphorylation of their target proteins within the consensus motif Arg–P–Ser–X–X–Ser–Pro (X is any amino acid and P indicates phosphorylation) for binding (106).

GTP–Ras, binding to Raf-1, orchestrates the changes in 14–3–3 binding that coincide with Raf-1 activation. Thus, inactive Raf-1 residues P–Ser259 and P–Ser621 bind a homodimer of 14–3–3ζ. GTP–Ras appears to be required to displace the 14–3–3ζ dimer from P–Ser259 leaving the 14–3–3ζ bound only to P–Ser621. GTP–Ras also unmasks a set of as yet incompletely identified phosphorylation sites in the Raf-1 CR-2 hinge and CR-3 catalytic regions. These sites are then phosphorylated by upstream kinases which have not yet been completely characterized. The free end of the 14–3–3 dimer, which had dissociated from Ser259, then binds to one of the newly phosphorylated sites generating a stable, active Raf-1 conformation (105) (Fig. 6).

Recent insight into the possible identity of one relevant Raf-1 activating kinase came from studies conducted by Marshall and colleagues. Again, regulation of Raf-1 by this kinase is ultimately Ras-dependent (107). Ser339 or 339 in the Raf-1 catalytic domain are known to undergo mitogen-induced phosphorylation. A kinase activity selective for these sites was purified approx. 20 000-fold from rat spleen and identified as p21-activated kinase (PAK)-3. PAKs are a family of Ser/Thr kinases that are effectors for Rho-family GTPases (discussed in Sections 3.8.1 and 3.8.2). PAK3-catalysed phosphorylation of Raf-1 at Ser338 or 339 contributes to Raf-1 activation (107). Activation of PAK3 requires binding to GTP–Rac or Cdc42; and GTP loading of Rac and Cdc42 is controlled by Rho-GEFs. These polypeptides contain pleckstrin homology domains and, accordingly, are thought to be regulated in part by Ras, through PI3-kinase (Fig. 6, see also Sections 3.8.1 and 3.8.2 and Chapter 1) (109). The significance of Ser338/339 phosphorylation to 14–3–3 binding is still unclear.

Raf-1 also undergoes regulatory Tyr phosphorylation; and the conditions under which Raf-1 is Tyr-phosphorylated differ from those under which Ser/Thr phosphorylation occurs. Thus, Raf-1 expressed from a baculovirus can be activated upon coexpression with either v-Src or v-Ras; and Src and Ras can synergistically activate Raf-1 (110, 111). Expression of Raf-1 baculovirus with oncogenic Ras baculovirus results primarily in Ser338 phosphorylation, the site phosphorylated by PAK3, while expression with v-Src baculovirus results in Tyr340 phosphorylation. Mutagenesis of Ser338 phosphorylation abolishes Ras activation of Raf-1, while mutagenesis of Tyr340 abolishes Src activation of Raf-1. Both phosphorylations require the Ras-dependent translocation of Raf-1 to the membrane; and mutant Raf constructs that do not bind Ras cannot undergo Ser338 or Tyr340 phosphorylation (110, 111).

Final components in Raf-1 activation are the molecular chaperones Cdc37 and Hsp90. These polypeptides are probably necessary to preserve the fidelity of Raf-1 structure during the conformational changes, described above, that accompany activation. Endogenous Raf-1 isolated from resting mammalian cells consistently copurifies with stoichiometric levels of Cdc37 and Hsp90 (112). Genetic screens of *Drosophila*

Sevenless Tyr-kinase signalling first identified Cdc37 as an essential component of the receptor Tyr-kinase/MAPK signalling pathway (113). Grammatikakis *et al.* observed that coinfection of Sf9 cells with Raf-1 and Cdc37 baculovirus resulted in an association between the Cdc37 and Raf-1 as well as dose-dependent Raf-1 activation (114). Cdc37 is also the primary recruiter of Hsp90 to Raf, and mutant Cdc37 constructs that cannot bring Hsp90 to the Raf-1 complex inhibit Raf-1 and ERK activation by mitogens in transfected cells (114). Thus, in addition to Ras binding and the aforementioned differential 14–3–3 binding, Raf-1 activation requires its association with Cdc37 and Hsp90 (114).

Thus Raf-1 activation is triggered upon binding GTP–Ras, and involves oligomerization, as well as Ser/Thr and Tyr phosphorylation and contributions from 14–3–3 proteins and heat-shock proteins. Although Ras binding is clearly the first step in Raf-1 activation, it remains to be determined in just what order these Ras-dependent elements act.

2.6 Differential regulation of Raf-1 and B-Raf by Rap1

B-Raf, which is expressed predominantly in the brain and CNS is a second member of the Raf family. B-Raf is a critical component in Ras-dependent ERK activation in certain cell systems. As with Raf-1, B-Raf is activated by Ha-Ras in response to mitogenic signals that recruit Tyr kinase receptors. In this instance, in contrast to Raf-1, a complex of B-Raf and 14–3–3 proteins can be directly activated by Ras in the apparent absence of additional polypeptides or phosphorylations catalysed by upstream kinases (115). B-Raf does not contain a Tyr residue analogous to Tyr340 of Raf-1. Moreover, phosphorylation of Ser445 of B-Raf, which corresponds to Ser338/339 of Raf-1 is constitutive, and not stimulated by Ras (111). However, phosphorylation of B-Raf Ser445 is necessary for activation of B-Raf. Thus, B-Raf, in the resting state is 'pre-phosphorylated' and is activated upon binding Ras (111). The mechanism of Ser445 phosphorylation is still unclear, and we still do not understand how B-Raf is activated by Ras.

Rap1 (also called K-Rev1) is a second member of the Ras superfamily and shares about 50% identity with Ha-Ras. Rap1 was originally isolated as a gene that could suppress Ras transforming activity *in vivo* (116). This inhibitory activity is probably due in part to the ability of Rap1 to bind Raf-1 directly and in a GTP-dependent manner, and to sequester it from Ras, thereby preventing the activation of Raf-1-dependent Ras effectors. Accordingly, Val12–Rap1 is a potent inhibitor of the Ras activation of ERK (92, 117).

The antagonism by Rap1 of the Ras activation of Raf-1 may partly account for the ability of cAMP to block ERK activation *in vivo*. cAMP has long been known to antagonize many of the effects of growth factors; and agents that elevate cAMP are often potent inhibitors of the Ras →Raf-1 → ERK pathway (118, 119). This inhibition was originally attributed to PKA-catalysed phosphorylation of the Raf-1 polypeptide at Ser43 (119). However, the levels of PKA needed to catalyse an appreciable phosphorylation of Raf-1 are exorbitant (119), and the significance of this phosphorylation is unclear. By contrast, agents that elevate cAMP have been shown to activate Rap1B GTP-loading (120). It is plausible to propose that this activation of Rap1 may trigger the binding of Rap-1 to Raf-1, thereby preventing Raf-1 activation.

In certain neuronal cell lines, however, cAMP agonists can activate the ERKs. Thus, for example, forskolin can foster PC12 cell differentiation (neurite outgrowth). While it was initially believed that the sole function of Rap1 was to act as a suppressor of Ras, a positive role for Rap1 in signal transduction from cAMP to the ERKs has recently been identified. Coexpression with constitutively active (Val12)–Rap1B results in the strong activation of B-Raf, but not Raf-1 (121). Moreover, B-Raf can associate directly, and in a GTP-dependent manner, with Rap1B. Two distinct functions of the Rap1 → B-Raf signalling axis have been identified. Both functions are observed in PC12 phaeochromocytoma cells (122). As was noted above, cAMP has been observed to trigger the activation of Rap1. Rap1 activation may be mediated by one of two processes. The first involves a novel cAMP effector, the exchange factor directly activated by cAMP (Epac, also called cAMP-GEF). Epac is a Rap1 GEF that directly binds and is activated by cAMP. Thus, agonists that generate the production of cAMP and activate the ERKs, such as isoproterenol, might recruit B-Raf by stimulating the Epac-dependent GTP loading of Rap-1. A critical caveat of this model is that Epac is relatively poorly expressed in brain, where B-Raf predominates (123, 124). Alternatively, Rap1 is itself a substrate for PKA, and it has been suggested that this phosphorylation promotes GTP loading through a mechanism that is poorly understood (120, 125). The physiological significance of cAMP activation of B-Raf was made evident when it was shown that Asn17 dominant inhibitory Rap1, but not Asn17-Ras could block forskolin-mediated B-Raf activation and, in parallel, PC12 cell neurite outgrowth (122).

Nerve growth factor (NGF) can also promote PC12 cell neurite outgrowth through engagement of the Trk family of Tyr kinase receptors. NGF treatment results in the sustained activation of ERK. By contrast, ERK activation by EGF is quite transient in PC12 cells. Engineering PC12 cells to express higher levels of EGF receptors enables EGF to promote sustained ERK activation; and, by this process, it was demonstrated that sustained ERK activation is critical to fostering PC12 cell differentiation (2, 126). NGF activation of ERK involves Ras- and Rap1-dependent processes, both of which are necessary for sustained ERK activation (127). The Ras-dependent mechanism is a conventional pathway mediated by Grb2–mSOS recruitment to Trk followed by Ras activation which, in turn, engenders Raf-1 activation. Rap1 activation involves a second SH2/SH3 adapter protein Crk. Crk recruits C3G, a GEF that can catalyse Rap1 GDP–GTP exchange (109). This Rap1 activation mechanism is Ras-independent and is not blocked with Asn17–Ras (127). When examined at a single intermediate time point (15–20 min), the expression of either Asn17–Ras or –Rap1 significantly blunts NGF activation of ERK. However, if ERK activation is examined over a 62-min time course, it is observed that activation is blocked by Asn17–Ras only in the first 10 min. Conversely, Asn17–Rap1 inhibits ERK activation only at later time points (20–62 min) (127). Thus, Rap1 is involved in propagating sustained activation of ERK in response to NGF.

2.7 Regulation of the ERK pathway by the scaffold protein MEK partner-1 (MP1)

The specificity and efficiency of many eukaryotic MAPK pathways is greatly enhanced by scaffold proteins. These proteins can be either intrinsic signalling elements within MAPK core modules, or distinct polypeptides; and serve to nucleate MAPK components into discrete complexes (Sections 1.1, 3.10) (3) (Figs 2 and 5). MEK binding partner-1 (MP1) was identified by Weber and colleagues as a polypeptide that could interact in a yeast two-hybrid screen, in transfected cells, and, *in vitro*, with MEK1 (16). MP1 can bind MEK2 *in vitro* but does not associate with MEK2 *in vivo*. MEK1 and MEK2 have a short proline-rich segment, the PRS (amino acids 261–307 of MEK1, Fig. 5A) (55). This domain is not present in other MEKs. Downstream signalling by MEK1 requires the PRS; and mutation/truncation analysis indicates that MP1 interacts with a portion of the MEK1 polypeptide that includes the PRS. Deletion of the PRS abolishes MP1 binding (16).

In *in vitro* assays, addition of MP1 enhances the phosphorylation of MEK1 by B-Raf, and enhances the ability of mutationally activated MEK1 to activate ERK in the absence of B-Raf. This would suggest that MP1 merely facilitates MEK1 regulation and function by binding to the MEK1 PRS; however, the situation is more complicated. Thus, the addition of MP1 to wild-type MEK1 previously activated by B-Raf is without effect, suggesting that MP1 preferentially interacts with non-phosphorylated MEK. The differential *in vivo* and *in vitro* association of MP1 with MEK2 also indicates that additional factors may interfere with MEK2 binding to MP1 *in vivo* (16).

Consistent with the ability of MP1 to enhance MEK activation of ERK, MP1 also interacts with ERK1, but, curiously, not with ERK2 (Fig. 5B). Thus, the expression of MP1 with ERK1 significantly enhances serum-dependent ERK activation *in vivo* (16). Although B-Raf activation of MEK1 is potentiated by MP1 (16), a possible interaction between B-Raf or other Raf family kinases and MP1 has not been reported.

2.8 Regulation of ERK function by nucleocytoplasmic shuttling

The ERKs were originally identified as cytosolic kinases; and, indeed, a substantial fraction of ERK activity can be recovered from cytosolic extracts (19). A critical question arising from this observation was how ERKs (and, indeed other MAPKs which are localized in the cytosol) could phosphorylate nuclear proteins such as transcription factors.

Immunocytochemical examination of the subcellular localization of ERKs revealed that upon mitogen stimulation, a significant portion of the pool of ERK immunoreactivity translocates to the nucleus. This translocation is reversible, and terminates upon cessation of stimulus (128). The potential importance of this nucleocytoplasmic shuttling became evident when it was observed that ERK-dependent *de novo* gene expression correlated with prolonged ERK activation—conditions that coincided with ERK nuclear translocation (129).

Pouysségur and colleagues have shown that sequestration of ERK in the cytosol, by coexpression with a catalytically inactive form of the MAPK phosphatase MKP3, has no effect on ERK-dependent Rsk activation or phosphorylation of an engineered cytosolic mutant of Elk-1, but strongly inhibits ERK-dependent gene transcription and entry into S phase (129).

The molecular mechanism by which ERK nuclear localization is regulated has only recently begun to be elucidated. Several novel findings suggest that inactive ERK is retained in the cytosol as part of a complex with MEK1. Formation of this complex requires the ERK binding domain (the amino terminal 32 amino acids of MEK1, Fig. 5) (130). MEK1 itself is retained in the cytosol by a consensus nuclear export signal (amino acids 33–44). Dissociation of the MEK–ERK complex requires MEK-catalysed phosphorylation of ERK at the Thr–Glu–Tyr phosphoacceptor region. Once dissociation of the complex occurs, activated ERK translocates to the nucleus (130) (Fig. 5B).

Whereas ERK nuclear translocation requires phosphorylation, ERK catalytic activity is not required for translocation. Thus, Cobb and colleagues demonstrated that both wild-type and kinase-inactive (Lys45Arg) ERK2, phosphorylated in bacteria upon coexpression with activated MEK1, could translocate to the nucleus upon microinjection into cells. Similarly, the unphosphorylated form of this kinase, expressed without MEK, could translocate upon treatment of the injected cells with mitogen. By contrast, mutation of the ERK2 phosphoacceptor sites to non-phosphorylatable residues (Thr185Ala/Tyr187Phe) completely abrogated the ability of these kinases to translocate upon injection into fibroblasts (131) (Fig. 5B).

ERK2 activation-dependent dimerization is also critical for translocation. Thus, the crystal structure of activated ERK2 revealed a dimer with the interface occurring to the rear of the catalytic cleft between the two kinase lobes (131, 132). Specifically, dimerization requires four Leu residues (333, 336, 341, 344) plus His176 which forms a salt bridge with Glu343 (131, 132). Coinjection of bacterially phosphorylated (expressed with active MEK1) wild-type ERK2 or Lys54Arg ERK2 with the phosphoacceptor mutant ERK2 (Thr185Ala/Tyr187Phe) enables translocation of the phosphoacceptor mutant (131). Mutation or deletion of the critical residues of the dimer interface has no effect on the *in vitro* kinase activity or MEK-catalysed phosphorylation of the ERK2, but completely abolishes ERK2 nuclear translocation (132). From these findings, it can be concluded that phosphorylation and activation of ERK2 results in dissociation from MEK1 and dimerization, both of which are necessary for translocation (Fig. 5). It is noteworthy that both the SAPKs and p38s have similar dimerization motifs and show similar ligand-dependent translocation. Indeed, as is described in Section 3.10.3, SAPKs form reversible, activation-dependent associations with the SAPK-specific MEK SEK1, complexes that may partly serve to retain inactive SAPKs in the cytosol (17). Inasmuch as the ERKs (as weil as the SAPKs and p38s) do not possess consensus nuclear localization motifs, the mechanism by which ERKs traverse the nuclear pore complex is unclear. It is possible that MAPKs, once liberated from elements restraining them in the cytosol, 'piggy back' into the nucleus by interacting with nuclear substrate proteins (Section 3.5.3).

3. The stress-activated protein kinase (SAPK)/c-Jun NH$_2$-terminal kinase (JNK), p38 MAPKs, and ERK5/big MAPK-1 (BMK1) pathways: mammalian MAPK signalling pathways activated by environmental stress and inflammatory cytokines

3.1 General considerations

The cellular and physiological responses to inflammation are central to the pathology of a number of important clinical conditions including ischaemic and hypertension injury, endotoxin shock, arthritis, inflammatory bowel disease, and diabetic nephropathy. In addition, these responses are central to the function of a number of treatment modalities commonly employed to combat cancer including chemotherapeutic agents and radiation. In recent years, it has become clear that, as with yeast, multiple, parallel, mammalian MAPK pathways exist; and that most of these, in conjunction with the nuclear factor-κB (NF-κB) pathway, are pivotal to the inflammatory response. As mammalian stress-activated signalling pathways are elucidated, it is becoming evident that these pathways will be important novel targets for anti-inflammatory drugs.

3.2 The stress-activated protein kinases/c-Jun NH$_2$-terminal kinases (SAPKs), a family of MAPKs activated by environmental stresses and inflammatory cytokines

Indications that mammalian cells possessed several MAPK pathways came shortly after the identification of the ERKs. The protein synthesis inhibitor cycloheximide, when administered to rats, can elicit the *in vivo* activation of ribosomal S6 phosphorylation; and, in fact, this strategy was used to activate p70 S6 kinase *in vivo* prior to purification (26). That cycloheximide could recruit p70 led to the notion that several protein kinase signalling pathways might be activated by cycloheximide.

This hypothesis was proved correct when it was demonstrated that injection of cycloheximide into rats activated a Ser/Thr kinase activity that could phosphorylate microtubule-associated protein-2 (MAP2). Purification of this kinase revealed a 54-kDa polypeptide, and the kinase was initially named p54-MAP2 kinase, or p54 (1, 133). p54 could be inactivated with Tyr or Ser/Thr phosphatases, indicating that, like the ERKs, this kinase required concomitant Tyr and Ser/Thr phosphorylation for activity (1, 134). Further support for the notion that p54 was a novel MAPK came from the demonstration that p54 was proline-directed (135).

However, the physiological substrate specificity of p54 clearly differed from that of the ERKs. In particular, p54 was unable to activate MAPKAP-K1/Rsk *in vitro* under conditions wherein ERK-mediated activation of MAPKAP-K1/Rsk (Section 2.2.1) was observed (133). More importantly, p54 was able to phosphorylate the c-Jun transcription factor at two sites (Ser63 and Ser73) implicated in the regulation of c-Jun and AP-1 *trans* activation function (136) (Section 3.5.2).

The SAPKs were cloned independently by two groups: one using degenerate oligonucleotides based on the sequence of purified p54, as PCR primers (137); and the other using a pure PCR strategy employing degenerate primers derived from regions conserved among all MAPKs (140). The availability of recombinant p54 enabled the generation of specific antibodies that could be used to analyse p54 regulation by extracellular stimuli. Immunoprecipitation of endogenous p54 from extracts of cells subjected to various treatments revealed that, in most cells, p54 was not strongly activated by mitogens such as insulin, EGF, PDGF, or FGF. By contrast, p54 was vigorously activated by environmental stresses such as heat shock, ionizing radiation, oxidant stress, DNA damaging chemicals (topoisomerase inhibitors and alkylating agents), reperfusion injury, mechanical shear stress, and, of course, protein synthesis inhibitors (cycloheximide and anisomycin). In addition, p54 could be activated by vasoactive peptides (endothelin and angiotensin-II) and inflammatory cytokines of the tumour necrosis factor (TNF) family (TNF, interleukin-1, CD40 ligand, CD27 ligand, Fas ligand, receptor activator of NF-κB, RANK ligand, etc.) (1, 137–145). p54 has since been renamed; and the nomenclature of this family of kinases is somewhat confusing (Table 1). Two systems are generally accepted: stress-activated protein kinase (SAPK) in reference to the regulation of these kinases by environmental stress and inflammation; and c-Jun NH_2-terminal kinase (JNK) in reference to the phosphorylation by these kinases of the c-Jun amino-terminal *trans* activation domain.

The SAPKs are encoded by at least three genes: *SAPKα/JNK2*, *SAPKβ/JNK3*, and *SAPKγ/JNK1* (137, 146) (Table 1). Like the ERKs, each contains a characteristic Thr–X– Tyr phosphoacceptor loop in subdomain VIII of the protein kinase catalytic domain. Whereas the ERK sequence is Thr–Glu–Tyr, that of the SAPKs is Thr183–Pro–Tyr185. The SAPK genes are further diversified into up to 12 polypeptides by differential hnRNA splicing. hnRNA splicing within the catalytic domain at a region spanning subdomains IX and X results in type-1 and -2 SAPKs (β-and α-JNKs, respectively, Table 1). Splicing at the extreme carboxyl terminus yields 54- (p54) and 46-kDa (p46) polypeptides (type-2 and -1 JNKs, respectively, Table 1). Whereas the significance of the carboxyl-terminal isoforms is not clear, there is some evidence that the type-1 and -2 kinases differ in their substrate-binding affinities (8, 137, 145).

3.3   The p38 MAPKs, a second stress-activated MAPK family

The p38 MAPKs are a third mammalian MAPK family. p38 (the α isoform) was originally described as a 38-kDa polypeptide that underwent Tyr phosphorylation in response to endotoxin treatment and osmotic shock (147). p38 was purified by antiphosphotyrosine immunoaffinity chromatography; and cDNA cloning revealed that p38 was the mammalian MAPK homologue most closely related to *HOG1*, the osmosensing MAPK of *S. cerevisiae* (Chapter 3). Most notably, the p38s, like Hog1p, contain the phosphoacceptor sequence Thr–Gly–Tyr (3, 147). An identical p38 isoform was identified independently by two groups as a kinase activated by stress and

IL-1 that could phosphorylate and activate MAPKAP kinase-2 (Section 3.5.1), a novel Ser/Thr kinase implicated in the phosphorylation and activation of the small heat-shock protein Hsp27 (148, 149).

Of particular interest, this same p38 isoform was purified and cloned as a polypeptide that could bind to a class of experimental pyridinyl-imidazole anti-inflammatory drugs, the cytokine-suppressive anti-inflammatory drugs (CSAIDs) (150). CSAIDs were originally characterized as compounds that could inhibit the transcriptional induction of TNF and IL-1 during endotoxin shock (150). The basis for the efficacy of these compounds as anti-inflammatory agents was their ability to bind and directly inhibit a subset of the p38s, thereby blocking the p38-mediated activation of AP-1, a *trans* acting factor crucial to TNF and IL-1 induction (150). Finally, p38 was isolated as a kinase that could interact with the Myc binding-partner Max (151) (Section 3.5.2). With the identification of additional p38 isoforms, four p38 genes are now known (Table 1): the original isoform, here referred to as p38α (also called CSAIDs binding protein (CSBP) and, somewhat confusingly, SAPK2a); p38β (also called SAPK2b and p38–2); p38γ (also called SAPK3 and ERK6); and p38δ (also called SAPK4).

Interestingly, only p38α and p38β are inhibited by CSAIDs, p38γ and p38δ are completely unaffected by these drugs *in vitro* or in transfected cells (156). The basis for this inhibition was revealed in the crystal structure of p38α complexed with the CSAID SB203580. Thr106 in the hinge of the p38α ATP binding pocket interacts with a fluorine atom in the SB203580 structure. This positions the drug to interact with His107 and Leu108 of the ATP binding pocket (159, 160). Substitution of Thr106, alone or in combination with His107 or Leu108, with the corresponding, more bulky residues from p38γ or p38δ (Met, and Pro or Phe, respectively, in both cases) abolishes SB203580 binding. Conversely, if the amino acid of p38γ, p38δ, or even SAPKγ/JNK1, which corresponds to p38α Thr106 is replaced with Thr, the resulting mutants display at least partial sensitivity to SB203580 (159, 160).

In further similarity with the SAPKs, the p38s are activated *in vivo* by environmental stresses and inflammatory cytokines and are only poorly activated by insulin and growth factors. In almost all instances the same stimuli that recruit the SAPKs also recruit the p38s (1). One exception is ischaemia–reperfusion. SAPKs are selectively activated during reperfusion, whereas the p38s are activated during ischaemia and remain active during reperfusion (1, 161, 162). The basis for this difference is unknown.

3.4   ERK5/big MAP kinase-1 (BMK1), a third class of stress-activated MAPK

The novel MEK MEK5 was cloned by degenerate PCR as part of an effort to identify new MAPK pathways and regulators (163, 164). ERK5, a putative MEK5 target, was cloned as part of a two-hybrid screen that employed MEK5 as bait (163). ERK5 is an approx. 90-kDa MAPK of which only one mammalian homologue is known. ERK5 has the sequence Thr–Glu–Tyr in its phosphoacceptor loop. The amino-terminal kinase domain of ERK5 is followed by an extensive carboxyl-terminal tail of unknown function that contains several consensus polyproline SH3 binding sites (163). The stimuli that recruit ERK5 have not been comprehensively characterized; however, ERK5 appears to be activated predominantly by environmental stresses in so far as oxidant stress (peroxide) and osmotic shock (sorbitol), but not mitogens, phorbol esters, vasoactive peptides, or inflammatory cytokines (TNF), can substantially activate ERK5 *in vivo* (165).

3.5 SAPK, p38, and ERK5 substrates

As with the ERKs, the SAPKs, p38s, and ERK5 phosphorylate both transcription factors and other protein kinases. These reactions are important to the inflammatory response.

3.5.1 Protein kinases

MAPKAP kinases-2 and -3

MAPKAP kinase-2 (MAPKAP-K2) and the structurally related MAPKAP-K3 (also called three-pathway regulated kinase or 3PK) are a small family of Ser/Thr kinases that are unrelated to MAPKAP-K1/Rsk. Each consists of an amino-terminal regulatory domain and a carboxyl-terminal kinase domain (166–169). MAPKAP-K2 and MAPKAP-K3 are responsible, at least in part (see below), for the phosphorylation of the small heat-shock protein Hsp27 (166–169). Non-phosphorylated Hsp27 normally exists in high molecular-weight multimers that serve as molecular chaperones, assisting in maintaining protein conformation. Phosphorylation of Hsp27 by MAPKAP kinase-2/3 at residues Ser15, Ser78, Ser82, and Ser90 correlates with the dissociation of Hsp27 into monomers and dimers and with the redistribution of Hsp27 to the actin cytoskeleton (170). In peroxide-treated, human umbilical-vein endothelial cells, this redistribution of Hsp27 may contribute to fostering the reorganization of F-actin into stress fibres, thereby affecting cell motility (170, 171). MAPKAP-K2/3-catalysed phosphorylation of Hsp27 at Ser90 appears necessary for this process and mutation of Ser90 to Ala prevents stimulus-induced changes in Hsp27 oligomerization (171).

Activation of MAPKAP-K2 requires phosphorylation at residues Thr25, Thr222, and Ser272 (172) (Fig. 7). Early studies had indicated that purified ERK1 could phosphorylate and activate MAPKAP-K2 *in vitro* (166). However, ERKs probably do not represent the physiological MAPKAP-K2 kinases. MAPKAP-K2 is activated not by insulin or mitogens, but by stresses and inflammatory cytokines, conditions wherein ERKs are not appreciably activated (148, 149). Instead, MAPKAP-K2 is phosphorylated and activated by p38α and p38β (but not by p38γ or p38δ) (173) (Fig. 8). Phosphorylation, catalysed by p38α and p38β of Thr25 gates subsequent phosphorylation, again catalysed by p38α and p38β, of Thr222 and Ser272 which reside in the kinase activation loop. Together, these phosphorylations result in activation of MAPKAP-K2 accompanied by an additional autophosphorylation at Thr334 (172) (Fig. 7). Consistent with regulation by p38α and p38β, MAPKAP-K2 activation and Hsp27 phosphorylation are inhibited by CSAIDs (168, 170–172) (Fig. 8).

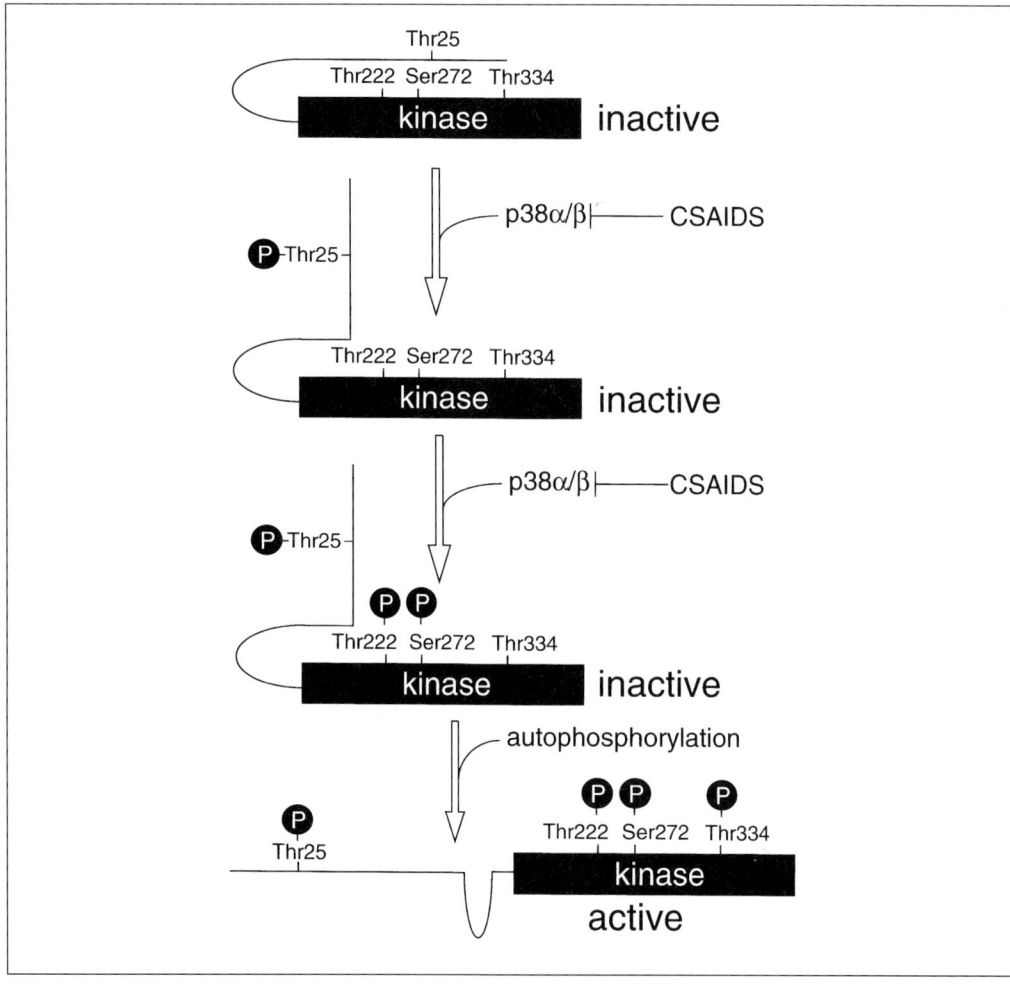

Fig. 7 Stepwise activation of MAPKAP-K2 by p38. Note that phosphorylation of Thr25 gates remaining phosphorylation events. Black circles with a white P indicate phosphorylation.

MAPKAP-K3 can also phosphorylate Hsp27 (168). Whereas MAPKAP-K3 can be activated *in vivo*, in overexpression experiments, by the ERK, SAPK, and p38 pathways (174), endogenous MAPKAP-K3 is only significantly activated by stressful stimuli and inflammatory cytokines; and in a manner that can be completely inhibited with CSAIDs, suggesting that MAPKAP-K3 is, in fact, stress- and not mitogen-activated, and that p38α and/or p38β are the major MAPKAP-K3 kinases *in vivo* (168) (Fig. 8).

PRAK

p38-regulated/activated kinase (PRAK) is an approx. 50-kDa Ser/Thr kinase with a similar overall structure to MAPKAP-K2 and -K3 and the MNKs, consisting of an

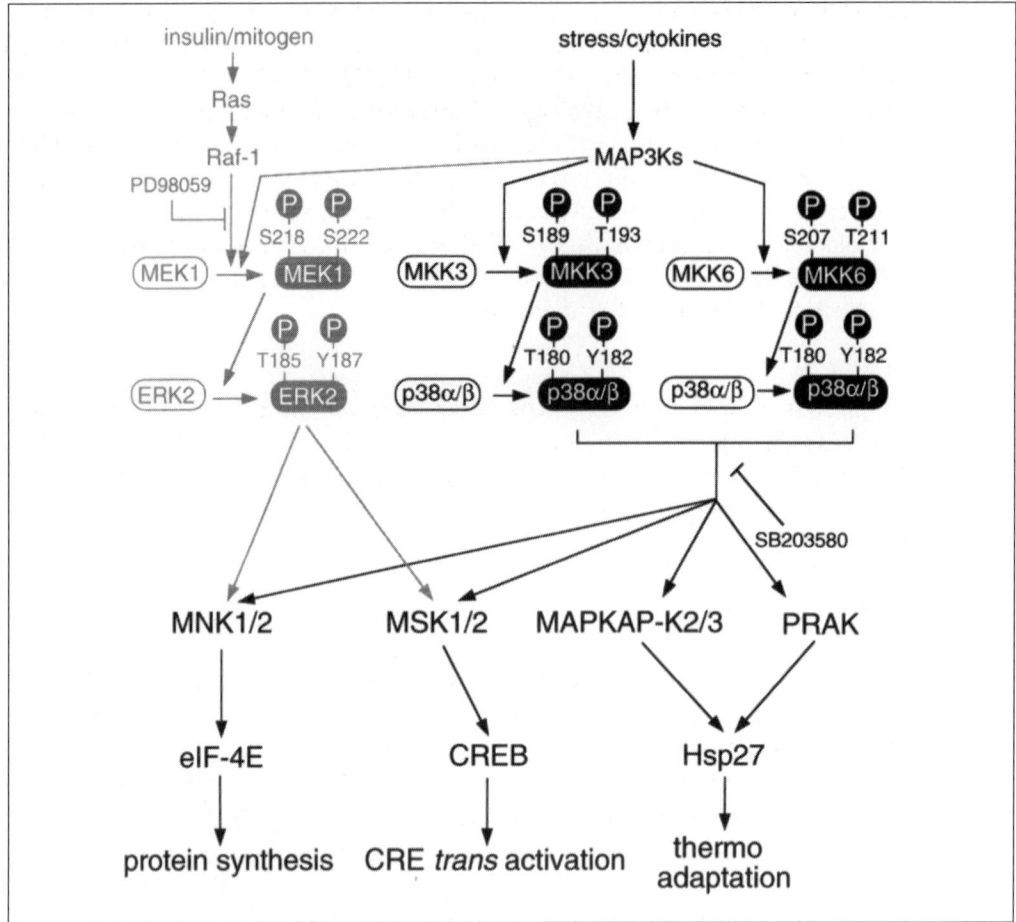

Fig. 8 Regulation of multiple protein kinases (MNKs, MSKs, MAPKAP-K2/3, and PRAK) by p38s, parallel regulation of MNKs and MSKs by ERKs. Black circles with a white P indicate phosphorylation. See text for details.

amino-terminal regulatory domain and a carboxyl-terminal kinase domain. PRAK is activated in response to stress and inflammatory cytokines and is not detectably activated by mitogens. PRAK activation can be blocked with CSAIDs (175). Consistent with this, PRAK can be activated *in vivo* and *in vitro* by p38α and p38β. Phosphorylation of PRAK catalysed by p38s is at Thr182 in the kinase domain activation loop (175). Once activated, PRAK can phosphorylate Hsp27 at the physiologically relevant sites; and in-gel kinase assays indicate that PRAK is an important stress-activated Hsp27 kinase (175) (Fig. 8).

MSK1/2
In addition to the ERKs, the MSKs (Section 2.2.2) can be phosphorylated and activated by the CSAID-sensitive p38s p38α or p38β. By contrast, neither SAPK, p38γ nor p38δ

can activate MSK1 *in vitro* (32). Consistent with this, MSK1 is activated *in vivo* not only by mitogens (Section 2.2.2), but by environmental stresses (arsenite, UV, peroxide), in a manner inhibited by the CSAID SB203580 (32). MSK1 is also activated by TNF; however, the pharmacological inhibition of TNF activation of MSK1 is complex. TNF activates the p38s (and the SAPKs) within 5 min. In HeLa cells, TNF can also activate HeLa cell ERKs; however, this activation is slower, reaching an apparent maximum at 15 min. Accordingly, TNF activation of MSK1 in HeLa cells can be completely suppressed with SB203580 after a 5-min stimulation. By 15 min, however, SB203580 inhibition is only partial and MSK1 can now also be partially inhibited with PD98059 (32) (Section 2.4 and Fig. 8).

MNKs

As discussed in Section 2.2.3, the MNKs are substrates for both the ERKs and p38 MAPKs (43, 44). The regulation of the MNKs by both the ERKs and p38s indicates that, as with AP-1 (see below), MNKs are a site of integration of stress and mitogenic signalling pathways (Fig. 8).

3.5.2   Transcription factors
CHOP/GADD153

CREB homologous protein (CHOP)/growth arrest and DNA damage-153 (GADD153) is a bZIP transcription factor of the CREB family (40). CHOP/GADD153 is transcriptionally induced in response to genotoxic and inflammatory stresses (176). These stimuli can also activate the transcriptional regulatory functions of CHOP/GADD153 through agonist-induced phosphorylation of Ser78 and Ser81. CHOP/GADD153 acts as a transcriptional repressor of certain cAMP-regulated genes and a transcriptional activator of stress-induced genes. Recruitment of CHOP/GADD153 correlates with cell-cycle arrest at G_1/S. This cell-cycle arrest is an important consequence of DNA damage inasmuch as it allows for DNA repair prior to DNA replication, thereby preserving genomic integrity. p38α can phosphorylate CHOP/GADD153 at Ser78 and Ser81 *in vivo* and *in vitro* and is a likely stress-activated regulator of CHOP/GADD153 function (176).

NFAT4

The nuclear factor of activated T cells (NFAT) family of transcription factors is distantly related to Rel/NF-κB (reviewed in ref. 177). NFATs are retained in the cytosol as a consequence of phosphorylation (catalysed by casein kinase-Iα and, possibly, glycogen synthase kinase (GSK)-3) at five to six sites (within NFAT4, this comprises a region spanning amino acids 204–215). This phosphorylation engenders a conformation that masks the nuclear localization signal. Agonist-induced Ca^2+ entry recruits the Ca^2+-dependent phosphatase calcineurin which dephosphorylates NFATs thereby enabling NFAT nuclear translocation. Dephosphorylation of NFATs also enhances DNA binding affinity (177). NFATs bind and *trans* activate genes with an NFAT *cis* acting element (consensus sequence: $^T/_A$GGAAAAT). NFAT sites are often located close to AP-1 sites in many promoters, allowing for the cooperative binding and synergistic *trans* activation of numerous genes (IL-2, IL-4, IL-5, and CD40L are examples) (177). Inasmuch as calcineurin is a major target of the immunosuppressants FK506 and cyclosporin-A, potent inhibition of NFAT activity is an important consequence of FK506 and cyclosporin-A action (101, 177).

Serum factors can markedly blunt the nuclear translocation of NFATs mediated by calcium ionophores acting through calcineurin. Casein kinases and GSK3 are not serum-stimulated; and this serum-dependent inhibition of NFAT activation implied that serum-responsive kinase cascades might contribute to NFAT inhibition (177, 178). Davis and colleagues showed that the SAPKs can phosphorylate the NFAT family member NFAT4 at Ser163 and Ser165. This phosphorylation correlates with an inhibition of agonist-induced NFAT4 nuclear translocation; and it has been proposed therefore that the SAPK pathway antagonizes NFAT4 action (178). However, the significance of this finding is unclear. McKeon *et al.* showed that NFAT4 mutants, wherein the putative SAPK phosphoacceptor sites (Ser163 and Ser165) have been changed to Ala, still show serum-stimulated inhibition of nuclear translocation in many instances. These investigators demonstrated that the SAPK-specific MAP3K MEK-kinase-1 (Section 3.7.2) can indirectly, and independently of SAPK, block NFAT4 dephosphorylation and activation. MEKK1 inhibition of NFAT translocation occurs whether or not the SAPK phosphorylation sites have been mutated to alanine; and the effect of MEKK1 is not reversed by dominant inhibitors of the SAPK pathway. Apparently, MEKK1 fosters inhibition of NFAT nuclear translocation by stabilizing the association between NFAT4 and the inhibitory NFAT kinase casein kinase-Iα (179).

Max

Max is a 12-kDa helix–loop–helix (HLH) polypeptide that interacts with the transcription factor c-Myc rendering Myc competent to signal. c-Myc is essential as a regulator of cell proliferation, differentiation, and apoptosis; and its biological functions are thought to partly depend upon the polypeptide binding partners with which c-Myc interacts, and the regulation of these binding partners (reviewed in ref. 180). p38α was isolated in a yeast two-hybrid screen for Max interactors. Max and p38α form a tight complex *in vivo* and *in vitro*; and the p38–Max interaction is thought to be mediated by the HLH domain of Max and a similar loop on p38. Max is a good substrate for p38α and for a truncated, constitutively active splicing isoform of p38α, Max interactor-2 (Mxi2) (151). Interestingly, there are no consensus proline-directed sites on Max, suggesting that the absolute proline requirement for MAPKs may be less pronounced for p38. The functional significance of p38-catalysed Max phosphorylation is unclear (151).

AP-1

The SAPKs and p38s are the dominant Ser/Thr kinases responsible for the recruitment of the activator protein-1 (AP-1) transcription factor in response to environmental stresses and inflammatory stimuli (1, 46). ERK5 may also play a role in AP-1 regulation (181). AP-1 comprises bZIP transcription factors—typically c-Jun and JunD, along with members of the *fos* (usually c-Fos) and ATF (usually ATF2) families. All bZIP transcription factors contain leucine zippers that enable homo- and hetero-dimerization; and, accordingly, AP-1 components are organized into Jun–Jun, Jun–Fos, or Jun–ATF dimers (46).

The presence of Jun family members enables AP-1 to bind to *cis* acting elements containing the tetradecanoyl phorbol myristate acetate (TPA) response element (TRE—consensus sequence: $TGA^C/_GTCA$). ATFs, including ATF2, are members of the CREB subfamily of bZIP transcription factors; and AP-1 heterodimers containing ATF transcription factors can also bind to the CRE (40, 46). AP-1 is an important *trans* activator of a number of stress-responsive genes including the genes for interleukins-1 and -2, CD40, CD30, TNF, and c-Jun itself. In addition, AP-1 participates in the transcriptional induction of proteases and cell-adhesion proteins (e.g. E-selectin) important to inflammation (46).

Activation of AP-1 involves both the direct phosphorylation/dephosphorylation of AP-1 components, as well as the phosphorylation and activation of transcription factors that induce elevated expression of c-*jun* or c-*fos*. Both events can be activated independently by several signalling pathways (Fig. 9). Phosphorylation of c-Jun or ATF2 within their *trans* activation domains correlates well with enhanced *trans* activating activity (136, 137, 179, 183). The SAPKs can phosphorylate the c-Jun *trans* activating domain at Ser63 and Ser73 (136). These residues are phosphorylated *in vivo* under conditions wherein the SAPKs are activated; and depletion of SAPK from cell extracts removes all stress-activated c-Jun kinase (137, 140). Thus, SAPKs are the dominant kinases responsible for c-Jun phosphorylation (Fig. 9). JunD is also phosphorylated by SAPKs, albeit less effectively than is c-Jun. This phosphorylation occurs at Ser90 and Ser100, a region of the JunD *trans* activation domain similar to the phosphoacceptor domain of c-Jun (7).

The SAPKs, and p38s can phosphorylate ATF2 at Thr69 and Ser71 in the *trans* activation domain. Again, these residues are phosphorylated under circumstances

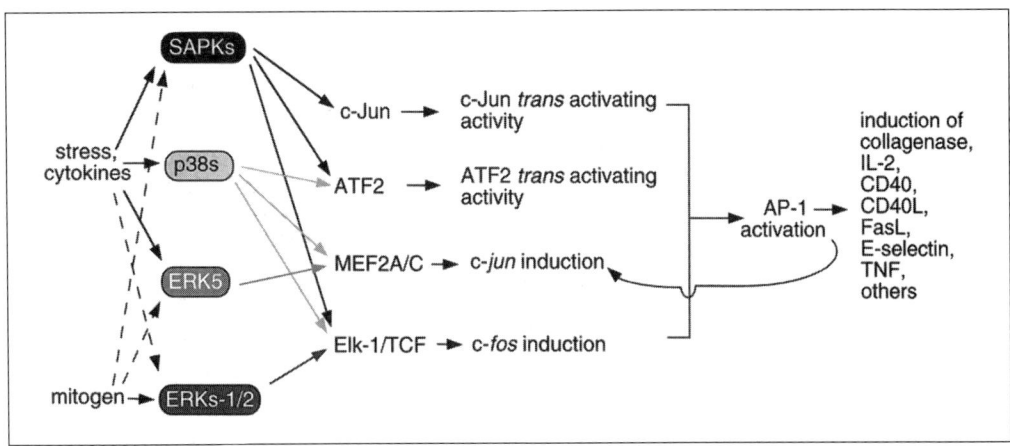

Fig. 9 Regulation of AP-1 by the integration of multiple MAPK pathways. Dashed lines indicate weak inputs. Note that AP-1 can *trans* activate the expression of c-*jun*.

wherein the SAPKs, and p38s are activated. Phosphorylation of ATF2 at Thr69 and Ser71 correlates with activation of ATF2 *trans* activating activity (182) (Fig. 9). Whether the SAPKs or p38s represent the dominant ATF2 kinases depends on the cell type and stimulus used. During reperfusion of an ischaemic kidney, for example, the SAPKs are the only detectable ATF2 kinases (183). By contrast, in response to IL-1, the p38s are the major ATF2 kinases activated in KB keratinocytes (173).

The SAPKs and p38s also contribute to AP-1 activation by stimulating the transcription of genes encoding AP-1 components (46) (Fig. 9). One of the earliest transcriptional events known to occur in response to mitogen stimulation is the induction of c-*fos* expression. The *fos* promoter contains a *cis* acting element, the serum response element (SRE) that mediates the recruitment of transcription factors which induce *fos* expression. The SRE binds a heterodimeric transcription factor containing two polypeptides, the serum response factor (SRF) and the ternary complex factors (TCFs) (reviewed in ref. 45) (Fig. 9).

The TCFs are a family of Ets-domain transcription factors that includes Elk-1 and Sap-1a (45). The SAPKs, and ERKs (Section 2.2.4), can phosphorylate two critical residues in the Elk-1 C-terminus (Ser383, Ser389), while the p38s can efficiently phosphorylate the analogous residues (Ser381 and Ser387) on Sap-1a (9, 47–50). This phosphorylation enhances the binding of TCFs to the SRF and thereby triggers *trans* activation at the SRE. By these processes MAPKs activated by both stress and mitogens can convergently contribute to c-*fos* induction (Fig. 3B).

The p38s and ERK5 can also phosphorylate the transcription factors myocyte enhancer factor-2A and -2C (MEF2A and C), members of the MEF subgroup of the MCM1-agamous and deficiens-SRF (MADS) box transcription-factor family (178, 184, 185). As with p38-catalysed ATF2 phosphorylation, phosphorylation of MEF2C by p38s and ERK5 correlates well with enhanced MEF2C *trans* activating activity (179, 184, 185). MEF2C was originally identified as a transcription factor that bound to AT-rich sequences (consensus: CTAAAAATAA) and *trans* activated a number of genes involved in myoblast differentiation; however, MEF2C is widely expressed and may regulate numerous transcriptional regulatory events (ref. 184 and references therein). Interestingly, the p38s and ERK5 phosphorylate different sets of sites on the MEF2C polypeptide. p38s phosphorylate Thr293 and Thr300, whereas ERK5 phosphorylates Ser387 (181, 184). All three sites are phosphorylated in response to serum or stress; however, Thr293/Thr300 phosphorylation is sufficient for p38 activation of MEF2C, while Ser387 phosphorylation is sufficient for ERK5 activation of MEF2C (181, 184). A *cis* element for MEF2C resides in the c-*jun* promoter; thus, p38 and ERK5 activation can contribute to the induction of c-*jun* expression (181). Indeed MEF2A or C, once activated by p38s can *trans* activate the c-Jun promoter (179, 182) (Fig. 9).

3.5.3 The physiological substrates of the SAPKs and p38s contain docking sites that confer specificity for distinct MAPK subgroups

The phosphorylation of c-Jun by the SAPKs illustrates an important point about the mechanism of substrate recognition by members of the MAPK family. All MAPKs are 'proline-directed'—phosphorylating Ser/Thr residues only if followed immediately by proline (Section 1.1 and Fig. 10A). However, the specificity of MAPKs for their physiological substrates is often largely dictated by the presence of binding sites, distal from the phosphoacceptor sites, that are specific for distinct MAPK subgroups. These MAPK binding sites allow for the selective interaction between MAPKs and their true *in vivo* substrates. As discussed in Section 2.2.1, MAPKAP-K1/Rsk has a specific binding site for ERKs that is at the extreme carboxyl terminus, distal from the ERK phosphoacceptor site (39).

The phosphorylation of Jun family transcription factors by the SAPKs is a second example of MAPK docking sites. Thus, the SAPKs, but not the ERKs or p38s, bind c-Jun quite strongly (7, 8). The SAPK binding site on c-Jun lies between residues 32 and 52, well away from Ser63 and Ser73, the sites of phosphorylation (7, 8, 46). This binding site lies within the so-called δ-domain (amino acids 30–57), a hydrophobic region initially implicated in the regulation of c-Jun oncogenicity due to its deletion in oncogenic v-Jun (for this reason, v-Jun does not bind SAPK and is not a SAPK substrate) (46) (Fig. 10A).

The presence of the SAPK binding site, coupled with the ability of c-Jun to hetero-dimerize with other members of the Jun family enables the SAPKs to phosphorylate other AP-1 constituents *in vivo* that, as monomers or homodimers, ordinarily are poor SAPK substrates (7, 46). Thus, while JunD possesses a phosphoacceptor region and even a SAPK binding pocket homologous to those of c-Jun, JunD binds SAPK poorly (7) (Fig. 10). Accordingly, JunD is generally not a SAPK substrate *in vitro*; however, when heterodimerized with c-Jun, JunD can undergo efficient SAPK-catalysed phos-phorylation and activation *in vitro*, and in transfection experiments *in vivo* (7) (Fig. 10). JunB, by contrast, possesses a SAPK binding pocket and binds SAPK well, but does not possess the proline-directed phosphoacceptor sites that are prerequisite for SAPK phosphorylation (7, 46) (Fig. 10A, and Section 1.1). Thus, JunB can bind SAPK, but is not phosphorylated *in vivo* or *in vitro* by SAPK (7) (Fig. 10B). However, in transfection experiments, JunB can heterodimerize with c-Jun mutants missing the δ-domain and foster SAPK phosphorylation of these mutants (7) (Fig. 10B). Interest-ingly, JunB may function as a negative regulator of c-Jun (46).

ATF2 also contains a hydrophobic pocket (amino acids 20–60) similar to the c-Jun δ-domain that binds SAPKs and p38s. As with the c-Jun δ-domain, the ATF2 MAPK binding site lies amino-terminal to the phosphoacceptor sites (Thr69 and Thr71) (182). Likewise, Elk-1 has a MAPK docking site, the D-domain (amino acids 312–334) that lies amino-terminal to the phosphoacceptor sites (Ser383, Ser389) (9, 50). Interest-ingly, the Elk-1 D-domain is essential for ERK and SAPK binding and phosphoryla-tion; however, p38 binding and phosphorylation appears not to require this domain (9).

3.5.4 Multiple genes and differential hnRNA splicing in the SAPK catalytic domain sequences may generate SAPK isoforms with different substrate affinities

As noted above (Section 3.2), the SAPKs are encoded by three genes that undergo differential hnRNA splicing to generate at least 10, and possibly up to 12, polypeptide

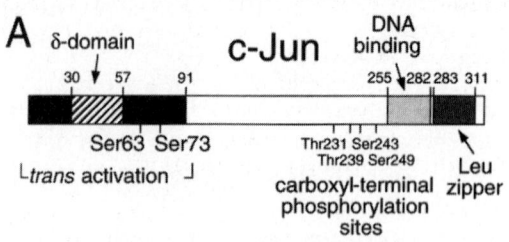

A

δ-domain

c-Jun

DNA binding

30 ↓ 57   91

255 ↓ 282 283 311

Ser63 Ser73

Thr231 Ser243
Thr239 Ser249

⌐trans activation⌐

carboxyl-terminal phosphorylation sites

Leu zipper

Amino-terminal phosphoacceptor motifs

c-Jun	58	SDLLTSPDVGLL KLASPELER	78
JunD	85	DGLLASPDLGLL KLASPELER	105
JunB	69	FSGQGS DTGASLKLASTELER	89

Domains implicated in SAPK/JNK binding

c-Jun-δ	25	AYGYSNPKILKQSMTLNLADP VGSLK	50
JunD	43	APPTSSM LKKDALTLSLADEGAAGLK	68
JunB	26	SLSLHDYKLLKPTLALNLADP YRGLK	51

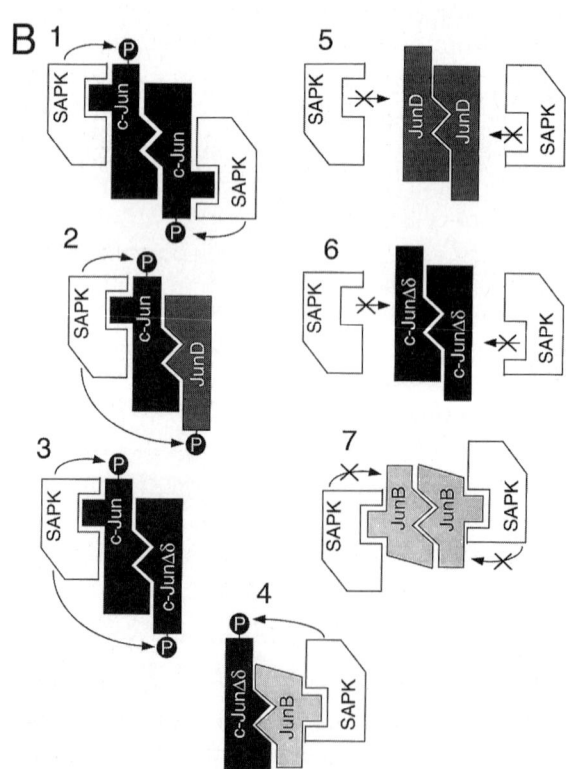

Fig. 10 The c-Jun δ-domain is a docking site for SAPKs and illustrates the principle that docking sites enable selective phosphorylation by MAPKs of their target proteins. (A) Comparison of the phosphoacceptor and putative docking sites of three Jun family members. Only c-Jun and JunD are SAPK substrates as only they have Pro-directed phosphoacceptor motifs. In spite of the similarities between docking sites, only c-Jun and JunB can interact with SAPKs. (B) Phosphorylation of Jun family proteins by SAPKs mediated by Jun–Jun dimerization, the presence of docking sites and Pro-directed phosphoacceptor sites. Black circles with a white P indicate phosphorylation. (1) c-Jun homodimers are SAPK substrates due to the presence of SAPK docking sites. (2) c-Jun–JunD heterodimerization enables JunD phosphorylation by SAPK bound to the docking site of c-Jun. JunD cannot bind SAPK. (3) Same as (2), except that c-Jun is heterodimerized with a c-Jun construct wherein the δ domain is mutated. (4) JunB can bind SAPK, and, although not a SAPK substrate, can heterodimerize with a Jun construct which is itself unable to bind SAPK (c-JunΔδ), thereby enabling phosphorylation of c-JunΔδ. (5–7) JunD (5), c-JunΔδ (6), or JunB (7) homodimers are not SAPK substrates due to a lack of a SAPK docking site (5) and (6) or a Pro-directed phosphoacceptor site (7).

---

species (137, 146). The significance of this heterogeneity is largely unknown; however, there is some evidence that the hnRNA splicing that generates the type-1 and -2 kinases (Table 1) may affect the affinity of different SAPK splicing isoforms for substrate. Thus, SAPK-p54α2 (JNK2α2, Table 1) binds more strongly to c-Jun than does SAPK-p54α1 (JNK2β2, Table 1). Moreover, SAPKα type-2 (JNK2 type α, Table 1) isoforms bind more strongly to c-Jun than to ATF2, whereas SAPKα type-1 (JNK2 type β) isoforms bind more strongly to ATF2 than to c-Jun (146).

In addition to differences in substrate binding among splicing isoforms of the same SAPK gene, the different *sapk* gene products themselves, SAPKα, β, and γ (JNKs-2, -3, and -1, respectively, Table 1), may also display differential substrate selectivity. Thus, whereas all SAPK isoforms preferentially interact with c-Jun as compared to ATF2, SAPKα/JNK2 isoforms are overall higher affinity c-Jun kinases than are SAPKβ/JNK3 or SAPKγ/JNK1 (8, 146). It should be noted, however, that these differences, as well as those among the splicing isoforms, are modest (two- to fivefold *in vitro*); and it is unclear what the *in vivo* biochemical significance of these differences is. By contrast, as we shall see (Section 3.14), the more dramatic differences among SAPK isoforms are observed when the biological functions of these enzymes are examined.

3.5.5 The regulation of AP-1 involves the integration of several MAPK pathways and is mediated by divergent AP-1 component transcription-factor subunits

How does the complex regulation of AP-1 constituent transcription factors translate into the recruitment of AP-1 by diverse extracellular stimuli? The different aspects of AP-1 regulation—activation of constituent transcription-factor expression and direct phosphorylation/activation of constituent transcription factors—can be independently regulated by several pathways (Fig. 9). Thus, mitogenic stimuli, which preferentially recruit the ERKs (and inhibit GSK3 via PKB/Akt, Chapter 1) will preferentially activate AP-1, respectively, through enhancement of the expression of AP-1 components (via ERK phosphorylation of Elk-1, resulting in c-*fos* expression, for example) and through the relief of GSK3-mediated inhibition of c-Jun DNA binding through dephosphorylation of the carboxyl-terminal phosphorylation sites on c-Jun which are phosphorylated by GSK3 (Fig. 9).

By contrast, stresses and inflammatory cytokines such as TNF, which preferentially activate the SAPKs and p38s, can recruit AP-1 through the direct phosphorylation of AP-1 components (c-Jun by the SAPKs and ATF2 by both the SAPKs and p38s). However, stress pathways can also promote enhanced expression of AP-1 components through the recruitment of Elk-1 (mediated by SAPK phosphorylation), which results in elevated c-*fos* expression; and through p38-catalysed phosphorylation of MEF2A/C (Fig. 9). MEF2A/C can bind and *trans* activate the promoter for c-*jun*. Stresses can also modestly recruit PKB/Akt, which can inhibit GSK3 thereby blocking its negative regulation of c-Jun DNA binding (Chapter 1). Finally, the c-Jun promoter also contains an AP-1 site; thus, c-*jun* expression can be autoregulated by any pathway that activates AP-1 (Fig. 9).

3.6 MEKs upstream of the SAPKs and p38s

3.6.1 Activation of the SAPKs by SEK1 and MKK7

As with the ERK pathway and all MAPK pathways, the p38 and SAPK pathways are organized into three-tiered, MAP3K → MEK → MAPK core signalling modules. The SAPKs are activated upon concomitant phosphorylation at Thr183 and Tyr185. SAPK/ERK kinase-1 (SEK1)/MAPK-kinase-4 (MKK4, Table 2) was cloned independently by two groups who employed degenerate PCR to identify novel MAPK signalling components. Two SEK1 isoforms are known; these represent hnRNA splicing variants that differ at their extreme amino termini. The striking homology shared by SEK1 and MEK1 and -2 (as well as yeast MEKs) indicated that this kinase lay upstream of MAPKs. It was subsequently shown by Sánchez *et al.*, and later by Dérijard *et al.*, that SEK1 could phosphorylate and activate all three SAPK isoforms *in vivo* and *in vitro* (186, 187). Interestingly, although SEK1 is a dual-specificity kinase, it does not phosphorylate the Tyr and Thr residues on SAPK at equal rates (see below). Thus, Tyr185 is preferentially phosphorylated by SEK1 as compared to Thr183 (186, 188). Dérijard *et al.* also showed that SEK1 could phosphorylate and activate p38 *in vivo*, when overexpressed, and *in vitro* (187). The significance of p38 activation by SEK1 is unclear. Targeted disruption of *sek1* in mice has no effect on p38 activity in ES cells (189). Moreover, if SEK1 concentrations in *in vitro* assays are adjusted to initial rate conditions for SAPK activation, little or no p38 activation is observed (174, 190).

Substantial biochemical evidence indicated that SEK1 was not the only SAPK-activating MEK. These studies revealed that the spectrum of SAPK activators recruited by different stimuli depended on the stimulus used and on the cell type. Thus, hydroxyapatite fractionation of extracts of osmotically shocked 3Y1 fibroblasts demonstrated a broad peak of SAPK activating activity that eluted at a position separate from SEK1 immunoreactivity (191). Similarly Mono-S chromatography of KB cell extracts showed that IL-1 failed to activate SEK1, but stimulated a broadly eluting peak of SAPK activating activity distinct from SEK1. Similar multiple peaks of SAPK activating activity were observed in extracts of PC-12 cells treated with arsenite or osmotic shock; and in KB cells treated with osmotic shock, UV radiation, or anisomycin (173, 190). In osmotically shocked 3Y1 fibroblasts, SEK1 represented a comparatively minor peak of SAPK activating activity (191). By contrast, SEK1 was more strongly activated by osmotic shock, UV, and arsenite in PC-12 cells (173, 190). In KB cells, IL-1 failed to activate SEK1; however, SEK1 and other SAPK activators were activated by anisomycin, osmotic shock, and UV radiation (191).

Further support for the idea of multiple SAPK-specific MEKs came from genetic studies. Thus, targeted disruption of *sek1*, while embryonically lethal, results in only a partial ablation of SAPK activation; *sek1–/–* ES cells are refractile to anisomycin and heat-shock activation of SAPK, while osmotic shock and UV activation of SAPK are unaffected (189). Finally, *hemipterous* is a *Drosophila* MEK required for dorsal closure during embryogenesis; and deletion/mutagenesis of *hemipterous* is lethal (Section 3.14). Although *hemipterous* is significantly homologous to SEK1, SEK1 cannot rescue *Drosophila* mutants wherein *hemipterous* is deficient (192, 193).

MKK7 (Table 2) was isolated contemporaneously by several laboratories. The two strategies used, database mining or degenerate PCR, sought mammalian MEKs with close homology to *hemipterous* (194–199). Alternatively, MKK7 was cloned in a two-hybrid screen as a polypeptide that could associate *in vivo* with MEK1 (193). The significance of this association is unclear. Thus, it is not surprising that, structurally, MKK7 is more similar to *hemipterous* than is SEK1; and MKK7 can effectively rescue *hemipterous* lethality (193–199). MKK7 displays a strong preference for SAPK, even under conditions of high expression. This contrasts with SEK1 which can, under conditions of high overexpression, activate p38 (186, 193–199). MKK7 can activate all SAPK isoforms tested equally well (193–199). MKK7, like SEK1, is subject to alternative hnRNA splicing—yielding enzymes with three different amino termini (α, β, and γ) and two different carboxyl termini (types 1 and 2). The β- and γ-isoforms bind directly to SAPKs via an amino-terminal extension missing in the α-isoforms. Upon over-expression, the α-isoforms exhibit a lower basal activity and higher-fold activation by upstream stimuli (200).

MKK7 is strongly activated by TNF and IL-1, conditions which, at best, induce modest SEK1 activation. In contrast, SEK1 and MKK7 are activated with equal potency by osmotic shock, while MKK7 is more weakly activated by anisomycin than is SEK1 (193–199). Thus it is plausible to argue that MKK7 represents at least a substantial portion of the biochemically detected SAPK activating activity present in 3Y1, PC-12, or KB cells subjected to osmotic shock or anisomycin, or in KB cells treated with IL-1.

While the activation patterns of SEK1 and MKK7 appear not to overlap entirely, there is evidence that SEK1 and MKK7 actively cooperate in the activation of SAPK. Thus, whereas SEK/MKK4 is not strongly activated by TNF, TNF cannot activate SAPK in fibroblasts derived from *sek1–/–* early mouse embryos, in spite of the fact that MKK7 is expressed in these cells (201, 202). Evidence for the basis of this differ-ence comes from biochemical studies of the mechanisms of SEK1 and MKK7 catalysis. As was mentioned above, SEK1 preferentially targets SAPK Tyr185 and only weakly phosphorylates Thr183 (188). By contrast, MKK7 preferentially phosphorylates SAPKs at Thr183 rather than Tyr185 (188). Accordingly, when employed individually at low concentrations, SEK1 and MKK7 are comparatively poor SAPK activators *in vitro*—each stimulating, at best, a fivefold activation of SAPK. When added together, however, SEK1 and MKK7 synergistically activate SAPK, resulting in an approx. 102-fold activation *in vitro*. This synergistic activation is accompanied by an equal phosphorylation of SAPK Thr183 and Tyr185 (188). *In vivo*, activation of SAPK by, say, TNF may involve the combined effect of modestly activated SEK1 acting primarily on SAPK Tyr185 and strongly activated MKK7 acting primarily on Thr183. Because phosphorylation of both Thr183 and Tyr185 is required for full SAPK activation, deletion of *sek1* would, by sharply decreasing SAPK Tyr phosphorylation, substantially compromise SAPK activation by TNF (188).

3.6.2 Activation of the p38s by MKK3 and MKK6

MKK3 and MKK6 (Table 2) were cloned by degenerate PCR using conserved MEK sequences as templates (187, 203, 204). Both enzymes are highly selective for p38 and do not activate SAPK under all conditions tested (187, 203, 204). MKK3 and MKK6 differ more substantially in their substrate selectivity. MKK3 preferentially activates p38α and p38β while MKK6 can strongly activate all known p38 isoforms. Similarly, MKK3 appears to be more restricted with regard to activation by upstream stimuli. Whereas MKK6 is activated by all known p38 activators, MKK3, like SEK1, is more strongly activated by physical and chemical stresses (204).

3.6.3 MEK5, a potential regulator of ERK5

ERK5 was originally isolated in a two-hybrid screen for interactors with MEK5 (163). Independently, MEK5 (Table 2) was cloned by degenerate PCR using conserved MEK sequences as primers (163, 164). Four MEK5 polypeptides arise from differential hnRNA splicing. Splicing near the 5′ end shifts the reading frame and generates long (450 amino acids) and short (359 amino acids) polypeptides. The long polypeptide localizes primarily to the particulate fraction while the shorter polypeptide is cytosolic. A second splicing event substitutes two 10-amino acid cassettes in a region between subdomains IX and XI. The latter splicing event gives rise to α- and β-MEK5 isoforms and is analogous to that which results in type-1 and -2 SAPKs (137, 146, 164). Inasmuch as this region is important to substrate binding, differential hnRNA splicing here may alter the substrate specificity of the resulting MEK5 enzymes. Given the observed interaction between MEK5 and ERK5 (163), it is plausible to speculate that MEK5 is a bona fide upstream activator of ERK5. However, activation of ERK5 by MEK5 has not yet been demonstrated.

3.7 The SAPKs and p38s are activated by several divergent families of MAP3Ks

3.7.1 General considerations

As with MEK1 and -2, the stress-activated MEKs are probably regulated by Ser/Thr phosphorylation within the P-activation loop of subdomain-VIII of the kinase domain, inasmuch as mutation of these residues ablates stress-activated MEK activation *in vivo* in transfection experiments. Thus, the SEK1 phosphorylation sites are Ser257 and Thr261 (186, 187), those for MKK7 are Ser206 and Thr210 (α-isoforms) (192–200), those for MKK3 are Ser189 and Thr193 (187), and those for MKK6 are Ser207 and Thr211 (187, 203, 204) (see Fig. 12A).

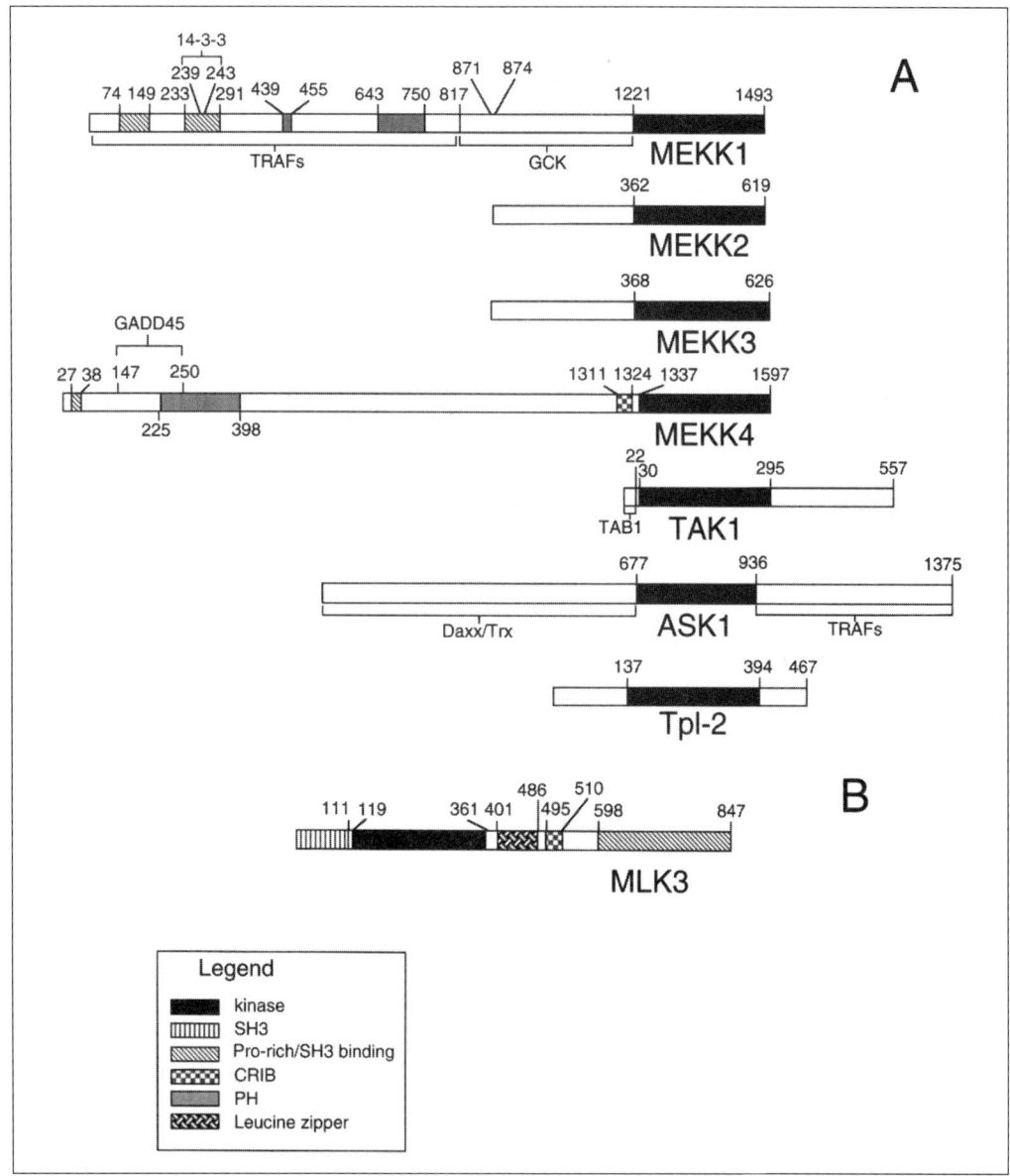

Fig. 11 Structure of mammalian MEKKs and MLKs. (A) Structure of mammalian MEKKs. Note that outside the conserved catalytic domains, the structures are widely divergent. Defined binding domains for putative upstream regulators are indicated. (B) Structure of MLK3, a canonical MLK.

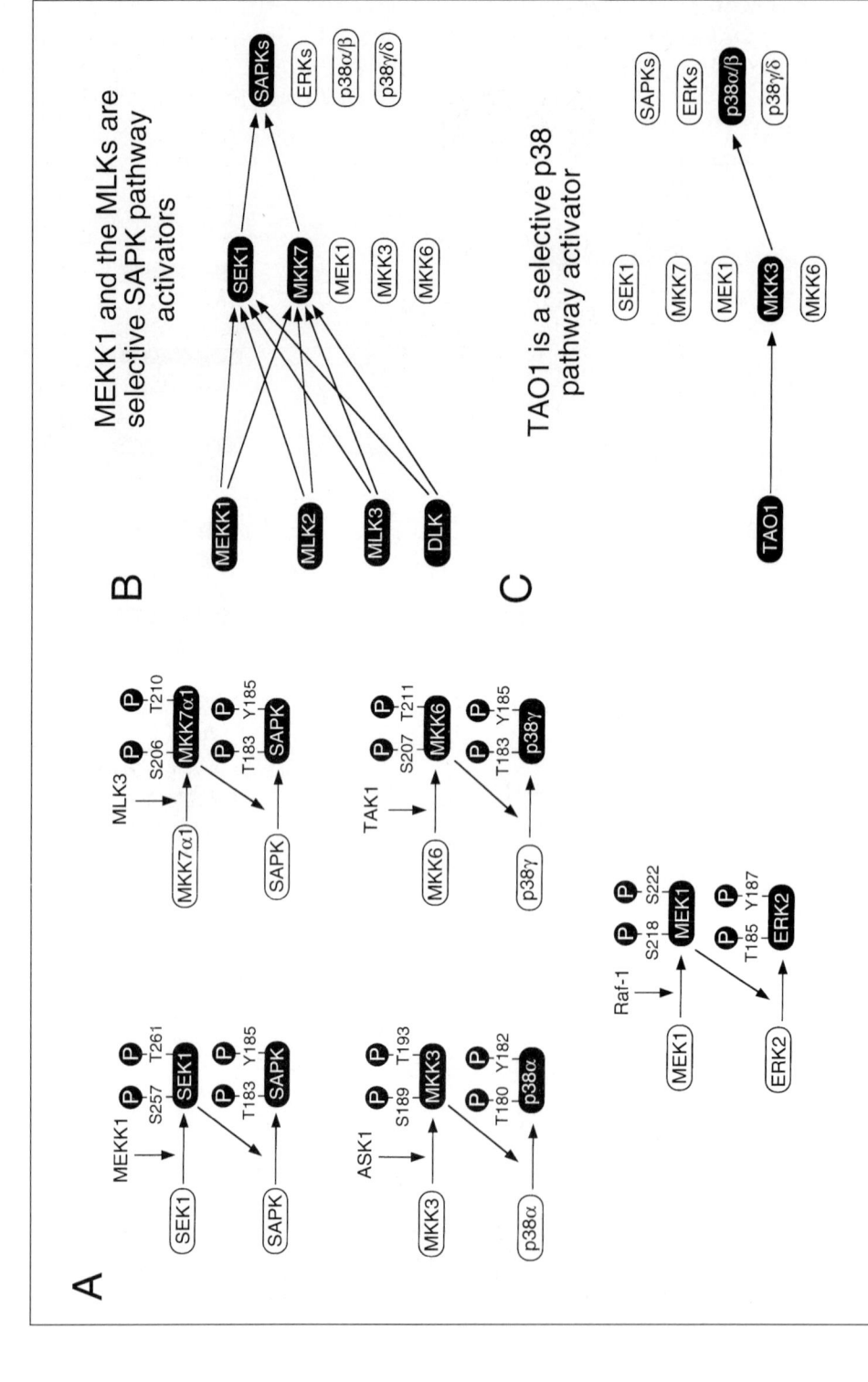

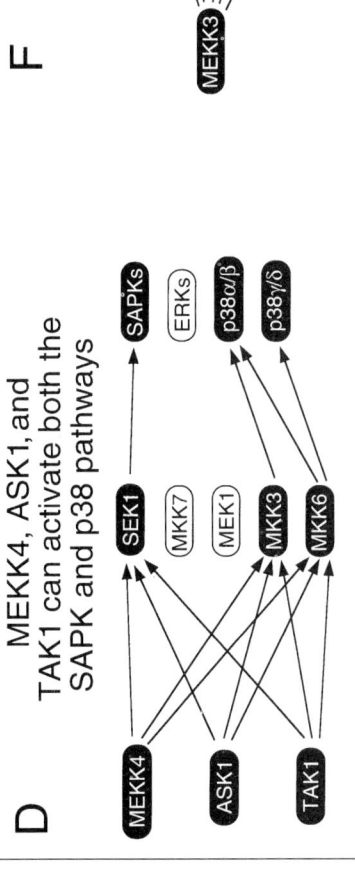

D MEKK4, ASK1, and
   TAK1 can activate both the
   SAPK and p38 pathways

E MEKK2 and Tpl-2 can
   activate both the SAPK and
   ERK pathways

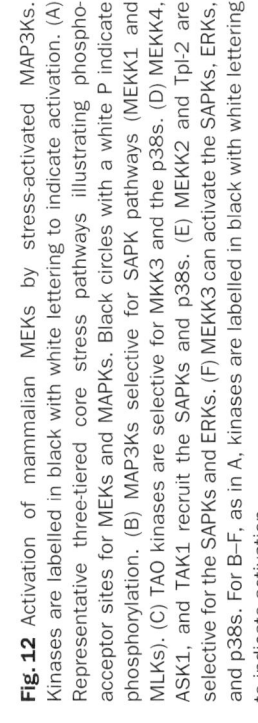

F MEKK3 can activate the SAPK,
   ERK, and p38 pathways

Fig. 12 Activation of mammalian MEKs by stress-activated MAP3Ks. Kinases are labelled in black with white lettering to indicate activation. (A) Representative three-tiered core stress pathways illustrating phospho-acceptor sites for MEKs and MAPKs. Black circles with a white P indicate phosphorylation. (B) MAP3Ks selective for SAPK pathways (MEKK1 and MLKs). (C) TAO kinases are selective for MKK3 and the p38s. (D) MEKK4, ASK1, and TAK1 recruit the SAPKs and p38s. (E) MEKK2 and Tpl-2 are selective for the SAPKs and ERKs. (F) MEKK3 can activate the SAPKs, ERKs, and p38s. For B–F, as in A, kinases are labelled in black with white lettering to indicate activation.

While, as yet, no MAP3Ks upstream of ERK5 have been identified, the number and diversity of Ser/Thr kinases that act as MAP3Ks upstream of the SAPKs and p38s is daunting. This heterogeneity is consistent with the multiple different stimuli that recruit these MAPK pathways. The MAP3Ks upstream of SAPK and p38 fall into three, broad, protein kinase families (Fig. 11 and 12B-F).

1. *The MEK-kinases (MEKKs)*: These enzymes are significantly homologous within their catalytic domains to the catalytic domain of *S. cerevisiae* Ste11p, a MAP3K that regulates both the yeast mating-pheromone and osmosensing pathways (3, 10) (Chapter 3). By contrast, MEKK extracatalytic sequences are quite divergent, and these enzymes can in some instances catalyse the activation of multiple different MAPK pathways and interact with a wide array of putative regulatory proteins (Fig. 11). Mammalian MEKKs include MEKK1–4, apoptosis signal-regulating kinase-1 (ASK1), transforming growth factor-β (TGFβ)-activated kinase-1 (TAK1), Tpl-2 (the product of the *cot* proto-oncogene), and NF-κB-inducing kinase (NIK) which is a specific activator of the NF-κB pathway (11, 205–216) (Table 3).

2. *Mixed lineage kinases (MLKs)*: These kinases bear structural homology to both Ser/Thr and Tyr kinases and also contain leucine zippers and SH3 binding sites as well as, in the case of MLK2 and -3, SH3 domains (Fig. 11B). MLKs are SAPK-specific (200, 217–222) (Table 3).

3. *Thousand and one (TAO)-1 and -2*: These comprise a novel MAP3K group that bears structural homology both to *S. cerevisiae* Ste11p and Ste20p (223).

Whereas Raf-1 is highly specific for the Ras → ERK pathway, the MAP3Ks upstream of the SAPKs and p38s, while in some instances showing distinct preferences for a particular MAPK pathway, are often markedly promiscuous, activating several classes of MEKs with substantial alacrity. The ability of stress-regulated MEKs to be

Table 3 MAP3K nomenclature

Name	Alternative names	Substrates/effectors
Raf-1		MEK1, MEK2
A-Raf		MEK1, MEK2
B-Raf		MEK1, MEK2
MEKK1		SEK1, MKK7
MEKK2		SEK1, MEK1
MEKK3		SEK1, MEK1, ?MKK3, ?MKK6
MEKK4	MTK1	SEK1, MKK3, MKK6
ASK1	MAPK kinase-kinase-5 (MAPKKK5)	SEK1, MKK3, MKK6
TAK1		SEK1, MKK3, MKK6
Tpl-2 (in rats)	Cot (in humans)	MEK1, SEK1
MLK2	MST	SEK1, MKK7
MLK3	SPRK, PTK1	SEK1, MKK7
DLK	MUK, ZPK	SEK1, MKK7
TAO1/2		MKK3

activated by several MAP3Ks, and of MAP3Ks to regulate several MEKs is reminiscent of signalling in *S. cerevisiae* and is an illustration of the general principle of multifarious regulation of MAPK signalling components (Section 1.1). Many stress-activated MAP3Ks also appear to be regulated by several distinct mechanisms; thus, stress-regulated MAP3Ks are theoretically able to receive divergent inputs and translate these into distinct sets of MAPK pathway activation events.

3.7.2 The MEKKs

MEKK1, a comparatively selective activator of the SAPKs

MEKK1 (Table 3) was cloned by Johnson and colleagues as part of an effort to identify MAP3Ks that were upstream of the ERKs. The cloning strategy exploited the existing knowledge of yeast signalling pathways (the mating-factor pathway of *S. cerevisiae* in particular, see Chapter 3). Accordingly, degenerate PCR primers based on conserved elements of the *STE11* sequence and the related *S. pombe* MAP3K *byr2* (Chapter 3) were used to amplify mammalian homologues of the yeast kinases. A partial cDNA fragment, designated MEKK1, was generated that corresponded to amino acids 817–1493 of the full-length enzyme (205). A complete MEKK1 cDNA was subsequently cloned by Cobb and co-workers (206). The full-length MEKK1 is a 150-kDa polypeptide that consists of a carboxyl-terminal kinase domain (amino acids 1221–1493) and an extensive amino-terminal domain (amino acids 1–1221) that includes two proline-rich segments (amino acids 74–149 and 233–291) containing putative binding sites for proteins with SH3 domains, a consensus binding site for 14–3–3 proteins (amino acids 239–243), two PH domains (amino acids 439–455 and 643–750), and an acid-rich motif (amino acids 817–1221) which contains two sites for cleavage by proapoptotic caspase cysteine proteases (Asp871, Asp874) (205, 206, 224, 225). Within the kinase domain is a binding site for Ras, the exact location of which has not been clearly defined (226). MEKK1 can also bind Rho family GTPases (227) (see Sections 3.8 and 3.13.2).

The 80-kDa MEKK1 fragment originally isolated (amino acids 817–1493) was shown in overexpression experiments to activate the ERKs. Of note, however, in comparison with Raf-1, much larger amounts of MEKK1 were required for MEK1/2 activation. Based on these results, it was proposed that MEKK1 was part of a distinct ERK activation mechanism (205). Inasmuch as Ste11p is recruited *in vivo* as a consequence of activation by mating factor of the heterotrimeric G-protein complex Gpa1p (α)–Ste4p (β)–Ste18p (γ) (3), it was suggested that MEKK1 routed signals from trimeric G protein-coupled receptors to the ERKs (205).

When MEKK1 was stably introduced into fibroblasts, however, it was observed that expression of the kinase caused growth inhibition and, in some instances apoptosis—effects that were unexpected for a bona fide activator of the ERKs. Stable cell lines expressing an isopropyl β-D-thiogalactoside (IPTG)-inducible, truncated MEKK1 construct (catalytic domain only) were generated by Templeton and co-workers in order to bypass the toxic effects of constitutive MEKK1 expression (228). With increasing doses of IPTG, MEKK1 selectively activated the endogenous SAPK pathway; ERK activation was not observed until massive overexpression of MEKK1 was achieved (228). Similar results were obtained from transient transfection experiments if MEKK1 expression was carefully titrated beginning at low levels of plasmid (229).

Accordingly, *in vivo*, MEKK1 is a highly selective activator of the SAPKs (Fig. 12B). This phenomenon is consistent with biochemical studies that indicate that MEKK1 can activate SEK1 *in vitro* and *in vivo* (228). MEKK1 can also activate MKK7α1, -α2, -β1, and -β2 as well as -γ1 *in vivo* (200). However, careful biochemical and kinetic analyses of MEKK1 activation of MKK7γ1 versus SEK1 indicate that, overall, SEK1 is a preferred MEKK1 substrate (other MKK7 isoforms were not tested). This preference may be due to the intrinsic scaffold properties of SEK1 (17) (see Section 3.10). By the same token, whereas overexpressed MEKK1 can activate the ERKs and even p38, kinetic studies of MEK1/2 and MKK3/6 activation by MEKK1 suggest that these MEKs are activated with a K_{cat} at least three orders of magnitude lower that that for activation of SEK1 and two orders of magnitude lower than that for activation of MKK7γ1 (17).

MEKK2 and -3 can activate both the SAPKs and the ERKs

MEKK2 and -3 (Table 3) were cloned by Johnson *et al.* using an identical strategy to that employed in the cloning of MEKK1 (207). MEKK3 was independently cloned by Siebenlist and colleagues using a similar approach (208). While these enzymes are clearly *STE11* homologues, as is MEKK1; MEKK2 and -3 are more closely related within their kinase domains to each other (> 90% identity) than they are to MEKK1 (~ 65% identity). Both MEKK2 and MEKK3 are smaller polypeptides than MEKK1 (~ 70 and 71-kDa, respectively). Both kinases have a carboxyl-terminal catalytic domain (amino acids 362–619 for MEKK2, amino acids 368–626 for MEKK3) and amino-terminal non-catalytic domains. The amino-terminal non-catalytic portions of MEKK2 and -3 are also significantly similar; however, these regions contain no motifs indicative of function or regulation (207, 208) (Fig. 11A).

In contrast to MEKK1, which is strongly selective for the SAPKs, MEKK2 and -3 can activate both the SAPK and ERK pathways with nearly equal potency (Figs 12E, F). p38 is not activated by MEKK2; however, MEKK3 can also activate the p38 pathway *in vivo* and can activate MKK3 and MKK6 *in vivo* and *in vitro* (230–232) (Fig. 12F). Consistent with concomitant ERK and SAPK activation, MEKK2 and -3 can each activate MEK1 and SEK1 *in vivo* and *in vitro* (207, 208). MEKK3 can also activate MKK7α1 and, to a much lesser extent, MKK7α2, -β1, and -β2 *in vivo* (200, 233). The reason for the selectivity of MEKK3 for different MKK7 isoforms is unclear.

MEKK4 can activate both the SAPKs and the p38s

MEKK4 (also called MAP three kinase-1, MTK1, Table 3) was isolated independently by Johnson *et al.* and Saito *et al.*, the former again employing a PCR strategy based on degenerate primers derived from *STE11* and *byr2* (209). Saito *et al.* cloned MEKK4 as a human cDNA that when expressed could rescue osmosensitive yeast missing the genes for all three MAP3Ks (*SSK2*, *SSK22*, and *STE11*) of the *HOG1* osmoadaptation pathway (Chapter 3) (3, 210). The kinase domain of MEKK4 shares about 55% amino acid homology with that of MEKK1, -2, and -3. An approx. 150-kDa polypeptide, MEKK4, contains a carboxyl-terminal kinase domain (amino acids 1337–1597) and an extensive amino-terminal regulatory region that includes a putative polyproline SH3 binding motif (amino acids 27–38) (209, 210), a binding site for growth arrest and DNA damage (GADD)-45 family proteins (amino acids 147–250, see Section 3.13.6) (234), a putative PH domain (amino acids 225–398), and a Cdc42/Rac interaction and binding (CRIB) domain (amino acids 1311–1324, see Section 3.13.6) (209, 210) (Fig. 11A). Johnson and colleagues have reported that MEKK4 is selective for the SAPKs and can activate only SEK1 *in vivo* and *in vitro* (209); however, Saito *et al.* used a titred expression of increasing levels of MEKK4 and unambiguously demonstrated that MEKK4 could activate the SAPKs (via SEK1) and the p38s (via MKK3 and MKK6) with equal potency *in vivo* and *in vitro* (210) (Fig. 12D).

TAK1 can activate both the SAPKs and the p38s

TGFβ-activated kinase-1 (TAK1) (Table 3) was cloned by Yamaguchi *et al.* in an intriguing screen for mammalian MAP3Ks. This screen employed a mutant yeast strain wherein *STE11* (Chapter 3) was deleted and a mutant form of the mating-factor pathway MEK *STE7*, $STE7^{P368}$ was expressed. $STE7^{P368}$ is a gain-of-function mutant that still requires Ste11p for full activity; however, constitutively active Raf-1 or the kinase domain of MEKK1, and, presumably other MAP3Ks, can substitute for Ste11p in $\Delta STE11/STE7^{P368}$ mutant cells (3, 211). Thus, the $\Delta STE11/STE7^{P368}$ mutant is an excellent platform for the identification of novel MAP3Ks. Accordingly, TAK1 was cloned as an additional MAP3K that could substitute for Ste11p in $\Delta STE11/STE7^{P368}$ cells. TAK1 is an approx. 60-kDa polypeptide with an amino-terminal kinase domain (amino acids 30–294) preceded by a short regulatory motif (amino acids 1–22) and followed by a carboxyl-terminal extension (amino acids 295–557) of unknown function (211) (Fig. 11A). The amino-terminal regulatory motif appears to serve an inhibitory role inasmuch as full-length TAK1 cannot substitute for Ste11p in $\Delta STE11/STE7^{P368}$ cells, whereas TAK1Δ1–22 can (211). Endogenous TAK1 is activated by TGFβ IL-1 and TNF; and the amino-terminal regulatory motif binds TAK1 binding proteins (TABs)-1 and -2, two regulatory proteins that couple TAK1 to upstream signals (see Section 3.13.5) (234–236). In overexpression experiments, TAK1 can activate both the SAPKs and p38s. *In vitro* and *in vivo*, TAK1 can phosphorylate and activate SEK1, MKK3, and MKK6 (211, 237) (Fig. 12D).

ASK1 can activate both the SAPKs and the p38s

Apoptosis signal-regulating kinase-1 (ASK1, also called MAPK kinase-kinase-5, MAPKKK5, Table 3) is a sixth MEKK that was cloned by Ichijo *et al.* using degenerate primers based on conserved elements of Ser/Thr kinase subdomains VI and VIII (211). ASK1 was also cloned by Wang *et al.* using a similar approach (212). ASK1 is an approx. 150-kDa polypeptide with a centrally located kinase domain (amino acids 677–936), an amino-terminal extension (amino acids 1–676) that includes segments that bind the Fas-associated adapter protein Daxx, and the redox-sensing enzyme thioredoxin (see Section 3.13.4) (211, 212, 238, 239) (Fig. 11A). In addition, ASK1 possesses a carboxyl-terminal extension (amino acids 937–1375) that has been implicated in binding polypeptides of the TNFR-associated factor (TRAF) family (see Section 3.13.4) (211, 212, 240) (Fig. 11A). ASK1 can activate the SAPKs (via SEK1) and the p38s (via MKK3 and -6) *in vivo* and *in vitro* (211, 212) (Fig. 12D). Endogenous ASK1 is activated by oxidant stress, TNF, and Fas. ASK1 is likely to be an important effector coupling these agonists to the SAPKs and p38s.

Expression of ASK1 from a Zn-inducible promoter induces apoptosis in several cell lines (211). The targets recruited by ASK1 (especially if the SAPKs or p38s are involved) that promote apoptosis are unknown. Dominant inhibitory, kinase-dead ASK1 mutants can block apoptosis stimulated by TNF or oxidant stress (211, 239). Thus, either ASK1 is a direct apoptogenic signalling component recruited by these ligands, or the dominant inhibitory ASK1 is sequestering a common upstream element that regulates both ASK1 and TNF/oxidant-induced apoptosis.

Tpl-2 can activate both the SAPKs and the ERKs

Tumour progression locus-2 (*Tpl-2*) was identified by Tsichlis and colleagues as a rat homologue of the human proto-oncogene *cot* (Table 3). The oncogenic potential of *Tpl-2* is activated in rat thymomas as a result of the continued additive, spontaneous proviral insertion of the Moloney leukaemia virus into the infected cell genome at the *Tpl-2* locus; the consequence of which is positive selection of tumour cells and enhanced tumour progression *in vivo* (214). Tpl-2 is an approx. 50-kDa protein Ser/Thr kinase with an amino-terminal domain of unknown function that is truncated in some mRNA splicing isoforms, a central catalytic domain (amino acids 139–394) that is significantly homologous to the *S. cerevisiae* MAP3K *STE11*, and a carboxyl-terminal regulatory domain (amino acids 395–467) (214, 215). Moloney leukaemia virus proviral insertions at *Tpl-2* always target the last intron of the gene and elicit the enhanced expression of a carboxyl terminally truncated protein, both through enhanced transcription and mRNA stabilization. It is likely, therefore, that the C-terminal regulatory domain of Tpl-2 exerts a negative effect on Tpl-2 activity. Thus, the free Tpl-2 kinase and carboxyl-terminal domains, when expressed together in Sf9 cells, interact and can be coimmunoprecipitated. Moreover, transgenic mice expressing the carboxyl terminally truncated protein develop T-cell lymphomas, whereas expression of the wild-type protein is without effect (215).

Tsichlis and colleagues demonstrated that transient expression of Tpl-2 activated the ERK pathway in parallel to Raf-1, and possibly downstream of Ras (241). Subsequently, Ley *et al.*, as well as Tsichlis *et al.*, showed that expression of Tpl-2 activated the SAPKs and ERKs with equal potency (11, 215). Ley *et al.* then demonstrated that, consistent with its homology to *STE11*, Tpl-2 was a MAP3K that could directly activate MEK1 and SEK1 *in vivo* and *in vitro* (11) (Fig. 12E). While overexpression of full-length Tpl-2 activates both the SAPKs and ERKs, expression of the oncogenic, carboxyl terminally truncated Tpl-2 results in substantially greater SAPK and ERK activation, further supporting the contention that the carboxyl-terminal domain negatively regulates Tpl-2 activity (215).

3.7.3 The mixed lineage kinases (MLKs): MLK3, MLK2, and DLK selectively activate the SAPKs

The mixed lineage kinases (MLKs) are a small family of protein Ser/Thr kinases that share a general structural configuration wherein an amino-terminal kinase domain is followed by one to two leucine zippers, a Cdc42/Rac interaction and binding (CRIB) domain (Section 3.8.2), and a carboxyl-terminal proline-rich domain with several consensus SH3 binding motifs. So far, four MLKs have been identified: MLK1; MLK2 (also called MKN28 cell-derived Ser/Thr kinase, MST, Table 3); MLK3 (also called SH3 domain-containing proline-rich kinase, SPRK or protein Tyr kinase-1, PTK1) and dual leucine zipper kinase (DLK, also called MAPK upstream kinase, MUK, or zipper-containing protein kinase, ZPK, Table 3). MLK2 and MLK3, in addition to the common features described above, contain SH3 domains amino-terminal to the kinase domains (217, 222, 242, 243) (Fig. 11B).

The name mixed lineage kinase is derived from the observation that the kinase domains of the MLKs bear structural similarities to both Ser/Thr and Tyr kinases. Thus, for example, MLK1 contains a Lys residue (Lys129 in the sequence His–Arg–Asp–Leu–Lys) in subdomain VIb that is characteristic of Ser/Thr kinases; however, two Trp residues in MLK1 subdomain IX (Trp192 and Trp199) are highly conserved among Tyr kinases, as is a motif in subdomain XI (Met241–Glu–Asp–Cys–Trp–Asn–Pro–Asp–Pro–His–Pro–Ser–Arg–Pro–Ser–Phe255) which conforms to a conserved region in Tyr kinases (Met–X–X–Cys–Trp–X–X–Asp/Glu–Pro–X–X–Arg–Pro–Ser/Thr–Phe, where X is any amino acid) (242). Biochemically, however, MLKs are clearly Ser/Thr-specific.

Out of the four known MLKs, three have been assayed for activation of MAPK pathways: MLK3, MLK2, and DLK. All three are potent activators of the SAPKs *in vivo* in transfection experiments (219–222) (Fig. 12B). MLK3, MLK2 and DLK are all established MAP3Ks. Thus, *in vitro* and *in vivo*, MLK3, MLK2, and DLK can activate SEK1 to a degree commensurate with that catalysed by MEKK1 (200, 217–220). In addition, MLK3 and DLK can activate MKK7 *in vivo* in transfection experiments, and *in vitro*, albeit with differential isoform selectivity *in vivo*. Thus, in both instances, MKK7β1 and -β2 are strongly activated *in vivo*, while α-isoforms are modestly activated (200). The significance and molecular basis of this selectivity are unknown. MLK2 can also activate MKK7α1. Indeed, MLK2 is a stronger MKK7α1 kinase than it is a SEK1 kinase (220).

3.7.4 TAO kinases are novel MAP3Ks that regulate the p38s and are structurally homologous to both Ste20 and MAP3Ks

The Thousand-and-one (TAO) kinases (TAO1 and TAO2) are a novel family of 1001 amino acid Ser/Thr kinases that were cloned by Cobb *et al.* as part of a strategy to identify novel mammalian kinases homologous to *S. cerevisiae STE20*, a proximal kinase thought to be important in the regulation of the Ste11p → Ste7p → Fus3p/Kss1p yeast mating-pheromone, MAPK core signalling module (3, 223). TAOs consist of amino-terminal kinase domains and very large (700 amino acids) carboxyl-terminal extensions of unknown function. The kinase domains of TAOs are signific-antly homologous to Ste20p (40% identity) and the germinal centre kinases (43% identity with GCK) (Section 3.11) established mammalian Ste20-like kinases (3, 12, 223). However, there is appreciable identity with MLK2 (33% overall). Notably, most of the identity with MLK2 resides within the substrate binding motifs of the kinase domain (223).

In vitro, the purified kinase domain of TAO1 will directly phosphorylate and acti-vate SEK1, MKK3, and MKK6; however, *in vivo* only MKK3 is activated. Moreover, when expressed in Sf9 cells from a recombinant baculovirus, TAO1 will interact *in vivo* with and coimmunoprecipitate with MKK3. Thus, although TAO1 bears signific-ant homology to Ste20s and GCKs, unlike kinases of the Ste20 and GCK groups, which are thought to regulate MAP3Ks (Sections 3.8.2, 3.11), TAO1 appears to be a direct MAP3K selective for MKK3 (223) (Fig. 12C). The ability of TAO1 to directly catalyse the activation of MEKs may be due to the kinase domain homology with MLK2 in the substrate binding region.

3.8 Regulation of the SAPKs and p38s by Rho family GTPases

3.8.1 General considerations

Sections 3.2–3.7 have described much of what we know of mammalian stress-activated MAP3K → MEK → MAPK core signalling modules. The following five sections (3.8–3.12) will describe some of the known proximal signalling components thought to couple to these core modules.

As noted above (Section 3.5), monomeric GTPases of the Ras superfamily are potent upstream activators of signal transduction pathways, and Ras is pivotal to the activation of the ERK pathway (Section 3.5). The activation of Ras by GEFs, and the inactivation of Ras by GAPs is discussed in Section 3.5.

In mammals, the Rho subfamily of the Ras superfamily of GTPases comprises the Rho (RhoA–E), Rac (Rac1 and -2), and Cdc42 (Cdc42Hs, G25K, Tc10, and Chp) sub-groups (77, 108, 244). As with other members of the Ras superfamily, Rho subfamily GTPases are active in the GTP-bound state (promoted by GEFs) and inactive in the GDP-bound state (promoted by GAPs) (77, 108, 109). Rac, Cdc42Hs, and other Rho subfamily GTPases share a large number of GEFs, belonging to the Dbl family, including: Ost, a GEF for RhoA and Cdc42Hs; Tiam-1, a GEF for Rac1 and Cdc42Hs; Lbc, a GEF for RhoA; and FGD1, a GEF selective for Cdc42Hs. All these proteins consist of a conserved Dbl homology (DH) domain that is necessary for promoting GDP dissociation, and a pleckstrin homology (PH) domain (Chapter 1). Because inositol phospholipids, including inositol-3′,4′,5′-trisphosphate, the product of the Ras effector PI3-kinase, are putative targeting signals for PH domain-containing polypeptides, Rho family GEFs themselves may be subject to regulation by Ras → PI3-kinase (109).

Among the first recognized physiological functions for the Rho family was regulation of the actin cytoskeleton (reviewed in refs 108, 109). Thus, fibroblasts in culture will form actin stress fibres that associate with the focal adhesions which, in turn, bind the cells to the substratum. Formation of these stress fibres requires RhoA

(108). Within minutes of mitogen treatment of cells, transient changes in the actin cytoskeleton occur. Notably, any dense perinuclear actin filaments present in the resting cells are disassembled and the cells form peripheral filopodia (actin microspikes) and lamellopodia (membrane ruffles) from the free actin (108). Both these morphological changes occur in the cortical actin cytoskeleton as a result of cyclical actin depolymerization and repolymerization. Filopodium formation requires Cdc42Hs which promotes actin depolymerization, creating the barbed ends from which actin polymerization into microspikes can occur. Lamellopodium formation requires Rac1-induced actin depolymerization parallel to the plane of the membrane. The formation of filopodia, lamellopodia, and stress fibres can also be induced by growth factors and activated Ras, responses inhibitable by dominant-negative mutants of Cdc42Hs, Rac, and Rho, respectively; suggesting that in response to mitogens, Cdc42Hs, Rac, and Rho are Ras effectors (108).

The physiological role of lamellopodia and filopodia in cellular growth and transformation is unclear; however, the ability of Rho family GTPases to trigger these cytoskeletal changes correlates, at least in part, with Rho family GTPase transforming capability. Thus, Rac, Rho, and Cdc42Hs all appear necessary for the entry of G_0 cells into the cell cycle, and expression of constitutively active, GTPase-deficient mutants of Rac, Rho, or Cdc42Hs will drive quiescent cells into the cell cycle in the absence of serum. Conversely, *ras* transformation is blocked by Asn17, exchange-deficient mutants of Rac or Cdc42, or by effector loop mutations that also abrogate cytoskeletal functions (108).

Constitutively active, GTPase-deficient (Val12) mutants of Rac1, Cdc42Hs, or Chp are potent activators of the SAPKs (245–247). In addition, Val12–Rac1 and Cdc42Hs can activate p38 (248, 249). RhoA generally does not activate SAPK or p38, but can in certain cell types (notably 293 cells) (250). Consistent with these findings, expression of the Dbl proto-oncogene product or FGD1, Rho GEFs that recruit Rac and Cdc42, also results in strong SAPK activation (245, 251). Transforming alleles of *ras* and EGF (through Ras) may signal to the SAPKs via Rac1 inasmuch as Asn17 Rac1 can block the activation of SAPK by these stimuli. In addition, Rac1 is also involved in SAPK activation in response to CD3–CD28 T-cell costimulation (246). Aside from this, no Ras-independent agonists that selectively target Rac1 or Cdc42Hs are known.

While Rac1 and Cdc42Hs can each activate both MAPK signalling pathways and reorganization of the actin cytoskeleton, it appears that these functions are mediated by distinct pathways. Support for this idea comes from the use of effector loop mutants (analogous to the effector loop mutants of Ras). Thus, Phe37Ala Rac1 is unable to stimulate membrane ruffling, but still activates SAPK. Conversely, Tyr40Cys Rac1 or Cdc42Hs can no longer activate SAPK, but are still able to induce, respectively, membrane ruffling and filopodium formation (252).

3.8.2 Rho GTPase targets that may couple to the SAPKs and p38s

Proteins of the Ras superfamily activate their targets through a direct binding interaction between the GTP-charged G-protein and the effector (78, 79). Thus, Ras can bind to the mitogenic MAP3K Raf-1 of the Ras-MAPK pathway; and this binding may mediate Raf-1 oligomerization and activation (Section 3.5). Most, but not all, direct targets for Rac1 and Cdc42Hs contain a so-called Cdc42/Rac interaction and binding (CRIB) domain which is necessary for Rac1 and Cdc42Hs binding (253). Several polypeptide species, including protein kinases and adapter proteins, are candidate Rac1 and Cdc42Hs effectors that couple to the SAPKs and p38s.

p21-activated kinases (PAKs)

STE20 encodes a *S. cerevisiae* Ser/Thr protein kinase with an amino-terminal regulatory domain containing a CRIB motif that binds Cdc42Sc, the yeast Cdc42Hs orthologue (3). Ste20p is thought to regulate the mating-pheromone MAPK core signalling module Ste11p → Ste7p → Fus3p/Kss1p, and genetic studies suggest that Ste20p is upstream of Ste11p; although the mechanism of Ste20p action in this process has yet to be defined completely (3) (Chapter 3). Genetic epistasis studies provisionally placing Ste20p upstream of the MAP3K Ste11p led to the notion that Ste20-like kinases in mammals might regulate MAPK core signalling modules (3).

The p21-activated kinases (PAKs, PAK1, also called αPAK; PAK2, also called γPAK or hPAK65; PAK3, also called βPAK; and PAK4, Table 4) are a family of mammalian kinases that are structural and functional orthologues of Ste20p (3, 254–258). PAK1 and PAK2 were originally purified as protein kinases that selectively bound GTP–Rac1 and GTP–Cdc42Hs (254, 255); and *in vitro* and *in vivo*, the kinase activity of PAKs is activated upon binding GTP–Rac1 or GTP–Cdc42Hs through a process that involves an obligatory autophosphorylation event (254, 256). PAK2 can also undergo activation during apoptosis through a mechanism that involves caspase-mediated cleavage at Asp212, a process that removes the amino-terminal regulatory domain (including the CRIB motif) (259).

A major function of the PAKs (PAK1 in particular) is to serve as effectors for Cdc42 and Rac in the regulation of the actin cytoskeleton. Thus, microinjection of activated PAK1 into quiescent Swiss 3T3 cells results in the formation of polarized filopodia and membrane ruffles (260, 261). Interestingly, this phenomenon can be reproduced

Table 4 GCK/PAK nomenclature

Name	Alternative names	Substrates/effectors
GCK		MEKK1, ?MLK3
GCKR	KHS[a]	MEKK1
GLK1		?
HPK1		MEKK1, MLK3
NckIK	NIK[b], HGK	MEKK1
Misshapen	Msn	?
PAK1	αPAK	?
PAK2	γPAK, hPAK65	?
PAK3	βPAK	Raf-1

[a]Kinase homologous to Ste20. [b]Not to be confused with NF-κB-inducing kinase. HGK, HPK1/GCK-related kinase.

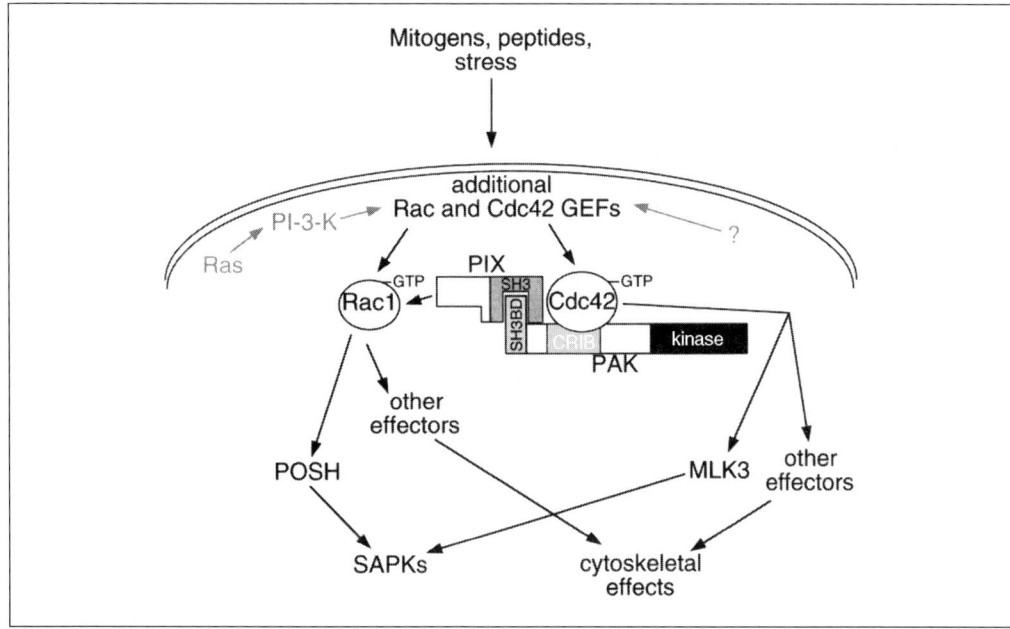

Fig. 13 Regulation of Rho family GTPases and PAKs by Cdc42 and PIX. GTP–Cdc42 recruits a PIX–PAK complex to the membrane, activating Rac1. Together these culminate in cytoskeletal changes (filopodia elicited by Cdc42 and lamellopodia elicited by Rac). Cdc42 and Rac also recruit upstream activators of the SAPKs such as MLK3 (a Cdc42 effector) or POSH (a Rac effector). Rac and Cdc42 can be themselves activated by additional Rho GEFs that are possibly under regulation by PI-3-kinase.

with a construct consisting solely of the free PAK amino-terminal regulatory domain —even an amino-terminal domain wherein the CRIB motif is mutated so as to abrogate Rac1 and Cdc42Hs binding (Leu83, Leu86) (260). The PAK1 amino-terminal regulatory domain (and indeed the regulatory domains of all the PAKs), in addition to a CRIB motif (amino acids 69–110), contains a consensus polyproline SH3 binding domain (amino acids 12–15) (254–258). Mutational inactivation of this SH3 binding site (Pro13Ala) results in a complete loss of PAK1 cytoskeletal regulation (260) (Fig. 13).

Lim and colleagues identified the PAK interacting exchange factors (PIXs), a family of SH3 domain-containing Rac GEFs, as species that could interact with the PAK SH3 binding site (at amino acids 12–15 in PAK1) (262). Transient expression of PIXs results in membrane ruffling, and PIXs can direct PAKs to membrane focal adhesions (262). It has been proposed, based on these results, that Cdc42Hs activation of PAK1 results in PIX binding to the PAK1 SH3 binding site, an event which, in turn, targets PAKs to focal adhesions and fosters Rac1 activation and membrane ruffling (262) (Fig. 13).

It is conceivable that Cdc42Hs, either directly or through PIXs and Rac1, could target the SAPKs via the PAKs. Indeed, several reports have indicated that constitutively activated forms of PAKs can activate the SAPKs and p38s (248, 249, 263, 264). Thus, for example, mutation of Leu107 to Phe, within the CRIB motif of human PAK1, results in a constitutively active construct that can activate coexpressed SAPK (263). Moreover, addition, to cell-free extracts of *Xenopus* oocytes, of a PAK1 mutant wherein the amino-terminal regulatory domain has been deleted also results in substantial SAPK activation (264). However, SAPK activation by coexpressed PAKs has not been universally observed. Moreover, the activation of SAPK by wild-type PAKs is generally small (227, 252); and expression of PAKs does not synergize with Rac or Cdc42 to activate further SAPK (227). Thus, it is not clear if PAKs are true effectors coupling Rho family GTPases to the SAPKs and p38s.

MAP3Ks

While activation of the SAPKs and p38s by PAKs is ambiguous, several established stress-sensitive MAP3Ks, including MEKK1 (227) (discussed in Section 3.13.2), MEKK4 (209, 227) (discussed in Section 3.13.6), MLK2, and MLK3 (253, 265) (discussed in Section 3.13.7), are putative effectors for Rac1 and Cdc42Hs (Fig. 13). However, whereas the PAKs are clearly activated by Rac1 and Cdc42Hs, the functional role of the interactions between these MAP3Ks and Rho family GTPases is, in most cases, unclear.

POSH

Plenty of SH3 domains (POSH) is a novel approx. 90-kDa polypeptide that consists of four SH3 domains (amino acids 139–190, 198–254, 457–511, and 838–892), a domain rich in putative polyproline SH3 binding sites (amino acids 368–405), but no CRIB motif. In spite of this, however, POSH interacts directly with Rac1 (but not Cdc42Hs, Ras, or Rho) in yeast two-hybrid assays and a GTP-dependent manner *in vitro*. The Rac binding domain has been localized to amino acids 292–362, a domain that appears thus far to be unique to POSH (266). Transient expression of POSH in COS1 cells results in potent SAPK activation. In addition, POSH expression triggers nuclear translocation of NF-κB and apoptosis (266). Significantly, in contrast to MLK3 and MLK2, POSH can interact with Phe37Ala–Rac1, a Rac1 effector mutant that cannot elicit lammelopodium formation but does activate coexpressed SAPK (252, 266). Moreover, in contrast to MLK2 and -3 (Section 3.13.7), POSH cannot interact with Tyr40Cys–Rac1, a second Rac1 effector mutant that does not activate the SAPKs *in vivo* (252, 266). These results suggest that POSH, but not the MLKs, is a likely effector for Rac1 activation of the SAPKs. However, as discussed in Section 3.13.7, MLK3 is a candidate Cdc42Hs effector.

Summary

In summary, PAKs, MEKK1 (Section 3.13.2) and -4 (Section 3.13.6) can interact with both Rac1 and Cdc42Hs. Constitutively active PAK mutants can in some, but not all, instances modestly activate the SAPKs *in vitro*. However, kinase-dead MEKKs cannot block this activation, calling into question the original hypothesis, based on studies of yeast signalling (3), that, in response to activated Rho family GTPases, PAKs would recruit MEKKs to activate the SAPKs and p38s. By contrast, MLK3 and MLK2 are probably effectors for Cdc42Hs (but not Rac1) (see Section 3.13.7), while POSH is a candidate adapter protein that couples Rac1 to the SAPKs.

3.9 Regulation of the SAPKs and p38s by heterotrimeric G-proteins

A number of ligands that bind to seven pass membrane receptors can strongly recruit the SAPKs. These include potent vasoactive compounds, suggesting a major role for the SAPKs in cardiovascular regulation (139, 140, 267, 268). Thus, angiotensin-II and endothelin-1 are strong SAPK agonists in several cell types including cardiomyocytes (angiotensin-II and endothelin-1), hepatocellular carcinomas (angiotensin-II), and pulmonary epithelia (endothelin-1) (139, 140, 267, 268). In addition, the SAPKs can be activated upon engagement of the M_1 muscarinic acetylcholine receptor and the thrombin receptor (267, 269).

The α- and $\beta\gamma$-subunits of heterotrimeric G-proteins couple seven pass receptors to intracellular effectors. Receptor engagement fosters GTP binding to the α-subunit which results in dissociation of the α-subunit from the $\beta\gamma$-complex. Each subunit then binds and regulates target proteins (thus, for example, $G_{s\alpha}$ activates, and $G_{i\alpha}$ inhibits adenylyl cyclase, while $G_{\beta\gamma}$ activates PLC-β2). The G_α-subunits have a modest intrinsic GTPase activity which is substantially accelerated by a novel family of GAPs, the regulators of G-protein signalling (RGS) proteins (reviewed in 77, 270, 271).

Trimeric G-protein α-subunits are divided into four subfamilies based on primary sequence conservation and shared effectors: $G_{i\alpha}$ ($G_{i\alpha1-3}$, $G_{z\alpha}$, $G_{o\alpha1/2}$, $G_{t\alpha}$, and $G_{gust\alpha}$); $G_{q\alpha}$ ($G_{q\alpha}$, $G_{11\alpha}$, $G_{14\alpha}$, and $G_{15/16\alpha}$); $G_{s\alpha}$ ($G_{s\alpha}$ and $G_{olf\alpha}$), and $G_{12\alpha}$ ($G_{12\alpha}$ and $G_{13\alpha}$) (77, 270, 271). Of the ligands that engage seven pass receptors and activate the SAPKs, not much is known regarding the subset of trimeric G-proteins to which these receptors couple. The angiotensin-II and thrombin receptors recruit primarily $G_{o\alpha}$ and, possibly, $G_{12\alpha}$ and $G_{13\alpha}$. Several of these α-subunits, in particular, $G_{q\alpha}$, $G_{i\alpha2}$, $G_{12\alpha}$, $G_{13\alpha}$, and $G_{16\alpha}$ have also been implicated in oncogenesis, in certain cell systems; and ectopic over-expression of transforming G protein-coupled receptors G-protein subunits results in ERK and SAPK activation (269–271).

There are five different mammalian β-subunits (β1–5) and ten different γ-subunits (γ1–10) known (270). Although the β- and γ-subunits do not pair completely indiscriminately, most β- and γ-subunits can form complexes *in vivo* and *in vitro*. By combining different α-, β-, and γ-subunits, a very large number of different heterotrimeric G-protein complexes can be assembled within a cell (270, 271).

The ability of selected G_α-subunits to transform cells and activate the ERKs, coupled with the observation that angiotensin-II, endothelin-I, and other ligands that bind to trimeric G protein-coupled receptors, could recruit the SAPKs, led to the examination of trimeric G-protein subunits as SAPK (and p38) activators. From these studies, it is evident that G-protein α- and $\beta\gamma$-subunits can both recruit the SAPKs, and that the degree of activation, as well as the mechanism (α- or $\beta\gamma$-driven) is cell- and stimulus-dependent.

Thus, transient overexpression of GTPase-deficient mutants of $G_{12\alpha}$, $G_{13\alpha}$, $G_{16\alpha}$, or

$G_{q\alpha}$ results in modest (PC12 cells) to strong (293 cells) SAPK activation. Strong activation of SAPK is also observed in NIH3T3 cells transformed by stable overexpression of GTPase-deficient $G_{\alpha12}$ and $G_{13\alpha}$ (272–274). p38 was not tested in these studies. However, activation of SAPK by G-protein α-subunits has not been universally observed; and it is evident that G-protein β- and γ-subunits can also activate coexpressed SAPK and may, in fact, represent a major mechanism by which trimeric G-proteins signal to the SAPKs (275). Thus, in COS-7 cells, SAPK is barely activated upon transient coexpression with GTPase-deficient $G_{13\alpha}$, $G_{12\alpha}$, or $G_{q\alpha}$. By contrast, coexpression of $G_{\beta1}$ plus $G_{\gamma1}$ results in striking SAPK activation (275). Furthermore, recruitment of SAPK by the M_1 or M_2 muscarinic acetylcholine receptor can be reversed with a construct expressing the G-protein β-subunit binding domain of the β-adrenergic receptor kinase (β-ARK), a scavenger for free $G_{\beta\gamma}$-subunits (275).

Several candidate mechanisms have been proposed for the coupling of heterotrimeric G-proteins to the SAPKs; however, there is little really definitive evidence in favour of any of these models. A role for Ras and Rac has been proposed for $G_{12\alpha}$ based on experiments that employed dominant inhibitory Asn17–Ras or –Rac. The role of Ras as an effector for $G_{12\alpha}$ may be cell-dependent, however. Thus, Asn17–Ras can block signalling from GTPase-deficient (Gln229Leu) $G_{12\alpha}$ to SAPK in NIH3T3 cells and in stably transfected COS cells but not in HEK293 cells (272, 273). By contrast, Asn17–Rac can inhibit Gln229Leu–$G_{12\alpha}$ activation of SAPK in both NIH3T3 and HEK293 cells (272, 273). Moreover, Asn17–Rac can also inhibit the activation of SAPK by free $G_{\beta1\gamma1}$ (275). $G_{13\alpha}$ signalling may also require Ras inasmuch as Asn17–Ras blocks SAPK activation by Gln226Leu GTPase-deficient $G_{13\alpha}$ in COS cells (272).

Non-receptor Tyr kinases may also couple trimeric G-proteins to the SAPKs. Thus, the Bruton's Tyr kinase (Btk) directly binds and is activated by $G_{12\alpha}$, a putative effector of the receptor for angiotensin-II a strong activator of the SAPKs (140, 270, 271, 276). Ectopic overexpression of Btk potently activates coexpressed SAPK (277). $G_{q\alpha}$- and free $G_{\beta\gamma}$-subunits can also activate phospholipase-Cβ (PLCβ). IP_3 generated as a consequence of PLCβ activity releases intracellular Ca^{2+} stores into the cytosol (270, 276). Pyk2 is a non-receptor Tyr kinase that is activated *in vivo* by elevations in intracellular free Ca^{2+} and, in some cells (hepatocellular carcinoma cells) by angiotensin-II (140, 278). Pyk2, acting in concert with Src has been implicated in the relay of signals from trimeric G-proteins to ERKs; and transient expression of Pyk2 and SAPK also results in significant SAPK activation (278, 279). Thus, agonists such as angiotensin-II could recruit the SAPKs by direct $G_{q\alpha}$-mediated activation of Btk and/or through PLCβ/Ca^{2+}-dependent activation of Pyk2. It is attractive to speculate that Btk and Pyk2 recruit Ras (Section 2.5) and, possibly Rac1, thereby triggering activation of SAPK; however, data in support of this hypothesis are not yet available. PLCβ activation also results in the generation of diacylglycerol and activation of classical PKCs; however, evidence implicating PKCs in the regulation of the SAPKs or p38s in response to the recruitment of seven pass receptors is unavailable (1, 137).

A final mechanism by which trimeric G-proteins might couple to the SAPKs and p38s comes from studies of planar polarity signalling in *Drosophila* (see Section 3.14.6). The establishment of planar polarity requires the seven pass receptor Frizzled which signals to Dishevelled, an adapter protein that couples to *Drosophila* Rho and Rac1, and, in turn, *Drosophila* SAPK (280). Three mammalian Dishevelled clones have been identified (281, 282). These may relay signals from mammalian trimeric G protein-coupled seven pass receptors to the SAPKs (and p38s).

3.10   Regulation of the SAPKs by scaffold proteins

Given the large number of MAP3Ks and additional upstream regulators of the SAPKs and p38s, sequestration of individual MAP3K → MEK → MAPK core signalling modules is vital to the maintenance of signalling speed and integrity. In yeast, much of this specificity is mediated by scaffolding proteins which do not activate MAPK pathways *per se*, but instead group specific MAP3K → MEK → MAPK core modules into organized clusters that can be efficiently and specifically regulated by distinct activators and can recruit selected effectors (3, 10). These scaffolding proteins may be distinct polypeptides such as Ste5p, or, as is the case with Pbs2p, the signalling components themselves may, in addition to signalling activity, possess intrinsic scaffold properties (3, 10) (see Chapter 3).

It is likely that mammalian MAPK signalling pathways operate along the same lines. Indeed, the ERK pathway employs the scaffold protein MP1 (Section 2.7). Scaffold proteins for stress-activated MAPK pathways are also being identified. As with yeast, these include proteins whose sole apparent function is that of a scaffold, others which are multipurpose binding proteins, and still others which are also components of MAP3K → MEK → MAPK core modules.

3.10.1   JIP1 is a distinct scaffold protein that may couple MLKs, MKK7, and SAPKs

JNK interacting protein-1 (JIP1) is a novel mammalian scaffold protein that was identified from a two-hybrid screen which sought polypeptides that could interact with SAPKγ/JNK1. JIP1 is an approx. 60-kDa polypeptide that consists of an N-terminal domain that binds SAPKs (amino acids 143–163) and a C-terminal SH3 domain (amino acids 491–600). In between is a proline-rich segment (amino acids 281–448) with several consensus putative SH3 binding sites (14). JIP1 can bind, in addition to SAPKs, MKK7 (but not SEK1), MLK3, and DLK (but not MEKK1). JIP1 can also bind the mammalian germinal-centre kinase (GCK) homologue, haemopoietic progenitor kinase-1 (HPK1) (but not GCK-related, another GCK homologue, see Section 3.11) (15). The binding of HPK1 is likely to be indirect—mediated by the binding of HPK1 to MLK3 (see Sections 3.11 and 3.13.7). MLK3 binds JIP1 via the JIP1 SH3 domain, whereas MKK7 binding to JIP requires a central region (amino acids 283–471) of the JIP1 polypeptide that overlaps with the proline-rich region (15) (Fig. 14A).

Coexpression of JIP with MLK3 or MKK7 enhances the ability of these proteins to foster SAPK activation *in vivo*. However, overexpression of JIP1 inhibits the activation of SAPK by extracellular stimuli including TNF and UV radiation (15). The significance of these paradoxical results is unclear; however, the *in vivo* binding data

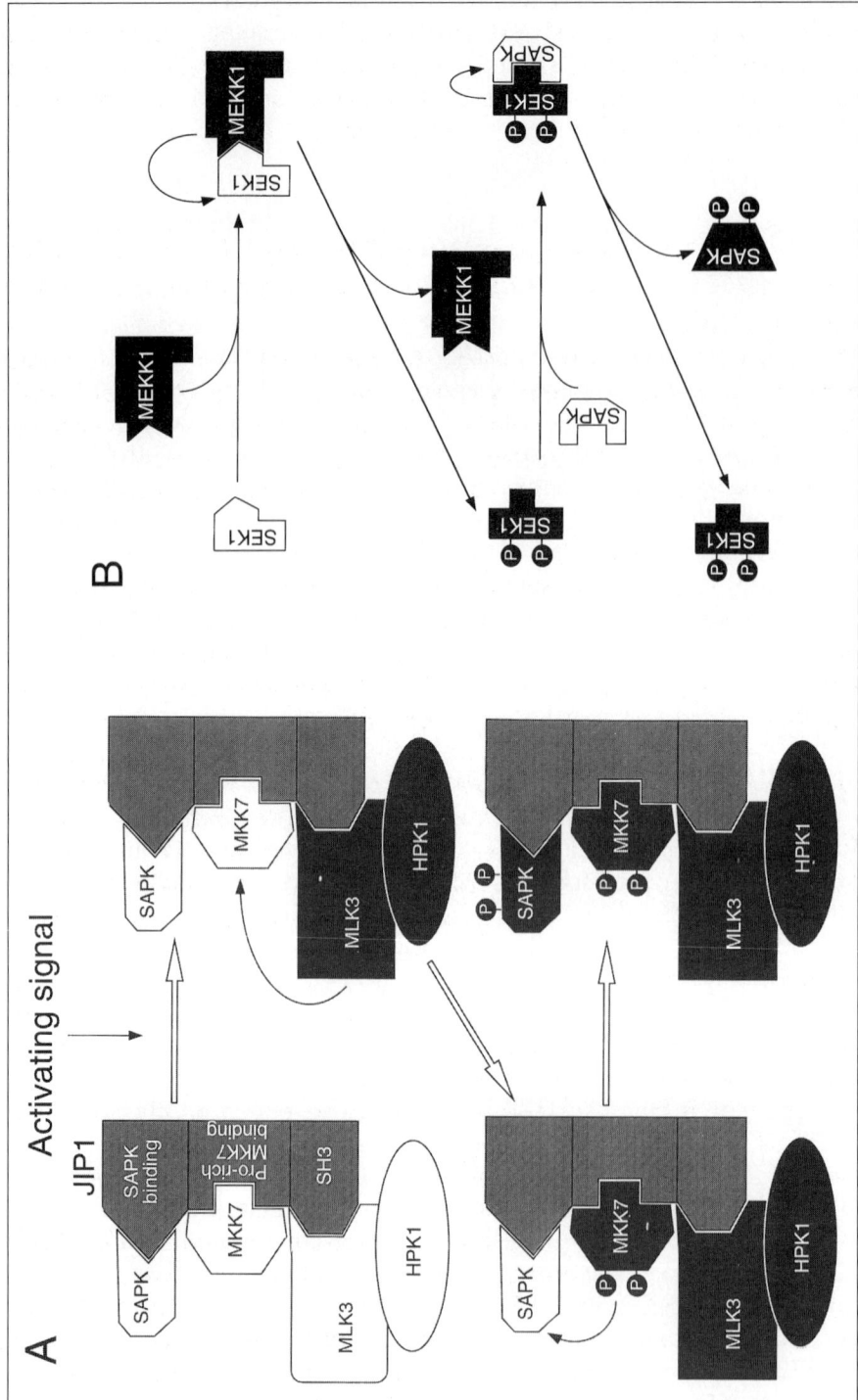

Fig. 14 Scaffold proteins and regulation of the SAPKs. (A) JIP1 forms a complex with MLK3/HPK1, MKK7, and SAPK allowing for efficient activation of the MLK3 → MKK7 → SAPK core module by relevant stimuli. (B) SEK1 has dynamic scaffold properties forming a complex with MEKK1 that dissociates upon phosphorylation of SEK1. A SEK1–SAPK complex then forms which itself dissociates upon phosphorylation of SAPK. Complex formation is mediated by an amino-terminal domain on the SEK1 polypeptide. See text for details. Kinases are labelled in black with white lettering to indicate activation. Black circles with a white P indicate phosphorylation.

indicate that JIP1 may serve to organize SAPK core signalling modules so as to permit regulation by selective sets of upstream stimuli. Accordingly, overexpression of JIP1 could non-specifically sequester endogenous MLK3, DLK, MKK7, or SAPK, enhancing the ability of recombinant, expressed MLK3, DLK, or MKK7 to activate the SAPKs, but inhibiting the ability of stimuli which normally recruit these proteins independently of JIP1 to activate the SAPKs.

3.10.2 The 280-kDa actin binding protein-280 (ABP-280) may be a scaffold protein for TNF and lysophosphatidic acid activation of the SAPKs in melanoma cells

ABP-280 is a 280-kDa protein that binds actin through its amino terminus, homodimerizes at its carboxyl terminus, and contains two flexible hinge regions, one two-thirds along its length and the other immediately adjacent to the homodimerization domain. Structurally, ABP-280 is rod-shaped; and, aside from the actin binding domain, the various protein interaction domains are distributed among 24 repeats of an approximately 98-amino acid structure. The hinge regions reside immediately upstream of repeats 16 and 24; and repeat 24 itself is the homodimerization domain. The structure of ABP-280 enables it to cross-link actin filaments into orthogonal arrays and to contribute to the structure of the cortical actin meshwork. This process may be regulated by the binding of the ABP-280 carboxyl-terminal region to cell-surface receptors such as those for integrins. ABP-280 may also be regulated by phosphorylation; and, indeed ABP-280 undergoes phosphorylation *in vivo* in response to serum, lysophosphatidic acid (LPA), and other treatments. However, the functional consequences of ABP-280 phosphorylation are unclear (ref. 283, and references therein). Recent studies indicate that ABP-280 may also serve as a signalling scaffold protein.

Thus, a yeast two-hybrid screen employed to identify novel SEK1 interactors identified ABP-280 as a SEK1 binding protein. SEK1 binds ABP-280 *in vivo* and *in vitro*; and the binding requires a carboxyl-terminal region of ABP-280 consisting of repeats 21–23C (amino acids 2282–2454) (283). Binding is quite specific; and ERK1, p38α, and MEK1 bind ABP-280 at best very modestly. Activation of SEK1 with MEKK1 has no effect on the binding of SEK1 to ABP-280; however, ABP-280 appears to be important for the activation of SEK1 and the SAPKs *in vivo* in melanoma cells, in response to discrete sets of extracellular stimuli (283).

M2 cells are a human melanoma cell line that has spontaneously lost ABP-280 expression. These cells show prolonged membrane-blebbing after plating, poor spreading in cell culture, and a deficit in directed cell motility. Reconstitution, in M2A7 cells, of physiological levels of ABP-280 corrects these defects (ref. 283 and references therein). SAPK activation by TNF and LPA is strikingly abrogated in M2 cells; while activation by arsenite, anisomycin, or hyperosmolarity are unaffected (283). By contrast, activation of SAPK by TNF and LPA is restored in M2A7 cells (283). Thus, ABP-280 may serve as a scaffold protein to coordinate the proper activation of the SAPKs by a select subset of activators (TNF and LPA) to the exclusion of others.

The mechanism by which ABP controls SAPK activation is unclear. However,

MEKK1 and SAPK can coprecipitate with ABP-280 if, and only if, SEK1 is present (283). As discussed below, SEK1 itself has scaffold properties (17) and may permit the formation of MEKK1–SEK1–SAPK–ABP-280 complexes that can be specifically recruited by TNF or LPA.

3.10.3 SEK1 has intrinsic scaffold properties

SEK1 possesses intrinsic scaffold properties. Thus, the N-terminus (amino acids 1–77) of inactive, dephosphorylated SEK1 can specifically bind one of its upstream activators, MEKK1. MEKK1 does not readily interact *in vivo* or *in vitro* with MEK1 or -2, MKK3 or -6, or with the γ1-isoform of MKK7 under conditions in which each MEK is expressed separately with MEKK1; and as a consequence, none of these MEKs is as good an *in vitro* substrate of MEKK1 as is SEK1 (17). However, while MKK7γ1 does not interact with MEKK1 and is a modest MEKK1 substrate *in vitro*, several MKK7 isoforms (α1, α2, β1, and β2) are strongly activated by MEKK1 *in vivo* (Section 3.6.1) (200, 201). These MKK7 isoforms are either amino-terminally deleted or truncated (201), and their binding to MEKK1 has not been ascertained.

The SEK1–MEKK1 complex can be recapitulated both *in vivo* and *in vitro*. *In vitro*, the addition of ATP to the SEK1–MEKK1 complex results in SEK1 activation which, in turn, triggers dissociation of the SEK1–MEKK1 complex (17) (Fig. 14B). Phosphorylated and activated SEK1 then forms a second specific complex, with SAPK, again mediated by the SEK1 amino-terminal MEKK1 binding domain. Upon phosphorylation and activation of SAPK, this second complex dissociates (17) (Fig. 14B). Thus, as is the case for the yeast MEK Pbs2p and its upstream activators and effectors (10), the intrinsic scaffold properties of SEK1 enable it to form specific, sequential dynamic complexes with both an upstream activator and a substrate.

3.11 Regulation of the SAPKs by the germinal centre kinases, a family of putative regulated scaffold kinases

3.11.1 General considerations

Germinal centre kinase (GCK) is the founding member of a novel family of protein kinases, some of which are selective activators of the SAPKs (12, 284). Mounting evidence suggests that the GCKs which recruit the SAPKs do so by binding MAP3Ks and elements upstream of MAP3Ks in a regulated manner reminiscent of the dynamic scaffold function of SEK1 (Section 3.10.3) (12, 18). To date, 11 mammalian GCKs have been cloned. In addition, there are *Drosophila*, *C. elegans*, and *Dictyostelium* homologues, as well as two *S. cerevisiae* genes with known phenotypes (reviewed in ref. 12). All GCKs possess amino-terminal kinase domains that are distantly related to those of the PAKs and Ste20p. The kinase domains are followed by extensive carboxyl-terminal regulatory domains (CTDs) (12). The distant sequence homology between the kinase domains of the GCKs, Ste20p, and the PAKs led to the initial placement of the GCKs in the Ste20 family. However, the domain arrangement of the GCKs, coupled with the fact that these kinases do not possess CRIB motifs and are not

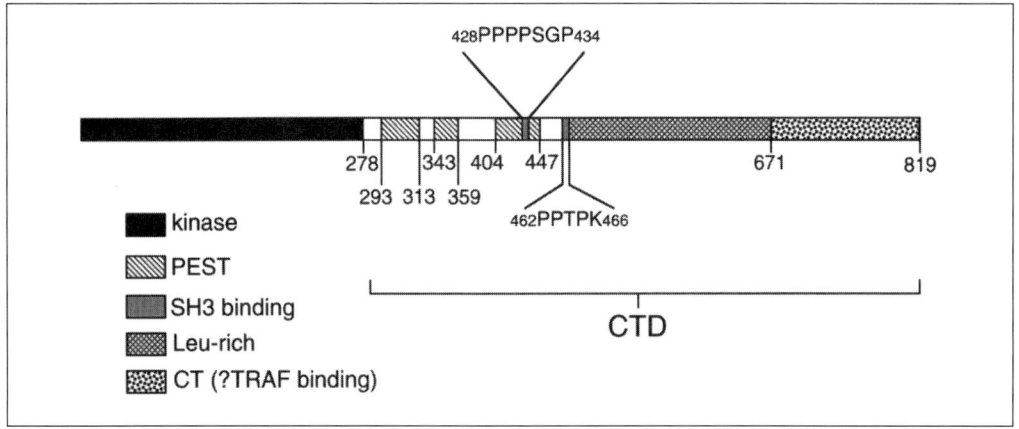

Fig. 15 Domain structure of GCK, a typical group-I GCK.

directly activated by Rho family GTPases, indicate that the GCKs should be considered a distinct protein kinase family (12).

GCKs can be subdivided into two groups based on overall sequence conservation and function: group-I GCKs are structurally similar within both the kinase and CTD regions to GCK itself and are specific upstream activators of the SAPKs (12) (Fig. 15). Group-II GCKs are more closely related to the *S. cerevisiae* GCK Sps1p and are activated *in vivo* by extreme stresses (12). No downstream effectors of group-II GCKs have been identified (12). This discussion will be limited to group-I GCKs.

So far, five mammalian group-1 GCKs have been identified: GCK itself; GCK-related (GCKR, also called kinase homologous to Ste20, KHS, Table 4); GCK-like kinase (GLK); haemopoietic progenitor kinase-1 (HPK1); and Nck-interacting kinase (NckIK, also called HPK1/GCK-like kinase, HGK. NckIK is also abbreviated to NIK; however, to avoid confusion with NF-κB inducing kinase, also abbreviated NIK, we will use the abbreviation NckIK) (284–291) (Table 4). There is striking conservation of the CTDs of group-I GCKs. The CTDs of all group-I GCKs include two or more proline/glutamic acid/serine/threonine (PEST) motifs and at least two polyproline putative SH3 domain binding sites. Of particular importance, all the group-I kinases possess a highly conserved approx. 350-amino acid region at the carboxyl-terminal end of the CTD that consists of a hydrophobic, leucine-rich domain and a 142–152 amino acid stretch, the carboxyl-terminal (CT) motif (284–291, reviewed in ref. 12). The leucine residues in the Leu-rich domains are not organized into leucine zippers, nor are these domains sufficiently hydrophobic for membrane insertion. The carboxyl terminal-most putative SH3 binding sites of GCK, GCKR, GLK, and HPK1 lie at the amino-terminal end of the cognate Leu-rich domains. Such a binding site is not evident in the Leu-rich segment of NckIK, which is the most divergent mammalian group-I GCK (12, 284–291).

GCK, GCKR, GLK, and NckIK are all activated *in vivo* by TNF (285, 287, 291, 292); and the CT domains of GCK, GCKR, and GLK along with that of HPK1 are quite homologous (284–291). The similarity of the CT motif of NckIK and those of the other mammalian group-1 GCKs, while apparent, is less dramatic. Studies of GCK and GCKR indicate that the CT motif is required for binding proteins of the TNF receptor-associated factor (TRAF) family, and, possibly, for gating the binding of MAP3Ks (12, 18, 285) (Sections 3.13.2, 3.13.3, and 3.13.7). The notion that GCKs can both potently activate the SAPKs as well as stably associate with SAPK core signalling module elements and their upstream activators in a manner dependent upon the kinase activity of the GCKs (Sections 3.13.2, 3.13.3, and 3.13.7) suggests that the group-I GCKs, like SEK1, possess both scaffolding and enzymatic signalling properties.

3.11.2 Group-1 GCKs: selective activators of the SAPKs

GCK

GCK (Table 4) was cloned as part of a subtractive screen to identify genes preferentially expressed in B-cell germinal centres; and, although GCK is present in all tissues examined, its distribution in B-lymphocyte follicular tissue is restricted largely to the germinal centre and not the surrounding mantle zone (284). Germinal centres are sites of B-cell maturation and selection, processes that are driven by ligands of the TNF family (CD40L, CD30L, TNF, in particular) (293, 294). Notably, mice in which the gene for the type-1 TNF receptor has been disrupted manifest little or no germinal centre formation (294). Accordingly, it was logical to propose that GCK, which is selectively expressed in the germinal centre, might be an effector for TNF family ligands and might signal to the SAPKs.

Indeed, endogenous GCK, while exhibiting substantial basal activity upon immunoprecipitation from several B-cell lines, is significantly activated *in vivo* by TNF (292). Moreover, transient expression of GCK results in potent SAPK activation in the absence of external stimulus. By contrast, expression of GCK does not result in activation of the ERKs or NF-κB and only activates the p38s if massively over-expressed (285, 292). Thus, like MEKK1 and the MLKs, GCK is a specific SAPK activator.

The high basal activity of both endogenous and recombinant GCK suggests that GCK is activated *in vivo* either by limiting concentrations of an inhibitor lost during immunoprecipitation or stoichiometrically titred-out upon GCK overexpression. Alternatively, GCK could be regulated by aggregation or oligomerization, a process that that could be mimicked by immunoprecipitation and/or overexpression. In either case, the CTD is the region of the GCK polypeptide most likely to mediate GCK regulation and substrate recognition (12). In support of this idea, transient expression of the free GCK-CTD, or a GCK construct (Lys44Met) devoid of kinase activity results in SAPK activation (18, 292). As we shall see (Sections 3.13.2, 3.13.3, and 3.13.7), the GCK-CTD binds both GCK regulators and effectors.

GCKR

GCKR (Table 4) was cloned by Kehrl and colleagues by DNA database mining (285). Independently, Blenis *et al.* cloned GCKR (referred to by them as KHS, Table 4) as part of a screen to identify novel mammalian homologues of *STE20* (286). GCKR is approximately 60% identical to GCK throughout its primary sequence. Most notably, the Leu-rich and CT motifs of the CTD are highly conserved (12, 284–286). Like GCK, GCKR is a selective activator of the SAPKs; the ERKs, p38s, and NF-κB are not activated upon expression of GCKR (285, 286). In contrast to GCK itself, transient expression of the GCKR CTD does not result in significant SAPK activation.

Endogenous GCKR is activated *in vivo* by TNF (285). In addition, GCKR is activated by ultraviolet radiation—a process that may be mediated by UV-induced clustering of TNF receptors (285, 295). It is likely that GCKR is an important effector for TNF, at least in some cell types, inasmuch as antisense constructs of GCKR, when expressed in 293 cells, can completely inhibit TNF activation of the SAPKs (285) (Section 3.13.3).

GLK

GLK (Table 4) is a third type-I GCK homologue that is, like GCK and GCKR, a selective activator of the SAPKs. The CTD of GLK is strikingly conserved with those of GCK and GCKR, especially within the Leu-rich and CT domains. Consistent with this close similarity, endogenous GLK is activated *in vivo* by TNF (287).

HPK1

HPK1 (Table 4) was cloned in a subtractive screen that sought genes selectively expressed in cells of the haemopoietic lineage. HPK1 shares many of the structural features of group-I GCKs, although it is more distantly related to GCK than is GCKR and GLK. Transient overexpression of HPK1 results in robust activation of the SAPKs. p38 is activated only very weakly and no ERK activation is observed upon expression of HPK1 (12, 288, 289).

HPK1 has four consensus SH3-domain binding sites (12, 288, 289). The carboxyl terminal-most SH3 binding site interacts with the SH3 domain of MLK3 (288) (discussed in Section 3.13.7). Two recent studies indicate that HPK1 may be regulated by SH2/SH3 adapter proteins that couple to receptor and/or non-receptor Tyr kinases. Thus, the amino-terminal two SH3 binding sites of HPK1 interact with the carboxyl-terminal SH3 domain of SH2/SH3 adapter protein Grb2, which is also a key regulator of Ras (Section 3.5). As with Grb2–SOS, the Grb2–HPK1 interaction is not affected by extracellular stimuli; instead, treatment of cells coexpressing Grb2 and HPK1 results in a translocation of the Grb2–HPK1 complex to the membrane (296).

The significance of the HPK1–Grb2 interaction is still somewhat unclear, however, given that coexpression with Grb2 has no effect on HPK1 activity. By contrast, Crk and CrkL, two additional SH2/SH3 adapter proteins, can bind to the second and fourth polyproline SH3 binding sites of HPK1. As with the Grb2–HPK1 interaction, the Crk(CrkL)–HPK1 interaction enables the EGF-dependent recruitment of HPK1 to the EGF receptor (297). The Crk(CrkL)–HPK1 interaction is likely to be physiologically significant inasmuch as coexpression of Crk or CrkL with HPK1 results in HPK1 activation, and synergistic SAPK activation (297). Although the SAPKs are usually only modestly activated by EGF and other mitogens that signal through receptor Tyr kinases, the Grb2–HPK1 and Crk/CrkL–HPK1 interaction could partly underlie the regulation of the SAPKs by any of several non-receptor Tyr kinases such as Btk and Pyk2 that strongly activate the SAPKs *in vivo* (Section 3.9) (see Fig. 17A).

In this regard, it is noteworthy that Btk and Pyk2 are putative effectors for trimeric G protein-coupled receptors (Section 3.9) (275, 278, 279). Several MAP3Ks upstream of the SAPKs (the SAPK-specific MEKK1, Section 3.13.2, and MLK3, Section 3.13.7, for example) bind Rac1 in a GTP-dependent manner. Rac1 has also been implicated in trimeric G-protein signalling to SAPK; and it is possible that one function of HPK1 translocation is to colocalize it with effector MAP3Ks simultaneously translocated to the plasma membrane (reviewed in ref. 12).

NckIK

Murine NckIK (Table 4) was cloned as part of a two-hybrid screen to identify novel polypeptides that interact with the SH2/SH3 domain-containing adapter protein Nck. A human homologue designated HPK1/GCK-like kinase (HGK) (Table 4) was cloned by degenerate PCR using primers based on sequences conserved in the GCK family (290, 291). Of the mammalian group-I GCKs, NckIK is the most distantly related. Still, functionally speaking, NckIK shares many of the same features of the group-I GCKs. Notably, NckIK transient overexpression strongly and specifically activates the SAPKs (291). Moreover, as with GCK, expression of the free NckIK CTD significantly activates the SAPKs, albeit to a lesser degree than does wild-type NckIK (291).

NckIK contains two consensus SH3 binding motifs, both of which can interact with the SH3 domains of Nck (291). The significance of the Nck–NckIK interaction is not clear; however, Nck has been implicated in the regulation of the actin cytoskeleton. Specifically, the SH3 domains of Nck can interact with the carboxyl-terminal SH3 binding site of PAK1, localizing PAK1 to the plasma membrane, an interaction that may, in conjunction with PIX–PAK (Section 3.8.2), underlie the directed membrane ruffling that occurs upon transient overexpression of PAK1 (260, 261, 262, 298). NckIK may also mediate changes in cell motility and shape. Indeed, a *Drosophila* orthologue of NckIK, Misshapen, is required for embryonic dorsal closure (Section 3.14.6) (299). NckIK can be activated by TNF (292), and recent studies of Misshapen signalling indicate that it (and, by extension, NckIK), like GCK and GCKR, is an effector for TNF receptor-associated factors (TRAFs) (18, 285, 300) (Section 3.13.3).

3.12 Regulation of the SAPKs and p38s by adapter proteins that couple to TNF family receptors

The tumour necrosis factor receptor (TNFR) family includes TNFR1, TNFR2, interleukin-1 (IL-1), and lymphotoxin-β (LT-β) receptors, CD40, CD27, Fas, receptor activator of NF-κB (RANK, also called osteoprotegerin, OPG, or TNF-related activation-induced cytokine, TRANCE, receptor), TNF-related, apoptosis-inducing ligand receptor (TRAILR), CD30, Ox40, death receptor (DR)-3, DR4, DR5, 4–1BB, the

THE STRESS-ACTIVATED PROTEIN KINASE

p75 neurotrophin receptor (p75NTR), and others (301–305). Several viral genes also encode TNFR family receptors; the latent membrane protein-1 (LMP1) of Epstein–Barr virus is an example. Ligands of the TNF family are structurally homologous, and, accordingly, the extracellular ligand-binding domains of these receptors are conserved. However, the intracellular extensions of these receptors are divergent. Upon binding ligand, TNFR family receptors can elicit a wide variety of inflammatory responses and are critical to immune-cell development, innate and acquired immunity, as well as the pathogenesis of a number of diseases such as arthritis, septic shock and, possibly, type-2 diabetes mellitus (301–305). Accordingly, this family of receptors is among the most important activators of the SAPKs, p38s, and NF-κB.

3.12.1 The protein recruitment model for TNFR family signalling

Receptors of the TNFR family possess no intrinsic enzymatic activity. Upon binding ligand, receptors of the TNF family homotrimerize or hetero-oligomerize with receptor accessory proteins. Receptor oligomerization is thought to trigger conformational changes that initiate signal transduction. The protein recruitment hypothesis posits that receptor oligomerization enables the binding of signal-transducing polypeptides that recruit downstream targets (302–304).

Death domains and signalling

Several TNFR family receptors (TNFR1, Fas, TRAILR, and the DRs, in particular) contain an 82–102-amino acid extension, the death domain (301–304). Death domains mediate homotypic and heterotypic protein–protein interactions and are critical for nucleating receptor–effector complexes and implementing several signalling programmes including, as their ominous name suggests, apoptosis (302–305). The type-1 TNFR (TNFR1) can recruit all the known signalling pathways activated by TNF. Upon ligand-induced TNFR1 trimerization, the TNFR1 death domain binds the death domain of the adapter protein TNFR-associated death-domain protein (TRADD) (302–304).

TRADD consists of a carboxyl-terminal death domain and an amino-terminal domain, the TNFR-associated factor (TRAF) interaction domain, that binds TRAF proteins (see below). TRADD is a pivotal component in the coupling of TNFR1 to downstream targets; and overexpression of TRADD can activate many TNF signalling pathways including apoptosis and NF-κB (306). Curiously, although TRADD can recruit TRAF2 and receptor interacting protein (RIP) (307, 308), two key upstream activators of the SAPKs and p38s (18, 309–311) (see below), TRADD overexpression does not activate the SAPKs (309–311). The interaction between the TRADD death domain and that of TNFR1 is thought to result in the binding of additional death-domain proteins to the TRADD death domain, and of TRAF proteins to the TRAF interaction domain (302–304).

TRAFs

The TRAFs are an emerging family of proinflammatory, signal-transducing adapter proteins that are important for the activation of a number of pathways in response

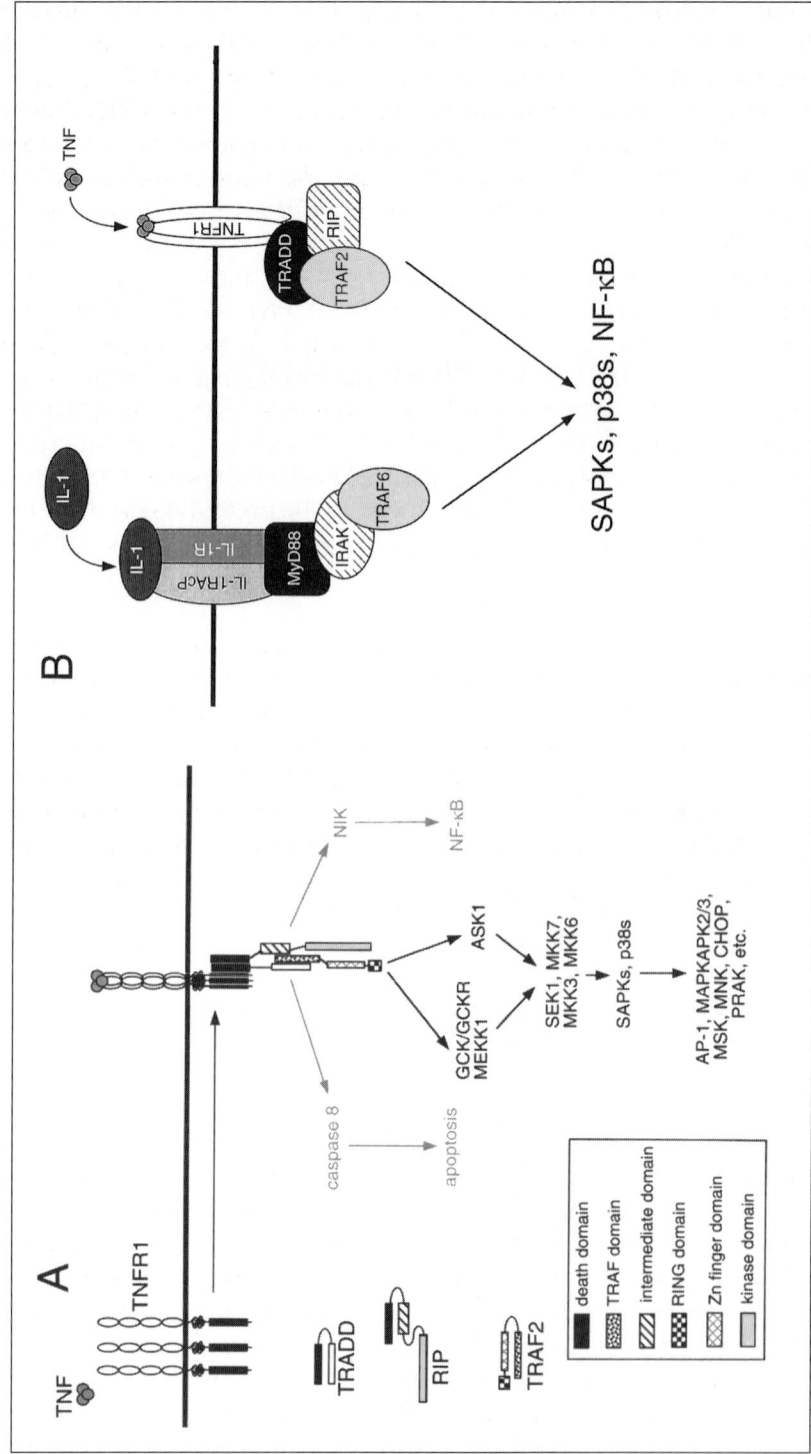

Fig. 16 Inflammatory cytokine signalling to SAPK and p38—the protein recruitment model. (A) TNFR1 signalling. Ligand-stimulated trimerization of TNFR1 fosters formation of a signalling complex consisting of the platform death-domain adapter protein TRADD (the death domain of which is associated with the TNFR1 death domain), RIP (the death domain which is also associated with the TRADD death domain), and TRAF2 (associated with both the TRADD effector domain and the RIP intermediate and kinase domains) This complex signals apoptosis (from TRADD), NF-κB activation (which requires RIP), and SAPK/p38 activation (which requires TRAF2. (B) Signalling by IL-1: comparison with TNFR1 signalling. IL-1 binding to its receptor results in the formation of a complex with IL-1 accessory protein (IL-1RAcP) This heteromer in turn recruits the platform death-domain adapter MyD88 which binds IRAK and TRAF6. IRAK and TRAF6 are necessary for SAPK/p38 and NF-κB activation.

to TNF family ligands (reviewed in refs 303, 304). So far, six mammalian TRAF polypeptides have been identified (TRAF1–6). The TRAFs consist of two, tandem, carboxyl-terminal TRAF domains (TRAF-N followed by TRAF-C), a central zinc-finger motif and, with the exception of TRAF1, an amino-terminal RING-finger motif (303, 304). The TRAF domains are responsible for binding upstream activators as well as TRAF effectors. The function of the Zn-finger domains is unclear; however, the RING fingers are important for the activation of downstream effectors (18, 240, 303–312) (Fig. 16A).

Details of the structural features required for TRAF binding to upstream proteins are emerging. TRAF1 and -2 were purified and cloned based on their interactions with the intracellular extension of TNFR2 (303, 304, 313). As with TRADD, the intracellular extension of TNFR2 contains a TRAF interaction domain. Within this domain is a common binding motif (consensus Pro–X–Gln–X–Ser/Thr, where X is any amino acid) that has been demonstrated to mediate TRAF binding in some but not all cases. Thus, many TNFR family receptors including CD40, CD27, LT-βR, RANK, and LMP1 contain Pro–X–Gln–X–Ser/Thr TRAF interaction domains. However, a consensus TRAF interaction domain is not always necessary for TRAF binding (144, 303, 304).

RIP

Receptor interacting protein (RIP) was cloned as a polypeptide that could interact in a two-hybrid screen with the death domain-containing receptor Fas. Indeed, ectopic expression of RIP results in rapid apoptosis and, in parallel, activation of NF-κB (309, 314, 315). Subsequent studies using cell lines in which expression of the endogenous RIP gene is ablated have shown that RIP is not involved in Fas proapoptotic signalling, but is instead essential for TNF activation of NF-κB, a result borne out by subsequent gene disruption studies (315, 316). RIP consists of a carboxyl-terminal death domain, an amino-terminal Ser/Thr kinase domain, and an intermediate domain which probably forms a coiled-coil structure (314).

RIP does not interact directly with TNFR1. Instead, the RIP death domain binds to that of TRADD in a TNF-dependent manner (309). The RIP kinase and intermediate domains can also bind to the TRAF domains of TRAF2 (312), and TNF treatment is thought to result in the formation of a TRADD–RIP–TRAF2 complex at the membrane (Fig. 16A).

3.12.2 TRAF2, -5, -6 and RIP can activate the SAPKs and p38s

Many conclusions as to the presumed biological and biochemical functions of the TRAFs and RIP are based on results from overexpression studies; and inasmuch as TRAFs and RIP appear to be constitutively active upon overexpression, non-specific artefacts of overexpression could prompt incorrect inferences. Still, in spite of this caveat, overexpression studies have provided much valuable insight into the possible mechanisms by which the TRAFs and RIP act.

Transient overexpression of TRAF2 results in potent SAPK and p38 activation (18, 309–311). The RING domain appears necessary for relaying signals from TRAFs to downstream effectors. Thus, deletion of the TRAF2 RING-finger domain completely abrogates the ability of TRAF2 to recruit the SAPKs and p38s, and ΔRING-TRAF2 can act as a dominant inhibitor of TNF activation of the SAPKs and p38s (18, 309–311). Targeted disruption of *traf2*, or transgenic expression of ΔRING-TRAF2 abolishes TNF activation of SAPK; however, IL-1 activation of SAPK is unaffected (p38 was not tested) (317, 318). Thus, TRAF2 is an essential signal transducer for coupling TNFRs to the SAPKs and, possibly, the p38s.

TRAF5 and -6, when expressed ectopically, can also strongly activate the SAPKs; however, overexpression of TRAF1, 3, and 4 does not result in SAPK activation (319). Inasmuch as TRAF2 appears to be the dominant TRAF coupling the TNFRs to the SAPKs, activation of the SAPKs by TRAF5 and -6 may couple the SAPKs to other signalling pathways. In support of this idea, coexpression experiments indicate that the intracellular extension of RANK binds TRAF1, -2, -3, -5, and -6 predominantly through two domains (amino acids 568–573 and 607–611) that conform to a consensus TRAF binding sequence (Pro–X–Gln–X–Ser/Thr). Interestingly, a second, novel motif (amino acids 340–421) that does not contain the TRAF binding motif associates with TRAF6 selectively, indicating that the so-called TRAF interaction domain may not be absolutely required for TRAF binding (144). Despite the apparent binding of multiple TRAFs to RANK, TRAF6 appears to be the major signal transducer from RANK to the SAPKs. Thus, deletion of the TRAF interaction motifs or RANK abolishes TRAF2 and TRAF5 binding, but has no effect on TRAF6 binding. Under these circumstances, SAPK and NF-κB activation triggered by RANK ligand is only modestly affected. Deletion of the TRAF6 binding site, however, severely impairs RANK activation of SAPK and NF-κB (144).

IL-1 probably requires TRAF6 to signal to the SAPKs; and targeted disruption of *traf6* severely blunts IL-1 activation of the SAPKs and NF-κB (320). IL-1 signalling follows the protein recruitment paradigm for inflammatory cytokine signalling. IL-1 binding to its receptor (IL-1R) triggers heterodimerization of the IL-1R with the IL-1R accessory protein (IL-1RAcP). The IL-1R–IL-1RAcP heterodimer then recruits the death-domain adapter protein MyD88. The death domain of receptor-bound MyD88 binds an amino terminal non catalytic domain of the Ser/Thr kinase IL-1R-associated kinase (IRAK). IRAK, in turn, binds TRAF6, forming a MyD88–IRAK–TRAF6 complex analogous to the TNFR-associated complex of TRADD, RIP, and TRAF2 (303, 304, 321). As with TRAF2, -5, and -6, IRAK can, upon overexpression, activate both NF-κB and the SAPKs (322) (Fig. 16B).

RIP, like TRAF2, is, when overexpressed, a potent activator of both SAPK and p38. The RIP intermediate domain is both necessary and sufficient for this function (18, 309). A dominant inhibitory construct of RIP wherein the intermediate domain is deleted (RIP-ΔID) effectively blocks TNF activation of both SAPK and p38 (18). RIP and TRAF2 can interact directly *in vivo*; however, gene disruption studies indicate that deletion of *rip* does not affect TNF activation of SAPK, but, instead abrogates the activation of NF-κB (312, 315, 316). Inasmuch as GCK, GCKR, ASK1, and MEKK1 also bind TRAFs (Sections 3.13.2, 3.13.3), it is likely that multiple, parallel mechanisms couple TRAF2 to the SAPKs and deletion of *rip* can be compensated for by the presence of these alternate pathways. A similar situation may exist for p38.

3.12.3 Daxx couples FasL to the SAPKs

Fas is a widely expressed cell death receptor that, among other things, is crucial to immune-cell regulation where it governs, in part, the ablation of autoreactive T lymphocytes. Engagement of Fas by FasL triggers apoptosis through a process that involves the adapter protein Fas-associated death-domain protein (FADD), the death domain of which interacts with that of activated Fas to signal apoptosis through activation of the caspase family of proapoptotic cysteine proteases (reviewed in refs 301–305, 323).

Daxx is a novel adapter protein that also interacts with Fas and couples Fas to the SAPKs (324). Daxx overexpression can also trigger apoptosis; however, Daxx does not possess a death domain. Instead, Daxx has a carboxyl-terminal, death-domain interaction motif (amino acids 625–739) that binds directly to the Fas death domain, but not to the death domain of FADD. The central portion of Daxx (amino acids 500–625) is apparently necessary for recruiting the SAPKs and for signalling apoptosis. The remainder of the Daxx polypeptide (the death-domain interaction motif, amino acids 626–739, and the amino-terminal segment of the molecule, amino acids 1–500) is thought to perform an autoinhibitory function that is relieved upon the interaction of Daxx with Fas. Expression of a construct consisting solely of the Fas binding region of Daxx (amino acids 625–739) blocks both Fas-induced SAPK activation and apoptosis (325). As is discussed in Section 3.13.4, Daxx may recruit the SAPKs by directly binding and activating ASK1 (328).

The SAPKs appear to be required for Daxx-induced cell death, but only in a subset of the cell lines tested (293 and L929 cells, but not HeLa cells) (324). Still, it is not clear what the relative significance of Daxx is to Fas-induced apoptosis. Fas mutants, wherein the death domain fails to interact with FADD, are completely unable to trigger apoptosis in response to FasL, in spite of the fact that Daxx binding and Fas-induced SAPK activation by these mutant receptors are unimpaired (325). More importantly, targeted disruption of *daxx* is embryonic lethal, with lethality occurring as a result of massive cellular apoptosis. Taken together, these results argue in favour of an antiapoptotic role for Daxx, rather than a proapoptotic role. Finally, the exact role of Daxx as an effector for Fas is also unclear. Several laboratories have cloned Daxx independently of its association with Fas; and in some instances Daxx has been isolated in yeast two-hybrid screens as an interactor with nuclear proteins such as DNA methyltransferase. Paradoxically, specific antisera were used to demonstrate a predominantly nuclear localization for Daxx, and Fas-dependent association of endogenous Daxx with endogenous, cytosolically localized ASK1 (238, 326). Clearly more work is needed before the role of Daxx in signalling to the SAPKs is established.

3.13 Coupling stress-activated MAP3Ks to upstream signals

3.13.1 General considerations

Sections 3.2–3.7 have outlined the basic elements of known mammalian stress-activated MAP3K → MEK → MAPK core modules, while Sections 3.8–3.12 have described upstream components thought to couple to these core pathways. Still, despite considerable progress in the identification of SAPK, and p38-activating MAP3Ks, and the identification of proximal signalling components, such as Ras superfamily and trimeric GTPases, PAKs, GCKs, and TRAFs, that are thought to couple to SAPK and p38 core pathways, the molecular mechanisms by which stress-activated MAP3Ks are regulated by events at the cell membrane remain obscure. What little is known suggests that stress-activated MAP3Ks either (1) bind directly to upstream activating regulatory proteins, or (2) couple to other protein kinases that, in turn, bind upstream regulatory molecules.

3.13.2 MEKK1 is a putative effector for Ras and Rho family GTPases, GCK family kinases, and TRAFs, and may be targeted to an apoptotic pathway upon caspase cleavage

Existing data suggest three mechanisms for the regulation of MEKK1: (1) recruitment by Ras and Rho family GTPases; (2) site-specific cleavage to generate an active fragment that triggers apoptosis; and (3) binding to TRAFs and GCK family kinases that, in turn, couple to complexes associated with TNFR family receptors.

Ras superfamily GTPases and elements coupled to receptor and non-receptor Tyr kinases

MEKK1 can interact with Ras, Rac1, and Cdc42Hs in a GTP-dependent manner (226, 227). Moreover, kinase-dead forms of the MEKK1 catalytic domain (Δ1–1221/ K1253M) can block the activation of SAPK by Rac1 and Cdc42Hs (interestingly, in these studies, kinase-dead MEKK1 mutants failed to block the modest activation of SAPK incurred by PAK1) (227). MEKK1 does not contain a consensus CRIB motif; the interaction of MEKK1 and Ras or Rho family GTPases requires the MEKK1 kinase domain inasmuch as deletion of the entire regulatory domain does not affect Rac1 or Cdc42Hs binding (226, 227).

As yet there is no evidence indicating that Ras superfamily GTPases actually regulate MEKK1 in a manner analogous to that of Raf-1 regulation by Ras; and the functional significance of the GTPase–MEKK1 interaction is unclear. Moreover, inhibition of G-protein signalling by K1253M MEKK1 could merely reflect sequestration of SEK1. Indeed, MEKK1 and SEK1 form a tight complex *in vivo* (17) (see Section 3.10.2). It is possible that one function of GTP-dependent Ras, Rac, or Cdc42 binding to MEKK1 is to translocate MEKK1 to sites of activation by other polypeptides, or to sites where substrates are located.

The GTP-dependent coupling of MEKK1 to Ras superfamily GTPases indicates that MEKK1 may be an effector for those agonists that recruit the SAPKs through Ras, Rac1, or Cdc42Hs-dependent mechanisms. As discussed above (Section 3.9), these include agonists coupled to Tyr kinases such as mitogens, and agents that signal through heterotrimeric G-proteins. In this regard, it is noteworthy that endogenous or recombinant MEKK1 can interact with the SH2/SH3 adapter protein Grb2 (327). This interaction requires the carboxyl-terminal SH3 domain of Grb2 and amino acids

1–301 of a 98-kDa fragment of MEKK1 (amino acids 233–291, Section 3.7.2) which includes the carboxyl-terminal of two SH3 binding sites (206, 327). EGF recruits the Grb2–MEKK1 complex to the Tyr phosphorylated form of the SH2 adapter protein Shc, already bound to the Tyr phosphorylated EGFR (327). EGF also recruits to the membrane complexes of either Grb2 or Crk/CrkL and HPK1 a (putative regulator of MEKK1, Section 3.11.2) as well as the Ras-activating Grb2–mSOS complex to the EGFR (2, 296, 297). Based on these results, it is plausible to suggest that activation of MEKK1 (and, perhaps by extension other MAP3Ks) by Tyr kinases may involve the coordinated recruitment of MEKK1 activators to the membrane, followed by their binding to MEKK1 and contribution to its activation (Fig. 17A).

Caspases

Although endogenous, full-length MEKK1 when assayed *in vitro* displays high basal activity, cleavage of the amino-terminal regulatory domain results in a modest but significant increase in activity, suggesting that the MEKK1 amino terminus exerts a negative regulatory effect on activity in the resting state that is relieved by upstream stimuli (206). There is evidence that MEKK1 is regulated by controlled proteolysis of the MEKK1 polypeptide at Asp871 and Asp874 (225). Many cells undergo apoptosis upon dissociation from the substratum, a process referred to as anoikis. The apoptosis characteristic of anoikis was recently shown to be regulated in part by MEKK1. In turn, the activation of MEKK1 during anoikis involves the proteolytic cleavage of MEKK1 to remove the inhibitory N-terminal domain; a reaction catalysed by caspases (225).

Caspases are a family of at least 11 related cysteine proteases that are vital to the regulation of apoptosis and inflammation. Caspases cleave at Asp residues within the consensus sequence Asp–X–X–Asp (reviewed in ref. 323). Caspases are divided into at least two, and possibly three subfamilies based on overall sequence homology. Although caspases probably have multiple roles in inflammation and apoptosis, evidence suggests that the interleukin-1 converting enzyme (ICE, caspase-1) subfamily of caspases (caspases-1, -4 and -5) plays more of a role in inflammation, while caspases homologous to *C. elegans* CED-3 (caspases-2, -3, -6, -7, -8, -9, -10, and -11) are preferentially involved in triggering apoptosis (323). Caspases-3 and -7 can be inhibited with the fluromethyl ketone-derivatized, synthetic peptide DEVD-fmk. Caspase-7 is also inhibited by the cowpox viral protein CrmA (323). A role for caspase-7 and, possibly, caspase-3 in MEKK1 cleavage during anoikis was inferred from studies showing inhibition of MEKK1 proapoptotic activity by the inhibitor peptide DEVD and CrmA (225). In addition, mutant MEKK1 polypeptides wherein the consensus caspase cleavage sites were mutated (Asp871/Asp874 → Glu871/Glu874) were resistant to activation during anoikis and conferred resistance to apoptosis. Finally, caspases-3 and -7 can cleave MEKK1 at Asp871/Asp874 *in vitro* (225).

Whereas caspase cleavage of MEKK1 may be necessary for the activation of MEKK1's proapoptotic function, evidence suggests that the activation of SAPK by MEKK1 does not require prior MEKK1 cleavage; and MEKK1 induction of apoptosis is SAPK-independent. Thus, expression of caspase-resistant MEKK1 still results in

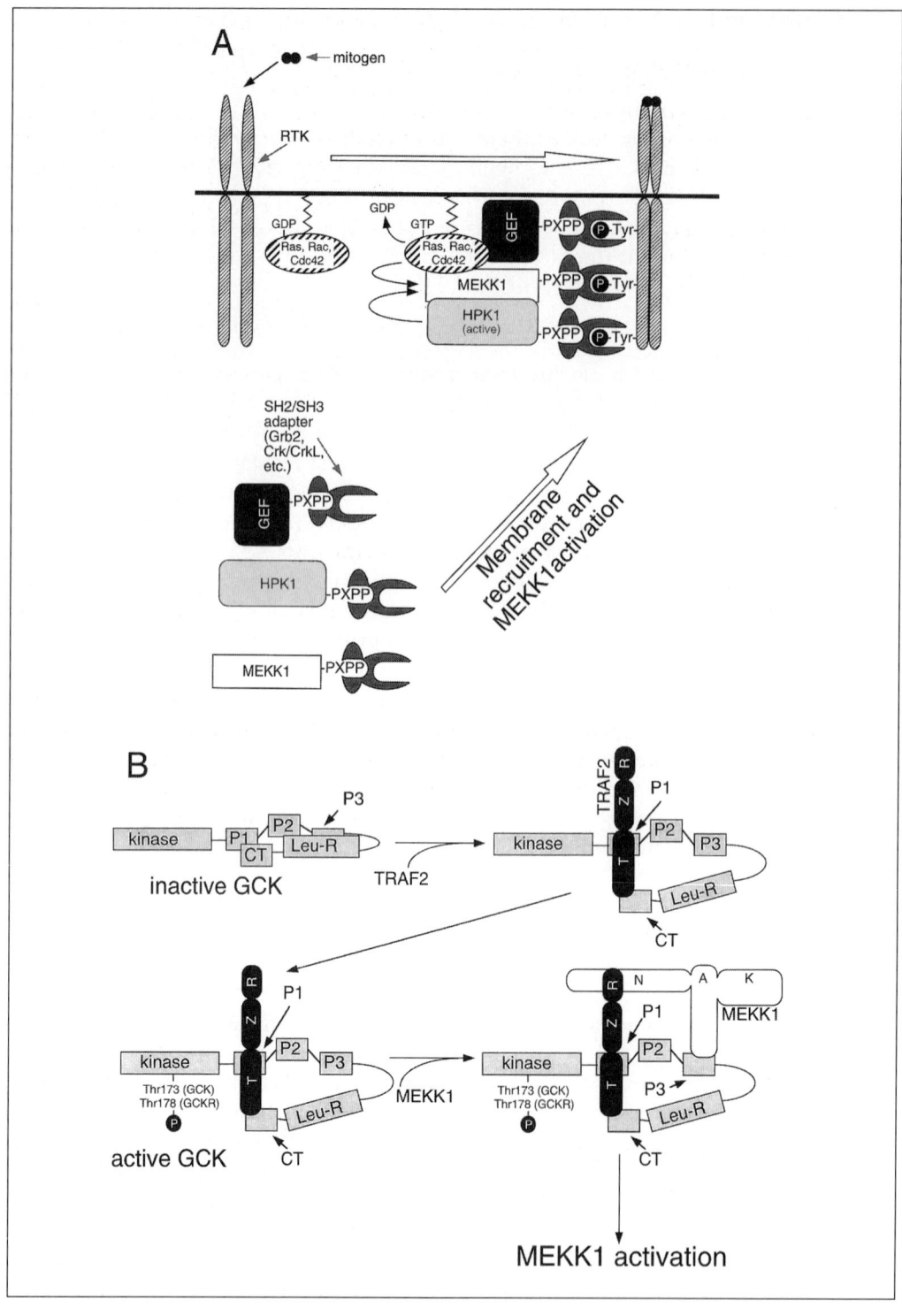

Fig. 17 Models for the complex regulation of stress-activated MAP3Ks by coordinated recruitment of upstream activators. MEKK1 is the example shown. (A) Regulation by Tyr kinase. A Rho or Ras–GEF, an upstream GCK (HPK1), and MEKK1 itself are constitutively associated with the SH3 domains of SH3/SH3 adapters (Grb2 for MEKK1 and HPK1, or Crk for HPK1 are examples) These complexes bind to phosphotyrosines on activated Tyr kinases via the SH2 domains of associated adapters. As a result MEKK1 is brought into close contact with HPK1 and with a Ras superfamily GTPase activated by the translocated GEF. The result is MEKK1 activation. Additional factors such as MAP3K aggregation or 14–3–3 binding may be important, but are not shown. (B) Model for coordinate activation of MEKK1 by TRAF2 and a group-1 GCK (GCK itself in this example). Activated TRAF2, at the receptor complex binds (via the TRAF domain of TRAF2 and PEST1 and the CT domains of GCK) and activates GCK (perhaps by fostering phosphorylation of Thr173 in the activation loop), GCK activation probably requires input from the TRAF2 RING domain. MEKK1 binds TRAF2 (via the amino-terminal domain of MEKK1 and the RING domain of TRAF2) and GCK (via the acid-rich domain of MEKK1 and the TRAF domain of TRAF2). Activated GCK and TRAF2 contribute to MEKK1 activation by mechanisms that are not understood. Additional factors such as MAP3K aggregation or 14–3–3 binding may be important, but are not shown. The fact that GCK activation (and, by extension, GCK activation of MEKK1) and MEKK1 binding requires the TRAF2 RING domain is consistent with the idea that the TRAF2 RING domain is indispensable for MEKK1 activation. P, PEST domain; Leu-R, leucine-rich domain; T, TRAF domain; Z, Zn-finger domain; R, RING domain; N, amino-terminal TRAF binding domain; A, acid-rich domain; K, kinase domain.

---

robust SAPK activation (328). Moreover, activation of the MAP3K activity of MEKK1 occurs before caspase cleavage (328). In addition, kinase-dead SEK1 does not reverse apoptosis incurred upon expression of the free MEKK1 catalytic domain (329). Caspase cleavage of MEKK1 results in a MEKK1 fragment that is free to migrate within the cell. Apparently, this diffusability is key to MEKK1's proapoptotic activity. The CAAX box of Ras (Section 2.5) was used to tether to the plasma membrane a recombinant MEKK1 fragment corresponding to the fragment generated upon caspase cleavage. This fragment was able to strongly activate SAPK but failed to induce apoptosis (328). Although ectopically overexpressed MEKK1 fragments can induce apoptosis, recent genetic studies call into question the significance of MEKK1 in promoting apoptosis: *mekk1–/–* ES cells have been generated. Surprisingly, these cells are more sensitive to proapoptotic environmental stresses, suggesting a protective, rather than a proapoptotic role for MEKK1 (330).

GCKs

Several group-I GCKs (GCK itself, HPK1 and NckIK) can interact with MEKK1 *in vivo* and *in vitro*; and there is evidence that MEKK1 is an effector for these GCKs (18, 287, 290). In all cases, the binding of MEKK1 requires the CTD of the GCK family kinase polypeptide. While the binding of MEKK1 to the HPK1 CTD has not been mapped in detail, the binding of MEKK1 to GCK and NckIK has been well characterized.

NckIK binds only full-length MEKK1. Deletion of the first 719 amino acids of MEKK1 abolishes NckIK binding (290). MEKK1 binding has been mapped to the carboxyl-terminal Leu-rich and CT domains (amino acids 948–1233) of NckIK and, accordingly, deletion of this domain prevents the binding of MEKK1 (290). A truncation construct of MEKK1 consisting solely of amino acids 1–719 will interact *in vivo* with amino acids 948–1233 of MEKK1 (290).

The GCK–MEKK1 interaction is somewhat more complex. GCK will interact with both endogenous or recombinant, full-length MEKK1 and with a MEKK1 construct consisting of amino acids 817–1493 (18). One GCK binding site on MEKK1 has been mapped to residues 817–1221 of the MEKK1 polypeptide. Deletion of the CT domain of GCK abolishes MEKK1[817–1221] binding, while subsequent deletion of the Leu-rich region restores binding. Deletion of the carboxyl-terminal PEST motif of GCK (amino acids 404–447) again abolishes binding (18).

The Leu-rich and CT motifs of group-1 GCKs are strikingly conserved. Why, then, does MEKK1 binding to NckIK and GCK appear to be so different? First, it is conceivable that the binding of MEKK1 to either GCK or NckIK involves multiple contact points both within and outside the Leu-rich segments. Although full-length MEKK1 interacts with both GCK and NckIK, the potential for binding of amino acids 1–719 of MEKK1 to the GCK Leu-rich domain was not tested (18). Second, NckIK is the most divergent of the group-1 GCKs. The Leu-rich domain of NckIK is considerably shorter than those of GCK, GCKR, GLK, and HPK1 (12, 284–290). Most notably, GCK, GCKR, GLK, and HPK1 possess an additional, highly conserved stretch of approx. 60 amino acids at the amino-terminal end of their Leu–rich motifs (284–289). This extension may perform an inhibitory role that is relieved upon the binding of GCK upstream activators to the CT region. Accordingly, deletion of the CT region, by abolishing the possibility of disinhibition of GCK, would prevent MEKK1 binding. In this regard, it is noteworthy that full-length, wild-type GCK binds MEKK1 less stably than does the free GCK CTD (18); and kinase-inactive GCK binds MEKK1 even more poorly than does wild-type GCK. Conversely, wild-type GCK is the best *in vivo* activator of the SAPK pathway (18). Activation of GCK's kinase activity could both permit MEKK1 binding and foster more efficient MEKK1 activation and turnover (12, 18).

It is possible that MEKK1 is an effector for GCK, NckIK, and HPK1 inasmuch as dominant inhibitory forms of MEKK1 can block GCK, NckIK, and HPK1 activation of SAPK—in particular, a fragment of MEKK1 that binds GCK but not SEK1 will inhibit GCK signalling without affecting the kinase activity of GCK (12, 18, 289, 290), suggesting that interfering with the binding of endogenous MEKK1 to GCK inhibits GCK signalling. Exactly how GCK family kinases might regulate MEKK1 is still unclear. GCK can phosphorylate MEKK1 *in vitro*, moreover the binding of MEKK1 to GCK may allow for the regulated targeting of MEKK1 to site(s) of activation. Alternatively, GCK's interaction with MEKK1 may foster MEKK1 by an oligomerization-dependent mechanism.

TRAFs

MEKK1 can also interact *in vivo* with TRAF2 and -6; and the interaction with TRAF2 is stimulated by TNF and requires the TRAF2 RING effector domain (Section 3.12.1) (322). It appears that oligomerization of TRAFs is sufficient to trigger MEKK1 binding. Fusion proteins were prepared comprising the RING- and Zn-finger loops of TRAF2 and -6 linked to the FK506 binding domain of FKBP12 (Section 2.5). In a manner analogous to the activation of ERK by forced dimerization of FKBP12–Raf-1 (Section 2.5), these FKBP–TRAF constructs, when expressed in transfected cells, could stimulate SAPK activity and AP-1-driven gene expression upon incubation with FK1012 (322).

Exposure of FKBP12–TRAF2 expressing cells to FK1012 results in the formation of insoluble FKBP–TRAF2 aggregates easily collected by centrifugation. Upon co-expression with MEKK1, it was observed that the MEKK1 coprecipitated with the FKBP–TRAF2 in a drug-dependent manner. Expression of a truncated FKBP12–TRAF2 construct missing the amino-terminal RING domain abolished this coprecipitation. Transiently expressed full-length MEKK1 generally undergoes progressive proteolysis, and it was observed that only full-length MEKK1, and not MEKK1 fragments coprecipitated with FKBP12–TRAF2 in response to FK1012. These results suggested that full-length MEKK1 could bind selectively to oligomerized TRAF2, *in vivo* (322).

Crystallographic studies indicate that the TRAF domains of TRAF2 exist as a trimer when complexed with a single molecule of the intracellular TRAF binding region of TNFR2 (331, 332). This result suggests that TRAFs oligomerize when complexed with their upstream activators; and it has been suggested that TRAFs are activated to bind their effectors upon oligomerization. The FK1012-dependent FKBP12–TRAF2–MEKK1 interaction (322) is consistent with this idea.

Further support for the idea that MEKK1 is a TRAF2 effector comes from the observation that transiently expressed MEKK1 and TRAF2 can associate *in vivo* in a TNF-dependent manner. Again, this association requires the TRAF2 RING domain. Moreover, coexpression of TRAF2 and MEKK1 also results in significant MEKK1 activation (322). These results imply a comparatively simple, linear mechanism whereby TRAF2, through its RING effector domain, directly binds and activates MEKK1. However, as discussed in the next section, the situation is likely to be more complex.

3.13.3 GCKs may couple TRAFs to MAP3Ks including MEKK1

In vivo, four closely related mammalian group-I GCKs, GCK itself, GCKR, GLK, and NckIK are activated by TNF (285, 287, 291, 292). There is emerging evidence that GCK, GCKR, and, possibly, NckIK (Section 3.14.6) are effectors for TRAF polypeptides (18, 285, 299, 300). In particular, GCKR is activated by TNF and upon coexpression with TRAF2 (285). Moreover, antisense GCKR constructs will block TNF activation of SAPK in 293 cells and kinase-dead MEKK1 can block GCKR activation of SAPK (285); suggesting that, in spite of the observation that TRAF2 and MEKK1 can associate apparently independently of coexpressed GCK family kinases (322), GCKR, perhaps acting through MEKK1, is an important TNF and TRAF2 effector in the SAPK pathway.

In further support of the idea that GCKs relay signals from TRAFs to the SAPKs, as with MEKK1, both GCK and GCKR can bind TRAF2 *in vivo* (12, 18). The CT and leucine-rich motifs of GCK and GCKR CTDs are strikingly conserved; and the binding of either GCK or GCKR to TRAF2 requires the TRAF domains of TRAF2 and the CT motifs of both kinase CTDs (12, 18). In addition, deletion of the amino-terminal PEST motif of GCK (Fig. 15) hampers TRAF2 binding (18). Coexpression with TRAF2 activates GCKR; and deletion of the TRAF2 RING domain cripples TRAF2's ability to activate coexpressed GCKR (285). From these results, it is plausible to speculate that the TRAF2 TRAF domains bind GCK and GCKR while the RING domains mediate GCK/GCKR activation (Fig. 17B). This result is consistent with genetic studies implicating the *Drosophila* group-I GCK Misshapen as an effector for *Drosophila* TRAF in the dorsal closure SAPK pathway (Section 3.14.6) (299, 300).

Thus, given the observed interactions between GCKs and both MAP3Ks and TRAFs, it is possible that GCKs, like SEK1 (17) and Pbs2p (10), serve as regulated scaffolds, assembling specific signalling components, in this instance to recruit the SAPKs.

Biochemical and genetic evidence (Section 3.14.6) suggest that GCKs relay signals from TRAFs to MEKK1 (18, 285, 299, 300), while biochemical evidence also indicates that TRAF2 and MEKK1 interact apparently independently of coexpressed GCKs (322). How might these conflicting results be reconciled? It is possible, of course, that both mechanisms operate separately and in parallel. Alternatively, the TRAF2–MEKK1 interaction may be comparatively weak, and might be stabilized or potentiated by GCKs once the GCKs themselves are activated by TRAFs. The TRAF–MEKK and GCK–MEKK binding results are consistent with the notion that a complex consisting of a group-I GCK and MEKK1 could interact with the TRAF2 RING domain. Accordingly, another possibility is that both TRAFs and GCKs might contribute to MEKK1 activation, requiring the obligate formation of a TRAF–group-I GCK family–MEKK1 complex.

Figure 17B is a speculative model, based on available data (18, 285, 299, 300, 322) of how TRAF2 and GCKs could collaborate to recruit and elicit activation of MEKK1. In this model, TRAF2, activated at the TNFR complex, binds the GCK at two sites (in the CT and the region containing PEST1 for GCK itself). This binding triggers GCK activation which, in turn, enables binding of MEKK1 to the GCK (at the region including PEST3 for GCK). The GCK–MEKK1 interaction, in turn, fosters a stable interaction between the TRAF2 RING domain and the MEKK1 amino terminus. MEKK1 is then activated by this process. The fact that GCK activation (and, by extension, GCK activation of MEKK1) and MEKK1 binding requires the TRAF2 RING domain is consistent with the idea that the TRAF2 RING domain is indispensable for MEKK1 activation.

Just what constitutes GCK and MEKK1 activation is still nebulous. The results with FKBP–MEKK1 implicate oligomerization in MEKK1 activation; but this has yet to be demonstrated. GCK, GCKR, and HPK1 contain a conserved Thr residue (Thr173 for GCK, Thr178 for GCKR, and Thr175 for HPK1) which lies just upstream of the Ala–Pro–Glu conserved residues in the subdomain VIII activation loop (284, 285, 286, 288). Mutagenesis of Thr178 of GCKR to Ala renders GCKR completely inactive and this construct can dominantly inhibit TNF- and TRAF2-activation of coexpressed SAPK (285). This finding suggests that Thr178 of GCKR (and, by extension, the analogous residues in GCK and HPK1) might be sites that must be phosphorylated for activation. Inasmuch as TRAFs are oligomerized upon interaction with receptors (330, 331), the clustering of GCKs at oligomerized TRAFs may elicit GCK phosphorylation (autophosphorylation?) and activation, triggering MEKK1 binding, and oligomerization-dependent (322) activation.

3.13.4 ASK1 is a TNF effector recruited by TRAF2, and a FasL effector recruited by Daxx: the role of cellular redox in ASK1 regulation

The coupling of group-I GCKs to MAP3Ks and TRAFs provides an attractive model for the regulation of the SAPKs by TNFR family receptors; however, the GCKs, as well as MEKK1 and MLK3 (Sections 3.7.2, 3.7.3, 3.11, 3.13.2, and 3.13.7) are highly selective for the SAPKs (1, 12), and, therefore, these models do not account for TNFR family regulation of the p38 pathway. Several recent studies indicate that the group-I GCK–TRAF and MEKK1–TRAF mechanisms are not the only means by which TNFR family receptors recruit the SAPKs. These studies also provide a mechanism for TRAF recruitment of p38.

ASK1 is a MAP3K that can activate both the SAPKs (through activation of SEK1) and the p38s (through MKK3 and MKK6) (211). Moreover, the MAP3K activity of ASK1 itself is activated by TNF (211). ASK1 is also activated upon Fas engagement (238); and ASK1 is emerging as an additional effector for TNFR family signalling to the SAPKs and p38s.

Recruitment by TRAF2

Expression of TRAFs2, 5, or 6 results in vigorous SAPK and p38 activation, and TRAF2 is required for SAPK activation by TNF (Section 3.12.2) (303, 304, 309–311, 317, 318). Recombinant ASK1 can interact *in vivo* with recombinant TRAF2, -5, and -6 (240). Endogenous TRAF2 and ASK1 interact *in vivo* in a TNF-dependent manner. The interaction between ASK1 and TRAF2 requires the TRAF domains of TRAF2 and a carboxyl-terminal regulatory domain of the ASK1 polypeptide (amino acids 940–1375). Coexpression of ASK1 and TRAF2 also results in activation of ASK1, and kinase-inactive ASK1 can block TRAF2 activation of the SAPKs, supporting the notion that ASK1 is a TRAF2 effector (240). Inasmuch as ASK1 is a potent activator of p38, the TRAF2–ASK1 interaction might also mediate the activation of p38 by TNF.

As discussed in Section 3.12.2, RIP is a potent activator of both the SAPKs and the p38s; and the RIP intermediate domain is both necessary and sufficient for this activity (18, 309). RIP, through its kinase and intermediate domains, interacts with the amino-terminal TRAF domain of TRAF2 (312). Inasmuch as deletion of *rip* does not abrogate TNF activation of the SAPKs (316), and RIP–ΔID fails to block TRAF2 activation of the SAPKs (18), multiple mechanisms of SAPK activation, such as the GCK and GCKR mechanisms described above, may emanate from TRAF2. By contrast, RIP appears to be a dominant effector for TRAF2 activation of p38 insofar as RIP–ΔID can effectively inhibit p38 activation by coexpressed TRAF2 (18). Immunoprecipitates of the RIP intermediate domain contain significant amounts of an endogenous MAP3K activity that can activate MKK6 (and p38) in a coupled assay *in vitro* (18). Given the TNF-dependent association of TRAF2 and ASK1, as well as TRAF2 and RIP, it is possible that this RIP-associated MAP3K is, in fact, a complex of TRAF2 and ASK1.

Recruitment by Daxx

As noted above (Section 3.12.3), Daxx is a novel adapter protein that couples Fas to SAPK activation (324). In certain cell types, Daxx-induced apoptosis requires activation of the SAPKs (324). Chang *et al.* observed that Fas engagement could activate endogenous ASK1. These investigators also observed that coexpression of Daxx and ASK1 results in substantial activation of ASK1. Moreover, both endogenous and recombinant Daxx can associate directly *in vivo* and coimmunoprecipitate with ASK1, a reaction that is Fas-dependent and requires the amino-terminal 648 amino acids of ASK1. Kinase-inactive ASK1 effectively blocks both Daxx-induced apoptosis and SAPK activation (238). Thus, ASK1 is a putative downstream target of Daxx that couples Fas to the SAPKs.

A critical caveat of these findings comes from recent examinations of the subcellular localization of Daxx. As was mentioned above (Section 3.12.3), Daxx has been isolated in yeast two-hybrid screens as an interactor with nuclear proteins such as DNA methyltransferase (326). Moreover, in apparent contrast to the ASK1 coimmunoprecipitation results described above (238), anti-Daxx antisera were used in cell fractionation studies to demonstrate a predominantly nuclear localization for Daxx (326).

Inhibition of ASK1 by thioredoxin

The exact molecular mechanisms by which TNF and Fas, through TRAF2 and Daxx, respectively, actually activate ASK1 have not been determined. Coimmuno-precipitation experiments indicate that GCK and ASK1 do not reliably interact *in vivo*; moreover, GCK is a selective SAPK activator, whereas ASK1 can activate both the SAPKs and p38s (211, 292). Thus, it appears unlikely that TRAF2 (or Daxx) activation of ASK1 proceeds through GCK. Yeast two-hybrid screening has revealed that the redox-sensing enzyme thioredoxin (Trx) is an endogenous inhibitor of ASK1 (239). This inhibition requires that Trx be in a reduced state. Thus, treatment of cells with oxidant stresses (H_2O_2) triggers the dissociation of Trx from ASK1 and the activation of ASK1 *in vivo* (239). TNF treatment is known to generate a pulse of reactive oxygen intermediates (ROIs) with slow kinetics (20 min–1 hour) that parallel the comparatively slow activation of ASK1 by TNF (333, 334). Furthermore, this ROI pulse may itself be TRAF2-dependent inasmuch as activation of the SAPKs by coexpressed TRAF2 is partially reversed with free radical scavengers (310). TNF fosters the dissociation of ASK1 from Trx by a process that can be blocked with free radical scavengers (239). Thus, TNF- (TRAF-) and, possibly, Fas-induced ROI formation could trigger the release of Trx from ASK1 and ASK1 activation (Fig. 18). Although TRAF2 is clearly necessary for TNF activation of ASK1, whether or not Trx dissociation from ASK1 precedes or is followed by TRAF2/Daxx binding remains to be determined.

Activation of ASK1 by dimerization

A likely consequence of Trx dissociation from ASK1 is dimerization-dependent ASK1 activation. Upon overexpression, ASK1 spontaneously dimerizes *in vivo*, and TNF promotes the dimerization of endogenous ASK1 by a mechanism that requires ROIs and which can be reversed with free radical scavengers (334). Moreover, expressed fusion proteins of ASK1 and DNA gyrase can be forced to dimerize *in vivo* upon

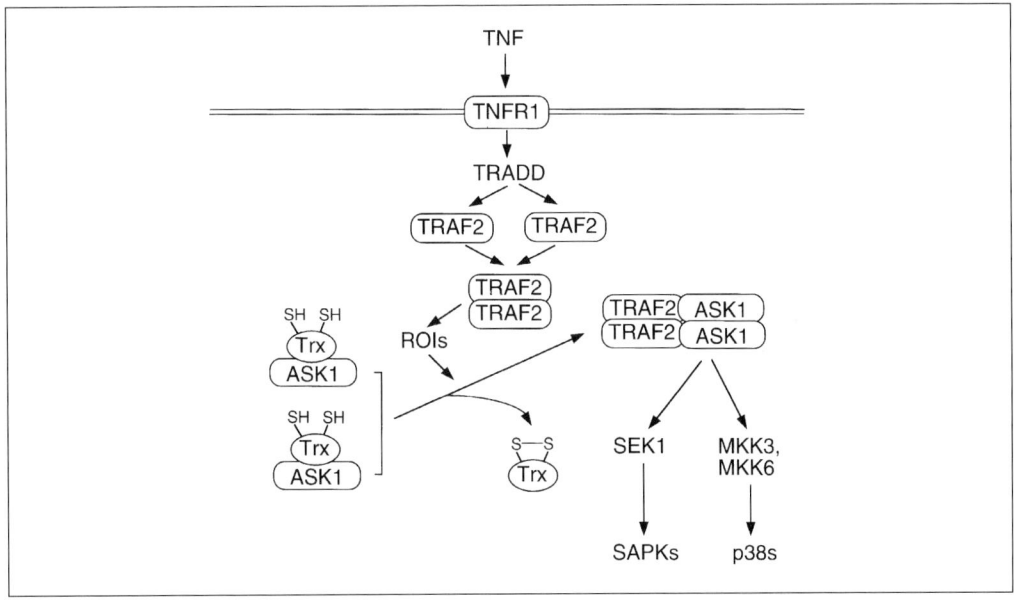

Fig. 18 Regulation of ASK1 by TNF. TRAF2-dependent ROI production leads to the dissociation of Trx from ASK1. TRAF2 also binds ASK1. Both TRAF2 binding and Trx dissociation may contribute to ASK1 oligomerization-dependent activation.

administration of the binary DNA gyrase-binding drug coumermycin. This coumermycin-induced dimerization results in substantial activation of coexpressed SAPK (334). TRAF2 is known to homodimerize *in vivo* (312, 331, 332); and subsequent to Trx dissociation from ASK1, one function of TRAF2 homodimerization could be to foster the dimerization/activation of associated ASK1 (Fig. 18).

3.13.5 TAK1 is a target for TGFβ and IL-1 through its association with TAB1

TAK1 was cloned as part of a screen to isolate novel mammalian MAP3Ks that could rescue *S. cerevisiae* deficient in *STE11*, a MAP3K of the mating-factor pathway (213) (Section 3.7.2). TAK1 can recruit both the SAPKs (via activation of SEK1) and the p38s (via activation of MKK3 and MKK6). Substantial evidence suggests that TAK1 couples the SAPKs to the transforming growth factor-β (TGFβ) receptors (213, 237).

TGFβ is a cytokine that arrests some cell types (Mv1-Lu mink lung-epithelial cells are an example) at G_1/S. In other cell types such as fibroblasts, TGFβ is growth-promoting, primarily due to the transcriptional induction of polypeptide mitogens. The TGFβ superfamily (TGFβ, the activins, bone morphogenic proteins (BMPs) and others) controls a broad array of morphogenic and developmental functions. TGFs exert their effects through a parallel family of receptor Ser/Thr kinases that can be subdivided into two classes—type-I and -II. Type-II receptors exist as dimers in resting cells and manifest constitutively active Ser/Thr kinase activity. Type-II receptor dimers bind TGF family ligands with high affinity, but are incompetent to signal. Type-I receptors are also dimerized in the resting cell, but are inactive in the absence of ligand and bind ligand with very low affinity (335). Ligand binding to type-II receptors triggers the formation of a heterotetramer, consisting of type-I and -II receptor dimers. Within the complex, type-I receptors are phosphorylated and activated by the type-II receptors. Activated type-I TGFβ family receptors then recruit, phosphorylate, and activate several types of signal transducing adapter/ transcriptional regulatory proteins including members of the similar to mothers against decapentaplegic (SMAD) family (335, 336).

TGFβ and BMP2 and -4 can activate the SAPKs in some cell lines. Furthermore, the MAP3K activity of TAK1 itself is activated by TGFβ and BMP4 in murine osteoblastic cells (MC3T3-E1); and ectopic expression of TAK1 can *trans* activate reporter genes driven by the TGFβ-sensitive plasminogen activator inhibitor-1 (PAI1) promoter (213, 237, 337). In resting cells, recombinant, ectopically expressed TAK1 is minimally active; however, deletion of the amino-terminal 22 amino acids from the TAK1 polypeptide results in constitutive activation. This suggests that the small amino-terminal extension of TAK1 plays an inhibitory role (213).

TAK1 binding protein-1 (TAB1) was cloned in a two-hybrid screen for interactors with the amino-terminal 22 amino acids of TAK1 (236). TAB1 is a 55-kDa polypeptide with no obvious structural features that indicate biochemical properties. Coexpression of full-length TAB1 with TAK1 results in TAK1 activation, assayed using the *S. cerevisiae STE11* rescue assay; or in transfected cells using either the PAI1 reporter system or p38 activation as a readout (236). TAB1 interacts with the amino terminus of TAK1—most notably, the amino-terminal 22-amino acid inhibitory domain is essential for TAB1 binding (236). Recently, a yeast two-hybrid screen to identify novel TAB1 interactors identified the human X chromosome-linked inhibitor of apoptosis (XIAP) as a TAB1 binding protein (337). XIAP also interacts directly with the type-I receptors for BMP2 and -4. The physiological significance of the XIAP–TAB1–TAK1 complex was demonstrated when it was observed that XIAP expression in *Xenopus* embryos could mimic the ventralizing effects of BMP, while expression of the TAB1 binding region of XIAP (see below) could block the expression of ventral mesoderm markers elicited by a constitutively active mutant type-I BMP receptor (337).

XIAP is a member of an emerging family of polypeptides, the inhibitors of apoptosis (IAPs); these were originally identified as baculovirally encoded proteins, and subsequently shown to exist in mammalian cells where they function as inhibitors of proapoptotic signalling from TNFR family receptors. All IAPs contain three baculoviral IAP repeat (BIR) motifs followed by a carboxyl-terminal RING domain. The amino-terminal-most of the three XIAP BIR motifs is required for binding TAB1, while the RING domain is required for interacting with type-I BMP receptors (337) (Fig. 19). Given that different domains of XIAP interact with TAB1 and BMPR type-I, it is not surprising that trimeric complexes of BMPR-I, XIAP, and TAB1 can be isolated (337) (Fig. 19). Other members of the IAP family can be recruited to TNFR complexes. Thus, cIAP1 and cIAP2 interact with TRADD and TNFR1 (303). The interaction between XIAP and TAB1 raises the possibility that IAPs may serve as adapters that recruit MAP3Ks to receptors of the TNFR family.

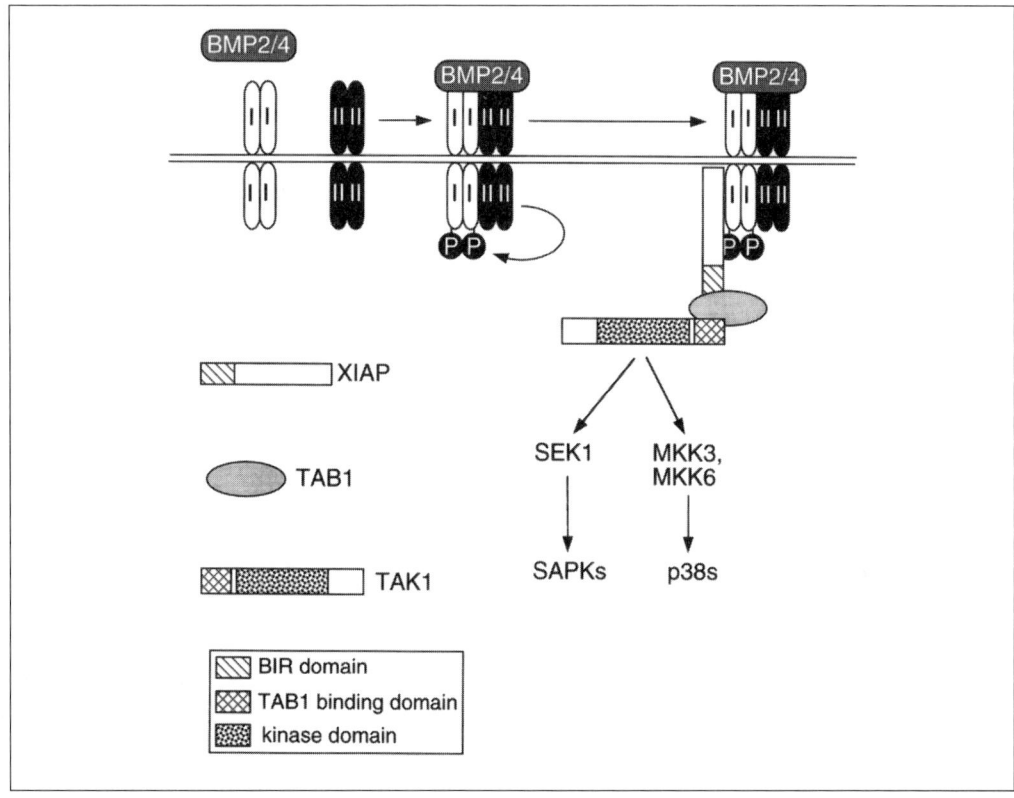

Fig. 19 Regulation of TAB1–TAK1 by TGFβ family receptors (BMP2 or BMP4). Type-II receptors, upon ligand binding, activate type-I receptors which, once activated, bind XIAP. XIAP, in turn, can recruit TAB1–TAK1, triggering TAK1 activation by mechanisms that are not yet understood.

In this regard, a recent study has also implicated TAK1 in IL-1 signalling to the SAPKs. Thus, both IL-1 and coexpressed TRAF6 can activate TAK1. In the latter instance, TAB1 is required for optimal activation. Moreover, coexpressed TRAF6 and TAK1 can coimmunoprecipitate in a reaction that requires TAB1 (235). The mechanism by which TRAF6 activates TAK1/TAB1 is unknown.

3.13.6 MEKK4 may be an effector for Rho family GTPases and is activated by DNA damage through a direct interaction with GADD45 homologues

Rho GTPases

MEKK4 is a MAP3K that can activate both the SAPK pathway (through activation of SEK1) and the p38 pathway (through MKK3 and MKK6) (209, 210) (Section 3.7.2). The MEKK4 polypeptide possesses a putative CRIB motif in its amino-terminal regulatory domain (amino acids 1311–1324) (Fig 11A), and there is some evidence that MEKK4 may be a target for Rho family GTPases (209, 210, 227). MEKK4 can bind Rac1 and Cdc42Hs *in vitro* and *in vivo*. The interaction with Rac1 appears to be GTP-independent, while the interaction with Cdc42Hs in more complex (209, 227). The binding of Cdc42Hs to a MEKK4 mutant consisting of the CRIB motif and the catalytic domain is GTP-dependent, while that to the full-length MEKK4 polypeptide is GTP-independent (209, 227). The significance of this discrepancy is unclear. Kinase-inactive MEKK4 can inhibit activation of SAPK by constitutively active Rac1 and Cdc42Hs (227). As with MEKK1 (227), there are no data indicating whether or not MEKK4 is actually regulated by Rac1 or Cdc42Hs, and, in the absence of these data, results using kinase-inactive MEKK4 need to be interpreted cautiously as these dominant inhibitors may merely sequester downstream effectors shared by MEKK4 and other MAP3Ks.

Ionizing radiation and other genotoxic stresses

Genotoxic stress requires rapid arrest of the cell cycle, and either the initiation of DNA and general cellular repair, or commitment to apoptosis, in order to maintain genomic integrity. Agents that cause DNA damage (UV- and γ-radiation, methyl methanesulfonate (MMS), cytosine arabinoside (AraC), N-acetoxy-2-acetylamino-fluorene, *cis*-platinum (CDDP), mitomycin-C, etc.) are often potent activators of the SAPKs, p38s, and other stress-activated signalling pathways (338–343). Past studies that have attempted to identify upstream components critical to genotoxic stress activation of the SAPKs have been controversial. Activation of the SAPKs by UV radiation is rapid (in this case most studies have employed UV-C, a component of the UV spectrum that does not penetrate the Earth's upper atmosphere). It is generally agreed that this activation involves UV-induced clustering and activation of cell-surface mitogen and cytokine receptors—a process that is probably driven by the formation of disulfide bridges between adjacent receptor proteins (295). It is unlikely that UV signalling to protein kinase cascades requires DNA damage *per se*; and, indeed, UV activation of NF-κB can proceed unaffected in enucleated cells (338).

By contrast, activation of the SAPKs by γ-radiation and chemical genotoxins proceeds slowly, and most investigators believe that recruitment of the SAPKs by these stimuli requires DNA damage as a triggering event (339–342). Thus, for example, the *ataxia telangiectasia mutated (ATM)* gene product, a polypeptide implicated in the cellular response to DNA damage, is necessary for SAPK activation in response to a number of radiant and chemical genotoxins (343–345). There is considerable disagreement, however, among investigators as to the degree of SAPK activation incurred by different genotoxins and the role played by the Abl Tyr kinase in genotoxin signalling. Thus, Kharbanda *et al.* see strong activation of the SAPKs by γ-radiation and chemical genotoxins (CDDP, mitomycin-C) which apparently requires c-Abl and does not occur in *abl–/–* mouse fibroblasts. These investigators were able to restore genotoxin activation of the SAPKs into *abl–/–* cells upon reintroduction of c-Abl cDNA (339, 341). By contrast, Liu *et al.* observe, at best, modest SAPK activation by γ-radiation and CDDP and substantial activation by mitomycin-C; however, activation of the SAPKs by these stimuli was unaffected in fibroblasts wherein c-*abl* was disrupted (340). The reason for this discrepancy is unclear.

However, a series of recent studies from Saito and colleagues provide important clues as to how genotoxins recruit the SAPKs. These findings clearly indicate that MEKK4 is recruited strongly by genotoxic stresses that transcriptionally induce proteins of the growth arrest and DNA damage-45 (GADD45) family. GADD genes are a set of transcripts that are rapidly induced by DNA damage and are thought to play a role in coordinating the cellular response to genotoxins (345). *GADD45* was one of the first GADD genes to be identified; although its function has remained obscure. Expression of GADD45 is the culmination of a signalling pathway that requires prior expression of the tumour suppressor protein p53 which *trans* activates the *gadd45* (*gadd45α*) gene. p53 expression is, in turn, driven by the activity of the *ATM* gene product (346). This is particularly noteworthy given the observation, cited above, that cell lines derived from patients with ataxia telangiectasia having null or inactivating *ATM* mutations are resistant to SAPK activation by genotoxins but not by TNF (344).

Saito and colleagues performed a yeast two-hybrid screen using MEKK4 as bait and identified three *gadd45* homologues (GADD45α, which is the original GADD45, plus GADD45β and -γ) as direct MEKK4 interactors (233). A broad segment in the middle of the GADD45 polypeptides (amino acids 24–147 of GADD45γ) binds to a small amino-terminal segment (amino acids 147–250) of the MEKK4 polypeptide (233) (Figs 11A and 20).

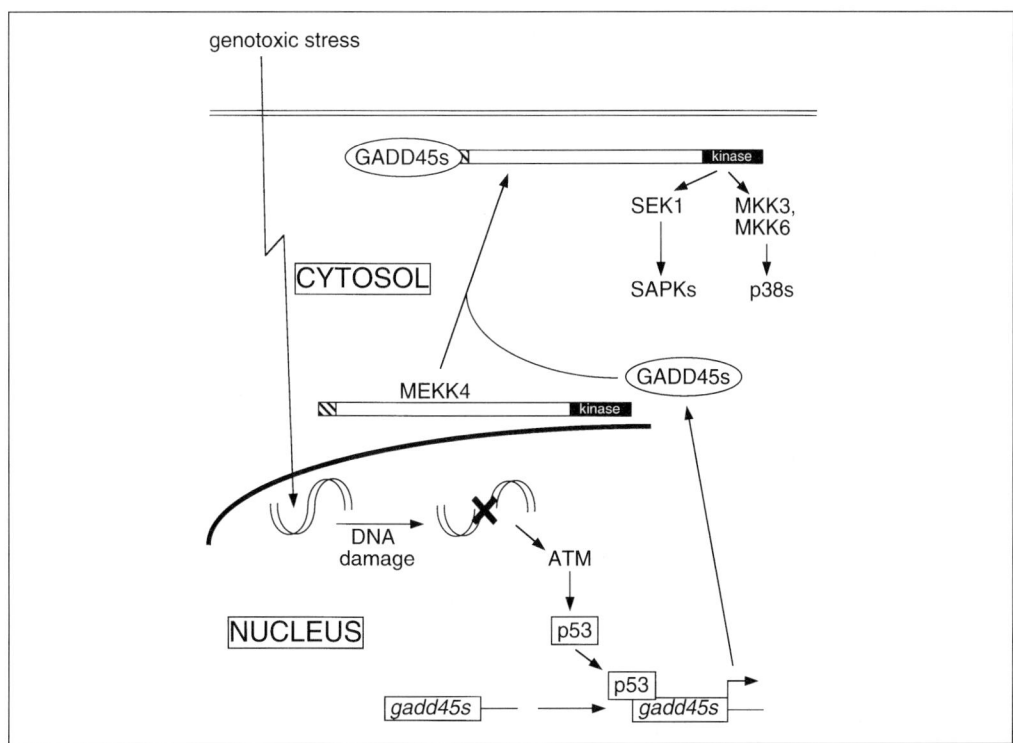

Fig. 20 MEKK4 activation by the GADD45 DNA damage-inducible protein family. DNA damage stimulates expression of GADD45s by a process that requires ATM and p53. GADD45s then bind and activate MEKK4.

GADD45 proteins are key to the regulation of MEKK4 signalling to the SAPKs and p38s in response to genotoxic stresses. Thus, all three *gadd45* genes are transcribed in response to the same genotoxic stimuli shown to upregulate *gadd45α* (233, 344). Moreover, ectopic expression of all three GADD45 polypeptides results in substantial activation of both the p38 and SAPK—a reaction that can be reversed upon co-expression with kinase-dead MEKK4 (233). Kinase-dead MEKK4 can also inhibit the activation of p38 by genotoxic stress that transcriptionally induce *gadd45* gene expression (233). In addition, expression of all three GADD45 polypeptides activates MEKK4 *in vivo* (233). Most importantly, however, addition of purified, recombinant GADD45 proteins to immunoprecipitated MEKK4 activates MEKK4 *in vitro*. Thus, the GADD45s represent a family of direct MEKK4 activators (233) (Fig. 20).

Ectopic expression of GADD45 proteins is proapoptotic in HeLa cells, a reaction that requires, in part, MEKK4 in so far as either kinase-dead MEKK4 or the GADD45 binding domain of MEKK4 (amino acids 147–250) significantly blocks apoptosis, while an in-frame deletion of the GADD45 binding domain of MEKK4 renders kinase-dead MEKK4 incapable of blocking apoptosis (233).

Taken together, these results implicate GADD45s as activators of the SAPKs, p38s, and apoptosis via an MEKK4-dependent mechanism. The requirement for GADD45 transcription also explains the slow kinetics of SAPK and p38 activation by many genotoxins (339–342). The fact that SAPK activation by genotoxins is ablated in *ATM* null or mutant cells (343) can also be explained by the observation that GADD45 expression (and, hence, subsequent activation of the SAPKs) is also blunted in these cells (345).

3.13.7 Regulation of MLK3 by group-I GCKs and Rho family GTPases: the role of dimerization through the Leu zippers

Regulation of MLK3 by group-I GCKs (HPK1 and GCK)

There is evidence that MLK3 is an effector for at least one mammalian group-1 GCKs: HPK1 (218, 288) (Section 3.11). Thus, not only can HPK1 associate *in vivo* with MEKK1, but it can bind MLK3 *in vivo* (288, 289). HPK1 binds MLK3 in an interaction that requires the amino-terminal SH3 domain of MLK3 (amino acids 1–93) and the carboxyl-terminal two of the four SH3 binding domains of HPK1 (amino acids 433–442 and 468–474) (218, 288). By contrast, the amino-terminal SH3 binding domains of HPK1 interact with Grb2 (Section 3.11) (296, 297). Expression of kinase-dead MLK3 abrogates activation of SAPK by coexpressed HPK1, suggesting that MLK3 is an HPK1 effector (288). In so far as expression of kinase-dead MLK3 abrogates the activation of SAPK by coexpressed GCK (288), MLK3 may also function as a GCK effector.

Regulation of MLK3 by Cdc42Hs

While the role of Rho family GTPases in the regulation of MEKK1 and MEKK4 is still nebulous (Section 3.8.2), MLK3 and MLK2, two mixed lineage kinase activators of the SAPKs, contain CRIB motifs and can interact directly with Cdc42Hs and Rac1 in a

GTP-dependent manner; and there is good evidence that at least MLK3 is an effector for Cdc42Hs (253, 265, 347). On the other hand, while MLK3 and MLK2 may be targets of Cdc42Hs, it is unlikely that they are Rac targets (265, 266). Phe37Ala–Rac1 is an effector loop mutant of Rac1 that retains its ability to recruit the SAPK pathway *in vivo* (252). However, neither MLK3 nor MLK2 can interact with Phe37Ala–Rac1 *in vivo* either in COS1 cells or in a yeast two-hybrid assay (265, 266). As discussed above (Section 3.8.2), Phe37Ala–Rac1 can bind POSH, and POSH is a likely candidate effector coupling Rac1 to the SAPKs (266).

MLK3 possesses a leucine zipper (amino acids 400–487) (243) (Fig. 11B). Leucine zippers frequently mediate protein–protein binding; for example the bZIP transcription factors in AP-1 dimerize via leucine zipper interactions. Lassam and colleagues have shown that MLK3, upon overexpression, will spontaneously dimerize *in vivo* (346). The leucine zipper domain of MLK3 mediates homodimerization, and deletion of this domain abrogates *in vivo* MLK3 homodimerization (346).

The MLK3 leucine zipper is also critical for signal transduction. Deletion of this domain, even in the presence of a fully competent kinase domain, completely inhibits the ability of MLK3 to recruit the SAPK pathway *in vivo* and converts the truncated construct into a dominant inhibitor of wild-type MLK3 signalling (346). Of particular importance, coexpression of MLK3 with Val12–Cdc42Hs substantially enhances the ability of MLK3 to homodimerize (346). From these results, it is plausible to speculate that GTP–Cdc42Hs binds and promotes MLK3 (or MLK2) homodimerization and activation.

Grb2, Crk, and CrkL can mediate the mitogen-stimulated recruitment of HPK1 to autophosphorylated Tyr kinase receptors at the plasma membrane (296, 297). Different regions of the HPK1 polypeptide mediate the interactions with SH2/SH3 adapters and MLK3 (296, 297); and one function of the recruitment of HPK1 to the membrane might be to position associated MLK3 in the vicinity of GTP–Cdc42, fostering MLK3 dimerization and activation. Alternatively, HPK1 (and GCK) may bind MLK3 and foster its dimerization-dependent activation independently of Cdc42Hs.

3.14 Biological functions of the SAPKs and p38s pathways

Studies of mammalian cell-culture and gene-disruption models, as well as the exploitation of lower eukaryotes that are amenable to genetic manipulation, have begun to reveal the biological functions of stress-regulated MAPK pathways. These results indicate a role for these pathways in apoptosis, progression to hypertrophy, cell-cycle arrest, and immune-cell development. Recent results also point to a critical role for the SAPKs in embryonic development.

3.14.1 The SAPKs can foster apoptosis in response to environmental stress: the role of transcriptional induction of FasL

Withdrawal of NGF from differentiated PC-12 cells or primary cultures of neurons results in apoptosis; and there is some evidence that this apoptotic programme requires the SAPKs and p38s (347, 348). Thus, nutrient withdrawal-induced apoptosis in PC-12 cells is coincident with the activation of the SAPKs and p38s, and a decrease in ERK activity. Overexpression of constitutively active MEKK1 in PC-12 cells both activates the SAPKs and promotes apoptosis even in the presence of NGF (347). Similarly, constitutively active forms of MKK3, when overexpressed with p38, can promote apoptosis in the presence of NGF. Reciprocally, and in apparent contrast to fibroblasts where SEK1-K129R fails to block MEKK1-induced apoptosis (329) (Section 3.13.2), non-phosphorylatable, dominant inhibitory mutants of c-Jun can prevent MEKK1-induced PC-12 cell apoptosis; and the expression of dominant interfering mutants of MKK3 can prevent apoptosis incurred by NGF withdrawal (347, 348). In so far as a constitutively active, oncogenic form of MEK1 can prevent apoptosis induced by NGF withdrawal, it is likely that the decision to initiate apoptosis in PC-12 cells depends, at least in part, on the balance of SAPK and p38 versus ERK activity (347, 348).

TR-4 cells are a thermotolerant subline of the murine fibroblast line RIF-1. Comparative studies of the cytocidal effects of heat-shock and genotoxic (cis-platinum) stress on these cells has implicated the SAPKs in the regulation of stress-induced cell death (349). Whereas TR-4 and RIF-1 cells express identical amounts of SAPK polypeptide, TR-4 cell SAPKs are not activated by heat shock, in spite of the ability of UV radiation to activate the SAPKs in both cell lines. Moreover, whereas heat shock and CDDP readily kill RIF-1 cells, these treatments have no cytocidal effects in TR-4 cells; RIF-1 and TR-4 cells, however, have comparable sensitivity to UV-C radiation (349). This finding is consistent with the observation that deletion of sek1 abolishes heat-shock activation of SAPK (189) (Section 3.6.1). Moreover, if the thermosensitive RIF-1 cells are stably transfected with a kinase-inactive mutant of SEK1 (SEK-Ser257Ala/Thr261Leu, SEK-AL) the activation of SAPK, but not p38, to all stimuli is inhibited. RIF-1 cells expressing SEK-AL acquire resistance to the cytocidal effects of UV-C as well as to heat shock and cis-platinum, suggesting that activation of the SAPK pathway is required for efficient cell death induced by these stress agonists (349).

Studies of heat-shock regulation of the SAPKs have revealed a novel mechanism of activation that appears to bypass the 'traditional' three-tiered core module. Although SEK1 is required for heat-shock activation of SAPK (189) (Section 3.6.1), SEK1 is not actually activated by heat shock. Instead heat shock suppresses a Tyr phosphatase that targets SAPK thereby allowing basal SEK1 Tyr kinase activity (SEK1 and SAPK activity is never entirely absent, even in the resting state) to slowly activate SAPK (350). A consequence of heat stress is the transcriptional induction of the chaperone protein Hsp70. Hsp70 expression reactivates this SAPK-targeted Tyr phosphatase resulting in SAPK inactivation (350). In this regard, it is noteworthy that TR-4 cells constitutively express high levels of Hsp70 which are not transcriptionally regulated by heat stress, accounting perhaps for the inability of heat shock to activate SAPK in these cells (349).

Clues as to the mechanism by which the SAPKs might promote apoptosis in response to genotoxins, high salt conditions, or nutritional withdrawal came from the results of Green and colleagues. Jurkat cells treated with proapoptotic genotoxins (etoposide, tenoposide, UV) rapidly undergo apoptosis. Apoptosis incurred by these agonists coincides with the transcriptional induction and cell-surface expression of FasL (351). FasL expression triggers an autocrine apoptotic programme mediated by the binding of newly expressed FasL to existing constitutively expressed Fas. The FasL promoter contains AP-1 and NF-κB *cis* acting elements, and inhibition of AP-1 activation (with a dominant-negative c-Jun construct, TAM67 wherein the transactivation δ domain is deleted, or with dominant inhibitory SAPK to block AP-1 activation) prevents FasL induction (350). Similarly, overexpression of a non-degradable IκB mutant prevents FasL induction. Taken together, these results indicate that genotoxic stresses promote apoptosis by recruiting the SAPKs and IKKs which, in turn, activate AP-1- and NF-κB-mediated induction of FasL (350). Induction of FasL is not restricted to genotoxic stresses. Nutrient withdrawal in cultured cerebellar granule neurons, or PC-12 cells, also results in induction of FasL by a SAPK-dependent mechanism (348, 350).

3.14.2 SAPKβ/JNK3 is required for kainate seizure-induced neuronal apoptosis

Kainic acid is a glutamate receptor agonist that rapidly induces seizures similar to those of epilepsy. In addition, kainate also causes widespread hippocampal neuronal apoptosis. Whereas SAPKα/JNK2 and SAPKγ/JNK1 are ubiquitously expressed, SAPKβ/JNK3 is selectively expressed in brain and heart. Flavell and colleagues disrupted the gene for SAPKβ/JNK3 in mice and observed that the knockout animals were strikingly resistant to kainate seizures (352). More importantly, however, deletion of *sapkβ/jnk3* protected these animals from kainate-induced neuronal apoptosis (352). Hippocampal cell death is a major clinical indicator in stroke and Alzheimer's disease. It will be important to determine if ischaemia or Alzheimer's-induced cell death correlates with SAPKβ/JNK3 activation. It will also be important to determine if hippocampal cell death induced by kainate involves upregulation of FasL.

3.14.3 The SAPKs and p38s are required for the progression of cardiomyocyte hypertrophy in response to pressure overload and vasoactive peptides

Pressure-overload cardiac hypertrophy (a consequence of hypertension) and cardiomyocyte apoptosis and necrosis arising from ischaemic injury are the leading causes of heart disease morbidity and mortality in the developed world (267). At the cellular level, hypertension subjects cardiomyocytes to mechanical shear stress. In addition, hypertension elicits the release of the vasoactive peptides endothelin-1 and angiotensin-II. Both shear stress and vasoactive peptides, in turn, cause cardiomyocyte hypertrophy, as manifested by cell enlargement as well as the formation of rigid cytoskeletal actomyosin fibrillar networks and the expression and release of atrial natriuretic factor (ANF). Cardiomyocyte contractility is compromised as a result of these changes, leading ultimately to heart failure (267).

Recent results from Olson and co-workers have implicated the Ca^{2+}/calmodulin-dependent phosphatase calcineurin (CaN/PP2B), signalling through NFAT3, in cardiac hypertrophy. Thus, transgenic expression of a Ca^{2+}-independent, constitutively active mutant of CaN/PP2B in mice significantly enhances cardiac hypertrophy and associated mortality in a response that can be reversed with the CaN/PP2B inhibitor FK-506 (353). However, the situation is likely to be much more complicated.

Chien and colleagues have observed that markers of cardiomyocyte hypertrophy (cell enlargement, actomyosin fibrils, and ANF expression) could be induced in cultured neonatal rat cardiomyocytes upon ectopic expression from recombinant adenoviruses of constitutively active forms of MKK6 and MKK7. By contrast, expression of active MKK3 mutants caused apoptosis (354, 355). MKK6 activates all forms of p38, whereas MKK3 preferentially recruits p38α and p38β. These results suggest that the SAPKs, p38γ, and p38δ are critical for promoting cardiomyocyte hypertrophy, while p38α and p38β are more important for eliciting cardiomyocyte apoptosis (355).

Force and colleagues extended these results to known physiological activators of cardiomyocyte hypertrophy. Thus, expression of kinase-dead Lys129Arg–SEK1 from recombinant adenoviruses can completely block cardiomyocyte hypertrophy (cell enlargement, actomyosin fibrils, and ANF expression) incurred by endothelin-1 in neonatal rat cardiomyocytes (268).

3.14.4 The p38s may act to promote cell-cycle arrest at G_1/S as part of a signalling pathway in late-G_1 mediated by Cdc42Hs

The Rho family GTPases RhoA, Rac1, and Cdc42Hs are important components in $G_0 \rightarrow G_1$ cell-cycle transition and in the oncogenesis arising from transforming alleles of *ras*. Thus, microinjection, in G_0, of active mutants of Rac1 or Cdc42Hs can drive quiescent Swiss 3T3 fibroblasts from G_0 into G_1; and continuous, ectopic expression of Cdc42Hs can transform NIH3T3 and Rat-1 fibroblasts. Moreover, inhibition of Cdc42Hs or Rac1 significantly reduces Ras-mediated transformation of Rat-1 fibroblasts (247, 356, 357).

However, Cdc42Hs, signalling through p38, in contrast to its role in quiescent cells, may have a role in cell-cycle inhibition—if the Cdc42Hs is selectively activated in the mid to late stages of G_1, in cells already committed to the cell cycle (358). Thus, microinjection of p38α, MKK3, or MKK6, but not SAPK, into NIH3T3 cells synchronized in early G_1 (with expression first detectable in mid-G_1) arrests cells at the G_1/S boundary. p38α-induced cell-cycle arrest is prevented by kinase-dead MKK3, implicating active p38 in this process (358). In addition, activation of p38 upon expression of an oestrogen-inducible, oestrogen-receptor MEKK3 fusion construct blocks cell proliferation incurred by activated Ras (230). Expression of Cdc42Hs, but not Rac1 or RhoA, in early G_1 also potently arrests cells at G_1/S in several cell lines (NIH3T3 fibroblasts, Mv1-Lu mink lung-epithelial cells, and MG63 osteosarcoma cells); and this G_1/S cell-cycle arrest can be reversed with kinase-dead MKK3, again implicating the p38s in Cdc42Hs-mediated cell-cycle arrest (358). Thus Cdc42Hs may serve as a transforming/growth-promoting protein in quiescent cells, or as a growth suppressor in cells already in the cell cycle. Inasmuch as p38 is a common target of Rac1 and Cdc42Hs, different Cdc42Hs targets may collaborate with p38 in G_1/S growth arrest versus growth promotion.

3.14.5   The SAPKs are key regulators of T-cell maturation, activation, and protection from FasL apoptosis

Two key gene-disruption studies indicate a complex role for the SAPKs in T-cell maturation and function. When stimulated with antigen by antigen-presenting cells (APCs), CD4+ T-helper cells (T_H cells) undergo clonal proliferation. In addition, they begin to produce IL-2. T_H cells so activated may then become T_{H1} or T_{H2} effector cells. T_{H1} cells mediate inflammatory responses while T_{H2} cells regulate humoral responses. Differentiation to T_{H1} or T_{H2} cells, and hence the type of immune response, are partly governed by the cytokine milieu to which the immature T_H cells are exposed. Thus, IL-12, which is a product of APCs, promotes differentiation to T_{H1} cells while IL-4, produced by mature T cells leads to T_{H2} cells.

APCs present antigen to the T-cell receptor (TCR) which recruits the TCR/CD28 costimulatory pathway. The T-cell receptor (TCR)/CD28 pathway mediates the initial stimulus that culminates in T_H cell maturation. While stimulation of either the TCR or CD28 alone is insufficient for SAPK activation, TCR/CD28 costimulation results in strong SAPK activation. AP-1, in turn, is necessary for cytokine production, including IL-2 and, possibly other cytokines involved in T-cell maturation (46, 359).

Flavell and colleagues deleted the gene for SAPKγ/JNK1 in mice (360). Immature T_H cells from these mice were refractory to costimulatory SAPK activation, in spite of the continued presence of SAPKα/JNK2. Thus, in the T_H stage, prior to differentiation into T_{H1}/T_{H2} cells, SAPKγ/JNK1 is the only Jun kinase recruited by costimulation. Costimulation of the *sapkγ/jnk1* knockout cells elicited indistinguishable levels of IL-2 production, suggesting that SAPK is not a rate-limiting pathway in TCR/CD28-stimulated IL-2 production (360). On the other hand, costimulation triggered enhanced proliferation in the knockout cells as compared to controls. TCR engagement did stimulate IFN-γ production, a hallmark of T_{H1} differentiation; however, CD28, which in combination with TCR recruits SAPK (359), failed to enhance IFN-γ production (360). By contrast, TCR engagement alone stimulated the production of exaggerated levels of the T_{H2} markers IL-4, IL-5, and IL-10. Thus, deletion of *sapkγ/jnk1* renders T_H cells hyperresponsive to TCR engagement in the absence of costimulation; and favours the production of T_{H2} marker cytokines. These results suggest that SAPKγ/JNK1 suppresses the differentiation of T_H lymphocytes into T_{H2} cytokine-producing cells (360).

SAPKα/JNK2 is apparently also required for peripheral T-cell activation. Mice were generated wherein *sapkα/jnk2* was deleted (361). The requirement for SAPKα/JNK2 differs from that for SAPKγ/JNK1. Whereas deletion of *sapkγ/jnk1* enhanced costimulatory proliferation and TCR-stimulated IL-4 production, while having no effect on TCR-stimulated IL-2 production; deletion of SAPKα/JNK2 reduced TCR-mediated proliferation and the production of IL-2, IL-4, and IFN-γ, markers of peripheral T-cell activation (360, 361). B-cell activation in the *sapkα/jnk2* knockout mouse was unimpaired. In addition, CD3-induced apoptosis of immature (CD4+ CD8+) T cells was reduced in *sapkα/jnk2–/–* animals. Thus, in apparent contrast to SAPKγ/JNK1, which suppresses T-cell differentiation, SAPKα/JNK2 is required for efficient peripheral T-cell activation (360, 361).

Early in T- and B-lymphocyte development, ablation of autoreactive cells is mediated, in part, by FasL-stimulated apoptosis. Penninger and colleagues deleted the gene for SEK1 in mice, and observed that this deletion abrogated SAPK activation by anisomycin and heat shock in cultured ES cells (189) (Section 3.6.1). However, deletion of *sek1* was early embryonic lethal. In order to circumvent this problem, and examine the role of SEK1 (and, hence, the SAPKs) in immune-cell function, *rag2*-deficient blastocysts were microinjected with *sek1–/–* ES cells to produce chimeric mice bearing a SEK1-deficient immune system against a wild-type background. Surprisingly, T lymphocytes isolated from the chimeras were strikingly hypersensitive to FasL-induced apoptosis, suggesting that SEK1 and its substrates actually protect T cells from FasL-induced apoptosis (189).

The lethality incurred upon deletion of *sek1* is probably due to impaired hepatogenesis. Thus, *sek1–/–* embryos die early in embryogenesis, subsequent to the establishment of the primitive vasculature, coincident with early hepatogenesis. Death is accompanied by excessive apoptotic death in the liver, again indicating an antiapoptotic role for SEK1 (199, 200).

3.14.6 *Drosophila* SAPK and the regulation of dorsal closure, planar polarity, and immunity

Recent studies of *Drosophila* SAPK signalling mechanisms implicates these pathways in the control of key developmental steps. Dissection of these mechanisms will provide invaluable clues as to the mechanisms and functions of mammalian SAPK and p38 pathways.

Dorsal closure

During *Drosophila* embryogenesis, establishment of a contiguous dorsal epidermis occurs after germ band retraction and organogenesis. The epidermal primordia of early *Drosophila* embryos is ventral, and the dorsal surface is covered by a primitive tissue, the aminoserosa. Dorsal closure involves a concerted ventrodorsal cell sheet movement that positions and eventually fuses the lateral epidermal primordia over the aminoserosa (reviewed in ref. 362). Fusion of the dorsal epidermis is followed by degeneration of the aminoserosa. Several key components of a *Drosophila* SAPK pathway are required for dorsal closure, including the products of *basket* (*bsk*, also called DJNK, a *Drosophila* SAPK), *hemipterous* (*hep*, a *Drosophila* orthologue of MKK7), *Djun* (*Drosophila* c-Jun), and *Dfos* (*Drosophila* c-Fos). Biochemical and genetic evidence indicate that Hep, Bsk, and DJun are directly linked in a signalling pathway. Thus, epistasis studies indicate that a constitutively active mutant of DJun, wherein the phosphoacceptor sites are mutated to Asp, can rescue the dorsal closure defects caused by ablation of *hep* or *bsk* function (192, 362). DRac1, DCdc42, and DPAK are also important to dorsal closure; and there is evidence that DRac functions upstream of Bsk and DJun inasmuch as the active DJun mutant can also rescue the dorsal closure defects incurred by mutations in *Drac1* (362). The ultimate consequence of recruitment of DJun is expression of *decapentaplegic* (*dpp*), a member of the TGFβ family that is similar to BMP4. This *dpp* is expressed in leading-edge epidermal cells during dorsal closure and deletion of *dpp* leads to dorsal closure defects. Expression of *dpp* in the primordial epidermis is abolished in *hep*, *bsk*, or *Djun* mutant backgrounds, clearly implicating *dpp* as a target for Bsk, Hep, and DJun (362).

Recent studies from Skolnik and colleagues implicate *misshapen* (*msn*) in dorsal closure signalling and suggest that Msn acts in parallel with dRac to recruit Bsk. *msn* is a *Drosophila* orthologue of NckIK (290, 299). Deletion or mutational inactivation of *msn* results in dorsal closure defects that are strikingly similar to the *bsk* phenotype (299, 362). Consistent with the idea that Msn signals to Bsk, a significant percentage of either doubly heterozygous *msn–/+*, *bskt–/+*, or *msn–/+*, *hep–/+* flies exhibit a dorsal-open phenotype—with the severity of the phenotype correlating well with the strength of the *bsk* or *hep* allele (299). Moreover, constitutively active Djun rescues not only the *bsk* phenotype, but the *msn* phenotype as well (299). Biochemically, expression of *msn* in mammalian cells activates coexpressed SAPK (299). Of particular interest, Skolnik *et al.* recently identified a *Drosophila* TRAF (*Dtraf*) that interacts with Msn in a yeast two-hybrid screen. DTRAF also activates coexpressed SAPK in mammalian cells; and kinase-dead Msn can block DTRAF activation of SAPK in mammalian cells (300). DTRAF binds to the Msn CTD at a region of the Msn polypeptide (amino acids 293–587) that corresponds to the PEST1 sequence of GCK (amino acids 293–313), a domain of GCK which, in addition to the GCK-CT extension, is required for TRAF2 binding (12, 18, 300) (Section 3.13.3). Although inactivating *dtraf* mutants exhibit a dorsal-open phenotype, genetic evidence placing DTRAF upstream of Msn is not yet available (300). However, these results suggest that a signalling pathway strikingly similar to the TRAF2 → GCK/GCKR → (MEKK1) → SAPK pathway (Sections 3.13.2, 3.13.3) exists in *Drosophila* to regulate, in part, dorsal closure (12, 18, 285, 299, 300). Moreover, given the close homology between NckIK and Msn, plus the observation that NckIK is activated by TNF (291), it is plausible to speculate that NckIK is also an effector for mammalian TRAFs.

Planar polarity

Planar polarity phenotypes in *Drosophila* are characterized by the misorientation of primordial epithelial cells within the wings, eyes, and legs. Thus, for example, in the eye, planar polarity defects are manifested in a disruption of the mirror symmetry of ommatidial units, with individual ommatidia adopting random chirality (reviewed in refs 363, 364).

The product of *dishevelled* (*dsh*) is required for two developmental steps: establishment of planar polarity and establishment of patterning programmes that give rise to proper embryonic segmentation and imaginal disc development during wing formation (*wingless* phenotype) (280, 363, 364). Consistent with these two roles, Dsh serves as an effector for two Wnt family (Chapter 3) receptors: DFrizzled2 (DFz2), a receptor for the *wingless* pathway, and Frizzled (Fz), a receptor required for proper planar polarity (363, 364). Two major challenges, then, are (1) the dissection of the mechanisms by which Dsh discriminates between two upstream signals, and (2) characterization of how Dsh generates two quite different developmental programmes.

Some answers to the latter question come from recent studies that indicate that different portions of the Dsh polypeptide are required for recruiting pathways which culminate in regulation of the planar polarity vs *wingless* developmental programmes (280).

Dsh can signal downstream to two parallel pathways: (1) a GSK3/β-catenin pathway that employs the products of the *Drosophila* GSK3 homologue *zeste-white-3* (*zw3*), the β-catenin homologue *armadillo* (*arm*), and the TCF homologue *pangolin* (*pan*) (363, 364) (see Chapter 3), and (2) a SAPK pathway that employs Bsk. Fz is a seven pass receptor that probably signals by recruiting heterotrimeric G-proteins (Section 3.9) that have yet to be identified. Defects in Fz and Dsh cause planar polarity defects; and genetic epistasis studies indicate that Dsh is an effector for Fz. In addition, constitutively active forms of the *Drosophila* RhoA orthologue *DrhoA* can rescue the ommatidial planar polarity defects incurred by the *dsh* defect (280). Similarly, overexpression of additional components of the *Drosophila* SAPK pathway including *bsk*, *hep*, and *Djun* can dominantly suppress the inactive *dsh* phenotype (280). These results suggest a pathway regulating planar polarity wherein Fz signals to Dsh which then recruits RhoA. RhoA, in turn, signals to Bsk and DJun.

Dsh consists of an amino-terminal Dsh/Axin (DIX) domain, a central PSD95/disks large (Dlg)/Z0–1 (PDZ) domain, and a carboxyl-terminal Dsh/Egl-10/pleckstrin (DEP) domain (280–282, 363, 364). These structures are indicative of a protein that serves as a docking site for signalling components; although the binding partners of Dsh are unknown. Biochemically, Dsh can activate the SAPKs in transfected mammalian cells (280). The DEP domain is conserved among several proteins implicated in signalling, and deletion or inactivating mutations of the DEP domain abolish SAPK activation (280). Thus, *dsh^1* is a point mutation of a Lys residue (Lys417Met) in the *dsh* DEP domain. This mutation abolishes Dsh activation of coexpressed SAPK in mammalian cells; and the *dsh^1* mutation is sufficient to cause planar polarity defects in the *Drosophila* eye (280). These *dsh^1* mutants can be rescued with ΔPDZ or ΔDIX Dsh, but not by Dsh^1 or Dsh–ΔDEP, clearly implicating the DEP domain in the establishment of planar polarity (280). By contrast, ΔDEP mutants of *dsh* have no inhibitory effect on signalling to Arm and are, in fact, sufficient to rescue *dsh* mutants that display a *wingless* phenotype (280, 363, 364). Thus, the DEP domain of Dsh recruits *Drosophila* SAPK to establish planar polarity, while the PDZ and DIX domains of Dsh signal to *Drosophila* GSK3 and β-catenin in the *wingless* pathway (280, 363, 364). As described in Section 3.9, signalling from seven pass receptors such as Fz (or angiotensin-II or endothelin-1) to the SAPKs is poorly understood. Three mammalian *dsh* homologues have been described (281, 282). The studies of *Drosophila* Dsh suggest that mammalian *dsh* homologues and Rho family GTPases may link seven pass receptors to the SAPKs.

Immunity

Insects possess an innate immune system which employs signalling pathways that are recruited to fight microbial infection. These pathways lead to the expression of a number of antimicrobial peptides that bind and perforate bacterial pathogens, thereby inactivating the infection (reviewed in ref. 362). These pathways involve *Drosophila* homologues of NF-κB and AP-1 (DJun/DFos). Biochemical studies indicate that Bsk is activated during the *Drosophila* innate immune response (362). Further genetic studies are needed to identify the physiological significance of this activation of Bsk, as well as the pathway that recruits Bsk during the *Drosophila* immune response.

3.14.7 Convergent TAK1 and Wnt signalling regulate TCF HMG domain-containing transcription factors in *C. elegans* and mammals

A further example of the convergence of SAPK and Wnt signalling comes from studies of the genetics of *C. elegans* embryonic asymmetry. In *C. elegans*, POP-1 is a TCF orthologue that acts as a transcriptional repressor which distinguishes the fates of anterior daughter cells from their posterior sister cells throughout the developmental process. *C. elegans* Wnt signalling down-modulates POP-1 in one posterior daughter cell referred to as E, thereby establishing E posterior localization in the developing embryo. Bowerman and colleagues identified *mom-4* and *lit-1* as additional genes required for the downregulation of POP-1 and manifestation of posterior cell morphology (365). Thus, as with a loss-of-function mutation of the *C. elegans* β-catenin homologue *wrm-1*, *mom-4* and *lit-1* mutant embryos display a fully penetrant loss of POP-1 asymmetry (365).

mom-4 is a *C. elegans* orthologue of TAK1 (213, 365); *lit-1* is a *C. elegans* orthologue of Nemo-like kinase (NLK). NLK is a member of a MAPK subgroup of previously unknown function, the founding member of which is the *Drosophila* kinase Nemo (not to be confused with the IKK-subunit NEMO). Thus, in mammalian cells, expressed Mom-4 can activate the SAPK pathway, when coexpressed with mammalian TAB1 (365, 366). Moreover, Mom-4 associates with a *C. elegans* orthologue of TAB1, TAP-1 (365, 366).

In a further demonstration of the conservation of convergent Wnt and MAPK signalling, Matsumoto and colleagues demonstrated that mammalian TAK1 activates coexpressed, mammalian NLK and downregulates transcriptional induction mediated by β-catenin and TCF (366). In mammalian cells, Wnt signalling stabilizes β-catenin which translocates to the nucleus and converts TCF family transcriptional repressors into transcriptional activators (363, 364). By contrast, in *C. elegans*, Wrm1, the β-catenin orthologue, appears to block the repressor function of POP-1 instead of converting it into a transcriptional activator (363, 364). This difference in function may account for the observation that the *C. elegans* Mom-4 → Lit-1 pathway positively regulates Wnt signalling, whereas the mammalian TAK → NLK pathway negatively regulates it (365, 366).

4. Concluding remarks

In the intervening 6 years since the original publication of this volume our understanding of MAP kinase signalling pathways has expanded dramatically. It has become apparent in the intervening time, that these pathways can no longer be considered simple and linear. Instead, these mechanisms form complex signalling networks. At this time, the two most compelling issues in the study of MAPK pathways are MAP3K → MEK → MAPK core pathway regulation and biological function.

4.1 Oligomerization, adapter/G-protein binding, and translocation as themes in MAP3K regulation

While a large number of potential MAP3K → MEK → MAPK core pathways, and their possible upstream activators and downstream targets have been identified, there is still a troubling lack of clarity concerning MAP3K regulation. In the case of stress-regulated MAP3Ks, this is due in large part to the high basal activity of these enzymes, making analysis of their activation difficult. It is clear that many MAP3Ks (Raf-1, ASK1, MLK3) are regulated, in part, by oligomerization. MEKK1 selectively associates with oligomerized TRAF2, a process that could foster MEKK1 oligomerization-dependent activation. Whether this oligomerization triggers *trans* autophosphorylation in a manner analogous to receptor Tyr kinases has not been determined. Alternatively, oligomerization could also render these kinases more attractive targets for additional upstream activators.

There are some MAP3Ks for which putative upstream kinases have been identified (PAK3 for Raf-1, GCKs for MEKK1 and MLK3). Inasmuch as both the PAKs and, possibly, GCKs reversibly associate with membrane/receptor-associated polypeptides (Rho GTPases in the case of PAKs, TRAFs in the case of GCKs), the parallel recruitment of target MAP3Ks to membrane receptor complexes could enable the formation of MAP3K activating complexes (Fig. 17).

G-proteins, and adapter proteins, are clearly required for both the regulation of MAP3K oligomerization and activation by upstream kinases, and may be a general paradigm of MAP3K regulation. Thus, Ras-dependent Raf-1 oligomerization and Ras → Rac1 regulation of PAKs is an example of a single G-protein (Ras) regulating both MAP3K aggregation and phosphorylation by upstream kinases. The binding of TRAF2 oligomers to both MEKK1 and the group-I GCKs, putative MEKK1 upstream kinases, is a second example. Finally, MLK3 is activated, at least in part, by Cdc42-dependent oligomerization. HPK1, a putative upstream activator of MLK3, is activated by the SH2/SH3 adapter protein Crk/CrkL, which has also been implicated in the regulation of Rho family GTPases. Thus, Tyr kinases might recruit MLKs by fostering Rho GTPase-dependent and group-1 GCK-dependent activation. Clearly it will be important to sort out the different roles of oligomerization, adapter protein binding, and upstream kinases in MAP3K regulation.

4.2 MAPK pathway biology

Genetic models were indispensable in the dissection of the Ras–ERK pathway. Although until recently there has been a comparative paucity of genetic models from which to draw conclusions as to stress-activated MAPK pathway regulation, new emerging genetic models such as the POP1 regulation mechanism in C. *elegans* as well as the dorsal closure and planar polarity pathways in *Drosophila*, coupled with the completion of the *C. elegans* and other genome sequencing projects, should make it possible to understand the epistatic relationships between MAP3Ks and their upstream activators.

Placing these pathways in the context of human disease pathology will be more difficult. In this instance, pharmacological inhibitors such as PD98059 or SB203580, plus mouse knockout and transgenic approaches will enable a greater understanding of how mammalian MAPK pathways relate to disease, and will identify important new drug targets.

References

1. Kyriakis, J. M. and Avruch, J. (1996). Sounding the alarm: protein kinase cascades activated by stress and inflammation. *J. Biol. Chem.*, **271**, 24313.
2. Marshall, C. J. (1995). Specificity of receptor tyrosine kinase signaling: transient versus sustained extracellular signal regulated kinase activation. *Cell*, **80**, 179.
3. Herskowitz, I. (1995). MAP kinase pathways in yeast: for mating and more. *Cell*, **80**, 187.
4. Countaway, J. L., Northwood, I. C., and Davis, R. J. (1989). Mechanism of phosphorylation of the epidermal growth factor receptor on threonine 669. *J. Biol. Chem.*, **264**, 10828.
5. Erickson, A. K., Payne, D. M., Martino, P. A., Rossomando, A. J., Shabanowitz, J., Weber, M. J., Hunt, D. F., and Sturgill, T. W. (1990). Identification by mass spectrometry of threonine 97 in bovine myelin basic protein as a specific phosphorylation site for mitogen-activated protein kinase. *J. Biol. Chem.*, **265**, 19728.
6. Clark-Lewis, I., Sanghera, J. S., and Pelech, S. L. (1991). Definition of a consensus sequence for peptide substrate recognition by p44mpk, the meiosis-activated myelin basic protein kinase. *J. Biol. Chem.*, **266**, 15180.
7. Kallunki, T., Deng, T., Hibi, M., and Karin, M. (1996). c-Jun can recruit JNK to phosphorylate dimerization partners via specific docking interactions. *Cell*, **87**, 929.
8. Dai, T., Rubie, E., Franklin, C. C., Kraft, A., Gillespie, D. A., Avruch, J., Kyriakis, J. M., and Woodgett, J. R. (1995). Stress-activated protein kinases bind directly to the delta domain of c-Jun in resting cells: implications for repression of c-Jun function. *Oncogene*, **10**, 849.
9. Yang, S.-H., Whitmarsh, A. J., Davis, R. J., and Sharrocks, A. D. (1998). Differential targeting of MAP kinases to the ETS-domain transcription factor Elk-1. *EMBO J.*, **17**, 1740.
10. Posas, F. and Saito, H. (1997). Osmotic activation of the HOG MAPK pathway via Ste11p MAPKKK: scaffold role of Pbs2p MAPKK. *Science*, **276**, 1702.
11. Salmerón, A., Ahmad, T. B., Carlile, G. W., Pappin, D., Narsimhan, R. P., and Ley, S. C. (1996). Activation of MEK-1 and SEK-1 by Tpl-2 proto oncoprotein, a novel MAP kinase kinase kinase. *EMBO J.*, **15**, 817.
12. Kyriakis, J. M. (1999). Signaling by the germinal center kinase family of protein kinases. *J. Biol. Chem.*, **274**, 5259.
13. Pawson, T. and Scott, J. D. (1997). Signaling through scaffold, anchoring, and adaptor proteins. *Science*, **278**, 2075.
14. Dickens, M., Rogers, J. S., Cavanagh, J., Raitano, A., Xia, Z., Halpern, J. R., Greenberg, M. E., Sawyers, C. L., and Davis, R. J. (1997). A cytoplasmic inhibitor of the JNK signal transduction pathway. *Science*, **277**, 693.
15. Whitmarsh, A. J., Cavanagh, J., Tournier, C., Yasuda, J., and Davis, R. J. (1998). A mammalian scaffold complex that selectively mediates MAP kinase activation. *Science*, **281**, 1671.

16. Schaeffer, H. J., Catling, A. D., Eblen, S. T., Collier, L. S., Krauss, A., and Weber, M. J. (1998). MP1: a MEK binding partner that enhances enzymatic activation of the MAP kinase cascade. *Science*, **281**, 1668.

17. Xia, Y., Wu, Z., Su, B., Murray, B., and Karin, M. (1998). JNKK1 organizes a MAP kinase module through specific and sequential interactions with upstream and downstream components mediated by its amino terminal extension. *Genes Dev.*, **12**, 3369.

18. Yuasa, T., Ohno, S., Kehrl, J. H., and Kyriakis, J. M. (1998). Tumor necrosis factor signaling to stress-activated protein kinase (SAPK)/Jun NH2-terminal kinase (JNK) and p38. *J. Biol. Chem.*, **273**, 22681.

19. Sturgill, T. W. and Ray, L. B. (1986). Muscle proteins related to microtubule associated protein-2 are substrates for an insulin-stimulatable kinase. *Biochem. Biophys. Res. Commun.*, **134**, 565.

20. Rossomando, A. J., Payne, D. M., Weber, M. J., and Sturgill, T. W. (1988). Evidence that pp42, a major tyrosine kinase target, is a mitogen-activated serine/threonine protein kinase. *Proc. Natl Acad. Sci. USA*, **86**, 6940.

21. Boulton, T. G., Yancopoulos, G. D., Gregory, J. S., Slaughter, C., Moomaw, C., Hsu, J., and Cobb, M. H. (1990). An insulin-stimulated protein kinase homologous to yeast kinases involved in cell cycle control. *Science*, **249**, 64.

22. Boulton, T. G., Nye, S. H., Robbins, D. J., Ip, N. Y., Radziejewska, E., Morgenbesser, S. D., DePinho, R. A., Panayotatos, N., and Cobb, M. H. (1991). ERKs: a family of protein-serine/threonine kinases that are activated and tyrosine phosphorylated in response to insulin and NGF. *Cell*, **65**, 663.

23. Jones, S. W., Erikson, E., Blenis, J., Maller, J. L., and Erikson, R. L. (1988). A *Xenopus* ribosomal protein S6 kinase has two apparent kinase domains that are each similar to distinct protein kinases. *Proc. Natl Acad. Sci. USA*, **85**, 3377.

24. Erikson, R. L. (1991). Structure, expression and regulation of protein kinases involved in phosphorylation of ribosomal protein S6. *J. Biol. Chem.*, **266**, 6007.

25. Cohen, P. (1988). Muscle glycogen synthase. In *The enzymes* (ed. P. Boyer and E. G. Krebs), p. 461. Academic Press, New York.

26. Avruch, J. (1998). Insulin signal transduction through protein kinase cascades. *Mol. Cell. Biochem.*, **182**, 31.

27. Cross, D. A. E., Alessi, D. R., Cohen, P., Andjelkovich, M., and Hemmings, B. A. (1995). Inhibition of glycogen synthase kinase-3 by insulin mediated by protein kinase B. *Nature*, **378**, 785.

28. Dent, P., Lavoinne, A., Nakielny, S., Caudwell, F. B., Watt, P., and Cohen, P. (1990). The molecular mechanism by which insulin stimulates glycogen synthesis in mammalian skeletal muscle. *Nature*, **348**, 302.

29. Sassone-Corsi, P., Mizzen, C. A., Cheung, P., Crosio, C., Monaco, L., Jacquot, S., Hanauer, A., and Allis, C. D. (1999). Requirement of Rsk-2 for epidermal growth factor-activated phosphorylation of histone H3. *Science*, **285**, 886.

30. Chadee, D. N., Hendzel, M. J., Tylipski, C. P., Allis, C. D., Bazett-Jones, D. P., Wright, J. A., and Davie, J. R. (1999). Increased Ser-10 phosphorylation of histone H3 in mitogen-stimulated and oncogene-transformed mouse fibroblasts. *J. Biol. Chem.*, **274**, 24914.

31. Xing, J., Ginty, D. D., and Greenberg, M. E. (1996). Coupling of the RAS-MAPK pathway to gene activation by RSK2, a growth factor-regulated CREB kinase. *Science*, **273**, 959.

32. Deak, M., Clifton, A. D., Lucocq, J., and Alessi, D. R. (1998). Mitogen- and stress-activated protein kinase-1 (MSK1) is directly activated by MAPK and SAPK2/p38, and may mediate activation of CREB. *EMBO J.*, **17**, 4426.

33. Ghoda, L., Lin, X., and Greene, W. C. (1997). The 90-kDa ribosomal S6 kinase (pp90rsk) phosphorylates the N-terminal regulatory domain of IκBα and stimulates its degradation *in vitro*. *J. Biol. Chem.*, **272**, 21281.

34. Sturgill, T. W., Ray, L. B., Erikson, E., and Maller, J. L. (1988). Insulin-stimulated MAP-2 kinase phosphorylates and activates ribosomal protein S6 kinase II. *Nature*, **334**, 715.

35. Dalby, K. N., Morrice, N., Caudwell, F. B., Avruch, J., and Cohen, P. (1998). Identification of regulatory phosphorylation sites in mitogen-activated protein kinase (MAPK)-activated protein kinase-1a/p90rsk that are inducible by MAPK. *J. Biol. Chem.*, **273**, 1496.

36. Richards, S. A., Fu, J., Romanelli, A., Shimamura, A., and Blenis, J. (1999). Ribosomal S6 kinase 1 (RSK1) activation requires signals dependent on and independent of the MAP kinase ERK. *Curr. Biol.*, **9**, 810.

37. Alessi, D. R., Deak, M., Casamayor, A., Caudwell, F. B., Morrice, N., Norman, D. G., Gaffney, P., Reese, C. B., MacDougall, C. N., Harbison, D., Ashworth, A., and Bownes, M. (1997). 3-phosphoinositide-dependent kinase-1 (PDK1): structural and functional homology with the *Drosophila* DSTPK61 kinase. *Curr. Biol.*, **7**, 776.

38. Alessi, D. R., Kozlowski, M. T., Weng, Q. P., Morrice, N., and Avruch, J. (1997). 3-phosphoinositide-dependent protein kinase 1 (PDK1) phosphorylates and activates the p70 S6 kinase *in vivo* and *in vitro*. *Curr. Biol.*, **8**, 69.

39. Gavin, A.-C. and Nebreda, A. R. (1999). A MAP kinase docking site is required for phosphorylation and activation of p90rsk/MAPKAP kinase-1. *Curr. Biol.*, **9**, 281.

40. Habener, J. F. (1990). Cyclic AMP-response element binding proteins: a cornucopia of transcription factors. *Mol. Endocrinol.*, **4**, 1087.

41. Tan, Y., Rouse, J., Zhang, A., Cariati, S., Cohen, P., and Comb, M. (1996). FGF and stress regulate CREB and ATF1 via a pathway involving p38 MAP kinase and MAPKAP kinase-2. *EMBO J.*, **15**, 4629.

42. Sonenberg, N. and Gingras, A.-C. (1998). The mRNA 5′ cap-binding protein eIF-4E and control of cell growth. *Curr. Opin. Cell Biol.*, **10**, 268.

43. Waskiewicz, A. J., Flynn, A., Proud, C. G., and Cooper, J. A. (1997). Mitogen-activated protein kinases activate the serine/threonine kinases Mnk1 and Mnk2. *EMBO J.*, **16**, 1909.

44. Fukunaga, R. and Hunter, T. (1997). MNK1, a new MAP kinase-activated protein kinase, isolated by a novel expression screening method for identifying protein kinase substrates. *EMBO J.*, **16**, 1921.

45. Treisman, R. (1996). Regulation of transcription by MAP kinase cascades. *Curr. Opin. Cell Biol.*, **8**, 205.

46. Karin, M., Liu, Z.-G., and Zandi, E. (1997). AP-1 function and regulation. *Curr. Opin. Cell Biol.*, **9**, 240.

47. Gille, H., Sharrocks, A. D., and Shaw, P. E. (1992). Phosphorylation of transcription factor p62TCF by MAP kinase stimulates ternary complex formation at c-fos promoter. *Nature*, **358**, 414.

48. Marais, R., Wynne, J., and Treisman, R. (1993). The SRF accessory protein Elk-1 contains a growth factor-regulated transcriptional activation domain. *Cell*, **73**, 381.

49. Janknecht, R. and Hunter, T. (1997). Convergence of MAP kinase pathways on the ternary complex factor Sap-1a. *EMBO J.*, **16**, 1620.

50. Whitmarsh, A. J., Shore, P., Sharrocks, A. D., and Davis, R. J. (1995). Integration of MAP kinase signal transduction pathways at the serum response element. *Science*, **269**, 403.

51. Anderson, N. G., Maller, J. L., Tonks, N. K., and Sturgill, T. W. (1990). Requirement for integration of signals from two distinct phosphorylation pathways for activation of MAP kinase. *Nature*, **343**, 651.

52. Payne, D. M., Rossomando, A. J., Martino, P., Erickson, A. K., Her, J., Shabanowitz, J., Hunt, D. F., Weber, M. J., and Sturgill, T. W. (1991). Identification of the regulatory phosphorylation sites in pp42/mitogen-activated protein kinase (MAP kinase). *EMBO J.*, **10**, 885.

53. Ahn, N. G., Seger, R., Bratlien, R. L., Diltz, C. D., Tonks, N. K., and Krebs, E. G. (1991). Multiple components in an epidermal growth factor-stimulated protein kinase cascade. *In vitro* activation of a myelin basic protein/microtubule–associated protein 2 kinase. *J. Biol. Chem.*, **266**, 4220.

54. Ahn, N. G., Seger, R., and Krebs, E. G. (1992). The mitogen-activated protein kinase activator. *Curr. Opin. Cell Biol.*, **4**, 992.

55. Crews, C., Alessandrini, A. A., and Erikson, R. L. (1992). The primary structure of MEK, a protein kinase that phosphorylates the ERK gene product. *Science*, **258**, 478.

56. Seger, R., Seger, D., Lozeman, F. J., Ahn, N. G., Graves, L. M., Campbell, J. S., Ericsson, L., Harrylock, M., Jensen, A. M., and Krebs, E. G. (1992). Human T-cell MAP kinase–kinases that are related to yeast signal transduction kinases. *J. Biol. Chem.*, **267**, 25628.

57. Wu, J., Harrison, J. K., Vincent, L. A., Haystead, C., Haystead, T. A. J., Michel, H., Hunt, D. F., Lynch, K. R., and Sturgill, T. W. (1993). Molecular structure of a protein-tyrosine/threonine kinase activating p42 mitogen-activated protein (MAP) kinase: MAP kinase kinase. *Proc. Natl Acad. Sci. USA*, **90**, 173.

58. Zheng, C. F. and Guan, K.-L. (1994). Activation of MEK family kinases requires phosphorylation of two conserved Ser/Thr residues. *EMBO J.*, **13**, 1123.

59. Alessi, D. R., Saito, Y., Campbell, D. G., Cohen, P., Sithanandam, G., Rapp, U., Ashworth, A., Marshall, C. J., and Cowley, S. (1994). Identification of the sites in MAP kinase kinase-1 phosphorylated by p74raf-1. *EMBO J.*, **13**, 1610.

60. Cowley, S., Paterson, H., Kemp, P., and Marshall, C. J. (1994). Activation of MAP kinase kinase is necessary and sufficient for PC12 differentiation and for transformation of NIH 3T3 cells. *Cell*, **77**, 841.

61. Mansour, S. J., Matten, W. T., Hermann, A. S., Candia, J. M., Rong, S., Fukasawa, K., Vande Woude, G. F., and Ahn, N. G. (1994). Transformation of mammalian cells by constitutively active MAP kinase kinase. *Science*, **265**, 966.

62. Fukuda, M., Gotoh, Y., and Nishida, E. (1997). Interaction of MAP kinase with MAP kinase kinase: its possible role in the control of nucleocytoplasmic transport of MAP kinase. *EMBO J.*, **16**, 1901.

63. Dang, A., Frost, J. A., and Cobb, M. H. (1998). The MEK1 proline-rich insert is required for efficient activation of the mitogen-activated protein kinases ERK1 and ERK2 in mammalian cells. *J. Biol. Chem.*, **273**, 19909.

64. Duesbery, N. S., Webb, C. P., Leppla, S. H., Gordon, V. M., Klimpel, K. R., Copeland, T. D., Ahn, N. G., Oskarsson, M. K., Fukasawa, K., Paull, K. D., and Vande Woude, G. F. (1998). Proteolytic inactivation of MAP-kinase-kinase by anthrax lethal factor. *Science*, **280**, 734.

65. Avruch, J., Zhang, X.-F., and Kyriakis, J. M. (1994). Raf meets Ras: completing the framework of a signal transduction pathway. *Trends Biochem. Sci.*, **19**, 279.

66. Gallego, C., Gupta, S. K., Heasley, L. E., Quian, N. X., and Johnson, G. L. (1992). Mitogen-activated protein kinase activation resulting from selective oncogenic expression in NIH3T3 and rat la cells. *Proc. Natl Acad. Sci. USA*, **89**, 7355.

67. Bruder, J. T., Heidecker, G., and Rapp, U. R. (1992). Serum-, TPA-, and Ras-induced expression from Ap-1/Ets-driven promoters requires Raf-1 kinase. *Genes Dev.*, **6**, 545.

68. Kyriakis, J. M., App, H., Zhang, X.-F., Banerjee P., Brautigan, D. L., Rapp, U. R., and Avruch, J. (1992). Raf-1 activates MAP kinase-kinase. *Nature*, **358**, 417.

69. Dent, P., Haser, W., Haystead, T. A. J., Vincent, L. A., Roberts, T. M., and Sturgill, T. W. (1992). Activation of mitogen-activated protein kinase kinase by v-Raf in NIH 3T3 cells and *in vitro*. *Science*, **257**, 1404.

70. Howe, L. R., Leevers, S. J., Gomez, N., Nakielny, S., Cohen, P., and Marshall, C. J. (1992). Activation of the MAP kinase pathway by the protein kinase raf. *Cell*, **71**, 335.

71. Kyriakis, J. M., Force, T. L., Rapp, U. R., Bonventre, J. V., and Avruch, J. (1993). Mitogen regulation of c–Raf–1 protein kinase activity toward mitogen-activated protein kinase-kinase. *J. Biol. Chem.*, **268**, 16009.

72. Pang, L., Sawada, T., Decker, S. J., and Saltiel, A. R. (1995). Inhibition of MAP kinase kinase blocks the differentiation of PC-12 cells induced by nerve growth factor. *J. Biol. Chem.*, **270**, 13585.

73. Alessi, D. R., Cuenda, A., Cohen, P., Dudley, D. T., and Saltiel, A. R. (1995). PD 098059 is a specific inhibitor of the activation of mitogen-activated protein kinase kinase *in vitro* and *in vivo*. *J. Biol. Chem.*, **270**, 27489.

74. Azpiazu, I., Saltiel, A. R., DePaoli-Roach, A. A., and Lawrence, J. C. (1996). Regulation of both glycogen synthase and PHAS-I by insulin in rat skeletal muscle involves mitogen-activated protein kinase-independent and rapamycin-sensitive pathways. *J. Biol. Chem.*, **271**, 5033.

75. Sagata, N. (1997). What does Mos do in oocytes and somatic cells? *Bioessays*, **19**, 13.

76. Posada, J., Yew, N., Ahn, N. G., Vande Woude, G. F., and Cooper, J. A. (1993). Mos stimulates MAP kinase in *Xenopus* oocytes and activates a MAP kinase kinase *in vitro*. *Mol. Cell. Biol.*, **13**, 2546.

77. Bourne, H. R., Sanders, D. A., and McCormick, F. (1991). The GTPase superfamily: conserved structure and molecular mechanism. *Nature*, **349**, 117.

78. McCormick, F. and Wittinghofer, A. (1996). Interactions between Ras proteins and their effectors. *Curr. Opin. Biotechnol.*, **7**, 449.

79. McCormick, F. (1994). Activators and effectors of ras p21 proteins. *Curr Opin. Genet. Dev.*, **4**, 71.

80. Downward, J., Graves, J. D., Warne, P. H., Rayter, S., and Cantrell, D. A. (1990). Stimulation of p21ras upon T-cell activation. *Nature*, **346**, 719.

81. Satoh, T., Endo, M., Nakafuku, M., Nakamura, S., and Kaziro, Y. (1990). Platelet-derived growth factor stimulates formation of active p21ras·GTP complex in Swiss mouse 3T3 cells. *Proc. Natl Acad. Sci. USA*, **87**, 5993.

82. Hattori, S., Fukuda, M., Yamashita, T., Nakamura, S., Gotoh, Y., and Nishida, E. (1992). Activation of mitogen-activated protein kinase and its activator by ras in intact cells and in a cell-free system. *J. Biol. Chem.*, **267**, 20346.

83. Shibuya, E. K., Polverino, A. J., Chang, E., Wigler, M., and Ruderman, J. V. (1992). Oncogenic Ras triggers the activation of 42-kDa mitogen-activated protein kinase in extracts of quiescent *Xenopus* oocytes. *Proc. Natl Acad. Sci. USA*, **89**, 9831.

84. Leevers, S. and Marshall, C. J. (1992). Activation of extracellular signal-regulated kinase, ERK2, by p21ras oncoprotein. *EMBO J.*, **11**, 569.

85. Simon, M. A., Bowtell, D. D. L., Dodson, G. S., Laverty, T. R., and Rubin, G. M. (1992). Ras1 and a putative guanine nucleotide exchange factor perform crucial steps in signaling by the sevenless protein tyrosine kinase. *Cell*, **67**, 701.

86. Ambrosio, L., Mahowald, A. P., and Perrimon, N. (1992). Requirement of the *Drosophila raf* homologue for *torso* function. *Nature*, **342**, 288.

87. Steinberg, P. W. and Horvitz, H. R. (1991). Signal transduction during *C. elegans* vulval induction. *Trends Genet.*, **7**, 366.

88. Thomas, S. M., DeMarco, M., D'Arcangelo, G., Halegoua, S., and Brugge, J. S. (1992). Ras is essential for nerve growth factor- and phorbol ester–induced tyrosine phosphorylation of MAP kinases. *Cell*, **68**, 1031.

89. Wood, K. W., Sarnecki, C., Roberts, T. M., and Blenis, J. (1992). *ras* mediates nerve growth factor receptor modulation of three signal-transducing protein kinases: MAP kinase, Raf-1, and RSK. *Cell*, **68**, 1041.

90. Moodie, S. A., Willumsen, B. M., Weber, M. J., and Wolfman, A. (1993). Complexes of Ras·GTP with Raf-1 and mitogen-activated protein kinase kinase. *Science*, **260**, 1658.

91. Van Aelst, L., Barr, M., Marcus, S., Polverino, A., and Wigler, M. (1993). Complex formation between RAS and RAF and other protein kinases. *Proc. Natl Acad. Sci. USA*, **90**, 6213.

92. Zhang, X. F., Settleman, J., Kyriakis, J. M., Takeuchi-Suzuki, E., Elledge, S. J., Marshall, M. S., Bruder, J. T., Rapp, U. R., and Avruch, J. (1993). Normal and oncogenic p21ras bind to the amino-terminal regulatory domain of c-Raf–1. *Nature*, **364**, 308.

93. Vojtek, A. B., Hollenberg, S. M., and Cooper, J. A. (1993). Mammalian Ras interacts directly with the serine/threonine kinase Raf. *Cell*, **74**, 205.

94. Warne, P. H., Viciana, P. R., and Downward, J. (1993). Direct interaction of Ras and the amino terminal region of Raf-1 *in vitro*. *Nature*, **364**, 352.

95. Luo, Z., Diaz, B., Marshall, M. S., and Avruch, J. (1997). An intact Raf zinc finger is required for optimal binding to processed Ras and for ras-dependent Raf activation in situ. *Mol. Cell. Biol.*, **17**, 46.

96. Barnard, D., Diaz, B., Hettich, L., Chuang, E., Zhang, X.-F., Avruch, J., and Marshall, M. (1995). Identification of the sites of interaction between c-Raf-1 and Ras-GTP. *Oncogene*, **10**, 1283.

97. Leevers, S. J., Paterson, H. F., and Marshall, C. J. (1994). Requirement for Ras in Raf activation is overcome by targeting Raf to the plasma membrane. *Nature*, **369**, 411.

98. Luo, Z., Tzivion, G., Belshaw, P. J., Vavvas, D., Marshall, M., and Avruch, J. (1996). Oligomerization activates c-Raf-1 through a Ras-dependent mechanism. *Nature*, **383**, 181.

99. Farrar, M. A., Alberol-Ila, J., and Perlmutter, R. M. (1996). Activation of the Raf-1 kinase cascade by coumermycin-induced dimerization. *Nature*, **383**, 178.

100. Force, T., Bonventre, J. V., Heidecker, G., Rapp, U., Avruch, J., and Kyriakis, J. M. (1994). Enzymatic characteristics of the c-Raf-1 protein kinase. *Proc. Natl Acad. Sci. USA*, **91**, 1270.

101. Schreiber, S. L. and Crabtree, G. R. (1992). The mechanism of action of cyclosporin A and FK506. *Immunol. Today*, **13**, 136.

102. Luo, Z., Zhang, X.-F., Rapp, U., and Avruch, J. (1995). Identification of the 14.3.3ζ domains important for self-association and Raf binding. *J. Biol. Chem.*, **270**, 23681.

103. Freed, E., Symons, M., Macdonald, S. G., McCormick, F., and Ruggieri, R. (1994). Binding of 14–3–3 proteins to the protein kinase Raf and effects on its activation. *Science*, **265**, 1713.

104. Fantl, W., Muslin, A. J., Kikuchi, A., Martin, J. A., MacNicol, A. M., Gross, R. W., and Williams, L. T. (1994). Activation of Raf-1 by 14–3–3 proteins. *Nature*, **371**, 612.

105. Tzivion, G., Luo, Z., and Avruch, J. (1998). A dimeric 14–3–3 protein is an essential cofactor for Raf kinase activity. *Nature*, **394**, 88.

106. Muslin, A. J., Tanner, J. W., Allen, P. M., and Shaw, A. S. (1996). Interaction of 14–3–3 with signaling proteins is mediated by the recognition of phosphoserine. *Cell*, **84**, 889.

107. King, A. J., Sun, H., Diaz, B., Barnard, D., Miao, W., Bagrodia, S., and Marshall, M. S. (1998). The protein kinase Pak3 positively regulates Raf-1 through phosphorylation of serine 338. *Nature*, **396**, 180.

108. Hall, A. (1998). Rho GTPases and the actin cytoskeleton. *Science*, **279**, 509.

109. Cerione, R. A. and Zheng, Y. (1996). The Dbl family of oncogenes. *Curr. Opin. Cell Biol.*, **8**, 216.

110. Fabian, J. R., Daar, I. O., and Morrison, D. K. (1993). Critical tyrosine residues regulate the enzymatic and biological activity of Raf-1 kinase. *Mol. Cell. Biol.*, **11**, 7170.

111. Mason, C. S., Springer, C. J., Cooper, R. G., Superti-Furga, G., Marshall, C. J., and Marais, R. (1999). Serine and tyrosine phosphorylations cooperate in Raf-1, but not B-Raf activation. *EMBO J.*, **18**, 2137.

112. Wartmann, M. and Davis, R. J. (1994). The native structure of the activated Raf protein kinase is a membrane-bound multi-subunit complex. *J. Biol. Chem.*, **269**, 6695.

113. Cutforth, T. and Rubin, G. M. (1994). Mutations in HSP83 and cdc37 impair signaling by the sevenless receptor tyrosine kinase in *Drosophila*. *Cell*, **77**, 1027.

114. Grammatikakis, N., Lin, J.-H., Grammatikakis, A., Tsichlis, P. N., and Cochran, B. H. (1999). p50cdc37 acting in concert with Hsp90 is required for Raf-1 function. *Mol. Cell. Biol.*, **19**, 1661.

115. Yamamori, B., Kuroda, S., Shimizu, K., Fukui, K., Ohtsuka, T., and Takai, Y. (1995). Purification of a Ras-dependent mitogen-activated protein kinase kinase kinase from bovine brain cytosol and its identification as a complex of B-Raf and 14–3–3 proteins. *J. Biol. Chem.*, **270**, 11723.

116. Kitayama, H., Sugimoto, Y., Matsuzaki, T., Ikawa, Y., and Noda, M. (1989). A ras-related gene with transformation suppressor activity. *Cell*, **56**, 77.

117. Cook, S. J., Rubinfeld, B., Albert, I., and McCormick, F. (1993). RapV12 antagonizes Ras-dependent activation of ERK1 and ERK2 by LPA and EGF in Rat-1 fibroblasts. *EMBO J.*, **12**, 3475.

118. Cook, S. J. and McCormick, F. (1993). Inhibition by cAMP of Ras-dependent activation of Raf. *Science*, **262**, 1069.

119. Wu, J., Dent, P., Jelinek, T., Wolfman, A., Weber, M. J., and Sturgill, T. W. (1993). Inhibition of the EGF-activated MAP kinase signaling pathway by adenosine 3', 5'-monophosphate. *Science*, **262**, 1065.

120. Altschuler, D. L., Peterson, S. N., Ostrowski, M. C., and Lapetina, E. G. (1995). Cyclic AMP-dependent activation of Rap1b. *J. Biol. Chem.*, **270**, 10373.

121. Ohtsuka, T., Shimizu, K., Yamamori, B., Kuroda, S., and Takai, Y. (1996). Activation of brain B-Raf protein kinase by Rap1B small GTP binding protein. *J. Biol. Chem.*, **271**, 1258.

122. Vossler, M. R., Yao, H., York, R. D., Pan, M.-G., Rim, C. S., and Stork, P. J.S. (1997). cAMP activates MAP kinase and Elk-1 through a B-Raf- and Rap1-dependent pathway. *Cell*, **89**, 73.

123. de Rooij, J., Zwartkruis, F. J., Verheijen, M. H., Cool, R. H., Nijman, S. M., Wittinghofer, A., and Bos, J. L. (1998). Epac is a Rap1 guanine-nucleotide-exchange factor directly activated by cyclic AMP. *Nature*, **396**, 474.

124. Kawasaki, H., Springett, G. M., Mochizuki, N., Toki, S., Nakaya, M., Matsuda, M., Housman, D. E., and Graybiel, A. M. (1998). A family of cAMP-binding proteins that directly activate Rap1. *Science*, **282**, 2275.

125. Altschuler, D. L. and Lapetina, E. G. (1993). Mutational analysis of the cAMP-dependent protein kinase-mediated phosphorylation site of Rap1b. *J. Biol. Chem.*, **268**, 7527.

126. Traverse, S., Seedorf, K., Paterson, H., Marshall, C. J., Cohen, P., and Ullrich, A. (1994). EGF triggers neuronal differentiation of PC12 cells that overexpress the EGF receptor. *Curr. Biol.*, **4**, 694.

127. York, R. D., Yao, H., Dillon, T., Ellig, C. L., Eckert, S. P., McCleskey, E. W., and Stork, P. J.S. (1998). Rap1 mediates sustained MAP kinase activation induced by nerve growth factor. *Nature*, **392**, 622.

128. Chen, R. H., Sarnecki, C., and Blenis, J. (1992). Nuclear localization and regulation of erk- and rsk-encoded protein kinases. *Mol. Cell. Biol.*, **12**, 915.

129. Brunet, A., Roux, D., Lenormand, P., Dowd, S., Keyse, S., and Pouysségur, J. (1999). Nuclear translocation of p42/p44 mitogen-activated protein kinase is required for growth factor-induced gene expression and cell cycle entry. *EMBO J.*, **18**, 664.

130. Fukuda, M., Gotoh, Y., and Nishida, E. (1997). Interaction of MAP kinase with MAP kinase kinase: its possible role in the control of nucleocytoplasmic transport of MAP kinase. *EMBO J.*, **16**, 1901.

131. Khokhlatchev, A. V., Canagarajah, B., Wilsbacher, J., Robinson, M., Atkinson, M., Goldsmith, E., and Cobb, M. H. (1998). Phosphorylation of the MAP kinase ERK2 promotes its homodimerization and nuclear translocation. *Cell*, **93**, 605.

132. Canagarajah, B. J., Khokhlatchev, A., Cobb, M. H., and Goldsmith, E. J. (1997). Activation mechanism of the MAP kinase ERK2 by dual phosphorylation. *Cell*, **90**, 859.

133. Kyriakis, J. M. and Avruch, J. (1990). pp54 MAP-2 kinase. A novel serine/threonine protein kinase regulated by phosphorylation and stimulated by poly-L-lysine. *J. Biol. Chem.*, **265**, 17355.

134. Kyriakis, J. M., Brautigan, D. L., Ingebritsen, T. S., and Avruch, J. (1991). pp54 micro-tubule-associated protein-2 kinase requires both tyrosine and serine/threonine phosphorylation for activity. *J. Biol. Chem.*, **266**, 10043.

135. Mukhopadhyay, N. K., Price, D. J., Kyriakis, J. M., Pelech, S., Sanghera, J., and Avruch, J. (1992). An array of insulin-activated, proline-directed (Ser/Thr) protein kinases phosphorylate the p70 S6 kinase. *J. Biol. Chem.*, **267**, 3325.

136. Pulverer, B. J., Kyriakis, J. M., Avruch, J., Nikolakaki, E., and Woodgett, J. R. (1991). Phosphorylation of c-jun mediated by MAP kinases. *Nature*, **353**, 670.

137. Kyriakis, J. M., Banerjee, P., Nikolakaki, E., Dai, T., Rubie, E. A., Ahmad, M. F., Avruch, J., and Woodgett, J. R. (1994). The stress-activated protein kinase subfamily of c-Jun kinases. *Nature*, **369**, 156.

138. Dérijard, B., Hibi, M., Wu, I.-H., Barrett, T., Su, B., Deng, T., Karin, M., and Davis, R. J. (1994). JNK1: a protein kinase stimulated by UV light and Ha-Ras that binds and phosphorylates the c-Jun transactivation domain. *Cell*, **76**, 1025.

139. Shapiro, P. S., Evans, J. N., Davis, R. J., and Posada, J. A. (1996). The seven transmembrane-spanning receptors for endothelin and thrombin cause proliferation of airway smooth muscle cells and activation of the extracellular signal regulated kinase and c-Jun NH2-terminal kinase groups of mitogen-activated protein kinases. *J. Biol. Chem.*, **271**, 5750.

140. Zohn, I. E., Yu, H., Li, X., Cox, A., and Earp, H. S. (1995). Angiotensin II stimulates the calcium-dependent activation of c-Jun N-terminal kinase. *Mol. Cell. Biol.*, **15**, 6160.

141. Berberich, I., Shu, G., Siebelt, F., Woodgett, J. R., Kyriakis, J. M., and Clark, E. A. (1996). Cross-linking CD40 on B cells preferentially induces stress-activated protein kinases rather than mitogen-activated protein kinases. *EMBO J.*, **15**, 92.

142. Bird, T. A., Kyriakis, J. M., Tyshler, L., Gayle, M., Milne, A., and Virca, G. D. (1994). Interleukin-1 activates p54 mitogen-activated protein (MAP) kinase/stress-activated protein kinase by a pathway that is independent of p21ras, Raf-1 and MAP kinase-kinase. *J. Biol. Chem.*, **269**, 31836.

143. Lenczowski, J. M., Dominguez, L., Eder, A. M., King, L. B., Zacharchuk, C. M., and Ashwell, J. D. (1997). Lack of a role for Jun kinase and AP-1 in Fas-induced apoptosis. *Mol. Cell. Biol.*, **17**, 170.

144. Galibert, L., Tometsko, M. E., Anderson, D. M., Cosman, D., and Dougall, W. C. (1998).

The involvement of multiple tumor necrosis factor receptor (TNFR)-associated factors in the signaling mechanisms of receptor activator of NF-κB, a member of the TNFR superfamily. *J. Biol. Chem.*, **273**, 34120.

145. Akiba, H., Nakano, H., Nishinaka, S., Shindo, M., Kobata, T., Atsuta, M., Morimoto, C., Ware, C. F., Malinin, N., Wallach, D., Yagita, H., and Okumura, K. (1998). CD27, a member of the tumor necrosis factor receptor superfamily, activates NF-κB and stress-activated protein kinase/c-Jun N-terminal kinase via TRAF2, TRAF5 and NF-κB-inducing kinase. *J. Biol. Chem.*, **273**, 13353.

146. Gupta, S., Barrett, T., Whitmarsh, A. J., Cavanagh, J., Sluss, H. K., Dérijard, B., and Davis, R. J. (1996). Selective interaction of JNK protein kinase isoforms with transcription factors. *EMBO J.*, **15**, 2760.

147. Han, J., Lee, J.-D., Bibbs, L., and Ulevitch, R. J. (1994). A MAP kinase targeted by endotoxin and hyperosmolarity in mammalian cells. *Science*, **265**, 808.

148. Rouse, J., Cohen, P., Trigon, S., Morange, M., Alonso-Llamazares, A., Zamanillo, D., Hunt, T. and Nebreda, A. (1994). A novel kinase cascade triggered by stress and heat shock that stimulates MAPKAP kinase-2 and phosphorylation of the small heat shock proteins. *Cell*, **78**, 1027.

149. Freshney, N. W., Rawlinson, L., Guesdon, F., Jones, E., Cowley, S., Hsuan, J., and Saklatvala, J. (1994). Interleukin-1 activates a novel protein kinase cascade that results in the phosphorylation of Hsp27. *Cell*, **78**, 1039.

150. Lee, J. C., Laydon, J. T., McDonnell, P. C., Gallagher, T. F., Kumar, S., Green, D., McNulty, D., Blumenthal, M. J., Heys, J. R., Landvatter, S. W., Strickler, J. E., McLaughlin, M. M., Siemens, I. R., Fisher, S. M., Livi, G. P., White, J. R., Adams, J. L., and Young, P. (1994). A protein kinase involved in the regulation of inflammatory cytokine biosynthesis. *Nature*, **273**, 739.

151. Zervos, A. S., Faccio, L., Gatto, J. P., Kyriakis, J. M., and Brent, R. (1995). Mxi2, a mitogen-activated protein kinase that recognizes and phosphorylates Max protein. *Proc. Natl Acad. Sci. USA*, **92**, 10531.

152. Jiang, Y., Chen, C., Li, Z., Guo, W., Gegner, J. A., Lin, S., and Han, J. (1996). Characterization of the structure and function of a new mitogen-activated protein kinase (p38β). *J. Biol. Chem.*, **271**, 17920.

153. Lechner, C., Zahalka, M. A., Giot, J. F., Moller, N. P., and Ullrich, A. (1996). ERK6, a mitogen-activated protein kinase is involved in C2C12 myoblast differentiation. *Proc. Natl Acad. Sci. USA*, **93**, 4355.

154. Mertens, S., Craxton, M., and Goedert, M. (1996). SAP kinase-3, a new member of the family of mammalian stress-activated protein kinases. FEBS Lett., **383**, 273.

155. Stein, B., Yang, M. X., Young, D. B., Janknecht, R., Hunter, T., Murray, B. W., and Barbosa, M. (1997). p38-2, a novel mitogen-activated protein kinase with distinct properties. *J. Biol. Chem.*, **272**, 19509.

156. Goedert, M., Cuenda, A., Craxton, M., Jakes, R., and Cohen, P. (1997). Activation of the novel stress-activated protein kinase SAPK4 by cytokines and cellular stresses is mediated by SKK3 (MKK6); comparison of its substrate specificity with that of other SAP kinases. *EMBO J.*, **16**, 3563.

157. Wang, X. S., Diener, K., Manthey, C. L., Wang, S.-W., Rosenzweig, B., Bray, J., Delaney, J., Cole, C. N., Chan-Hui, P.-Y., Mantlo, N., Lichenstein, H. S., Zukowski, M., and Yao, Z. (1997). Molecular cloning and characterization of a novel p38 mitogen-activated protein kinase. *J. Biol. Chem.*, **272**, 23688.

158. Jiang, Y., Gram, H., Zhao, M., New, L., Gu, J., Feng, L., Di Padova, F., Ulevitch, R. J., and

Han, J. (1997). Characterization of the structure and function of the fourth member of the p38 group mitogen-activated protein kinases, p38δ. *J. Biol. Chem.*, **272**, 30122.

159. Eyers, P. A., Craxton, M., Morrice, N., Cohen, P., and Goedert, M. (1998). Conversion of SB 203580-insensitive MAP kinase family members to drug-sensitive forms by a single amino-acid substitution. *Chem. Biol.*, **5**, 321.

160. Gum, R. J., MaLauthlin, M. M., Kumar, S., Wang, Z., Bower, M. J., Lee, J. C., Adams, J. L., Livi, G. P., Goldsmith, E. J., and Young, P. R. (1998). Acquisition of sensitivity of stress-activated protein kinases to the p38 inhibitor, SB 203580, by alteration of one or more amino acids within the ATP binding pocket. *J. Biol. Chem.*, **273**, 15605.

161. Pombo, C. M., Bonventre, J. V., Avruch, J., Woodgett, J. R., Kyriakis, J. M., and Force, T. (1994). The stress-activated protein kinases are major c-Jun amino-terminal kinases activated by ischaemia and reperfusion, *J. Biol. Chem.*, **269**, 26546.

162. Bogoyevitch, M. A., Gillespie-Brown, J., Ketterman, A. J., Fuller, S. J., Ben-Levy, R., Ashworth, A., Marshall, C. J., and Sugden, P. H. (1996). Stimulation of the stress-activated mitogen-activated protein kinase subfamilies in perfused heart. p38/RK mitogen-activated protein kinases and c-Jun N-terminal kinases are activated by ischaemia/reperfusion. *Circ. Res.*, **79**, 162.

163. Zhou, G., Bao, Z. Q., and Dixon, J. E. (1995). Components of a new human protein kinase signal transduction pathway. *J. Biol. Chem.*, **270**, 12665.

164. English, J. M., Vanderbilt, C. A., Xu, S., Marcus, S., and Cobb, M. H. (1995). Isolation of MEK5 and differential expression of alternatively spliced forms. *J. Biol. Chem.*, **270**, 28897.

165. Abe, J.-I., Kusuhara, M., Ulevitch, R. J., Berk, B. C., and Lee, J.-D. (1996). Big mitogen-activated protein kinase 1 (BMK1) is a redox-sensitive kinase. *J. Biol. Chem.*, **271**, 16586.

166. Stokoe, D., Campbell, D. G., Nakielny, S., Hidaka, H., Leevers, S., Marshall, C., and Cohen, P. (1992). MAPAKP kinase-2; a novel protein kinase activated by mitogen-activated protein kinase. *EMBO J.*, **11**, 3985.

167. Stokoe, D., Engel, K., Campbell, D. G., Cohen, P., and Gaestel, M. (1992). Identification of MAPKAP kinase 2 as a major enzyme responsible for the phosphorylation of the small mammalian heat shock proteins. *FEBS Lett.*, **313**, 307.

168. McLaughlin, M. M., Kumar, S., McDonnell, P. C., Van Horn, S., Lee, J. C., Livi, G. P., and Young, P. R. (1996). Identification of mitogen-activated protein (MAP) kinase-activated protein kinase-3, a novel substrate of CSBP p38 MAP kinase. *J. Biol. Chem.*, **271**, 8488.

169. Sithanandam, G., Latif, F., Smola, U., Bernal, R. A., Duh, F.-M., Li, H., Kuzmin, I., Wixler, V., Geil, L., Shresta, S., Lloyd, P. A., Bader, S., Sekido, Y., Tartof, K. D., Kashuba, V. I., Zabarovsky, E. R., Dean, M., Klein, G., Zbar, B., Lerman, M. I., Minna, J. D., Rapp, U. R., and Allikmets, A. (1996). 3pK, a new mitogen-activated protein kinase-activated protein kinase located in the small cell lung cancer tumor suppressor gene region. *Mol. Cell. Biol.*, **16**, 868 [published erratum appears in *Mol. Cell. Biol.* (1996). **16**, 1880].

170. Huot, J., Houle, F., Marceau, F., and Landry, J. (1997). Oxidative stress-induced actin re-organization mediated by the p38 mitogen-activated protein kinase/heat shock protein 27 pathway in vascular endothelial cells. *Circ. Res.*, **80**, 383.

171. Lambert, H., Charette, S. J., Bernier, A. F., Guimond, A., and Landry, J. (1999). HSP27 multimerization mediated by phosphorylation-sensitive intermolecular interactions at the amino terminus. *J. Biol. Chem.*, **274**, 9378.

172. Ben-Levy, R., Leighton, I. A., Doza, Y. N., Attwood, P., Morrice, N., Marshall, C. J., and Cohen, P. (1995). Identification of novel phosphorylation sites required for activation of MAPKAP kinase-2. *EMBO J.*, **14**, 5920.

173. Cuenda, A., Cohen, P., Buee-Scherrer, V., and Goedert, M. (1997). Activation of stress-

activated protein kinase-3 (SAPK3) by cytokines and cellular stresses is mediated via SAPKK3 (MKK6); comparison of the specificities of SAPK3 and SAPK2 (RK/p38). *EMBO J.*, **16**, 295.

174. Ludwig, S., Engel, K., Hoffmeyer, A., Sithanandam, G., Neufeld, B., Palm, D., Gaestel, M., and Rapp, U. R. (1996). 3pK, a novel mitogen-activated protein (MAP) kinase-activated protein kinase is targeted by three MAP kinase pathways. *Mol Cell Biol.*, **16**, 6687.

175. New, L., Jiang, Y., Zhao, M., Liu, K., Zhu, W., Flood, L. J., Kato, Y., Parry, G. C. N., and Han, J. (1998). PRAK, a novel protein kinase regulated by the p38 MAP kinase. *EMBO J.*, **17**, 3372.

176. Wang, X. and Ron, D. (1996). Stress-induced phosphorylation and activation of the transcription factor CHOP (GADD153) by p38 MAP kinase. *Science*, **272**, 1347.

177. Rao, A., Luo, C., and Hogan, P. G. (1997). Transcription factors of the NFAT family: regulation and function. *Annu. Rev. Immunol.*, **15**, 707.

178. Chow, C.-W., Rincón, M., Cavanagh, J., Dickens, M., and Davis, R. J. (1997). Nuclear accumulation of NFAT4 opposed by the JNK signal transduction pathway. *Science*, **278**, 1638.

179. Zhu, J., Shibakashi, F., Price, R., Guillemot, J.-C., Yano, T., Dötch, V., Wagner, G., Ferrara, P., and McKeon, F. (1998). Intramolecular masking of nuclear import signal on NF-AT-4 by casein kinase I and MEKK1. *Cell*, **93**, 851.

180. Blackwood, E. M., Kretzner, L., and Eisenman, R. N. (1992). Myc and Max function as a nucleoprotein complex. *Curr. Opin. Genet. Dev.*, **2**, 227.

181. Kato, Y., Kravchenko, V. V., Tapping, R. I., Han, J., Ulevitch, R. J., and Lee, J.-D. (1997). BMK1/ERK5 regulates serum-induced early gene expression through transcription factor MEF2C. *EMBO J.*, **16**, 7054.

182. Gupta, S., Campbell, D., Dérijard, B., and Davis, R. J. (1995). Transcription factor ATF2 regulation by the JNK signal transduction pathway. *Science*, **267**, 389.

183. Morooka, H., Bonventre, J. V., Pombo, C. M., Kyriakis, J. M. and Force, T. (1995). Ischaemia and reperfusion enhance ATF-2 and c-Jun binding to cAMP response elements and to an AP-1 binding site from the c-Jun promoter. *J. Biol. Chem.*, **270**, 30084.

184. Han, J., Jiang, Y, Li, Z., Kravchenko, V. V., and Ulevitch, R. J. (1997). MEF2C participates in inflammatory responses via p38-mediated activation. *Nature*, **386**, 563.

185. Zhao, M., New, L., Kravchenko, V. V., Kato, Y., Gram, H., di Padova, F., Olson, E. N., Ulevitch, R. J., and Han, J. (1999). Regulation of the MEF2 family of transcription factors by p38. *Mol. Cell. Biol.*, **19**, 21.

186. Sánchez, I., Hughes, R. T., Mayer, B. J., Yee, K., Woodgett, J. R., Avruch, J., Kyriakis, J. M., and Zon, L. I. (1994). Role of SAPK/ERK kinase-1 in the stress-activated pathway regulating transcription factor c-Jun. *Nature*, **372**, 794.

187. Dérijard, B., Raingeaud, J., Barrett, T., Wu, L.-H., Han, J., Ulevitch, R. J., and Davis, R. J. (1995). Independent human MAP kinase signal transduction pathways defined by MEK and MKK isoforms. *Science*, **267**, 682.

188. Lawler, S., Fleming, Y., Goedert, M., and Cohen, P. (1998). Synergistic activation of SAPK1/JNK1 by two MAP kinase kinases *in vitro*. *Curr. Biol.*, **8**, 1387.

189. Nishina, H., Fischer, K. D., Radvanyi, L., Shahinian, A., Razqallah, H., Rubie, E. A., Bernstein, A., Mak, T. W., Woodgett, J. R., and Penninger, J. M. (1997). Stress-signalling kinase Sek1 protects thymocytes from apoptosis mediated by CD95 and CD3. *Nature*, **385**, 350.

190. Meier, R., Rouse, J., Cuenda, A., Nebreda, A., and Cohen, P. (1996). Cellular stresses and cytokines activate multiple mitogen-activated-protein kinase kinase homologues in PC12 and KB cells. *Eur. J. Biochem.*, **236**, 796.

191. Moriguchi, T., Kawasaki, H., Matsuda, S., Gotoh, Y., and Nishida, E. (1995). Evidence for multiple activators for stress-activated protein kinases/c-Jun amino terminal kinases. *J. Biol. Chem.*, **270**, 12969.

192. Glise, B., Bourbon, H., and Noselli, S. (1995). Hemipterous encodes a novel *Drosophila* MAP kinase kinase, required for epithelial cell sheet movement. *Cell*, **83**, 451.

193. Holland, P. M., Suzanne, M., Campbell, J. S., Noselli, S., and Cooper, J. A. (1997). MKK7 is a stress-activated mitogen-activated proetin kinase kinase functionally related to hemipterous. *J. Biol. Chem.*, **272**, 24994.

194. Tournier, C., Whitmarsh, A. J., Cavanagh, J., Barrett, T., and Davis, R. J. (1997). Mitogen-activated protein kinase kinase 7 is an activator of the c-Jun NH_2-terminal kinase. *Proc. Natl Acad. Sci. USA*, **94**, 7337.

195. Yao, Z., Deiner, K., Wang, X. S., Zukowski, M., Matsumoto, G., Zhou, G., Mo, R., Sasaki, T., Nishina, H., Hui, C. C., Tan, T.-H, Woodgett, J. R., and Penninger, J. M. (1997). Activation of stress-activated protein kinases/c-Jun N-terminal kinases (SAPKs/JNKs) by a novel mitogen-activated protein kinase kinase (MKK7). *J. Biol. Chem.*, **272**, 32378.

196. Foltz, I. N., Gerl, R. E., Wieler, J. S., Luckach, M., Salmon, R. A., and Schrader, J. W. (1998). Human mitogen-activated protein kinase kinase 7 (MKK7) is a highly conserved c-Jun N-terminal kinase/stress-activated protein kinase (JNK/SAPK) activated by environmental stresses and physiological stimuli. *J. Biol. Chem.*, **273**, 9344.

197. Moriguchi, T., Toyoshima, F., Masuyama, N., Hanafusa, H., Gotoh, Y., and Nishida, E. (1997). A novel SAPK/JNK kinase, MKK7, stimulated by $TNF\alpha$ and cellular stresses. *EMBO J.*, **16**, 7045.

198. Wu, Z., Wu, J., Jacinto, E., and Karin, M. (1997). Molecular cloning and characterization of human JNKK2, a novel Jun NH_2-terminal kinase-specific kinase. *Mol. Cell. Biol.*, **17**, 7407.

199. Lu, X., Nemoto, S., and Lin, A. (1997). Identification of c-Jun NH2-terminal protein kinase (JNK)-activating kinase 2 as an activator of JNK but not p38. *J. Biol. Chem.*, **272**, 24751.

200. Tournier, C., Whitmarsh, A. J., Cavanagh, J., Barrett, T., and Davis, R. J. (1999). The MKK7 gene encodes a group of c-Jun NH_2-terminal kinase kinases. *Mol. Cell. Biol.*, **19**, 1569.

201. Ganiatsas, S., Kwee, L., Fujiwara, Y., Perkins, A., Ikeda, T., Labow, M. A., and Zon, L. I. (1998). SEK1 deficiency reveals mitogen-activated protein kinase cascade crossregulation and leads to abnormal hepatogenesis. *Proc. Natl Acad. Sci. USA*, **95**, 6881.

202. Nishina, H., Vaz, A., Billia, P., Nghiem, M., Sasaki, T., de la Pompa, J. L., Furlonger, K., Paige, C., Hui, C., Fischer, K. D., Kishimoto, H., Iwatsubo, T., Katada, T., Woodgett, J. R., and Penninger, J. M. (1999). Defective liver formation and liver cell apoptosis in mice lacking the stress signaling kinase SEK1/MKK4. *Development*, **126**, 505.

203. Raingeaud, J., Whitmarsh, A. J., Barett, T., Dérijard, B., and Davis, R. J. (1996). MKK3- and MKK6-regulated gene expression is mediated by the p38 mitogen-activated protein kinase signal transduction pathway. *Mol. Cell. Biol.*, **16**, 1247.

204. Cuenda, A., Alonso, G., Morrice, N., Jones, M., Meier, R., Cohen, P., and Nebreda, A. (1996). Purification and cDNA cloning of SAPKK3, the major activator of RK/p38 in stress- and cytokine-stimulated monocytes and epithelial cells. *EMBO J.*, **15**, 4156.

205. Lange-Carter, C. A., Pleiman, C. M., Gardner, A. M., Blumer, K. J., and Johnson, G. L. (1993). A divergence in the MAP kinase regulatory network defined by MEK kinase and Raf. *Science*, **260**, 315.

206. Xu, S., Robbins, D. J., Christerson, L. B., English, J. M., Vanderbilt, C., and Cobb, M. H. (1996). Cloning of Rat MEK kinase 1 cDNA reveals an endogenous membrane-associated 195-kDa protein with a large regulatory domain. *Proc. Natl Acad. Sci. USA*, **93**, 5291.

207. Blank, J. L., Gerwins, P., Elliot, E. M., Sather, S., and Johnson, G. L. (1996). Molecular cloning of mitogen activated protein/ERK kinase kinases (MEKK) 2 and 3. *J. Biol. Chem.*, **271**, 5361.

208. Ellinger-Ziegelbauer, H., Brown, K., Kelly, K., and Siebenlist, U. (1997). Direct activation of the stress-activated protein kinase and extracellular signal-regulated protein kinase (ERK) pathways by an inducible mitogen-activated protein kinase/ERK kinase kinase 3 derivative. *J. Biol. Chem.*, **272**, 2668.

209. Gerwins, P., Blank, J. L., and Johnson, G. L. (1997). Cloning of a novel mitogen-activated protein kinase-kinase-kinase, MEKK4, that selectively regulates the c-Jun amino terminal kinase pathway. *J. Biol. Chem.*, **272**, 8288.

210. Takekawa, M., Posas, F., and Saito, H. (1997). A human homolog of the yeast Ssk2/Ssk22 MAP kinase kinase kinases, MTK1, mediates stress-induced activation of the p38 and JNK pathways. *EMBO J.*, **16**, 4973.

211. Ichijo, H., Nishida, E., Irie, K., ten Dijke, P., Saitoh, M., Moriguchi, T., Takagi, M., Matsumoto, K., Miyazono, K., and Gotoh, Y. (1997). Induction of apoptosis by ASK1, a mammalian MAPKKK that activates SAPK/JNK and p38 signaling pathways. *Science*, **275**, 90.

212. Wang, X. S., Diener, K., Jannuzzi, D., Trollinger, D., Tan, T.-H., Lichenstein, H., Zukowski, M., and Yao, Z. (1996). Molecular cloning and characterization of a novel protein kinase with a catalytic domain homologous to mitogen-activated protein kinase kinase kinase. *J. Biol. Chem.*, **271**, 31607.

213. Yamaguchi, K., Shirakabi, K., Shibuya, H., Irie, K., Oishi, I., Ueno, N., Taniguchi, T., Nishida, E., and Matsumoto, K. (1995). Identification of a member of the MAPKKK family as a potential mediator of TGFβ signal transduction. *Science*, **270**, 2008.

214. Patriotis, C., Makris, A., Bear, S. E., and Tsichlis, P. N. (1993). Tumor progression locus 2 (Tpl-2) encodes a protein kinase involved in the progression of rodent T-cell lymphomas and in T-cell activation. *Proc. Natl Acad. Sci. USA*, **90**, 2251.

215. Ceci, J. D., Patriotis, C. P., Tsatsanis, C., Makris, A. M., Kovatch, R., Swing, D. A., Jenkins, N. A., Tsichlis, P. N., and Copeland, N. G. (1997). Tpl-2 is an oncogenic kinase that is activated by carboxy-terminal truncation. *Genes Dev.*, **11**, 688.

216. Malinin, N. L., Boldin, M. P., Kovalenko, A. V., and Wallach, D. (1997). MAP3K-related kinase involved in NF-κB induction by TNF, CD-95 and IL-1. *Nature*, **385**, 540.

217. Rana, A., Gallo, K., Godowski, P., Hirai, S.-I., Ohno, S., Zon, L. I., Kyriakis, J. M., and Avruch, J. (1996). The mixed lineage protein kinase SPRK phosphorylates and activates the stress-activated protein kinase activator, SEK1. *J. Biol. Chem.*, **271**, 19025.

218. Tibbles, L. A., Ing, Y. L., Kiefer, F., Chan, J., Iscove, N., Woodgett J. R., and Lassam, N. J. (1996). MLK-3 activates the SAPK/JNK and p38/RK pathways via SEK1 and MKK3/6. *EMBO J.*, **15**, 7026.

219. Hirai, S.-I., Katoh, M., Terada, M., Kyriakis, J. M., Zon, L. I., Rana, A., Avruch, J., and Ohno, S. (1997). MST/MLK2, a member of the mixed lineage kinase family, directly phosphorylates and activates SEK1, an activator of c-Jun N-terminal kinase/stress-activated protein kinase. *J. Biol. Chem.*, **272**, 15167.

220. Hirai, S.-I., Noda, K., Moriguchi, T., Nishida, E., Yamashita, A., Deyama, T., Fukuyama, K., and Ohno, S. (1998). Differential activation of two JNK activators, MKK7 and SEK1, by MKN28-derived nonreceptor serine/threonine kinase/mixed lineage kinase 2. *J. Biol. Chem.*, **273**, 7406.

221. Holtzman, L. B., Merritt, S. E., and Fan, G. (1994). Identification, molecular cloning and characterization of dual leucine zipper bearing kinase. *J. Biol. Chem.*, **269**, 30808.

222. Fan, G., Merritt, S. E., Kortenjann, M., Shaw, P. E., and Holtzman, L. B. (1996). Dual leucine zipper-bearing kinase (DLK) activates p46SAPK and p38mapk but not ERK2. *J. Biol. Chem.*, **271**, 24788.

223. Hutchinson, M., Berman, K. S., and Cobb, M. H. (1998). Isolation of TAO1, a protein kinase that activates MEKs in stress-activated protein kinase cascades. *J. Biol. Chem.*, **273**, 28625.

224. Fanger, G. R., Widmann, C., Porter, A. C., Sather, S., Johnson, G. L., and Vaillancourt, R. R. (1998). 14–3–3 proteins interact with specific MEK kinases. *J. Biol. Chem.*, **273**, 347.

225. Cardone, M. H., Salveson, G. S., Widmann, C., Johnson, G., and Frisch, S. (1997). The regulation of anoikis: MEKK-1 activation requires cleavage by caspases. *Cell*, **90**, 315.

226. Russell, M., Lange-Carter, C. A., and Johnson, G. L. (1995). Direct interaction between Ras and the kinase domain of mitogen-activated protein kinase kinase kinase (MEKK1). *J. Biol. Chem.*, **270**, 11757.

227. Fanger, G. R., Johnson, N. L., and Johnson, G. L. (1997). MEK kinases are regulated by EGF and selectively interact with Rac/Cdc42. *EMBO J.*, **16**, 4961.

228. Yan, M., Dai, T., Deak, J. C., Kyriakis, J. M., Zon, L. I., Woodgett, J. R., and Templeton, D. J. (1994). Activation of stress-activated protein kinase by MEKK1 phosphorylation of its activator SEK1. *Nature*, **372**, 798.

229. Minden, A., Lin, A., McMahon, M., Lange-Carter, C., Dérijard, B., Davis, R. J., Johnson, G. L., and Karin, M. (1994). Differential activation of ERK and JNK mitogen-activated protein kinases by Raf-1 and MEKK. *Science*, **266**, 1719.

230. Ellinger-Ziegelbauer, H., Kelly, K., and Siebenlist, U. (1999). Cell cycle arrest and reversion of Ras-induced transformation by a conditionally activated form of mitogen-activated protein kinase kinase kinase 3. *Mol. Cell. Biol.*, **19**, 3857.

231. Deacon, K. and Blank, J. L. (1997). Characterization of the mitogen-activated protein kinase kinase 4 (MKK4)/c-Jun NH_2-terminal kinase 1 and MKK3/p38 pathways regulated by MEK kinases 2 and 3. MEK kinase 3 activates MKK3 but does not cause activation of p38 kinase *in vivo*. *J. Biol. Chem.*, **272**, 14489.

232. Deacon, K. and Blank, J. L. (1999). MEK kinase 3 directly activates MKK6 and MKK7, specific activators of the p38 and c-Jun NH_2-terminal kinases. *J. Biol. Chem.*, **274**, 16604.

233. Takekawa, M. and Saito, H. (1998). A family of stress-inducible GADD45-like proteins mediate activation of the stress-responsive MTK1/MEKK4 MAPKKK. *Cell*, **95**, 521.

234. Shirakabe, K., Yamaguchi, K., Shibuya, H., Irie, K., Matsuda, S., Moriguchi, T., Gotoh, Y., Matsumoto, K., and Nishida, E. (1997). TAK1 mediates the ceramide signaling to stress-activated protein kinase/c-Jun N-terminal kinase. *J. Biol. Chem.*, **272**, 8141.

235. Ninomiya-Tsuji, J., Kishioto, K., Hiyama, A., Inoue, J.-I, Cao, Z., and Matsumoto, K. (1999). The kinase TAK1 can activate the NIK-IκB as well as the MAP kinase cascade in the IL-1 signalling pathway. *Nature*, **398**, 252.

236. Shibuya, H., Yamaguchi, K., Shirakabe, K., Tonegawa, A., Gotoh, Y., Ueno, N., Irie, K., Nishida, N., and Matsumoto, K. (1996). TAB1: an activator of the TAK1 MAPKKK in TGFβ signal transduction. *Science*, **272**, 1179.

237. Moriguchi, T., Kuroyanagi, N., Yamaguchi, K., Gotoh, Y., Irie, K., Kano, T., Shirakabe, K., Muro, Y., Shibuya, H., Matsumoto, K., Nishida, E., and Hagiwara, M. (1996). A novel kinase cascade mediated by mitogen-activated protein kinase kinase 6 and MKK3. *J. Biol. Chem.*, **271**, 13675.

238. Chang, H. Y., Nishitoh, H., Yang, X., Ichijo, H., and Baltimore, D. (1998). Activation of apoptosis signal-regulating kinase 1 (ASK1) by the adapter protein Daxx. *Science*, **281**, 1860.

239. Saitoh, M., Nishitoh, H., Fujii, M., Takeda, K., Tobiume, K., Sawada, Y., Kawabata, M., Miyazono, K., and Ichijo, H. (1998). Mammalian thioredoxin is a direct inhibitor of apoptosis signal-regulating kinase (ASK) 1. *EMBO J.*, **17**, 2596.

240. Nishitoh, H., Saitoh, M., Mochida, Y., Takeda, K., Nakano, H., Rothe, M., Miyazono, K., and Ichijo, H. (1998). ASK1 is essential for JNK/SAPK activation by TRAF2. *Mol. Cell*, **2**, 389.

241. Patriotis, C., Makris, A., Chernoff, J., and Tsichlis, P. N. (1994). Tpl-2 acts in concert with Ras and Raf-1 to activate mitogen-activated protein kinase. *Proc. Natl Acad. Sci. USA*, **91**, 9755.

242. Dorow, D. S., Devereux, L., Dietzsch, E., and De Kretser, T. (1993). Identification of a new family of human epithelial protein kinases containing two leucine/isoleucine-zipper domains. *Eur. J. Biochem.*, **213**, 701.

243. Gallo, K. A., Mark, M. R., Scadden, D. T., Wang, Z., Gu, Q., and Godowski, P. J. (1994). Identification and characterization of SPRK, a novel src-homology 3 domain-containing proline-rich kinase with serine/threonine kinase activity. *J. Biol. Chem.*, **269**, 15092.

244. Aronheim, A., Broder, Y. C., Cohen, A., Fritsch, A., Belisle, B., and Abo, A. (1998). Chp, a homologue of the GTPase Cdc42Hs, activates the JNK pathway and is implicated in reorganizing the actin cytoskeleton. *Curr. Biol.*, **8**, 1125.

245. Coso, O. A., Chiarello, M., Yu, J.-C., Teramoto, H., Crespo, P., Xu, N., Miki, T., and Gutkind, J. S. (1995). The small GTP binding proteins Rac1 and Cdc42 regulated the activity of the JNK/SAPK signaling pathway. *Cell*, **81**, 1137.

246. Minden, A., Lin, A., Claret, F.-X., Abo, A., and Karin, M. (1995). Selective activation of the JNK signaling cascade and c-Jun transcriptional activity by the small GTPases Rac and Cdc42Hs. *Cell*, **81**, 1147.

247. Olson, M. F., Ashworth, A., and Hall, A. (1995). An essential role for Rho, Rac, and Cdc42 GTPases in cell cycle progression through G1. *Science*, **269**, 1270.

248. Zhang, S., Han, J., Sells, M. A., Chernoff, J., Knaus, U. G., Ulevitch, R. J., and Bokoch, G. M. (1995). Rho family GTPases regulate p38 mitogen-activated protein kinase through the downstream mediator Pak1. *J. Biol. Chem.*, **270**, 23934.

249. Bagrodia, S., Dérijard, B., Davis, R. J., and Cerione, R. A. (1995). Cdc42 and PAK-mediated signaling leads to Jun kinase and p38 mitogen-activated protein kinase activation. *J. Biol. Chem.*, **270**, 27995.

250. Teramoto, H., Crespo, P., Coso, O. A., Igishi, T., Xu, N., and Gutkind, J. S. (1996). The small GTP binding protein RhoA activates c-Jun N-terminal kinases/stress-activated protein kinases in human kidney 293T cells. *J. Biol. Chem.*, **271**, 25731.

251. Zheng, Y., Fischer, D. J., Santos, M. F., Tigyi, G., Pasteris, N. G., Gorski, J. L., and Xu, Y. (1996). The factigenital dysplasia gene product FGD1 functions as a Cdc42Hs-specific guanine-nucleotide exchange factor. *J. Biol. Chem.*, **271**, 33169.

252. Lamarche, N., Tapon, N. Stowers, L., Burbelo, P., Aspenstrom, P., Bridges, T., Chant, J., and Hall, A. (1996). Rac and Cdc42 induce actin polymerization and G1 cell cycle progression independently of p65PAK and the JNK/SAPK MAP kinase cascade. *Cell*, **87**, 519.

253. Burbelo, P. D., Drechsel, D., and Hall, A. (1995). A conserved binding motif defines numerous candidate target proteins for both Cdc42 and Rac GTPases. *J. Biol. Chem.*, **270**, 29071.

254. Manser, E., Leung, T., Salihuddin, H., Zhao, Z.-S., and Lim, L. (1994). A brain serine/threonine protein kinase activated by Cdc42 and Rac1. *Nature*, **367**, 40.

255. Teo, M., Manser, E., and Lim, L. (1995). Identification and molecular cloning of a

p21cdc42/rac1-activated serine/threonine kinase that is rapidly activated by thrombin in platelets. *J. Biol. Chem.*, **270**, 26690.

256. Martin, G. A., Bollag, G., McCormick, F., and Abo, A. (1995). A novel serine kinase activated by rac1/CDC42Hs-dependent autophosphorylation is related to PAK65 and STE20. *EMBO J.*, **14**, 1970.

257. Manser, E., Chong, C., Zhao, Z.-S., Leung, T., Michael, G., Hall, C., and Lim, L. (1995). Molecular cloning of a new member of the p21-Cdc42/Rac-activated kinase (PAK) family. *J. Biol. Chem.*, **270**, 25070.

258. Abo, A., Qu, J., Cammarano, M. S., Dan, C., Fritsch, A., Baud, V., Belisle, B., and Minden, A. (1998). PAK4, a novel effector for Cdc42Hs, is implicated in the reorganization of the actin cytoskeleton and in the formation of filopodia. *EMBO J.*, **17**, 6527.

259. Lee, N., MacDonald, H., Reinhard, C., Halenbeck, R., Roulston, A., Shi, T., and Williams, L. T. (1997). Activation of hPAK65 by caspase cleavage induces some of the morphological and biochemical changes of apoptosis. *Proc. Natl Acad. Sci. USA*, **94**, 13642.

260. Sells, M. A., Knaus, U. G., Bagrodia, S., Ambrose, D. M., Bokoch, G. M., and Chernoff, J. (1997). Human p21-activated kinase (Pak1) regulates actin organization in mammalian cells. *Curr. Biol.*, **7**, 202.

261. Sells, M. A., Boyd, J. T., and Chernoff, J. (1999). p21-activated kinase 1 (Pak1) regulates cell motility in mammalian fibroblasts. *J. Cell Biol.*, **145**, 837.

262. Manser, E., Loo, T.-H., Koh, C.-G., Zhao, Z.-S., Chen, X.-Q., Tan, L., Tan, I., Leung, T., and Lim, L. (1998). PAK kinases are directly coupled to the PIX family of nucleotide exchange factors. *Mol. Cell*, **1**, 183.

263. Brown, J. L., Stowers, L., Baer, M., Trejo, J., Coughlin, S., and Chant, J. (1996). Human Ste20 homologue hPAK1 links GTPases to the JNK MAP kinase pathway. *Curr. Biol.*, **6**, 598.

264. Polverino, A., Frost, J., Yang, P., Hutchinson, M., Neiman, A. M., Cobb, M. H., and Marcus, S. (1995). Activation of mitogen-activated protein kinase cascades by p21-activated protein kinases in cell-free extracts of *Xenopus* oocytes. *J. Biol. Chem.*, **270**, 26067.

265. Nagata, K., Puls, A., Futter, C., Aspenstrom, P., Schaefer, E., Nakata, T., Hirokawa, N., and Hall, A. (1998). The MAP kinase kinase kinase MLK2 co-localizes with activated JNK along microtubules and associates with kinesin superfamily motor KIF3. *EMBO J.*, **17**, 149.

266. Tapon, N., Nagata, K., Lamarche, N., and Hall, A. (1998). A new Rac target POSH is an SH3-containing scaffold protein involved in the JNK and NF-κB signalling pathways. *EMBO J.*, **17**, 1395.

267. Force, T., Pombo, C. M., Avruch, J. A., Bonventre, J. V., and Kyriakis, J. M. (1996). Stress-activated protein kinases in cardiovascular disease. *Circ. Res.*, **78**, 947.

268. Choukroun, G., Bonventre, J. V., Kyriakis, J. M., Rosenzweig, A., and Force, T. (1998). Endothelin stimulation of cardiomyocyte hypertrophy requires the SAPK pathway. *J. Clin. Invest.*, **102**, 1311.

269. Coso, O. A., Chiariello, M., Kalinec, G., Kyriakis, J. M., Woodgett, J. R., and Gutkind, J. S. (1995). Transforming G protein-coupled receptors potently activate JNK (SAPK). *J. Biol. Chem.*, **270**, 5620.

270. Hamm, H. E. (1998). The many faces of G protein signaling. *J. Biol. Chem.*, **273**, 669.

271. Kehrl, J. H. (1998). Heterotrimeric G protein signaling: roles in immune function and fine-tuning by RGS proteins. *Immunity*, **8**, 1.

272. Prasad, M. V. V. S. V., Dermott, J. M., Heasley, L. E., Johnson, G. L., and Dhanasekaran, N. (1995). Activation of Jun kinase/stress-activated protein kinase by GTPase-deficient mutants of Gα12 and Gα13. *J. Biol. Chem.*, **270**, 18655.

273. Collins, L. R., Minden, A., Karin, M., and Heller Brown, J. (1996). Gα12 stimulates c-Jun NH2-terminal kinase through the small G proteins Ras and Rac. *J. Biol. Chem.*, **271**, 17349.

274. Heasley, L. E., Storey, B., Fanger, G. R., Butterfield, L., Zamarripa, J., Blumberg, D., and Maue, R. A. (1996). GTPase-deficient Gα16 and Gαq induce PC12 cell differentiation and persistent activation of c-Jun NH2-terminal kinases. *Mol. Cell. Biol.*, **16**, 648.

275. Coso, O. A., Teramoto, H., Simonds, W. F., and Gutkind, J. S. (1996). Signaling from G protein-coupled receptors to c-Jun kinase involves βγ subunits of heterotrimeric G proteins acting on a Ras and Rac1-dependent pathway. *J. Biol. Chem.*, **271**, 3963.

276. Jiang, Y., Ma, W., Kozasa, T., Hattori, S., and Huang, X. Y. (1998). The G protein Gα12 stimulates Bruton's tyrosine kinase and a rasGAP through a conserved PH/BM domain. *Nature*, **395**, 808.

277. Kawakami, Y., Miura, T., Bissonnette, R., Hata, D., Khan, W. N., Kitamura, T., Maeda-Yamamoto, M., Hartman, S. E., Yao, F., Alt, F. W., and Kawakami, T. (1997). Bruton's tyrosine kinase regulates apoptosis and JNK/SAPK kinase activity. *Proc. Natl Acad. Sci. USA*, **94**, 3938.

278. Dikic, I., Tokiwa, G., Lev, S., Courtenidge, S. A., and Schlessinger, J. (1996). A role for Pyk2 and Src in linking G-protein-coupled receptors with MAP kinase activation. *Nature*, **383**, 547.

279. Tokiwa, G., Dikic, I., Lev, S., Schlessinger, J. (1996). Activation of Pyk2 by stress signals and coupling with JNK signaling pathway. *Science*, **273**, 792.

280. Boutros, M., Paricio, N., Strutt, D. I., and Mlodzik, M. (1998). Dishevelled activates JNK and discriminates between JNK pathways in planar polarity and wingless signaling. *Cell*, **94**, 109.

281. Semënov, M. V., and Snyder, M. (1997). Human dishevelled genes constitute a DHR-containing multigene family. *Genomics*, **42**, 302.

282. Klingensmith, J., Yang, Y., Axelrod, J. D., Beier, D. R., Perrimon, N., and Sussman, D. J. (1996). Conservation of dishevelled structure and function between flies and mice: isolation and characterization of dvl2. *Mech. Dev.*, **58**, 15.

283. Marti, A., Luo, Z., Cunningham, C., Ohta, Y., Hartwig, J., Stossel, T., Kyriakis, J. M., and Avruch, J. (1997). Actin-binding protein-280 binds the stress-activated protein kinase (SAPK) activator SEK1 and is required for tumor necrosis factor-α activation of SAPK in mammalian cells. *J. Biol. Chem.*, **272**, 2620.

284. Katz, P., Whalen, G., and Kehrl, J. H. (1994). Differential expression of a novel protein kinase in human B lymphocytes: preferential localization in the germinal center. *J. Biol. Chem.*, **269**, 16802.

285. Shi, C.-S. and Kehrl, J. H. (1997). Activation of stress-activated protein kinase/c-Jun N-terminal kinase, but not NF-κB, by the tumor necrosis factor (TNF) receptor 1 through a TNF receptor-associated factor 2- and germinal center kinase related-dependent pathway. *J. Biol. Chem.*, **272**, 32102.

286. Tung, R. M. and Blenis, J. (1997). A novel human SPS1/STE20 homologue, KHS, activates Jun N-terminal kinase. *Oncogene*, **14**, 653.

287. Diener, K., Wang, X. S., Chen, C., Meyer, C. F., Keesler, G., Zukowski, M., Tan, T.-H., and Yao, Z. (1997). Activation of the c-Jun N-terminal kinase pathway by a novel protein kinase related to human germinal center kinase. *Proc. Natl Acad. Sci. USA*, **94**, 9687.

288. Kiefer, F., Tibbles, L. A., Anafi, M., Janssen, A., Zanke, B. W., Lassam, N., Pawson, T., Woodgett, J. R., and Iscove, N. R. (1996). HPK1, a hematopoietic protein kinase activating the SAPK/JNK pathway. *EMBO J.*, **15**, 7013.

289. Hu, M. C.-T., Qiu, W. R., Wang, X., Meyer, C. F., and Tan, T.-H. (1996). Human HPK1, a

novel human hematopoietic progenitor kinase that activates the JNK/SAPK kinase cascade. *Genes Dev.*, **10**, 2251.

290. Su, Y.-C., Han, J., Xu, S., Cobb, M. and Skolnik, E. Y. (1997). NIK is a new Ste20-related kinase that binds NCK and MEKK1 and activates the SAPK/JNK cascade via a conserved regulatory domain. *EMBO J.*, **16**, 1279.

291. Yao, Z., Zhou, G., Wang, X. S., Brown, A., Diener, K., Gan, H., and Tan, T.-H. (1999). A novel human STE20-related protein kinase, HGK, that specifically activates the c-Jun N-terminal kinase signaling pathway. *J. Biol. Chem.*, **274**, 2118.

292. Pombo, C. M., Kehrl, J. H., Sánchez, I., Katz, P., Avruch, J., Zon, L. I., Woodgett, J. R., Force, T., and Kyriakis, J. M. (1995). Activation of the SAPK pathway by the human STE20 homologue germinal center kinase. *Nature*, **377**, 750.

293. Cerutti. A., Schaffer, A., Shah, S., Zan, H., Liou, H.-C., Goodwin, R. G., and Casali, P. (1998). CD30 is a CD40-inducible molecule that negatively regulates CD40-mediated immunoglobulin class switching in non-antigen-selected human B cells. *Immunity*, **9**, 247.

294. Matsumoto, M., Mariathasan, S., Nahm, M. H., Baranyay, F., Peschon, J. J., and Chaplin, D. D. (1996). Role of lymphotoxin and the type I TNF receptor in the formation of germinal centers. *Science*, **271**, 1289.

295. Rosette, C. and Karin, M. (1996). Ultraviolet light and osmotic stress: activation of the JNK cascade through miltiple growth factor and cytokine receptors. *Science*, **274**, 1194.

296. Anafi, M., Kiefer, F., Gish, G. D., Mbamalu, G., Iscove, N. N., and Pawson, T. (1997). SH2/SH3 adaptor proteins can link tyrosine kinases to a Ste20-related protein kinase, HPK1. *J. Biol. Chem.*, **272**, 27804.

297. Ling, P., Yao, Z., Meyer, C. F., Wang, X. S., Oehrl, W., Feller, S. M., and Tan, T.-H. (1999). Interaction of hematopoietic progenitor kinase 1 with adapter proteins Crk and CrkL leads to synergistic activation of c-Jun N-terminal kinase. *Mol. Cell. Biol.*, **19**, 1359.

298. Lu, W., Katz, S., Gupta, R., and Mayer, B. J. (1997). Activation of PAK by membrane localization mediated by an SH3 domain from the adaptor protein Nck. *Curr. Biol.*, **7**, 85.

299. Su, Y.-C., Treisman, J. E., and Skolnik, E. Y. (1998). The *Drosophila* Ste20-related kinase misshapen is required for embryonic dorsal closure and acts through a JNK MAPK module on an evolutionarily conserved signaling pathway. *Genes Dev.*, **12**, 2371.

300. Liu, H., Su, Y.-C., Becker, E., Treisman, J., and Skolnik, E. Y. (1999). A *Drosophila* TNF-receptor-associated factor (TRAF) binds the Ste20 kinase Misshapen and activates Jun kinase. *Curr. Biol.*, **9**, 101.

301. Smith, C. A., Farrah, T., and Goodwin, R. G. (1994). The TNF receptor superfamily of cellular and viral proteins: activation, costimulation and death. *Cell*, **76**, 959.

302. Vandenabeele, P., Declercq, W., Beyaert, R., and Fiers, W. (1995). Two tumour necrosis factor receptors: structure and function. *Trends Cell Biol.*, **5**, 392.

303. Wallach, D., Varfolomeev, E. E., Malinin, N. L., Goltsev, Y. V., Kovalenko, A. V., and Boldin, M. P. (1999). Tumor necrosis factor receptor and fas signaling mechanisms. *Annu. Rev. Immunol.*, **17**, 331.

304. Arch, R. H., Gedrich, R. W., and Thompson, C. B. (1998). Tumor necrosis factor receptor-associated factors (TRAFs)—a family of adapter proteins that regulates life and death. *Genes Dev.*, **12**, 2821.

305. Tracey, K. J. and Cerami, A. (1993). Tumor necrosis factor, other cytokines and disease. *Annu. Rev. Cell Biol.*, **9**, 317.

306. Hsu, H., Xiong, J., and Goeddel, D. V. (1995). The TNF receptor-1-associated protein TRADD signals cell death and NF-κB activation. *Cell*, **81**, 495.

307. Hsu, H., Shu, H.-B, Pan, M.-G, and Goeddel, D. V. (1996). TRADD-TRAF2 and TRADD-FADD interactions define two distinct TNF receptor 1 signal transduction pathways. *Cell*, **84**, 299.

308. Hsu, H., Huang, J., Shu, H.-B., Baichwal, V., and Goeddel, D. V. (1996). TNF-dependent recruitment of the protein kinase RIP to the TNF receptor-1 signaling complex. *Immunity*, **4**, 387.

309. Liu, Z.-G., Hsu, H., Goeddel, D. V., and Karin, M. (1996). Dissection of TNF receptor-1 effector functions: JNK activation is not linked to apoptosis while NF-κB activation prevents cell death. *Cell*, **87**, 565.

310. Natoli, G., Costanzo, A., Ianni, A., Templeton, D. J., Woodgett, J. R., Balsano, C., and Levrero, M. (1997). Activation of SAPK/JNK by TNF receptor-1 through a noncytotoxic TRAF2-dependent pathway. *Science*, **275**, 200.

311. Reinhard, C., Shamoon, B., Shyamala, V., and Williams, L. T. (1997). Tumor necrosis factor-α-induced activation of c-jun N-terminal kinase is mediated by TRAF2. *EMBO J.*, **16**, 1080.

312. Takeuchi, M., Rothe, M., and Goeddel, D. V. (1996). Anatomy of TRAF2. *J. Biol. Chem.*, **271**, 19935.

313. Rothe, M., Wong, S. C., Henzel, W. J., and Goeddel, D. V. (1994). A novel family of putative signal transducers associated with the cytoplasmic domain of the 75-kDa tumor necrosis factor receptor. *Cell*, **78**, 681.

314. Stanger, B. Z., Leder, P., Lee, T.-H., Kim, E., and Seed, B. (1995). RIP: a novel protein containing a death domain that interacts with Fas/APO-1 (CD95) in yeast and causes cell death. *Cell*, **81**, 513.

315. Ting, A. T., Pimentel-Muiños, F.-X., and Seed, B. (1996). RIP mediates tumour necrosis factor receptor 1 activation of NF-κB but not Fas/APO-1-initiated apoptosis. *EMBO J.*, **15**, 6189.

316. Kelliher, M. A., Grimm, S., Ishida, Y., Kuo, F., Stanger, B. Z., and Leder, P. (1998). The death domain kinase RIP mediates the TNF-induced NF-κB signal. *Immunity*, **8**, 297.

317. Yeh, W.-C., Shahinian, A., Speiser, D., Kraunus, J., Billia, F., Wakeham, A., de la Pompa, J. L., Ferrick, D., Hum, B., Iscove, N., Ohashi, P., Rothe, M., Goeddel, D. V., and Mak, T. W. (1997). Early lethality, functional NF-κB activation, and increased sensitivity to TNF-induced cell death in TRAF2-deficient mice. *Immunity*, **7**, 715.

318. Lee, S. Y., Reichlin, A., Santana, A., Sokol, K. A., Nussenzweig, M. C., and Choi, Y. (1997). TRAF2 is essential for JNK but not NF-κB activation and regulates lymphocyte proliferation and survival. *Immunity*, **7**, 703.

319. Song, H. Y., Régnier, C. H., Kirschning, C. J., Goeddel, D. V., and Rothe, M. (1997). Tumor necrosis factor (TNF)-mediated kinase cascades: bifurcation of nuclear Factor-B and c-jun N-terminal kinase (JNK/SAPK) pathways at TNF receptor-associated factor 2. *Proc. Natl Acad. Sci. USA*, **94**, 9792.

320. Lomaga, M. A., Yeh, W.-C., Sarosi, I., Duncan, G. S., Furlonger, C., Ho, A., Morony, S., Capparelli, C., Van, G., Kaufman, S., van der Heiden, A., Itie, A., Wakeham, A., Khoo, W., Sasaki, T., Cao, Z., Penninger, J. M., Paige, C. J., Lacey, D. L., Dunstan, C. R., Boyle, W. J., Goeddel, D. V., and Mak T. W. (1999). TRAF6 deficiency results in osteopetrosis and defective interleukin-1, CD40, and LPS signaling. *Genes Dev.* **13**, 1015.

321. Wesche, H., Henzel, W. J., Shillihglaw, W., Li, S., and Cao, Z (1997). MyD88: an adapter that recruits IRAK to the IL-1 receptor complex. *Immunity*, **7**, 837.

322. Baud, V., Liu, Z.-G., Bennett, B., Suzuki, N., Xia, Y., and Karin, M (1999). Signaling by proinflammatory cytokines: oligomerization of TRAF2 and TRAF6 is sufficient for JNK

and IKK activation and target gene induction via an amino terminal effector domain. *Genes Dev.*, **13**, 1297.

323. Yuan, J. (1997). Transducing signals of life and death. *Curr. Opin. Cell Biol.*, **9**, 247.

324. Yang, X., Khosravi-Far, R., Chang, H. Y., and Baltimore, D. (1997). Daxx, a novel Fas-binding protein that activates JNK and apoptosis. *Cell*, **89**, 1067.

325. Chang, H. Y., Yang, X., and Baltimore, D. (1999). Dissecting Fas signaling with an altered-specificity death-domain mutant: requirement of FADD binding for apoptosis but not Jun N-terminal kinase activation. *Proc. Natl Acad. Sci. USA*, **96**, 1252.

326. Michaelson, J. S., Bader, D., Kuo, F., Kozak, C., and Leder, P. (1999). Loss of Daxx, a promiscuously interacting protein results in extensive apoptosis in early mouse development. *Genes Dev.*, **13**, 1918.

327. Pomerance, M., Multon, M. C., Parker, F., Venot, C., Blondeau, J. P., Tocque, B., and Schweighoffer, F. (1998). Grb2 interaction with MEK-kinase 1 is involved in regulation of Jun-kinase activities in response to epidermal growth factor. *J. Biol. Chem.*, **273**, 24301.

328. Widmann, C., Gerwins, P., Lassignal Johnson, N., Jarpe, M. B., and Johnson, G. L. (1998). MEK kinase 1, a substrate for DEVD-directed caspases, is involved in genotoxin-induced apoptosis. *Mol. Cell. Biol.*, **18**, 2416.

329. Johnson, N. L., Gardner, A. M., Diener, K. M., Lange-Carter, C. A., Gleavy, J., Jarpe, M. B., Minden, A, Karin, M., Johnson, L. I., and Johnson, G. L. (1996). Signal transduction pathways regulated by mitogen-activated /extracellular response kinase kinase kinase induce cell death. *J. Biol. Chem.*, **271**, 3229.

330. Yujiri, T., Sather, S., Fanger, G. R., and Johnson, G. L. (1998). Role of MEKK1 in cell survival and activation of JNK and ERK pathways defined by targeted gene disruption. *Science*, **282**, 1911.

331. Park, Y. C., Burkitt, V., Villa, A. R., Tong, L., and Wu, H. (1999). Structural basis for self-association and receptor recognition of human TRAF2. *Nature*, **398**, 533.

332. McWhirter, S. M., Pullen, S. S., Holton, J. M., Crute, J. J., Kehry, M. R., and Alber, T. (1999). Crystallographic analysis of CD40 recognition and signaling by human TRAF2. *Proc. Natl Acad. Sci. USA*, **96**, 8408.

333. Goossens, V., Grooten, J., De Vos, K., and Fiers, W. (1995). Direct evidence for tumor necrosis factor-induced mitochondrial reactive oxygen intermediates and their involvement in cytotoxicity. *Proc. Natl Acad. Sci. USA*, **92**, 8115.

334. Gotoh, Y. and Cooper, J. A. (1998). Reactive oxygen species- and dimerization-induced activation of apoptosis signal-regulating kinase 1 in tumor necrosis factor-α signal transduction. *J. Biol. Chem.*, **273**, 17477.

335. Massagué, J. (1998). TGFβ signal transduction. *Annu. Rev. Biochem.*, **67**, 753.

336. Attisano, L. and Wrana, J. (1998). Mads and smads in TGFβ signalling. *Curr. Opin. Cell Biol.*, **10**, 188.

337. Yamaguchi, K., Nagai, S.-I., Ninomiya-Tsuji, J., Nishita, M., Tamai, K., Irie, K., Ueno, N., Nishida, E., Shibuya, H., and Matsumoto, K. (1999). XIAP, a cellular member of the inhibitor of apoptosis protein family links the receptors to TAB1-TAK1 in the BMP signaling pathway. *EMBO J.*, **18**, 179.

338. Devary, Y., Rosette, C., DiDonato, J. A., and Karin, M. (1993). NF-κB activation by ultra-violet light not dependent on a nuclear signal. *Science*, **261**, 1442.

339. Kharbanda, S., Ren, R., Pandey, P., Shafman, T. D., Feller, S., Weichselbaum, R. R., and Kufe, D. W. (1995). Activation of the c-Abl tyrosine kinase in the stress response to DNA damaging agents. *Nature*, **376**, 785.

340. Liu, Z.-G., Baskaran, R., Lea-Chou, E. T., Wood, L. D., Chen, Y., Karin, M., and Wang, J. Y.

J. (1996). Three distinct signalling responses by murine fibroblasts to genotoxic stress. *Nature*, **384**, 273.

341. Pandey, P., Raingeaud, J., Kaneki, M., Weichselbaum, R., Davis, R. J., Kufe, D., and Kharbanda, S. (1996). Activation of p38 mitogen-activated protein kinase by c-Abl-dependent and -independent mechanisms. *J. Biol. Chem.*, **271**, 23755.

342. Chen, Y.-R., Meyer, C. F., and Tan, T.-H. (1996). Persistent activation of c-Jun N-terminal kinase-1 (JNK1) in γ radiation-induced apoptosis. *J. Biol. Chem.*, **271**, 631.

343. Shafman, T. D., Saleem, A., Kyriakis, J., Weichselbaum, R., Kharbanda, S., and Kufe, D. W. (1995). Defective induction of stress-activated protein kinase activity in ataxia-telangiectasia cells exposed to ionizing radiation. *Cancer Res.*, **55**, 3242.

344. Fornace, A. J., Jr., Alamo, I., Jr., and Hollander, C. (1988). DNA damage-inducible transcripts in mammalian cells. *Proc. Natl Acad. Sci. USA*, **85**, 8800.

345. Kastan, M. B., Zhan, Q., el-Diery, W. S., Carrier, F., Jacks, T., Walsh, W. V., Plunkett, B. S., Vogelstein, B., and Fornace, A. J., Jr. (1992). A mammalian cell cycle checkpoint pathway utilizing p53 and GADD45 is defective in ataxia-telangiectasia. *Cell*, **71**, 587.

346. Leung, I. W.-L., and Lassam, N. (1998). Dimerization via tandem leucine zippers is essential for the activation of the mitogen-activated protein kinase kinase kinase, MLK-3. *J. Biol. Chem.*, **273**, 32408.

347. Xia, Z., Dickens, M., Raingeaud, J., Davis, R. J., and Greenberg, M. E. (1995). Opposing effects of ERK and JNK-p38 MAP kinases on apoptosis. *Science*, **270**, 3126.

348. Le-Niculescu, H., Bonfoco, E., Kasuya, Y., Claret, F.-X., Green, D. R., and Karin, M. (1999). Withdrawal of survival factors results in activation of the JNK pathway in neuronal cells leading to Fas ligand induction and cell death. *Mol. Cell. Biol.*, **19**, 751.

349. Zanke, B. W., Boudreau, K., Rubie, E., Tibbles, L. A., Zon, L. I., Kyriakis, J., Liu, F.-F., and Woodgett, J. R. (1996). The stress-activated protein kinase pathway mediates cell death following injury-induced by cis-platinum, UV irradiation or heat. *Curr. Biol.*, **6**, 606.

350. Meriin, A., Yaglom, J. A., Gabai, V. L., Mosser, D. D., Zon, L., and Sherman, M. Y. (1999). Protein damaging stresses activate JNK via inhibition of its phosphatase: a novel pathway controlled by HSP72. *Mol. Cell. Biol.*, **19**, 2547.

351. Kasibhatla, S., Brunner, T., Genestier, L., Echeverri, F., Mahboubi, A., and Green, D. R. (1998). DNA damaging agents induce expression of Fas ligand and subsequent apoptosis in T lymphocytes via the activation of NF-κB and AP-1. *Mol. Cell*, **1**, 543.

352. Yang, D. D., Kuan, C. Y., Whitmarsh, A. J., Roncon, M., Zheng, T. S., Davis, R. J., Rakic, P., and Flavell, R. A. (1997). Absence of excitotoxicity-induced apoptosis in the hippocampus of mice lacking the Jnk3 gene. *Nature*, **389**, 865.

353. Molkentin, J. D., Lu, J.-R., Antos, C. L., Markham, B., Richardson, J., Robbins, J., Grant, S. R., and Olson, E. N. (1998). A calcineurin-dependent transcriptional pathway for cardiac hypertrophy. *Cell*, **93**, 215.

354. Wang, Y., Huang, S., Sah, V. P., Ross, J., Jr., Brown, J. H., Han, J., and Chien, K. R. (1998). Cardiac muscle cell hypertrophy and apoptosis induced by distinct members of the p38 mitogen-activated protein kinase family. *J. Biol. Chem.*, **273**, 2161.

355. Wang, Y., Su, B., Sah, V. P., Heller Brown, J., Han, J., and Chien, K. R. (1998). Cardiac hypertrophy induced by mitogen-activated protein kinase kinase 7, a specific activator for c-Jun NH2-terminal kinase in ventricular muscle cells. *J. Biol. Chem.*, **273**, 5423.

356. Qiu, R. G., Abo, A., McCormick, F., and Symons, M. (1997). Cdc42 regulates anchorage-independent growth and is necessary for Ras transformation. *Mol. Cell. Biol.*, **17**, 3449.

357. Qiu, R.-G., Chen, J., Kirn, D., McCormick, F., and Symons, M. (1995). An essential role for Rac in Ras transformation. *Nature*, **374**, 457.

358. Molnár, Á., Theodoras, A. M., Zon, L. I., and Kyriakis, J. M. (1997). Cdc42Hs, but not Rac1, inhibits serum-stimulated cell cycle progression at G1/S through a mechanism requiring p38/RK. *J. Biol. Chem.*, **272**, 13299.

359. Su, B., Jacinto, E., Hibi, M., Kallunki, T., Karin, M., and Ben-Neriah, Y. (1994). JNK is involved in signal integration during costimulation of T lymphocytes. *Cell*, **77**, 727.

360. Dong, C., Yang, D. D., Wysk, M., Whitmarsh, A. J., Davis, R. J., and Flavell, R. A. (1998). Defective T cell differentiation in the absence of Jnk1. *Science*, **282**, 2092.

361. Sabapathy, K., Hu, Y., Kallunki, T., Schreiber, M., David, J.-P., Jochum, W., Wagner, E. F., and Karin, M. (1999). JNK2 is required for efficient T-cell activation and apoptosis but not for normal lymphocyte development. *Curr. Biol.*, **9**, 116.

362. Ip, Y. T. and Davis, R. J. (1998). Signal transduction by the c-Jun N-terminal kinase (JNK) —from inflammation to development. *Curr. Opin. Cell Biol.*, **10**, 205.

363. Dierick, H. and Bejsovec, A. (1999). Cellular mechanisms of wingless/Wnt signal transduction. *Curr. Top. Dev. Biol.*, **43**, 153.

364. Wodarz, A. and Nusse, R. (1998). Mechanisms of Wnt signaling in development. *Annu Rev. Cell Dev. Biol.*, **14**, 59.

365. Meneghini, M. D., Ishitani, T., Carter, J. C., Hisamoto, N., Ninomiya-Tsuji, J., Thorpe, C. J., Hamill, D. R., Matsumoto, K., and Bowerman, B. (1999). MAP kinase and Wnt pathways converge to downregulate an HMG-domain repressor in *Caenorhabditis elegans*. *Nature*, **399**, 793.

366. Ishitani, T., Ninomiya-Tsuji, J., Nagai, S.-I., Nishita, M., Meneghini, M., Barker, N., Waterman, M., Bowerman, B., Clevers, H., Shibuya, H., and Matsumoto, K. (1999). The TAK1-NLK-MAPK-related pathway antagonizes signalling between β-catenin and transcription factor TCF. *Nature*, **399**, 798.

3 | Genetic approaches to protein kinase function in lower eukaryotes

SIMON E. PLYTE

1. Introduction

The study of protein kinase function in higher eukaryotes such as Mammalia has been generally restricted to biochemical analysis of purified proteins, due to the limitations of genetic approaches in these species. Most of these enzymes were first identified biochemically and their genes isolated subsequent to peptide sequence analysis. However, the promiscuous nature of protein kinases in phosphorylating proteins *in vitro* has complicated study of their function in cells. In contrast, the tractability of lower eukaryotes to genetic analysis has resulted in a distinct approach in which genes involved in a given process are initially identified via their phenotypic effects. These genetic strategies, of course, do not specifically target protein kinase genes but this family of proteins has surfaced at a surprising frequency during analysis of regulatory genes, which is a testimony to their adaptability and ubiquity as biological tools. Once identified, the relative ease of genetic manipulation in lower eukaryotes enables the investigation of kinase function *in vivo*. Thus, mutational analysis can be used to determine whether a gene product is required in a particular pathway. Additionally, genetic epistatic experiments, between two genes in a particular pathway, may be applied to determine the order of action of those gene products. During the last decade, functional homologues of many mammalian protein kinases have been identified in lower eukaryotes, allowing a degree of extrapolation in the workings of the proteins in different species. In fact many of these kinases seem to function in an identical manner in lower eukaryotes and participate in evolutionary conserved kinase cascades, vindicating the use of lower eukaryotes as a model system in which to study protein kinase function.

In this chapter I wish to give the reader a flavour of the type of techniques used in lower eukaryotic genetics and the sort of information that can be determined from such studies. In addition, I hope to convince you, in part, of the power of such studies in elucidating the mechanisms of signal transduction and how genetic experiments in lower eukaryotes are starting to shape the way we think about signal transduction specificity. For example, one of the major questions in signal transduction is how specificity arises from activation of a common pathway such as the Ras/MAP kinase pathway. Elegant work in yeast, where there are at least five MAP kinase pathways working in parallel, has shown that specificity can be achieved by recruiting only the correct kinase to a multimeric complex, thus ensuring that the signal is conveyed to its specific end point. Further, from work in *Caenorhabditis elegans*, another level of specificity has been demonstrated which shows that a combination of general and tissue-specific effectors can determine the outcome of activating a common signal transduction cassette.

In this chapter I will outline several examples of signal transduction pathways that have been elucidated by genetic means:

(1) signal transduction during R7 photoreceptor development in *Drosophila melanogaster* and during vulval induction in *C. elegans*, as examples of receptor tyrosine kinase signalling pathways;

(2) the role of cAMP-dependent protein kinase and GSK-3 during the life cycle of the slime mould, *Dictyostelium discoideum*;

(3) the pheromone signalling pathway in *Saccharomyces cerevisiae*, as an example of heterotrimeric G protein-coupled receptor signalling and the limitation of signalling cross-talk in parallel MAP kinase cascades.

2. Genetic manipulation in *Drosophila*

2.1 Signal transduction from the Sevenless receptor

Genetic dissection of the pathway responsible for the development of the *Drosophila* UV photoreceptor (R7 cell) has revealed a wealth of information regarding the mechanism of action of receptor tyrosine kinases. Since publication of the first edition of this book a further 12 genes have been identified as having a role in Sevenless signalling, demonstrating the power of genetic approaches in elucidating signal transduction processes (1–17). The fruit fly's compound eye comprises approximately 800 ommatidia. Each ommatidium is composed of a precise array of 20 cells: 8 neuronal photoreceptor cells (R1–8), 4 lens secreting cells, and 8 other accessory cells. During development, the R7 cell is the last of the eight photoreceptors to be recruited from the accessory cells and is responsible for UV detection (reviewed in refs 18 and 19). The *sevenless* mutation (encoding a receptor tyrosine kinase; 20) results in complete loss of the R7 photoreceptor from each ommatidium and adult flies are unable to respond to UV stimulation (21, 22). A subset of the accessory cells are competent to differentiate into the R7 photoreceptor, but only one of these cells adopts the neuronal cell fate, the rest developing into non-neuronal cone cells. R7 differentiation is thought to be controlled by the spatial and temporal expression of specific proteins on the R7 and R8 photoreceptors (reviewed in refs 6, 23, and 24). Dominant mutants that effect Sevenless signalling produce supernumerary R7 photoreceptors by per-

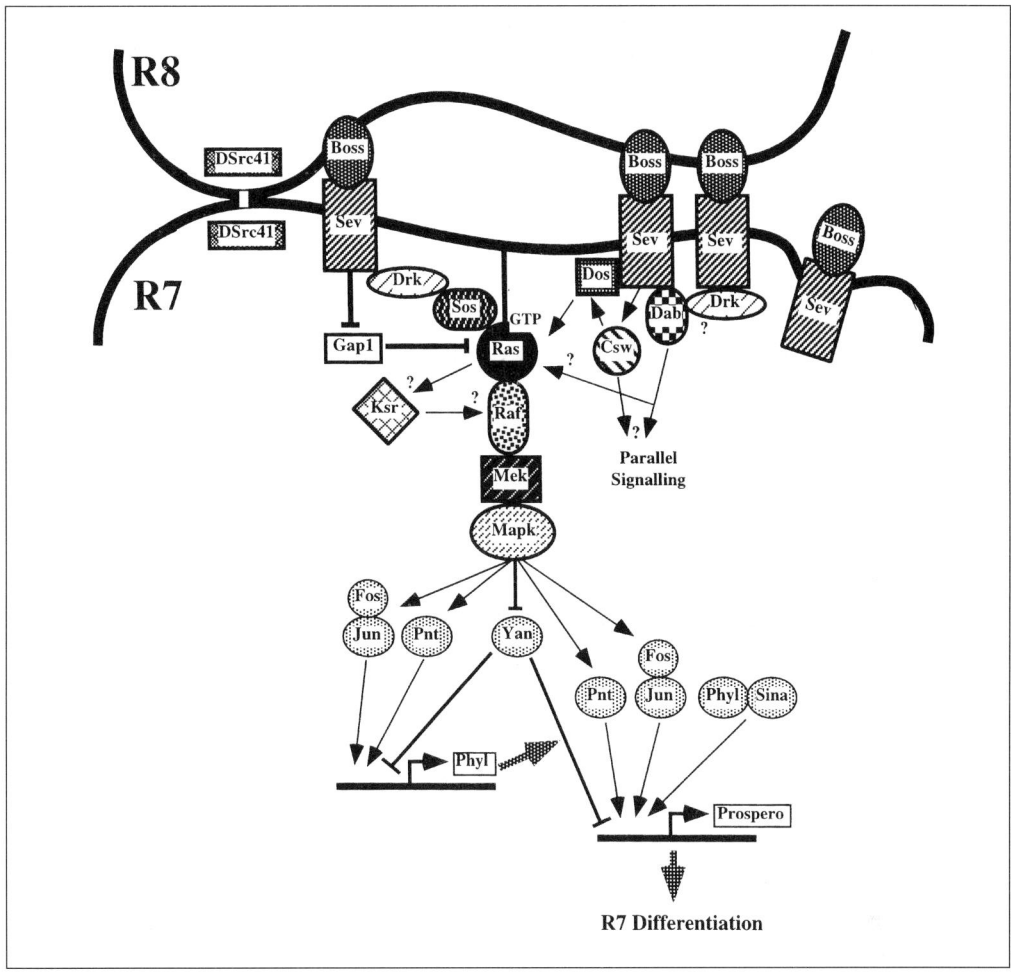

Fig. 1 The Sevenless signalling pathway. Interaction between the Bride of Sevenless (Boss) and Sevenless receptor (Sev) results in Ras/MAP kinase activation, transcriptional activation, and ultimately R7 photoreceptor development (for details see text). After an initial interaction the Boss–Sev complex is internalized. Sos, son of sevenless; Csw, corkscrew; Ksr, kinase supressor of ras; Pnt, pointed; Phyl, phyllopod; Sina, seven in absentia; Dab, disabled; Dos, daughter of sevenless.

mitting differentiation of the group of Sev-competent cells giving rise to a 'rough eye' phenotype. Figure 1 illustrates the genes which have thus far been implicated in signal transduction by the Sevenless receptor tyrosine kinase.

Interaction of the transmembrane protein encoded by *bride of sevenless* (*boss*) gene with the Sevenless receptor tyrosine kinase results in stimulation of the pathway and ligand–receptor internalization (25, 26). A *Drosophila* homologue of Src (DSrc41) has been shown to be involved in this process, and is thought to be associated with the formation of adherens junctions that facilitate the juxtaposition of *Boss* and *Sev* expressing cells (10). Upon Sevenless activation, the receptor becomes phosphorylated on tyrosine residues and creates docking sites for SH2 and PTB domain-containing proteins such as Drk (Grb2 homologue) and DAB (disabled) (7, 16, 27). The precise role of these adapter proteins is not clear but they are both required for efficient Sevenless receptor signalling. Drk can tether the Sevenless receptor to Sos (son of sevenless) via its SH2 and SH3 domains (16, 27, 28). The Sos protein is an activator of guanine nucleotide exchange related to the fission yeast *CDC25* gene (29) and this factor is thought to promote GTP binding and activation of Ras (30). Sev activation also leads to the inhibition of Gap1 which negatively regulates Ras activity (31). Dab can compete with Drk for binding to the same phosphorylated tyrosine residue on the Sev receptor and Drk has been shown to bind to a proline-rich region on Dab via one of its SH3 domains (7). This presents the possibility for the formation of a multimeric complex comprising various adapter proteins and aggregated receptors which may be required for efficient Sevenless signalling (indicated in Fig. 1). Interestingly, this situation is analogous to that seen upon T-cell receptor activation where Grb2 and Shc can bind the same phosphotyrosine residue on the ζ chain and/or bind each other via an SH3–proline interaction, ultimately leading to Ras activation. The PH domain-containing adapter protein Dos (daughter of sevenless) has also been shown to be involved in signalling from the Sev receptor, but it is not clear whether this is mediated by Ras activation or by a parallel pathway (8, 14). This protein is similar to IRS-1/2 and contains putative docking sites for various signalling molecules including Drk, Csw, PLCγ and PI3-kinase (14). In addition, the tyrosine phosphatase Csw (corkscrew) has been shown to be positively involved in the Sev signal transduction pathway, possibly by controlling the degree of tyrosine phosphorylation of several adapter proteins, thus regulating the types and number of protein complexes (15).

Ras activation leads to the recruitment of Raf (*D-Raf, polehole allele*) and activation of the MAP kinase cassette (reviewed in refs 9 and 32). The association of Ras with Raf is not sufficient for Raf activation as this kinase needs to be phosphorylated for full activity. One possible Raf-activating kinase is Ksr (kinase suppresser of *ras*) as it acts downstream of Ras in Sev signal transduction and upstream or in parallel to Raf (1, 13). The dual-specificity kinase, MEK (D-Sor1), is activated upon phosphorylation by Raf and this in turn activates MAP kinase (rolled/sevenmaker) by phosphorylation on both tyrosine and threonine residues (9, 33). MAP kinase is then possibly responsible for the direct or indirect regulation of several nuclear proteins/transcription factors including Yan, Pnt (pointed) jun, and sina (seven in absentia) (2–5, 11, 12, 17). Yan and Pnt are Ets-domain transcription factors and Yan negatively regulates the transcription of the *Phyllopod* (an immediate Sev-induced gene) and *Prospero* (a gene involved in axon connectivity) genes (2, 3). In contrast, Pnt is required for the transcription of these two genes, and it is thought that the balance in activity between these two transcription factors is important in R7 cell-fate determination. The induction of Phyl (a nuclear factor) is also dependent on AP-1 activity (Jun/fos). Phyl has been shown to physically associate with sina (another nuclear factor essential for R7 differentiation) and together with Pnt and AP-1 they control the expression of *prospero* (2, 3). Several other genes have been identified by genetic means that play a role in R7 cell-fate determination and these await full characterization.

2.1.1 Loss-of-function analysis

A variety of genetic techniques have been used to study the mechanism of receptor tyrosine kinase signalling in the Sevenless pathway. Screening for recessive, viable, loss-of-function mutations which result in the *sevenless* phenotype is a direct method for identifying genes in the pathway. Flies are mutagenized, either chemically or by UV irradiation, and the progeny are screened for the phenotype. The *boss*, *sina*, and *gap*1 genes were initially identified in this way (25, 34–36). Alternatively, mutations that give rise to the 'rough eye' phenotype can also be used to identify new genes involved in Sevenless signalling. A loss-of-function mutation in Yan and a gain-of-function mutation in MAPK (rolled/sevenmaker) resulted in flies with a rough-eye phenotype (12, 17). This approach only reveals genes dispensable for viability, i.e. that do not perform essential functions in other processes and pathways. Once a gene has been identified, mosaic analysis can be used to determine whether the gene acts in a cell autonomous- or non-autonomous manner. A cell-autonomous phenotypic marker must be identified which is proximal to the gene of interest. In studies with *boss*, the *chaoptic* (*chp*) gene, which effects rhombdomere morphology, was used as the phenotypic marker (36). In flies heterozygous for *chp* and *boss*, homozygous patches of cells were produced by X-ray-induced mitotic recombination. Cells mutant for *boss* were then identified by their *chaoptic* phenotype. The R7 photoreceptor failed to develop in ommatidia where the R8 cell expressed the *boss* mutation, suggesting that the Boss transmembrane protein, expressed exclusively on the R8 cell specifies the fate of the R7 photoreceptor cell. Mosaic analysis has shown that the *sevenless*, *sos*, *gap*1, *sina*, and other *E(sev)* genes are required in a cell-autonomous manner for R7 photoreceptor development (22, 30, 34).

2.1.2 P-element insertional mutagenesis

'Enhancer trap' techniques can be used to identify genes that give rise to a particular phenotype. Using transposable P-element vectors, the β-galactosidase gene fused to an enhancerless promoter can be randomly integrated into the *Drosophila* genome. The resultant expression of β-galactosidase may then reflect the expression pattern of a nearby gene (37–39). If the β-galactosidase construct also disrupts the host gene, this may result in an observable phenotypic effect. Genomic sequences flanking the site of P-element insertion can be recovered by plasmid rescue (40) and used to screen genomic libraries to reveal the disrupted gene. The *sos* and *seven-up* genes were identified in this way (31, 41). Seven-up is a member of the steroid receptor family and is required in the R1, R3, R4, and R6 photoreceptors for specification of their fate (41). These four photoreceptors develop R7 characteristics in *seven-up* mutant ommatidia (41).

2.1.3 Phenotypic rescue

Another approach to identify members of a particular signalling pathway is to look for second-site dominant suppressers of the phenotype. Flies already mutant for the phenotype of interest are fed a chemical mutagen and their progeny screened for rescue of that phenotype. For example, an activated Ras protein was able to rescue R7 photoreceptor development in *boss* and *sevenless* null flies (42). However, it failed to rescue the phenotype in flies carrying a mutation in *sina*, which suggested that Ras acts downstream of Boss and Sevenless and upstream of Sina. The *sos* gene was also identified as a second site suppresser of *sevenless* in flies expressing a mutant Seven-less receptor (43). However, rescue was only observed when Sevenless protein was produced, suggesting a Sos–Sevenless protein interaction. Similar experiments have demonstrated that Drk is required downstream of Sevenless and upstream of Ras (28). Additionally, this protein has been demonstrated to physically interact with the Sos and Sevenless proteins, consistent with genetic data associating these latter two gene products (28). Such genetic epistatic experiments have been widely used to study signal transduction pathways. However, they are limited to ordering the action of genes and cannot provide information about the number of steps that exist between identified genes. A dominant *sos* allele was able to enhance the 'rough eye' phenotype resulting from constitutive activation of the *ellipse*-encoded receptor tyrosine kinase (the fly EGF receptor homologue) (44), suggesting that the *sos* protein is common to both the Sevenless and EGF signalling pathways. A constitutively activated *raf* gene was able to rescue R7 development in *sevenless* mutant flies (45). In contrast, reduced levels of D-raf suppressed the effects of a constitutively activated Ras protein, suggesting that D-raf acts downstream of Sevenless and Ras in this pathway (45). Ras and D-raf are also required in the Torso receptor signalling pathway (46, 47, 50). Like Sevenless and Ellipse, Torso is a receptor tyrosine kinase (44) and is required for the determination of terminal embryonic structures (48, 49).

2.1.4 Pathway-restricted genetic screening

One of the problems with the techniques described above is the potential for the mutated protein to interfere with various other pathways, as demonstrated by the effect of mutations in *sos, ras, D-Sor1, rolled, Ksr, Dos,* and *D-raf* on Torso, DER and ellipse signal transduction (1, 8, 9, 12, 13, 31, 47, 50). When proteins common to more than one signal transduction pathway are mutated, the phenotype of interest may be obscured by secondary effects resulting from the disruption of several pathways. Simon *et al.* devised a sensitive genetic screen to circumvent the problems of signal-ling pathway cross-talk (30). They attempted to uncover genes that were required for signalling through a weakened Sevenless pathway by screening for second gene mutations that were functionally inefficient but not inactive, which therefore select-ively exacerbated the *sevenless* phenotype. This technique allows for the identification of genes common to more than one pathway. First, a temperature-sensitive mutant of *sevenless* was derived to enable manipulation of the efficiency of this signalling path-way. Flies were then mutagenized and progeny grown at a temperature where the Sevenless pathway was barely functional. Under these conditions a subtle mutation in another component of the pathway, resulting in as little as a twofold reduction in its efficiency, would be sufficient to block Sevenless signalling and prevent R7 de-velopment, but not affect the role of the secondary gene in other pathways. Seven independent gene mutations (termed *enhancers of sevenless, E(sev)*), including the *ras* and *sos* genes, were identified as having a role in Sevenless signalling (30). In addition, *ras*, *sos*, and two other *E(sev)* genes were shown to have a role in the Ellipse pathway.

Another way of circumventing signal transduction cross-talk between common effectors is the use of cell-specific promoters and enhancers permitting restricted controllable protein expression. For example, Karim *et al*. (1) generated flies that contained a dominant activated *ras* gene mutation under the control of a *sevenless* enhancer/promoter (*sev*-Ras^{V12}) thus restricting the expression of the dominantly active form of Ras to *Sev*-competent cells. The resulting flies had a 'rough eye' phenotype due to the differentiation of most *Sev*-competent cells into R7-type photoreceptors. A mutagenesis screen for second site suppressers or activators of this *sev*-Ras^{V12} 'rough eye' phenotype can then be performed to identify genes acting upstream or downstream of Ras in Sevenless signalling. Karim *et al*. identified 282 dominant suppressers and 577 dominant enhancers of the 'rough eye' phenotype in such a screen (1). However, there are at least four different classes of gene mutation that can enhance or suppress the 'rough eye' phenotype:

(1) bona fide genes truly involved in Sev/Ras signalling;
(2) genes which effect the postranslational modification of Ras but are not involved in Ras signalling *per se*;
(3) genes involved in the regulation of the *sev* promoter/enhancer thus affecting Ras^{V12} levels;
(4) non-specific mutations leading to suppression or enhancement of the 'rough eye' phenotype.

To confirm the participation of the isolated genes in Ras signalling two major types of experiment can be performed. The first is to cross flies carrying the mutation of interest with transgenic flies carrying a dominant-negative Ras mutation under control of the *sev* promoter/enhancer (*sev*-Ras^{N17}). *Sev*-Ras^{N17} flies have a phenotype complementary to the *sev*-Ras^{V12} 'rough eye' phenotype (having a *sevenless* phenotype). Mutation of a gene product involved in Sev/Ras signalling that suppressed the 'rough eye' phenotype of *sev*-Ras^{V12} should exacerbate the *sevenless* phenotype in *sev*-Ras^{N17} flies. The second test is to express other known Sevenless signalling gene mutations in the cell-specific expression system and compare the effect of the new gene mutation. This type of experiment will also provide information as to whether the newly identified gene acts upstream or downstream of the second 'tester' gene.

In such experiments using cell-specific expression of activated Ras, mutations in Raf, MEK, MAPK, Ksr, Phyl, and several uncharacterized genes were identified as suppressers of the 'rough eye' phenotype and act downstream of Ras (1). Similarly, both D-jun and Pnt were isolated in screens that could suppress an activated mutation in MAPK (*sevenmaker*) (5).The Dos and Dab genes were identified as dominant suppressers of an activated Sevenless receptor (in *sev*-Sev expressing transgenic flies) but were unable to suppress an activated Ras gene suggesting that they act downstream of Sev and upstream or in parallel to Ras (7, 8). Mutations in the tyrosine phosphatase Csw could suppress dominant active Sev and Ras transgenic effects, suggesting a role downstream of Ras (14, 15). However, a loss-of-function mutation in Dos blocks activated Sev and Csw thus placing Csw upstream of both Dos and Ras. It is thought that Csw acts in a pathway parallel to Ras but somehow dependent upon Ras expression. Genetic epistatic experiments between the various components of the Sevenless signalling pathway identified to date has led to the scheme presented in Fig. 1. Activation of Ras appears to be the primary response to Sevenless receptor signalling and may be a common mechanism for the activation of some tyrosine kinase receptor pathways. However, there are many more genes still to be characterized, and with the hint of parallel signals the sevenless pathway is by no means complete.

3. Vulval development in *C. elegans*

3.1 Vulval induction

The nematode *C. elegans* is a genetically manipulable multicellular organism which has yielded an abundance of signalling pathway information. Cells within the developing nematode are visible by microscopy, and fate maps of every cellular division have been compiled. With this information, the consequences of mutations on any of the processes occurring throughout development can be recognized at the single-cell level. Further, the use of Lac-Z and Green Fluorescent Protein reporter transgenes has enabled the cell-specific visualization of individual gene expression (51). Development of the nematode vulva has provided insights into the mechanism of action of receptor tyrosine kinases (reviewed in refs 52–57). The adult vulva develops from three of six embryonic cells termed vulval precursor cells (VPCs) (Fig. 2, top). These cells have equal potential to develop into one of three cell types denoted 1°, 2°, and 3° as shown by cell-ablation studies (58). The 1° and 2° cell types give rise to the vulva proper and usually arise from the P5.p (2°), P6.p (1°), and P7.p (2°) cells. The mechanism of induction and inhibition of the VPCs in vulval development has been used as a model system for investigating cell–cell signalling and is cartooned in Fig. 2 (bottom). An inductive signal from the gonadal anchor cell stimulates an EGF receptor-like tyrosine kinase (Let-23) on the surface of the VPC. The *lin-3* gene product is the inductive molecule and is similar to epidermal growth factor and transforming growth factor-α (elegant work in *C. elegans* on *lin-3* has provided significant information regarding the idea of morphogenic gradients in the specification of cell fate and is reviewed in refs 59 and 60). A newly identified kinase, Lin-2 (related to CAM kinase-II) is thought to act at the level of the Let-23 receptor and effects 1° cell differentiation, but it's kinase function appears to be dispensable for this role (61, 62). It is now thought that this protein, together with Lin-7 and Lin-12, is responsible for the correct processing and subcellular localization of the Let-23 receptor and does not play a role in signal transduction *per se* (61, 62). Stimulation of the Let-23 receptor is thought to override an inhibitory signal emanating from the syncytial hypoderm (acting via the Lin-15 protein) and to activate a Ras-like protein (Let-60). Activation of Let-60/Ras by Let-23 is mediated by the adapter protein Sem-5 (an SH2/SH3

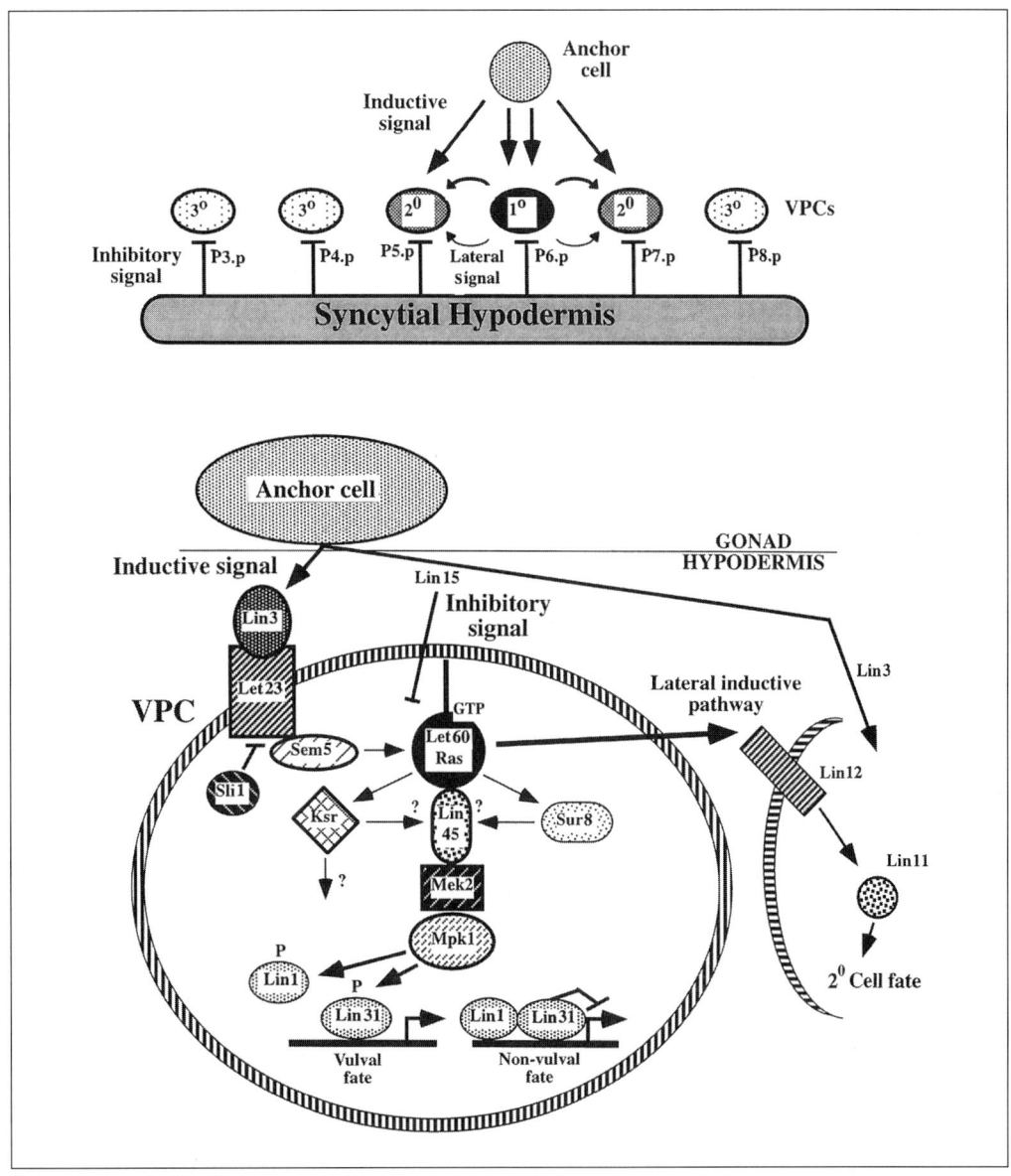

Fig. 2 Mechanism of vulval induction in *C. elegans*. (Top) Cell fates of the six vulval precursor cells (VPCs). An inductive signal from the anchor cell overcomes an inhibitory signal from the syncytial hypodermis and stimulates a 1° and 2° cell fate in three of the VPCs. A signal from the 1° cell helps to induce the two neighbouring cells to adopt the 2° cell fate. The three remaining VPCs adopt the 3° cell fate. (Bottom) The putative order of gene products thus far identified in the vulval inductive pathway (see text for details).

domain-containing protein that was first identified in C. *elegans*), which presumably activates Ras by association with a guanine nucleotide exchange factor (as yet un-identified in C. *elegans*) (see ref. 74). Another adapter-like molecule, Sli-1 (homologous to c-cbl), has been shown to negatively regulate Let-23 mediated vulval induction, possibly by direct interaction with the tyrosine kinase receptor itself (64, 65). Stimulation of Let-60/Ras leads to activation of the MAP kinase cassette involving Lin-45 (Raf), MEK2 (MEK), and mpk1/sur-1 (MAPK) and the regulation of at least two transcription factors (Lin-1 and Lin-31) resulting in 1° or 2° cell-fate deter-mination (66). In addition, the Ksr-1 kinase (a Raf-like serine/threonine kinase) and the leucine repeat protein, Sur-8, have been shown to act downstream of Ras and upstream or in parallel of Raf in vulval induction (67, 68). The cell closest to the anchor cell, which experiences the highest concentration of morphogen (Lin-3) usually adopts the 1° fate. The presumptive 1° cell, together with the Lin-3 morphogen, is then thought to induce the two flanking cells to adopt a 2° cell fate via a mechanism of lateral induction (mediated by a Notch-like protein encoded by *lin-12*) (56, 63, 69–74). The remaining three cells adopt the 3° fate and later fuse with the hypodermis.

Many more genes have been identified by mutation analysis that play a role in this signalling mechanism, but these await biochemical characterization (55, 75). Major dissection of this pathway has come about by the phenotypic analysis of various mutants. Mutations which effect the cell fates of these six VPCs can be classed into two types: vulvaless (vul) and multivulva (muv). In vulvaless mutants, all cells adopt the 3° cell fate as a result of disruption of the inductive signalling pathway. In multi-vulva mutants, all the cells develop as either 1° or 2° due to continual stimulation of the inductive signal. In some multivulva mutants, termed hypervulval, two 1° cells are found juxtaposed and are thought to result from disruption of the lateral inhibit-ory signal (57). Therefore, scoring of these phenotypes in mutational experiments has enabled several genes to be assigned to particular pathways.

Cell-specific laser ablation has revealed significant information regarding the equivalence of cells during development. Gonad ablation in loss-of-function *lin-15* mutants does not effect the resulting multivulval phenotype, which suggests that this gene does not act in the gonads or exert it's effect by inhibition of the inducing factor (55). Genetic mosaic and ablation analysis has shown that Lin-15 does not emanate from the gonads or VPCs but is now thought to arise from the surrounding syncytial hypodermis (76). The *let-23, let-60, let-45,* and *sem-5* mutations were identified as suppressors of the multivulva phenotype resulting from a *lin-15* mutation and act downstream of this signal (69, 74, 76–78). Genetic analysis of various *let-60* mutants has demonstrated that this Ras protein acts as the switch in determining the de-velopmental fate of the VPCs (72). Mutations that decrease the effectiveness of Let-60/Ras produce a vulvaless phenotype (3° cell fate), whereas those which increase its activity produce the multivulval phenotype (1° and 2° cell fates) (72, 73, 77). MEK2, *lin-45, Kkr-1,* and *sur-8* were all identified in a screen for second site suppressors of the multivulval phenotype resulting from transgenic animals carrying an activated *let-60/ras* gene (53, 54, 68). The *sur-8* mutant was able to suppress the multivulval pheno-type observed with an activated *let-60/ras* transgene or a loss-of-function *lin-15* mutation (68). However, it was unable to restore wild-type vulva to gain-of-function *lin-45/raf* and loss-of-function *lin-1* mutants, suggesting that Sur-8 acts downstream of Ras and upstream or in parallel to Raf (68). The multivulval phenotype observed with gain-of-function mutants in *lin-45/raf* can be blocked by mutations in *MEK2* (MEK) and *mpk1/sur-1* (MAPK) consistent with the conservation of the MAP kinase cascade. Strong mutants in *lin-45/raf* do not completely block Let-60/Ras signalling in vulval development, and this argues for the existence of a parallel pathway. One possible member of such a pathway is the newly identified kinase, Ksr-1 (67). *ksr-1* was initially identified as a suppresser of the multivulva phenotype produced by an activated *let-60/ras* transgene and has been shown to act downstream of Let-60/Ras and upstream of Lin-1 (67) The *ksr-1* null phenotype is much weaker than that of *lin-45/raf*, which suggests that it is unlikely to be the Raf activating kinase and participate in a linear pathway from Ras to Lin-1 (via Lin-45/Raf, MEK2, and mpk1/sur-1). Instead it is thought to contribute to the overall level of downstream signalling from Let-60/Ras, thus introducing another level of complexity to Ras-mediated signal transduction.

The *sem-5* gene was identified in a screen for suppressers of the multivulva phenotype in *lin-15* mutants. The protein was sequenced and shown to be composed of an SH2 domain flanked by two SH3 domains (74). This class of adapter protein was first identified in *C. elegans*, and since then numerous adapter proteins of this type have been identified that function in many diverse signalling processes (reviewed in ref. 88). These SH2/SH3 domain-containing proteins can tether phosphotyrosine-containing proteins (via the SH2 domain) and proline-rich proteins (via the SH3 domain) and are reviewed ref. 79. Another adapter molecule, Sli-1 was identified in a screen for mutations that could rescue the vulvaless phenotype resulting from a reduction-in-function mutation in the *let-23 receptor* (64, 65). This adapter molecule acts upstream of Sem-5 and Let-60/Ras and probably negatively regulates Let-23 activity by direct binding to the receptor. Interestingly, c-cbl (the mammalian homologue of Sli-1) has also been implicated in negative regulation of T-cell receptor stimulation but its mode of action is unclear (see ref. 89).

One of the major questions in cell signalling is how highly conserved signalling cascades such as the MAP kinase cassette (RAF–MEK–MAPK) can elicit distinct and specific cellular responses in a variety of different tissues. One attractive hypothesis has been the idea of tissue-specific targets for this signalling cascade. Lin-1 (ETS domain) and Lin-31 (winged helix) are both transcription factors involved in vulval development and have been shown, genetically, to act downstream of *mpk1/sur-1* (66, 80, 81). *lin-1* mutants are multivulval suggesting that this factor is a negative regulator of vulval cell fate. In addition, this factor has a role in other MAP kinase-mediated events in *C. elegans* such as development of the excretory system, the male tail, and the posterior ectoderm (81). *lin-31* null mutants exhibit both a vulvaless and multivulval phenotype (66, 80). In about 40% of *lin-31* null mutants, P5.p, P6.p, and P7.p adopt a 3° cell fate (vulvaless), whilst in 60% of animals P3.p, P4.p, and P8.p adopt 1° and 2° cell fates (multivulval). These observations suggest that Lin-1 could act as a general effector of MAPK kinase signalling, whilst Lin-31 could be a vulval-specific effector having two different activities: inhibition of 1° and 2° cell fates in P3.p, P4.p, and P8.p and induction of 1° and 2° cell fates in P5.p, P6.p, and P7.p. Tan *et al.* (66) have shown that both Lin-1 and Lin-31 are targets for mpk1/sur-1 in vulval development and have shown that these two proteins associated in a phosphory-lation-dependent manner. They propose that when these transcription factors are unphosphorylated they can physically associate and inhibit vulval development (see Fig. 2). However, upon stimulation of the MAPK cassette in p5.p, P6.p, and P7.p (by Let-23) these two factors become phosphorylated and dissociate from one another permitting a 'liberated' Lin-31 to induce 1° and 2° cell-fate differentiation. Thus from work in C. *elegans*, one level of specificity has been demonstrated which shows that a combination of general and tissue-specific effectors can determine the outcome of activating a common signal transduction cassette.

3.2 Lateral induction/inhibition

A hypervulval phenotype was observed in a screen for genes which changed the phenotype in *lin-15* mutants (71). In these animals 1° cells were found juxtaposed, suggesting disruption of a lateral inhibitory signal. The mutation was identified as *lin-12* (encoding another EGF receptor-like transmembrane tyrosine kinase) and was suggested to act in reception of the lateral inhibitory signal (82, 83). Further, over-expression of *lin-12* results in all cells adopting the 2° cell fate (84). This mechanism resembles the Notch-mediated lateral inhibition of sensory bristle formation in fruit flies (85, 86). However, 2° cells can form in the absence of any other VPCs, which suggests that the lateral signal is not essential for 2° cell fate. Katz *et al.*, using a controllable heat-shock transgene (expressing *lin-3*) in *lin-3* null animals, demon-strated that the concentration of Lin-3 experienced by the receiving cells is important for the specification of cell fate (59, 60). High expression levels of Lin-3 produced a multivulval phenotype (multiple 1° cells), whilst lower levels induced 2° cell-fate differentiation in recipient cells. Thus, low levels of Lin-3 can act to specify 2° cell fate. This implies that the lateral signal, acting via Lin-12/notch does not act to inhibit Lin-3/Let-23 signalling but rather synergizes with this pathway to specify the 2° cell fate. Further, mutations in *lin-11* (which encodes a homeodomain transcription factor) result in only one of the three possible progeny cell types found in 2° cell siblings (55, 70, 87). This suggests that Lin-11 acts in the Lin-12 signalling pathway to specify formation of the correct daughter cells.

Together, the fruit fly and nematode studies demonstrate the central importance of signal transducing proteins to processes such as development and differentiation. Several common principles recur, such as the role of receptor tyrosine kinases in signal reception, SH2 motif-containing proteins in recruiting transducing proteins, and protein–serine kinases acting to disseminate the signals throughout the cells. Equally important to these feed-forward pathways are those that antagonize or desensitize. Thus, inductive cycles can be established involving mutually dependent signals locking cells into complex relationships. By definition, disruption of these components yields phenotypic changes, suggesting there is very limited redundancy.

In view of the large number of protein kinase genes expressed per cell (probably > 200), these enzymes must be far more specific in their *in vivo* functions than is often apparent *in vitro*.

4. Signal transduction in slime mould

Cells of the slime mould *Dictyostelium discoideum* persist as single-celled amoeba until environmental conditions trigger aggregation and multicellular development (90). Cell growth and development are completely separate and can be controlled precisely in the laboratory (91). Thus, studying the role of developmentally important genes in *Dictyostelium* can circumvent the problems often encountered during embryogenesis (after genetic manipulation in higher eukaryotes), because multicellularity arises from aggregation of individual amoeboid cells and not from a process of cell division. Under conditions of starvation vast quantities of *Dictyostelium* synchronously initiate a developmental programme which is reversible until the last stages of maturation (91). Mutation and gene-disruption experiments are relatively straightforward in this slime mould as most of the life cycle is spent in the haploid state (for a guide to *Dictyostelium* biology and manipulation see ref. 91). cAMP is responsible for the coordination and regulation of many cellular processes during the life cycle of *Dictyostelium* (91–97). It is involved in the initial aggregation of amoeba by acting as an extracellular chemoattractant, as well as inducing chemotaxis and early gene expression (92). Additionally, during the multicellular phase of the life cycle, cAMP is required intracellularly for coordination of morphogenic movements and differential gene expression (97).

Upon starvation, a few cells start emitting pulses of cAMP and the surrounding amoeba respond by moving chemotactically towards the source. The cAMP pulses are received by heterotrimeric G protein-coupled receptors, which then activate three distinct pathways leading to chemotaxis, gene expression, and activation of cAMP-dependent protein kinase (PKA; reviewed in refs 94, 98, and 99). Genetic experiments have shown that receptor activation of phospholipase C and guanyl cyclase leads to gene expression and chemotaxis, respectively (100, 101). Receptor adaptation occurs within a few seconds, which facilitates the reception of the next flux of cAMP. This adaptation is thought to be mediated by specific kinases, as phosphorylated receptors are observed *in vivo* several seconds after cAMP stimulation and analogous protein kinases desensitize mammalian adrenergic receptors (93–107). After about 8 hours a tight aggregate is formed comprising approximately 10^5 cells, and it is at this stage that cell-fate decisions and cell proportioning occurs. There are three predominant cell types present in the aggregate, termed prestalk A (18%), prestalk B (3%), and prespore cells (79%) distinguishable by the cell-specific markers ecmA, ecmB, and Spi-1, respectively (Fig. 3). Under conditions of high humidity and low light levels, the aggregate will develop into a slug that can migrate until it reaches a more favourable environment where culmination is initiated. Here, cells redistribute and terminally differentiate to produce a mature fruiting body comprising four distinct structures (Fig. 3). Several serine/threonine and tyrosine kinases have now been isolated from

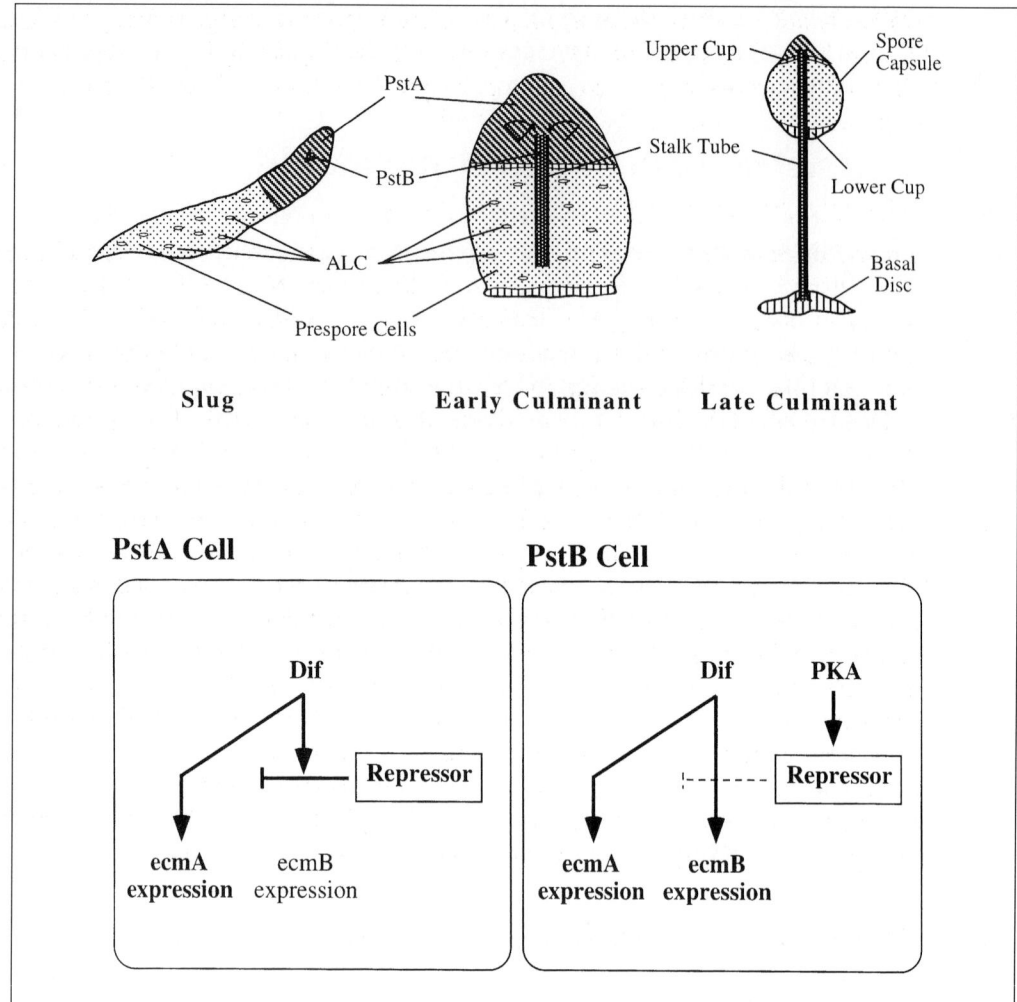

Fig. 3 (Top) Distribution of various cell types during different stages of *Dictyostelium discoideum* development. During the slugging phase, prestalk A (pstA) cells and a small node of prestalk B (pstB) cells are situated in the anterior region of the slug. The anterior-like cells (ALCs) are found distributed amongst the prespore cells. During culmination the pstA cells migrate to the region where the pstB cells are located. Upon arrival they are transformed into pstB cells and migrate into and form the stalk tube which lifts the spore-containing capsule into the air. (Bottom). One of the putative roles of cAMP-dependent protein kinase (PKA) during *Dictyostelium* development. The inductive action of Dif to promote pstB cell differentiation is inhibited in pstA cells by a repressor protein. Upon migration of the pstA cells to the mouth of the stalk tube this repression is lifted by the induced action of cAMP-dependent protein kinase, allowing pstB cell differentiation. This prevents ectopic differentiation of prestalk cells in other regions of the organism.

Dictyostelium and have been shown to be involved in developmental processes; many of which have developmentally important homologues in higher eukaryotes. Thus, *Dictyostelium* presents an attractive eukaryotic model system for the study of cell–cell communication and cellular differentiation.

4.1 Identification of protein kinases

Several protein kinase genes have been isolated by homology cloning techniques and include GSK-3, PKA, CKII, ERK1, and ERK2 (95, 97, 103, 108–111). A saturation mutagenesis screen (described in Section 4.4) has been developed in *Dictyostelium* to identify developmentally important genes that will undoubtedly uncover several new kinases. A multigene family comprising at least five putative serine/threonine kinases was identified by a PCR-based screen (112,113). Two putative serine/threonine kinases (termed Dd PK1 and Dd PK2) were identified by screening a cDNA library with oligonucleotides directed towards conserved regions of eukaryotic protein kinases (107). *Dd PK1* RNA levels were seen to decrease during aggregation and reaccumulate later in development, suggesting a role in culmination (107). Overexpression of *Dd PK2* caused rapid development and affected the intracellular level of cAMP in *Dictyostelium* (114). Tan and Spuddich (115) used antiphosphotyrosine antibodies to screen an expression library and isolated two fusion proteins that exhibited protein tyrosine kinase activity (termed *Dd PYK1* and *Dd PYK2*). These tyrosine kinases were the first to be identified in *Dictyostelium* and share homology with both mammalian serine/threonine and tyrosine kinases. This is consistent with the idea that the catalytic domains of tyrosine kinases were derived from an archetypal protein serine/threonine kinase, although it is presently unclear whether the Dd PYKs exclusively phosphorylate tyrosine (116).

4.2 The role of cAMP-dependent protein kinase

The cAMP-dependent protein kinase has a very complex role in *Dictyostelium* biology as cAMP, and thus PKA, appears to coordinate several developmental events (reviewed in ref. 117). There are four cAMP receptors (cAR1–4) expressed on the surface of *Dictyostelium* cells during development. Pulses of extracellular cAMP induce a chemotactic response and developmental gene expression. PKA is a mediator of both these early events. Stimulation of cAR1 leads to activation of adenyl cyclase and the MAP kinase homologue ERK2 (Fig. 4). Both these events lead to the elevation of intracellular cAMP, directly or by the inhibition of the phosphodiesterase (regA) resulting in activation of PKA. PKA stimulates developmental gene expression and downregulates ERK2 activity leading to signalling adaptation. Firtel and Chapman have demonstrated that overexpression of the regulatory subunit of PKA (PKA-Rm), mutated to prevent cAMP binding, results in a block to aggregation (95,118). These cells do not express genes which are normally induced in the multicellular stages of development. Additionally, cells overexpressing the catalytic subunit (PKA-cat) undergo accelerated growth, whilst those having a disrupted gene fail to aggregate

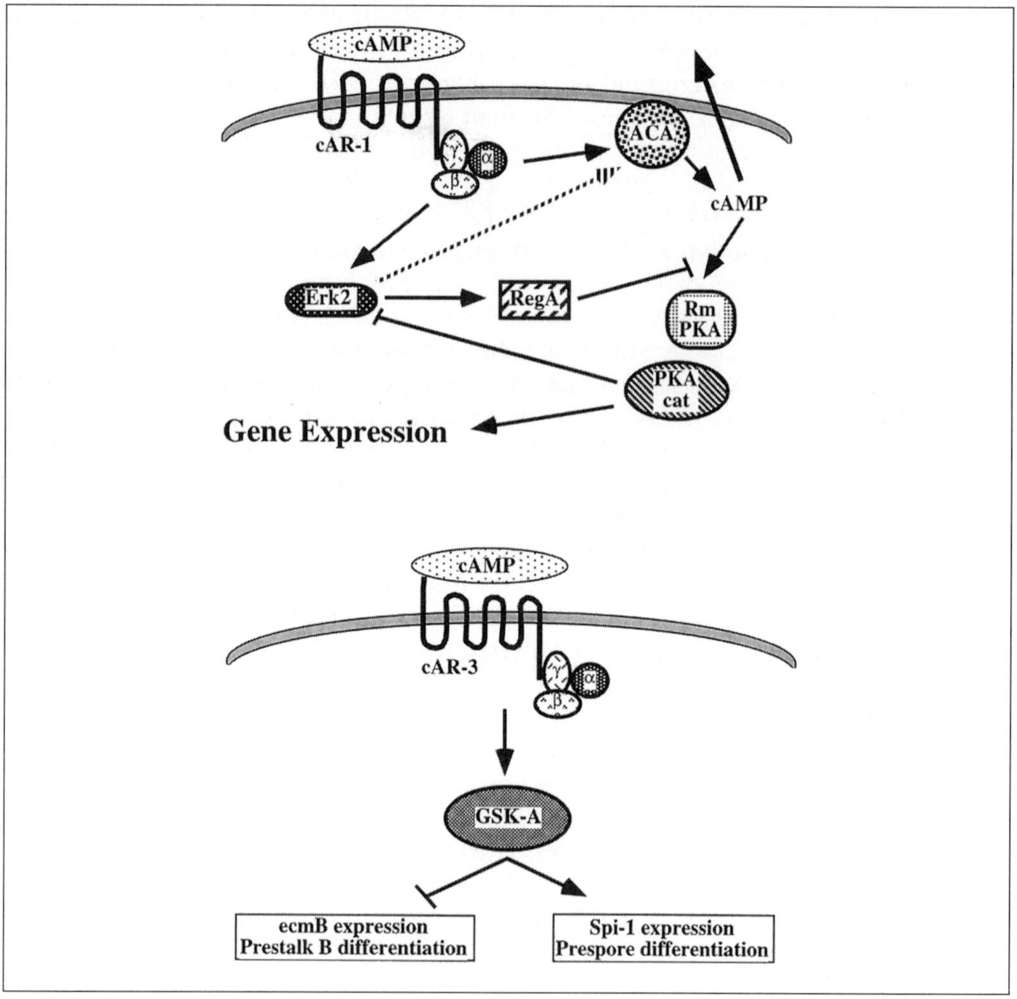

Fig. 4 (Top) Activation of PKA by extracellular cAMP. Binding of extracellular cAMP to the G protein-coupled receptor (cAR-1) leads to activation of ERK2 and adenyl cyclase (ACA). Intracellular cAMP then binds to the regulatory subunit of PKA (RmPKA) and activates the enzyme. ERK2 regulates the phosphodiesterase (RegA) and thus contributes to increased cAMP levels. It may also have a role in the stimulation of adenyl cyclase. (Bottom) Activation of GSK-3 by the serpentine receptor cAR-3. The level of GSK-3 activity can be regulated by the cAMP receptor cAR-3. GSK-3 activity inhibits prestalk cell formation and induces prespore cell differentiation. Thus, the level of GSK-3 activity is essential in determining the eventual cell fates in these precursor cell types.

(119). Further, disruption of the ERK2 gene results in a block to aggregation (120). ERK2 is thought to be required, at least in part, for the activation of adenyl cyclase and is possibly activated by the $G\alpha4$ subunit of the heterotrimeric G-protein complex (121).

More elaborate experiments, using stage-specific promoters, have provided further insights into the function of PKA during development (97). During the slugging and culmination stages of development, cells destined to form the stalk tube (pst cells) can be divided into several cell types. Two of these cell types can be distinguished on the basis of their differential gene expression. PstB cells express two genes encoding extracellular matrix proteins (termed *ecmA* and *ecmB*) whilst pstA cells only express one, *ecmA* (122–124). Both genes are induced by the stalk-cell morphogen, DIF (123, 124). During culmination the pstA cells migrate to the mouth of the stalk tube and are transformed into pstB cells by induction of the *ecmB* gene (Fig. 3). The pstB cells then migrate into and form the developing stalk tube, in a process likened to a 'reverse fountain'. A mutant regulatory subunit of PKA (acting as a dominant inhibitor) was fused to the stage-specific *ecmA* promoter to study the effect of blocking the action of this kinase during multicellular differentiation. *Dictyostelium* transformed with this vector (*ecmA-Rm*) exhibited prolonged migration and failed to culminate into mature fruiting bodies (97). This implied that PKA is required for the choice between continued migration and culmination. A more detailed analysis demonstrated that no pstB cells were produced as the cells failed to express the *ecmB* gene. Heterotypic aggregates, produced by mixing wild-type and mutant amoebae (tagged by β-gal expression), produced slugs that were able to culminate normally. However, none of the cells containing the *ecmA-Rm* construct were seen to migrate to the tip of the slug and were not found in the stalk tube (97). *Dictyostelium* transformed with either an *ecmB* promoter fused to the *Rm* gene, or with an *ecmA* or *ecmB* promoter fused to a constitutively active PKA gene, undergo normal development. These data suggested that protein kinase activity was required for *ecmB* gene induction and pstA cell migration. Harwood *et al.* have proposed that a repressor in the pstA cells, which blocks DIF induction, prevents precocious expression of the *ecmB* gene (Fig. 3) (97). When pstA cells migrate to the mouth of the stalk tube this repression is lifted, by the action of cAMP-dependent protein kinase, resulting in *ecmB* gene expression.

Overexpression studies of PKA-Rm and PKA-cat have indicated a role for PKA in the formation of spores (97, 114, 119, 125, 126). PKA activation is thought to be the developmental trigger for spore-cell terminal differentiation, and overexpression of PKA-cat in prespore cells leads to spore-cell formation in a cell-autonomous manner. This effect is independent of ERK2 activation, placing PKA either downstream of ERK2 or in a different pathway at a different developmental time point (127). Further, during culmination a secreted peptide (SDF-2) is probably responsible for the activation of PKA in spore cells and acts via a two-component histidine kinase cassette that leads to increased intracellular cAMP (128). Thus, PKA appears to play many roles during *Dictyostelium* development including chemotaxis, aggregation, developmental gene expression, migration, culmination, and terminal differentiation.

4.3 Glycogen synthase kinase-3

GSK-3 is a serine/threonine kinase implicated in various signalling pathways in higher eukaryotes including transcriptional regulation, cell-fate determination, and insulin signalling. However, determination of a role *in vivo* by genetic analysis has been hindered by its indispensability during embryogenesis (129, 130). GSK-3 was first identified in *Dictyostelium* by homology-cloning and subsequently surfaced in an insertional mutagenesis screen as a kinase required for cell-fate determination (108). GSK-3 null *Dictyostelium* have an altered proportion of cells at the aggregate stage, seen as a massive expansion of the prestalk-B cell population which are formed at the expense of the prespore cells. This suggests that GSK-3 acts to inhibit prestalk-B and induce prespore cell differentiation in normal aggregates; a function known to be controlled by extracellular cAMP and DIF. An increase in GSK-3 activity was shown to be associated with an induction and repression of prespore and prestalk-B specific gene expression, respectively (131) (Fig. 4). This increase in GSK-3 activity at a specific point during development was not observed in cells carrying a null mutation in cAR-3, suggesting that this receptor acts upstream of GSK-3 (131). Further, on closer examination of the cAR-3 null phenotype it was seen to be very similar to that of the GSK-3 null. Analysis of other REMI mutants having the same phenotype should rapidly identify the other components of this signalling pathway and perhaps identify novel regulators of GSK-3.

4.4 Restriction-enzyme mediated integration (REMI)

Kuspa *et al.* have devised a novel genetic technique of random insertional mutagenesis in *Dictyostelium* termed 'restriction-enzyme mediated integration', which is outlined in Fig. 5 (132). Basically, amoeba are transfected with an antibiotic-resistance plasmid linearized by a unique restriction enzyme together with that enzyme. The restriction enzyme then randomly 'cuts' the *Dictyostelium* genome and native enzymes repair the DNA, occasionally inserting the antibiotic-resistance plasmid into the genome. Cells are then grown in the presence of the antibiotic and clonal populations screened for phenotypes of interest. Once a phenotype has been scored the genomic DNA from that clone is isolated, cut with a second restriction enzyme that does not cut inside the plasmid, ligated, and transformed into *Escherichia coli*. In this way, the plasmid is isolated from the genome together with 5'- and 3'-flanking sequences of the gene responsible for the phenotype. In order to verify that the phenotype observed is due to a single insertional event at the identified locus the isolated plasmid containing the 5'- and 3'-flanking sequences can be used in homologous recombination gene disruption in normal cells to regenerate the phenotype. There are literally hundreds of REMI mutants in scores of phenotypic classes awaiting characterization. Thus, with the advent of such phenotypic screening techniques, the identification of mammalian kinase homologues, and the simplicity of its life cycle, the future for *Dictyostelium* as a model eukaryotic system for the study of kinase function looks bright.

5. Serine/threonine kinases in yeast

A wealth of data regarding kinase function in a diverse number of signalling pathways, including cell cycle, osmotic stress, mating, transcription, translation, and meiosis has been obtained from studies in both budding (*Saccharomyces cerevisiae*) and fission

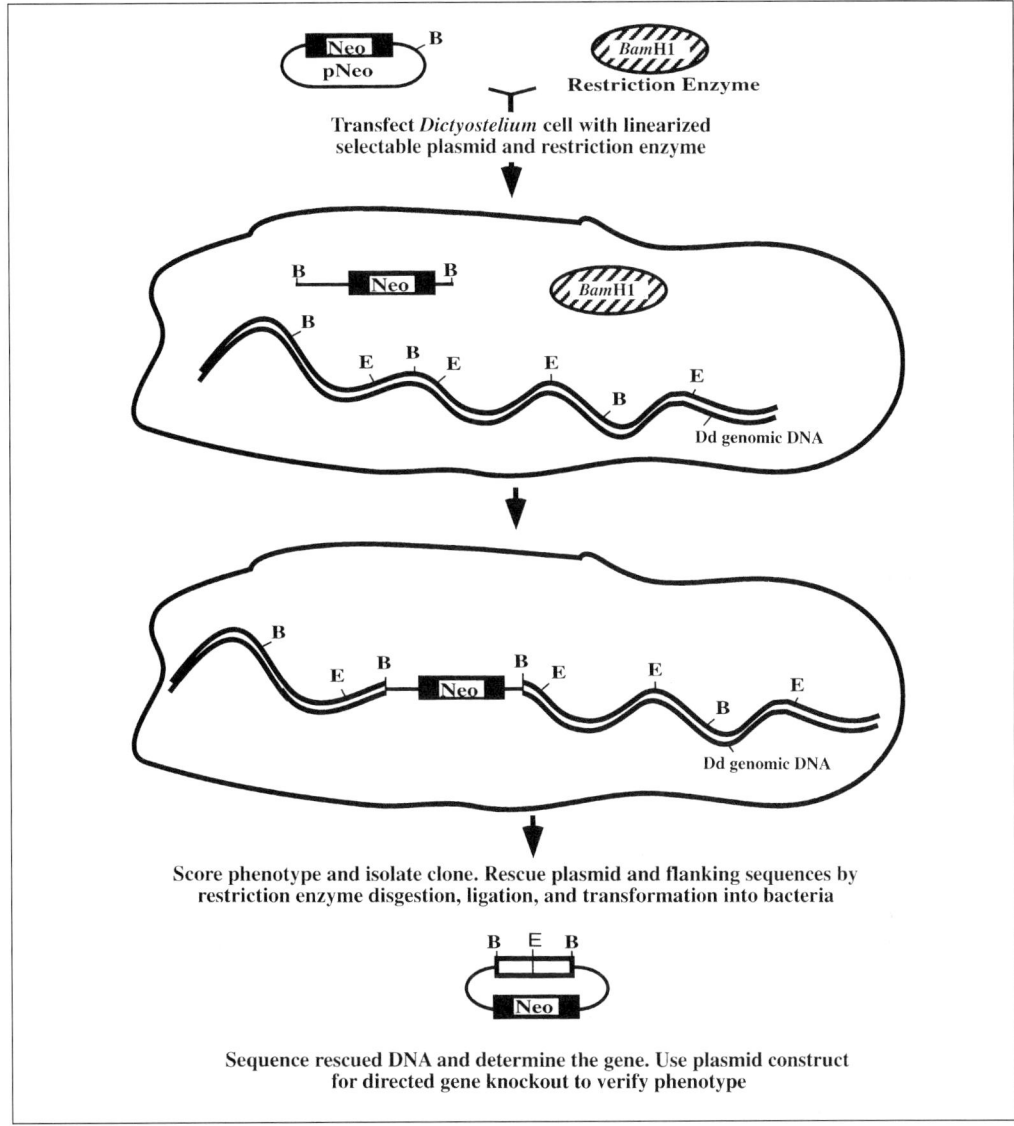

Fig. 5 Restriction-enzyme mediated integration (REMI). A random insertional mutagenesis technique that involves the random digestion of genomic DNA with a restriction enzyme, integration of a selectable plasmid, and repair of the genomic DNA. See text for full explanation. B, restriction site for *Bam*HI; E, restriction site for *Eco*RI.

(*Schizosaccharomyces pombe*) yeast (133, 134). A recurrent theme in many receptor-mediated signalling pathways is the activation of a MAPK kinase-like cascade (135). In budding yeast there are at least four distinct MAP kinase pathways that have been described, yet, despite the significant protein homologies, there appears very little cross-talk between these pathways, suggesting that kinase promiscuity is tightly controlled *in vivo*. In metabolic processes, the idea of substrate channelling has been accepted for a long time. This presents an attractive idea for limiting kinase promiscuity in signal transduction processes and ensuring efficient relay of a specific message down a not-so-specific cascade. In this section the yeast mating-response pathway will be described as an example of signalling from a heterotrimeric G protein-coupled receptor that acts via a MAP kinase cassette. Further, from genetic studies in yeast, possible methods for limiting signalling cross-talk and conveying specificity to MAP kinase pathways will be discussed.

5.1 The yeast mating response

In budding yeast, diploidy results from the fusion of two different haploid cell types termed a and α. Each cell secretes a mating factor (either a or α pheromone) which interacts with the corresponding receptor on the opposite cell type to promote mating. Receptor binding initiates a signal transduction pathway leading to cell-cycle arrest, conjugation, and ultimately meiosis (reviewed in refs 136 and 137). Mutational analysis produces two phenotypes: responsive and non-responsive (138). Here, the mating response is either constitutively activated or completely non-responsive to pheromone. Much of this pathway has been determined by combining these types of mutation in genetic epistasis experiments. Upon pheromone binding, both receptors elicit the same response by stimulation of a common pathway.

Binding of the pheromone to its corresponding heterotrimeric G protein-coupled receptor leads to activation of a signal transduction pathway that results in mating-specific gene expression, cell-cycle arrest, and polarized cell growth (Fig. 6) (136–150). Heterotrimeric G protein-coupled receptors can elicit similar effects in other systems (the T-cell response to an antigen-presenting cell and the *Dictyostelium* developmental programme, although the signalling pathways in these systems are poorly under-stood) (151, 152). Activation of the receptor leads to dissociation of the α-subunit (Gap) from the β/γ subunits (Ste4 and Ste18, respectively). In contrast to other known sys-tems, the α-subunit acts to inhibit signalling and its ligand-dependent dissociation from the β/γ-subunit relieves inhibition of signalling. The precise mechanism by which the β/γ-subunits activate signalling is not known, but it is thought that they act to localize signalling molecules to the site of pheromone action. Precise localiza-tion of the β/γ-subunits also provides positional information that is important for the orientation and initiation of polarized cell growth leading to cell fusion (153). The formation of a multiprotein 'transducersome' complex at the inner surface of the cytoplasmic membrane has been proposed and is supported by the biochemical and genetic evidence. Further, constitutive membrane localization of the β/γ-subunits is sufficient to activate the pheromone-response pathway (154, 155).

The Ste20 kinase appears to be the principal, but not sole, effector of signalling from the β/γ-subunits as *Ste20* null mutants still retain a weak pheromone response (144). Ste20 also has other roles in yeast signal transduction, and thus its localization to the 'transducersome' complex at the pheromone receptor probably serves to bring it in contact with specific substrates (149, 156). The kinase scaffold protein, Ste5, has been shown to be associated with the β/γ-subunits as well as with several other

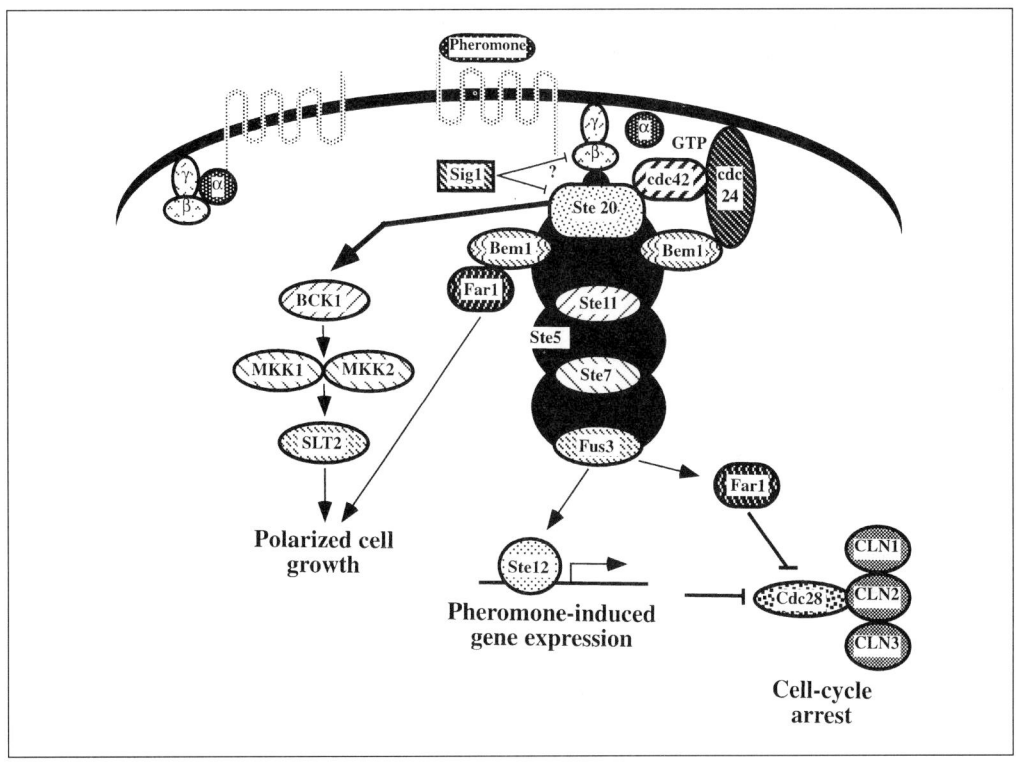

Fig. 6 Involvement of protein kinases in the mating response of *S. cerevisiae*. Pheromone binding stimulates guanine nucleotide exchange on the α-subunit of the G-protein–receptor complex which promotes dissociation of the β- and γ-subunits and activates the mating-response kinase cascade. The relative ordering of the components has been established by a combination of genetic and biochemical experiments. Some of the relationships remain ambiguous.

signalling molecules (154, 157, 158). By using a combination of genetic (including the yeast two-hybrid system, see Section 5.3) and biochemical analyses the following proteins have been shown to, or inferred to, constitute the multiprotein transducersome complex: the kinases, Ste20, Cdc42, Ste11, Ste7, Fus3, Far1; the adapter protein Bem1 and the guanidine nucleotide exchange factor Cdc24 (147, 148, 154, 157–161) Bem1 (an SH3 domain-containing protein) associates with Ste5 and Cdc24, possibly ensuring correct assembly and juxtaposition of the activated transducersome complex (162). Bem1 can also bind to the Far1 kinase and this interaction is important for the role of Far1 in polarized cell growth (145, 157). Interestingly, both Ste5 and Far1 contain a LIM domain (a protein–protein interaction domain) which possibly mediates their interaction with Bem1. Cdc24 is an activator of the GTP-dependent kinase Cdc42 and this kinase may be an upstream activator of Ste20 (147, 148, 159, 162, 163). Thus it is possible that Ste5 binds Cdc42 and Bem1. These interactions permit Bem1 to bind Cdc24 and activate Cdc42 which can then effect the activity of Ste20.

The components of the pheromone-specific MAP kinase pathway are physically associated with the kinase scaffold protein Ste5 (154, 160, 161). Ste11 is a known substrate of Ste20 and is activated upon phosphorylation in response to pheromone binding (149). Ste11 phosphorylates and activates the dual-specificity kinase, Ste7, which in turn activates the MAP kinase homologue Fus3 by phosphorylation on both threonine and tyrosine (164–166). The downstream target of this pathway appears to be a transcription factor encoded by *STE12* (167, 168). This factor is a phosphoprotein and binds to pheromone-responsive elements (PRE) found upstream of target genes (167, 169, 170). Ste12 is rapidly phosphorylated in response to pheromone binding and this results in increased transcription of pheromone-responsive genes (for example *fus1* and *fus2* which are required for cell fusion) (169). Fus3 can also phosphorylate and activate the Far1 kinase. The function of Far1 here is to promote cell-cycle arrest during mating by inhibiting the cyclin-dependent kinase Cdc28 and hence the cyclins CLN1, CLN2, and CLN3 (146, 152, 170–172, 180). The interaction between Far1 and the cyclin complex has been shown to be dependent on Fus3 activity, and thus a direct link between pheromone signalling and cell-cycle arrest has been established (172). Far1 also has a role in polarized cell growth which involves the reorganization of the actin cytoskeleton towards the point of cell fusion, and this function is dependent on its association with the adapter protein Bem1 (145, 146, 157). Efficient polarized cell growth and cell fusion requires the activation of another MAP kinase pathway (involving the kinases BCK1, MKK1/MKK2, and SLT2) and is also activated by Ste20 (173).

The identification of genes involved in this signal transduction process and their order of action has mainly been determined by genetic means. However, biochemical analysis has been invaluable in confirming specific protein interactions and clarifying points that were not clear from genetic analysis. For example, a dominant allele of *ste11* was able to rescue mating in *ste4* (Gβ) mutants but was unable to restore mating in cells carrying mutations in either *ste7*, *fus3*, or *kss1* (174, 175). This suggests that the product of *ste11* acts before these other genes in the mating-response pathway. In support of this, Ste7 has been shown to be a substrate for Ste11. However, a *ste11* null mutant, but not a *ste7* mutant, could block the growth defect produced by overexpression of *ste5* (155). From a genetic point of view, this does not place *ste7* downstream of *ste11* or *ste5*. Characterization of the *ste5* gene product showed it to be a kinase scaffold protein and thus it was possible to resolve the genetic incongruencies in the light of biochemical information. Overexpression of *ste4* (Gβ) can constitutively activate the pheromone-response pathway and this can be blocked by disruption of *ste5*, *ste7*, *ste11*, *ste18*, and *ste20*, suggesting that these genes act downstream of *ste4* (155). However, the growth defect observed by overexpression of *ste5* could be blocked by *ste4* and *ste18* (Gγ) null mutants which suggests a role downstream of *ste5* for these proteins. However, another interpretation of this information is that Ste5 requires the physical presence of Ste4 and Ste18 to function correctly. This is consistent with the idea of a large multiprotein transducersome complex in the vicinity of the heterotrimeric G protein-coupled receptor and the role of Gβ/Gγ in transducersome localization.

Screens for 'synthetic lethal' mutations (i.e. two temperature-sensitive (*ts*) mutations that result in lethality at the permissive temperature when combined) can often identify proteins that interact or act at the same step in a pathway. Akada *et al.* devised a 'synthetic sterile' screen to identify proteins that act close to *ste4* in the pheromone-response pathway (155). They generated sterile mutants in a yeast strain where *ste4* was replaced by a *ts ste4* mutation. They produced mutants and screened for cells that were sterile at the permissive temperature. The newly identified mutants were transformed with a plasmid containing wild-type *ste4* under the control of a galactose inducible promoter. Mutants were then assayed on galactose plates (i.e. in the presence of wild-type *ste4*) and those able to mate were termed 'synthetic sterile' mutants in which the sterile phenotype requires the presence of both the *ts ste4* mutation and the new *ts* mutation. The majority of mutants exhibited sterility in the presence of wild-type *ste4* and result from dominant-negative mutations in downstream signalling molecules. In this screen, Akada *et al.* identified *ste5*, *ste18*, and *ste20* as synthetic sterile mutations with respect to *ste4*, which demonstrated that this technique can indeed identify proteins acting close to the target protein in signalling cascades. Another technique, the yeast interaction trap assay, is a powerful genetic tool for identifying interacting proteins and will be discussed in more detail in Section 5.3.

5.2 Signalling cross-talk

The remarkable degree of conservation of intermediate elements of the pheromone pathway in budding yeast and pathways in fission yeast and mammals, including preservation of the order of action, suggests these interactions evolved before the divergence of these organisms' ancestors. There are at least four distinct MAP kinase cascades described in yeast effecting such processes as mating, invasive growth, osmotic sensing, and polarized cell growth (Fig. 7) (135, 173, 176, 179). Whereas the intermediate proteins, the protein kinases, are highly conserved, their input signals and output targets are quite different. This is a common theme of signal transducing systems. Many intermediary protein kinases are evolutionary dinosaurs, remnants from a bygone age perhaps trapped by their own efficiency in performing a transductory function that has been adapted as needs demand.

Remarkably, although there is considerable similarity between the kinases of the MAP kinase cassette (MAPKKK, MAPKK, and MAPK) in different systems, there is very little cross-talk, which suggests that their modes of action are tightly controlled *in vivo*. What mechanisms exist to restrict this cross-talk? Studies in yeast have shed some light on the processes of signal transduction specificity. The MAPKK kinase, Ste11, is involved in the activation of three distinct MAP kinase cascades (pheromone response, osmotic sensing, and invasive growth), yet when yeast cells are exposed to high osmolarity, Ste11 activity only induces HOG1 activation and the osmotic response. Using a *fus1–LacZ* reporter construct (that is specifically activated in response to pheromone signalling) O'Rouke *et al.* showed that when cells containing a mutation in *hog1* (or the HOG kinase, *pbs 2*) were subjected to high osmolarity the pheromone-response pathway was also activated (156). Further, they showed, using various mutant strains, that this activation required the osmolarity sensor, *sho1*, and the kinases *ste20*, *ste11*, *ste7*, *fus3*, and *Kss1*. This demonstrates that the osmolarity

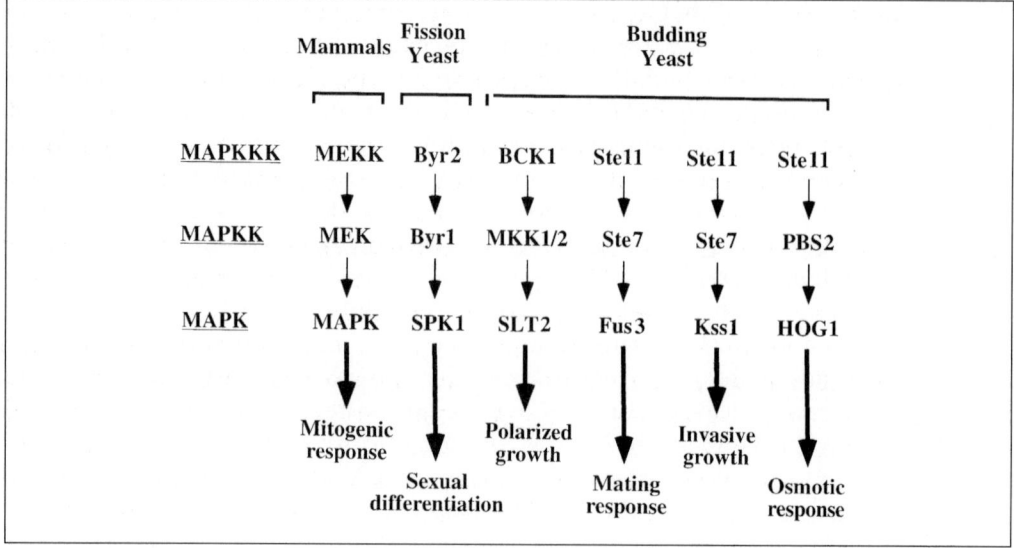

Fig. 7 Comparison of mammalian, fission yeast, and budding yeast MAP kinase-mediated signal transduction pathways.

sensor Sho1, acting via Ste11/Ste20 can inappropriately activate the pheromone-response pathway in the absence of HOG1. This genetic evidence suggests that HOG1 acts to prevent signalling cross-talk by regulating Sho1 in some way (possibly via a negative feedback loop).

In *fus3* null cells, pheromone signalling still occurs because the MAP kinase, Kss1, substitutes for Fus3 and activates Ste12, permitting expression of pheromone-responsive genes (140, 164). It was originally thought from this genetic evidence that Fus3 and Kss1 were functionally redundant. However, in *fus3* null cells, pheromone signalling also resulted in invasive growth. Subtle analysis of various mutants has determined that Kss1 is the MAP kinase required for invasive growth (177, 178). The transcription factor Ste12 is also required for expression of the genes responsible for invasive growth and is regulated both positively and negatively by Kss1 (177). Thus, in the absence of Fus3, Kss1 physically substitutes for this kinase and is inappropriately activated in response to pheromone which results in mating- and filamentation-specific gene expression. Moreover, using cells harbouring a kinase-dead *fus3* allele, *Kss1* was unable to rescue the pheromone-response defect as the Fus3 protein was still recruited to the signalling complex (178). Thus, from studies in yeast, two mechanisms of limiting signalling cross-talk have been proposed: channelling the signal to specifically assembled signalling cascades and pathway-specific feedback regulation.

5.3 Yeast interaction trap assays

The yeast two-hybrid system represents a powerful technique for the study of protein–protein interactions and for the identification of novel interacting proteins for a particular target protein. In the last two years alone, nearly 300 articles have been published in which the yeast two-hybrid system has been used to study protein kinase function. The technique affords the identification of new interacting proteins with a particular target protein or can be used to confirm protein–protein interactions in the context of a cellular environment. The principle of the technique (developed by Fields and Song, 181) relies on the fact that certain transcription factors comprise two distinct functional domains: a DNA binding domain (DBD) and a transactivation domain (TAD) (Fig. 8). Protein 'X' is fused to the DBD of a specific transcription factor (e.g. Gal-4) and protein 'Y' to its TAD and these are transfected into a yeast cell line harbouring a β-galactosidase reporter gene, under the control of the specific transcription factor. If the two proteins physically associate *in vivo* they will reconstitute the transcription factor, by virtue of proximity, and induce β-galactosidase expression (which can be scored as 'blue' yeast colonies). Thus the development of 'blue' yeast colonies in such an assay is an indication of a protein–protein interaction between the two target proteins. However, by far the most widely adopted use of this technique is in the screening of TAD fusion libraries with the 'bait-DBD' fusion protein in order to identify new interacting proteins. In such screens, reporter cell lines expressing the 'bait-DBD' protein are transfected with a TAD-fusion library and 'blue' colonies isolated to identify the 'bait-binding protein'. This technique has had considerable success in the field of signal transduction in last 7 years and a few examples are given below.

The ubiquitous 14–3–3 proteins were identified in a two-hybrid screen for proteins that interact with the cytosolic part of the type-1, insulin-like growth factor receptor (182). Interestingly, this protein has surfaced in several two-hybrid screens and is now thought to be a general phosphoserine binding protein that may act as a scaffold-like protein in signal transduction cascades, similar to the role determined for Ste5 in the pheromone-response pathway. These proteins were also shown (by two-hybrid analysis) to interact with the MEK kinases MEKK1, MEKK2, and MEKK3 (183) and GSK-3β (S. Plyte, unpublished observations). In addition to the 14–3–3 protein, Axin, Axil (axin-like protein) and human dynamin-like protein IV (HdynIV) were isolated in screens for proteins that interact with GSK-3*b* (184–186). A novel family of cofactors that interact with and regulate homeodomain transcription factors were identified in a two-hybrid screen (187). In this screen, three homeodomain-interacting protein kinases (HDPK1–3) were identified and shown to bind these transcription factors and act as corepressors. In contrast to this type of kinase DBP–protein interaction, a different DNA binding protein (dbpA) was isolated as a protein that could inhibit the cell-cycle dependent kinase, Cdk5 (188) Interestingly, this protein could also inhibit Cdk4 (but not CdK2) and this may reflect signalling cross-talk in dbpA-mediated pathways that regulate cell-cycle progression.

One of the drawbacks of this technique is that the protein–protein interaction needs to take place in the nucleus, and thus membrane-bound proteins are not ideally suited for this type of analysis. One possibility is to attach nuclear localization sequences to the proteins of interest, but this may lead to artefactual results due to inappropriate expression of a protein in the wrong cellular compartment. Another limitation is that several signalling proteins need to be activated (e.g., by phosphorylation) before they

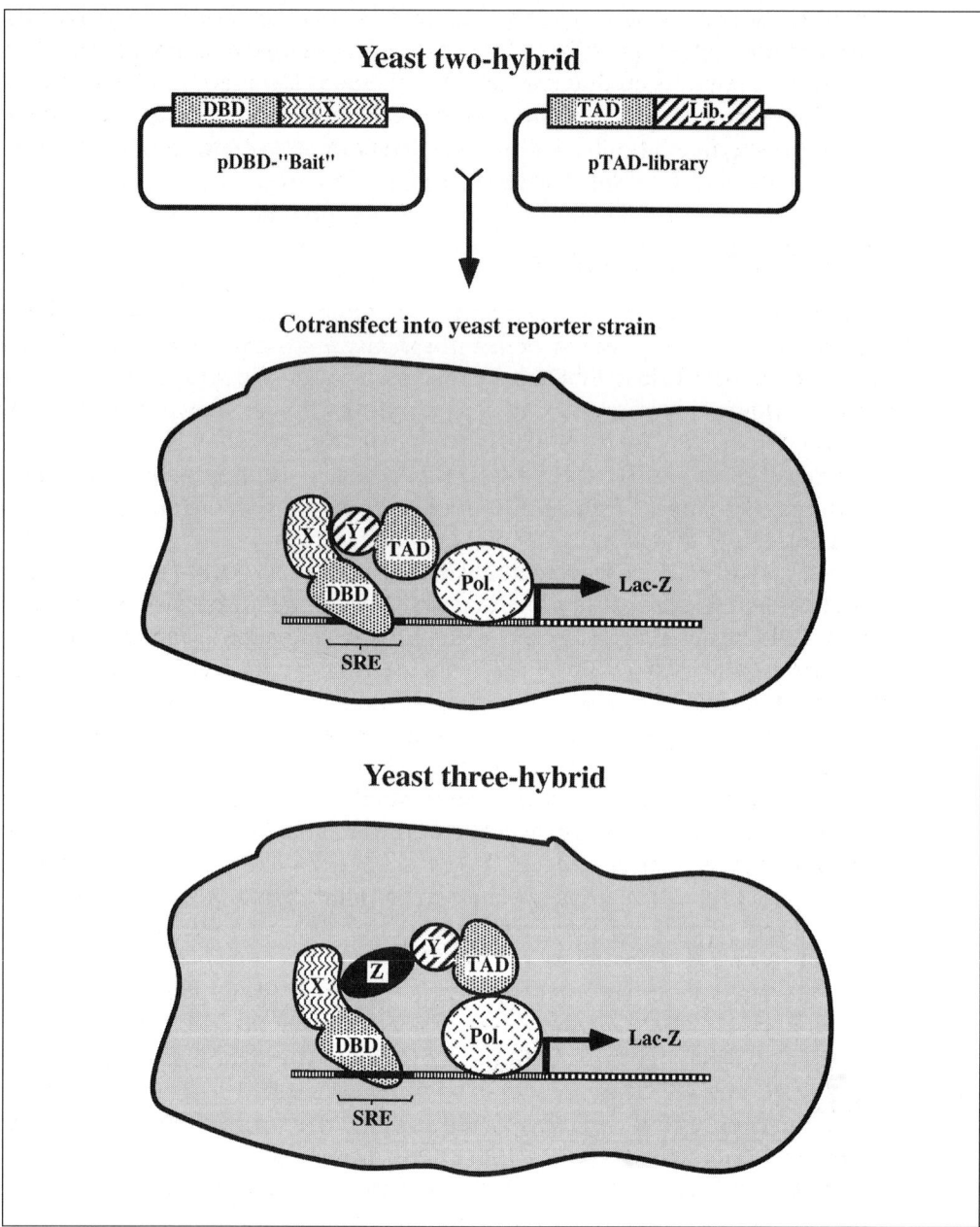

Fig. 8 The yeast interaction trap assay (yeast two-hybrid assay). A yeast strain, harbouring a reporter construct whereby the β-galactosidase gene is under the control of a specific promoter, is transfected with two fusion proteins: protein X fused to the DNA binding domain (DBD) of a specific transcription factor (that recognizes the promoter); protein Y (or a library of proteins) fused to the transactivation domain (TAD) of the same transcription factor. Physical association of proteins X and Y results in reconstitution of the transcription factor and reporter gene expression. (Pol., DNA polymerase.). In the three-hybrid assay a third linker protein is also expressed in these cells.

can bind other proteins. This can be overcome in some cases by coexpressing the activating protein together with the 'bait' protein. A further refinement of the technique is to coexpress an intermediate protein that is able to bind the 'bait' protein (yeast three-hybrid system, Fig. 8). This affords the identification of a third protein that binds the intermediate protein in the context of its association with the 'bait' protein and yields information about protein complex formation *in vivo*.

6. Concluding remarks

Reversible protein phosphorylation is a recurrent theme in biology, and it is evident that many signalling pathways present in lower eukaryotes persist in higher organisms. It seems evident that 'Nature' has taken this very efficient mechanism and employed it for a plethora of diverse activities. However, as we start to unravel the complicated knot of interconnected signal transduction pathways we seem to increasingly encounter several common proteins. Thus, it seems that 'Nature', possibly limited by the complexity of the genome or the need for efficiency, has been forced to utilize the same protein kinases for different functions. It is only now that we are starting to see the clever ways in which 'Nature' has been able to recycle old kinases for new functions and still maintain specificity. Theses glimpses are being afforded, in part, by the elegant studies that are being performed in lower eukaryotes.

Over the past decade, the realization that the fundamental processes of growth control, cell-cycle regulation, and signal transduction are similar in all eukaryotes has revolutionized these fields. The increased tractability of simpler eukaryotes to genetic analysis offers gene discovery based on function rather than purely structure. Genes identified in one organism can be rapidly isolated in others, thus stimulating the discovery of further genes. Conservation of function allows examination of the properties of one component in a heterologous system. Out of this enormous convergence of interests around the 'basic processes' of transduction, protein kinases have emerged as common denominators. They are implicated in every aspect of cellular regulation and represent the most varied class of proteins known. Application of both genetic and biochemical approaches to their study will undoubtedly reveal more of their manifold talents. With 'genome closure' (expressed genes) expected very soon the number of 'orphan kinases' will increase dramatically over the next couple of years. Homology cloning of such orphans or their expression as heterologous proteins in lower eukaryotes is one of the quickest ways to generate information relating to their function. I suspect that, with the arrival of many new kinases from the human genome sequencing programme, the genetic manipulation of such kinases in lower eukaryotes will become of paramount importance.

References

1. Karim, F. D., Chang, H. C., Therrien, M., Wasserman, D. A., Laverty, T., and Rubin, G. M. (1996). A screen for genes that function downstream of Ras1 during *Drosophila* eye development. *Genetics*, **143**, 315–29.

2. Kauffmann, R. C., Li, S., Gallagher, P. A., Zhang, J., and Carthew, R. W. (1996). Ras1 signalling and transcriptional competence in the R7 cell of *Drosophila*. *Genes Dev.*, 2167–78.

3. Chang, H. C., Solomon, N. M., Wasserman, D. A., Karin, F. D., Therrien, M., Rubin, G. M., and Wolff, T. (1995). *phyllopod* functions in the fate determination of a subset of photoreceptors in *Drosophila*. *Cell*, **80**, 463–72.

4. Brunner, D., Ducker, K., Oellers, N., Hafen, E., Scholz, H., and Klambt, C. (1994). The ETS domain protein Pointed-P2 is a target of MAP kinase in the *sevenless* signal transduction pathway. *Nature*, **370**, 386–9.

5. Bohmann, D., Ellis, M. C., Staszewski, L. M., and Mlodzik, M. (1994). *Drosophila* Jun mediates Ras-dependent photoreceptor development. *Cell*, **78**, 973–86.

6. Simon, M. A. (1994). Signal transduction during development of the *Drosophila* R7 photoreceptor. *Dev. Biol.*, **166**, 431–42.

7. Le, N. and Simon, M. A. (1998). Disabled is a putative adaptor protein that functions during signalling by the sevenless receptor tyrosine kinase. *Mol. Cell biol.*, **18**, 4844–54.

8. Raabe, T., Riesgo-Escovar, J., Liu, X., Bausenwein, B. S., Deak, P., Maroy, P., and Hafe, E. (1996). DOS, a novel pleckstrin homology domain-containing protein required for signal transduction between sevenless and Ras1 in *Drosophila*. *Cell*, **85**, 911–20.

9. Lim, Y., Tsuda, L., Inoue, Y. H., Irie, K., Adachi-Yamada, T., Hata, M., Nishi, Y., Matsumoto, K., and Nishida, Y. (1997). Dominant mutations of *Drosophila* MAP kinase kinase and their activities in *Drosophila* and yeast MAP kinase cascades. *Genetics*, **146**, 263–73.

10. Takahashi, F., Endo, S., Kojima, T., and Saigo, K. (1996). Regulation of cell–cell contacts in developing *Drosophila* eyes by Dsrc41, a new close relative of vertebrate c-src. *Genes Dev.*, **10**, 1645–56.

11. Kockel, L., Zeitlinger, J., Staszewski, L. M., Mlodzik, M., and Bohmann, D. (1997). Jun in *Drosophila* development: redundant and non redundant functions and regulation by two MAPK signal transduction pathways. *Genes Dev.*, **11**, 1748–58.

12. Brunner, D., Oellers, N., Szabad, J., Biggs III, W. H., Zipursky, S/L., and Hafen, E. (1994) A gain-of-function muatation in *Drosophila* MAP kinase activates multiple receptor tyrosine kinase signalling pathways. *Cell*, **76**, 875–88.

13. Therrien, M., Chang., H. C., Solomon, N. M., Karin, F. D., Wasserman, D. A., and Rubin, G. M. (1995). KSR, a novel protein kinase required for Ras signal transduction. *Cell*, **83**, 879–88.

14. Herbst, R., Carroll, P. M., Allard, J. D., Schilling, J., Raabe, T., and Simon, M. A. (1996). Daughter of sevenless is a substrate of the phosphotyrosine phosphatase corkscrew and functions during sevenless signalling. *Cell*, **85**, 899–909.

15. Allard, J. D., Chang, H. C., Herbst, R., McNeil, H., and Simon, M. A. (1996). The SH2 containing tyrosine phosphatase corkscrew is required during signalling by sevenless, Ras1 and Raf. *Development*, **122**, 1125–35.

16. Raabe, T., Olivier, J. P., Dickson, B., Liu, X., Gish, G. D., Pawson, T., and Hafen, E. (1995). Biochemical and genetic analysis of the DRK SH2-SH3 adaptor protein of *Drosophila*. *EMBO J.*, **14**, 2509–18.

17. Lai, Z-C. and Ribin, G. M. (1992). Negative control of photoreceptor development in *Drosophila* by the product of the *Yan* gene, an ETA domain protein. *Cell*, **70**, 609–20.

18. Tomlinson, A. and Ready, D. F. (1987). Cell fate in the *Drosophila* ommatidium. *Dev. Biol.*, **123**, 264.

19. Tomlinson, A. (1988). Cellular interactions in the developing *Drosophila* eye. *Development*, **104**, 183.

20. Hafen, E., Basler, K., Edstroem, J. E., and Rubin, G. M. (1987). *sevenless*, a cell-specific

homeotic gene of *Drosophila*, encodes a putative transmembrane receptor with a tyrosine kinase domain. *Science*, **236**, 55.

21. Campos-Ortega, J. A., Jurgens, G., and Hofbauer, A. (1979). Cell clones and pattern formation: studies on *sevenless*, a mutant of *Drosophila melanogaster*. *Roux's Arch. Dev. Biol.*, **186**, 27.

22. Harris, W. A., Stark, W. S., and Walker, J. A. (1976). Genetic dissection of the photoreceptor system in the compound eye of *Drosophila melanogaster*. *J. Physiol.* **256**, 415.

23. Rubin, G. M. (1991). Signal transduction and the fate of the R7 photoreceptor in *Drosophila*. *TIG*, **7**, 372.

24. Rubin, G. M. (1989). Development of the *Drosophila* retina: induction events studied at the single cell resolution. *Cell*, **57**, 519.

25. Kramer, H., Cagan, R. L., and Zipursky, S. L. (1991). Interaction of the *bride of sevenless* membrane-bound ligand and the *sevenless* tyrosine kinase receptor. *Nature*, **352**, 207.

26. Cagan, R. L., Kramer, H., Hart, A. C., and Zipursky, S. L. (1992). The *bride of sevenless* and *sevenless* interaction: internalisation of a transmembrane ligand. *Cell*, **69**, 393.

27. Simon, M. A., Dodson, G. S., and Rubin, G. M. (1993). An SH3–SH2–SH3 protein is required for p21Ras activation and binds to sevenless and sos proteins *in vitro*. *Cell*, **73**, 169–77.

28. Olivier, J. P., Raabe, T., Henkenmeyer, M., Dickson, B., Mbamalu, G., Margolis, B., Schlessinger, J., Hafen, E., and Pawson, T. (1993). A *Drosophila* SH2/SH3 protein couples the *sevenless* receptor tyrosine kinase to Sos, a putative guanine nucleotide exchange factor. *Cell*, **73**, 179–191.

29. Jones, S., Vignais, M. L., and Broach, J. R. (1991). The CDC25 protein of *S. cerevisiae* promotes exchange of guanine nucelotides bound to Ras. *Mol. Cell Biol.*, **11**, 2641.

30. Simon, M. A., Bowtell, D. D. L., Dodson, G. S., Laverty, T. R., and Rubin, G. M. (1991). Ras1 and a putative guanine nucleotide exchange factor perform crucial steps in signalling by the *sevenless* protein tyrosine kinase. *Cell*, **67**, 701.

31. Gaul, U., Mardon, G., and Rubin, G. M. (1992). A putative Ras GTPase activating protein acts as a negative regulator of signalling by the *sevenless* receptor tyrosine kinase. *Cell*, **68**, 1007.

32. Daum, G., Eisenmann-Tappe, I., Fries, H. W., Troppmair, J., and Rapp. U. R. (1994). The ins and outs of Raf kinases. *Trends Biochem. Sci.*, **19**, 474–80

33. Tsuda, L., Inoue, Y. H., Yoo, M. A., Mizuno, M., Hata, M., Lim, Y. M., Adachi-Yamada, T., Ryo, H., Masamune, Y. and Nishida, Y. (1993). A protein kinase similar to MAP kinase activator acts downstream of the raf kinase in *Drosophila*. *Cell*, **72**, 407–14.

34. Cathew, R. W. and Rubin, G. M. (1990). *seven in absentia*, a gene required for specification of the R7 cell fate in the *Drosophila* eye. *Cell*, **63**, 561.

35. Hart, A. C., Kramer, H., Van Vactor, D. L., Paidhungat, M., and Zipursky, S. L. (1990). Induction of cell identity in the *Drosophila* retina: the *bride of sevenless* protein is predicted to contain a large extracellular domain and seven transmembrane segments. *Genes Dev.*, **4**, 1835.

36. Reinke, R. and Zipursky, S. L. (1988). Cell–cell interaction in the *Drosophila* retina: the *bride of sevenless* gene is required in photoreceptor cell R8 for R7 cell development. *Cell*, **55**, 321.

37. Beir, E., Vaessin, H., Sheperd, S., Lee, K., McCall, K., Barbel, S., Ackerman, L., Carretto, R., Uemura, T., Grell, E. H., Jan, L. Y., and Jan, Y. N. (1989). Searching for pattern and mutations in the *Drosophila* genome with a p-*lacZ* vector. *Genes Dev.*, **3**, 1273.

38. Bellen, H., O'Kane, C. J., Wilson, C., Grossniklaus, U., Kurth-Pearson, R., and Gehring, W. J. (1989). P-element mediated enhancer detection: a versatile method to study development in *Drosophila*. *Genes Dev.*, **3**, 1288.

39. Wison, C., Kurth-Pearson, R., Bellen, H. J., O'Kane, C. J., Grossniklaus, U., and Gehring, W. J. (1989). P-element mediated enhancer detection: an efficient method for isolating and characterising developmentally regulated genes in *Drosophila*. *Genes Dev.*, **3**, 1301.

40. Hanahan, D., Lane, D., Lipsich, L., Wigler, M., and Botchan, M. (1980). Characteristics of an SV40-plasmid recombinant and its movement into and out of the genome of a murine cell. *Cell*, **21**, 127.

41. Mlodzik, M., Hiromi, Y., Weber, U., Goodman, C. S., and Rubin, G. M. (1990). The *Drosophila seven-up* gene, a member of the steroid receptor gene superfamily, controls photoreceptor cell fates. *Cell*, **60**, 211.

42. Fortini, M. E., Simon, M. A., and Rubin, G. M. (1992). Signalling by the *sevenless* protein tyrosine kinase is mimicked by Ras 1 activation. *Nature*, **355**, 559.

43. Rogge, R. D., Kariovich, C. A., and Banerjee, U. (1991). Genetic dissection of a neuro-developmental pathway: *son of sevenless* functions downstream of the *sevenless* and EGF receptor tyrosine kinases. *Cell*, **64**, 398.

44. Sprenger, F., Stevens, L. M., and Nusslein-Volhard, C. (1989). The *Drosophila* gene *torso* encodes a putative receptor tyrosine kinase. *Nature*, **338**, 478.

45. Dickson, B., Sprenger, F., Morrison, D., and Hafen, E. (1992). Raf functions downstream of Ras 1 in the Sevenless signal transduction pathway. *Nature*, **360**, 600.

46. Ambrosio, L., Mahowald, A. P., and Perrimon, N. (1989). Requirement of the *Drosophila raf* homologue for *torso* function. *Nature*, **342**, 288.

47. Lu, Z., Chou, T.-Z., Williams, N. G., Roberts, T., and Perrimon, N. (1993). Control of cell fate determination by $p21^{ras}$/Ras1, an essential component of *torso* signalling in *Drosophila*. *Genes Dev.*, **7**, 621.

48. Baker, N. E. and Rubin, G. M. (1989). Effect on eye development of dominant mutations in *Drosophila* homologue of the EGF receptor. *Nature*, **340**, 150.

49. Klingler, M., Erdelyi, M., Szabad, J., and Nusslein-Volhard, C. (1988). Function of *torso* in determining the terminal analgen of the *Drosophila* embryo. *Nature*, **335**, 275.

50. Doyle, H. J. and Bishop, J. M. (1993). Torso, a receptor tyrosine kinase required for embryonic pattern formation, shares substrates with Sevenless and EGF-R pathways in *Drosophila*. *Genes Dev.*, **7**, 633.

51. Dent, J. A. and Han., M. (1998) Post-embryonic expression of *C. elegans let-60 ras* reporter constructs. *Mech. Dev.*, **72**, 179–82.

52. Sternberg, P. W., Lesa, G., Lee, J., Katz, W. S., Yoon, C., Candinin, T. R., Huang, L. S., Chamberlin, H. M., and Jongeward, G. (1995). LET-23-mediated signal transduction during *Caenorhabditis elegans* development. *Mol. Reprod. Dev.*, **42**, 523–8.

53. Sundaram, M. and Han, M. (1996). Control and integration of cell signalling pathways during *C. elegans* vulval development. *Bioassays*, **18**, 473–80.

54. Kornfeld, K. (1997). Vulval development in *Caenorhabditis elegans*. *Trends Genet.*, **13**, 55–61.

55. Ferguson, E. L., Sternberg. P. W., and Horvitz, H. R. (1987). A genetic pathway for the specification of the vulval cell lineages of *Caenorhabditis elegans*. *Nature*, **326**, 259.

56. Horvitz, H. R. and Sternberg, P. W. (1991). Multiple intercellular signalling systems control the development of the *Caenorhabditis elegans* vulva. *Nature*, **351**, 535.

57. Sternberg, P. W. and Horvitz, H. R. (1991). Signal transduction during *C. elegans* vulval induction. *TIG*, **7**, 366.

58. Sternberg, P. W. and Horvitz, H. R. (1986). Pattern formation during vulval development in *Caenorhabditis elegans*. *Cell*, **44**, 761.

59. Katz, W. S., Hill, R. J., Clandinin, T. R., and Sternberg, P. W. (1995) Different levels of the *C. elegans* growth factor LIN-3 promote distinct vulval precursor fates. *Cell*, **82**, 297–307.

60. Kenyon, C. (1995). A perfect vulva every time: gradients and signalling cascades in *C. elegans*. *Cell*, **82**, 171–4.

61. Hoskins, R., Hajnal, A. F., Harp, S. A., and Kim, S. K. (1996). The *C. elegans* vulval induction gene lin-2 encodes a member of the MAGUK family of cell junction proteins. *Development*, **122**, 97–111.

62. Kaech, S. M., Whitfield, C. W., and Kim, S. K. (1998). The LIN-2/LIN-7/LIN-10 complex mediates basolateral membrane localization of the *C. elegans* EGF receptor LET-23 in vulval epithelial cells. *Cell*, **94**, 761–71.

63. Hill, R. J. and Sternberg, P. W. (1992). The gene *lin-3* encodes an inductive signal for vulval development in *C. elegans*. *Nature*, **358**, 470.

64. Yoon, C. H., Lee, J., Jongeward, G. D., and Sternberg, P. W. (1995). Sililarity of sli-1, a regulator of vulval development in *C. elegans*, to the mammalian proto-oncogene c-cbl. *Science*, **269**, 1102–6.

65. Jongeward, G. D., Clandinin, T. R., and Sternberg, P. W. (1995). sli-1, a negative regulator of Let-23 mediated signalling in *C. elegans*. *Genetics*, **139**, 1553–66.

66. Tan, P. B., Lackner, M. R., and Kin, S. K. (1998). MAP kinase signalling specificity mediated by the LIN-1 ETS/LIN-31 WH transcription factor complex during *C. elegans* vulval induction. *Cell*, **93**, 569–80.

67. Sundaram, M. and Han, M. (1995). The *C. elegans* ksr-1 gene encodes a novel raf-related kinase involved in Ras-mediated signal transduction. *Cell*, **83**, 889–901.

68. Sieburth, D. S., Sun, Q., and Han, M. (1998). SUR-8, a conserved Ras-binding protein with leucine rich repeats, positively regulates Ras-mediated signalling in *C. elegans*. *Cell*, **94**, 119–30.

69. Han, M., Golden, A., Han, Y., and Sternberg, P. W. (1993). *C. elegans lin-45 raf* gene participates in *let-60 ras*-stimulated vulval differentiation. *Nature*, **363**, 133.

70. Sternberg, P. W. (1988). Lateral inhibition during vulval induction in *Caenorhabditis elegans*. *Nature*, **335**, 551.

71. Sternberg, P. W. and Horvitz, H. R. (1989). The combined action of two intercellular signalling pathways specifies three cell fates during vulval induction in *C. elegans*. *Cell*, **58**, 679.

72. Beitel, G. J., Clark, S. G., and Horvitz, H. R. (1990). *Caenorhabditis elegans Ras* gene *let-60* acts as a switch in the pathway of vulval induction. *Nature*, **348**, 503.

73. Han, M. and Sternberg, P. W. (1990). *let-60*, a gene that specifies cell fates during *C. elegans* vulval induction, encodes a ras protein. *Cell*, **63**, 921.

74. Clark, S. G., Stern, M. J., and Horvitz, H. R. (1992). *C. elegans* cell-signalling gene *sem-5* encodes a protein with SH2 and SH3 domains. *Nature*, **356**, 340.

75. Stern, M. J. and DeVore, D.L. (1994). Extending and connecting signalling pathways in *C. elegans*. *Dev. Biol.*, **166**, 443–59.

76. Herman, R. K. and Hedgecock, E. M. (1990). Limitation of the size of the vulval primordium of *Caenorhabditis elegans* by *lin-15* expression in surrounding hypodermis. *Nature*, **348**, 169.

77. Han, M., Arioan, R. V., and Sternberg, P. W. (1990). The *let-60* locus controls the switch between vulval and nonvulval fates in *Caenorhabditis elegans*. *Genetics*, **126**, 899.

78. Aroian, R. V. and Sternberg, P. W. (1991). Multiple functions of *let-23*, a *Caenorhabditis elegans* receptor tyrosine kinase gene required for vulval induction. *Genetics*, **128**, 251.

79. Pawson, T. and Gish, G. D. (1992). SH2 and SH3 domains: from structure to function. *Cell*, **71**, 356.

80. Miller, L. M., Gallegos, M. E., Morisseau, B. A., and Kim, S. K. (1993). lin31, a *Caenorhabditis elegans* HNF-3/fprk head transcription factor homologue, specifies three alternative cell fates in vulval development. *Genes, Dev.*, **7**, 933–47.

81. Beitel, G. J., Tuck, S., Greewald, I., and Horvitz, H. R. (1995). The *Caenorhabditis elegans* gene lin-1 encodes and ETS domain protein and defines a branch of the vulval induction pathway. *Genes. Dev.*, **9**, 3149–62.

82. Greenwald, I. S. (1985). *lin-12*, a nematode homeotic gene, is homologous to a set of mamalian proteins that includes epidermal growth factor. *Cell*, **43**, 583.

83. Yochem, J., Weston, K., and Greenwald, I. S. (1988). The *Caenorhabditis elegans lin-12* gene encodes a transmembrane protein with overall homology to *notch*. *Nature*, **335**, 547.

84. Greenwald, I. S., Sternberg, P. W., and Horvitz, H. R. (1983). The *lin-12* locus specifies cell fates in *Caenorhabditis elegans*. *Cell*, **34**, 435.

85. Ruel, L., Bourouis, M., Heitzler, P., Pantesco, V., and Simpson, P. (1993). *Drosophila* shaggy kinase and rat glycogen synthase kinase-3 have conserved activities and act downstream of Notch. *Nature*, **362**, 557.

86. Simpson, P. and Carteret, C. (1990). Proneural clusters: equivalence groups in the epithelium of *Drosophila*. *Development*, **110**, 927.

87. Ferguson, E. and Horvitz, H. R. (1989). The multivulva phenotype of certain *Caenorhabditis elegans* mutants results from defects in two functionally redundant pathways. *Genetics*, **123**, 109.

88. Peterson, E. J., Clements, J. L., Fang, N., and Koretzky, G. A. (1998). Adaptor proteins in lymphocyte antigen receptor signalling. *Curr. Opin. Immunol.*, **10**, 337–44.

89. Boussiotis, V. A., Freeman, G. J., Berezovkaya, A., Barber, D. L., and Nadler, L. M. (1997). Maintenance of human T cell anergy: blocking of IL2 gene transcription by activated RAp1. *Science*, **278**, 124–8.

90. Devreotes, P. N. (1982). In *The development of* Dictyostelium discoideum (ed. W. Loomis), p. 117. Academic Press, San Diego, CA.

91. Spudich, J.A. (ed.) (1987). *Dictyostelium discoideum*: molecular approaches to cell biology. In *Methods in cell biology*, Vol. 28. Academic Press, London.

92. Dottin, R. P., Bodduluri, S. R., Doody, J. F., and Haribabu, B. (1991). Signal transduction and gene expression in *Dictyostelium discoideum*. *Dev. Genet.*, **12**, 2.

93. Anschultz, A., Um, H.-D., Tao, Y.-P., and Klein, C. (1991). Regulation of protein phosphorylation in *Dictyostelium discoideum*. *Dev. Genet.* **12**, 14.

94. Firtel, R. A. (1991). Signal transduction pathways controlling multicellular development in *Dictyostelium*. *TIG*, **7**, 381.

95. Simon, M., Driscoll, D., Mutzel, R., Part, D., Williams, J., and Veron, M. (1989). Overproduction of the regulatory subunit of the cAMP-dependent protein kinase blocks the differentiation of *Dictyostelium discoideum*. *EMBO J.*, **8**, 2039.

96. Pitt, G. S., Milona, N., Borleis, J., Lin., K. C., Reed, R. R., and Devreotes, P. N. (1992). Structurally, distinct and stage specific adenyl cyclase genes play different roles in *Dictyostelium* development. *Cell*, **69**, 305.

97. Harwood, A. J., Hopper, N. A., Simon, M. N., Driscoll, D. M., Veron, M., and Williams, J. G. (1992). Culmination in *Dictyostelium* is regulated by the cAMP-dependent protein kinase. *Cell*, **69**, 615.

98. Firtel, R. A., van Haarstert, P. J. M., Kimmel, A. R., and Devreotes, P. (1989). G protein linked signal transduction pathways in development: *Dictyostelium* as an experimental system. *Cell*, **58**, 235.

99. Soderbom, F. and Loomis, W. F. (1998). Cell–cell signalling during *Dictyostelium* development. *Trends Microbiol.*, **6**, 402–6.

100. Kumagi, A., Pupillo, M., Gundersen, R., Miake-Tye, R., Devreotes, P. N., and Firtel, R. A. (1989). Regulation and function of Ga proteins in *Dictyostelium*. *Cell*, **57**, 265.

101. Klein, P. S., Sun, T. J., Saxe, C. L., Kimmel, A. R., Johnson, R. L., and Devreotes, P. N. (1988). A chemoattractant receptor controls development in *Dictyostelium discoideum*. *Science*, **241**, 1467.

102. Johnson, R. L., Grundersen, R., Lilly, P., Pitt, G. S., Pupillo, M., Sun, T. J., Vaughan, R. A., and Devreotes, P. N. (1989). G-protein-linked signal transduction systems control development in *Dictyostelium*. *Development*, **107**, 75.

103. Kikkawa, U., Mann, S. K. O., Firtel, R. A., and Hunter, T. (1992). Molecular cloning of casein kinase II a subunit from *Dictyostelium discoideum* and its expression in the life cycle. *Mol. Cell. Biol.*, **12**, 5711.

104. Meisner, H. and Czech, M. P. (1991). Phosphorylation of transcriptional factors and cell cycle dependent proteins by casein kinase II. *Curr. Opin. Cell Biol.*, **3**, 474.

105. Hathaway, G. M. and Traugh, J. A. (1982). Casein kinases—multipotential protein kinases. *Curr. Top. Cell. Regul.*, **21**, 101.

106. Woodgett, J. R. (1991). A common denominator linking glycogen metabolism, nuclear oncogenes and development. *Trends Biochem. Sci.*, **16**, 177.

107. Burki, E., Anjard, C., Scholder, J. C., and Reymond, C. D. (1991). Isolation of two genes encoding putative protein kinases regulated during *Dictyostelium discoideum* development. *Gene*, **102**, 57.

108. Harwood, A. J., Plyte, S. E., Woodgett, J. R., Strutt, H., and Kay, R. R. (1995). Glycogen synthase kinase 3 regulates cell fate in *Dictyostelium*. *Cell*, **80**, 139–48.

109. Gaskins, C., Clark, A. M., Aubry, L., Segall, J. E., and Firtel, R. A. (1996). The *Dictyostelium* MAP kinase ERK2 regulates multiple independent devlopmental pathways. *Genes Dev.*, **10**, 118–28.

110. Aubry, L., Maeda, M., Insall, R., Devreotes, P. N., and Firtel, R. A. (1997). The *Dictyostelium* mitogen activated protein kinase ERK2 is regulated by Ras and cAMP-dependent protein kinase (PKA) and mediates PKA function. *J. Biol. Chem.*, **272**, 3883–6.

111. Kosaka, C. and Pears, C. J. (1997). Chemoattractants induce tyrosine phosphorylation of ERK2 in *Dictyostelium discoideum* by diverse signalling pathways. *Biochem. J.*, **324**, 347–52.

112. Haribabu, B. and Dottin, R. P. (1991). Homology cloning of protein kinase and phospho-protein phosphatase sequences of *Dictyostelium discoideum*. *Dev. Genet.*, **12**, 45.

113. Haribabu, B. and Dottin, R. P. (1991). Identification of a protein kinase multigene family of *Dictyostelium discoideum*: molecular cloning and expression of a cDNA encoding a developmentally regulated protein kinase. *Proc. Natl Acad. Sci. USA*, **88**, 1115.

114. Anjard, C., Pinaud, S., Kay, R. R., and Reymond, C. D. (1992). Overexpression of Dd PK2 protein kinase causes rapid development and affects the intracellular cAMP pathway of *Dictyostelium discoideum*. *Development*, **115**, 785.

115. Tan, J. L. and Spudich, J. A. (1990). Developmentally regulated protein-tyrosine kinase genes in *Dictyostelium discoideum*. *Mol. Cell. Biol.*, **10**, 3578.

116. Hunter, T. (1987). A thousand and one protein kinases. *Cell*, **50**, 823.

117. Loomis, W. F. (1998). Role of PKA in the timing of developmental events in *Dictyostelium* cells. *Microbiol. Mol. Biol. Rev.*, **62**, 684–94.

118. Firtel. R. A. and Chapman, A. L. (1990). A role for cAMP-dependent protein kinase A in early *Dictyostelium* development. *Genes Dev.*, **4**, 18.

119. Mann, S. K. O., Yonemoto, W. M., Taylor, S. S., and Firtel, R. A. (1992). DdPK3, which plays essential roles during *Dictyostelium* development, encodes the catalytic subunit of cAMP-dependent protein kinase. *Proc. Natl Acad. Sci. USA*, **89**, 10701.

120. Segall, J., Kuspa, A., Shaulsky, G., Ecke, M., Maeda, M., Gaskins, C., Firtel, R., and Loomis, W. (1995). A MAP kinase necessary for receptor mediated activation of adenyl cyclase in *Dictyostelium*. *J. Cell Biol.*, **128**, 405–13.

121. Maeda, M., Aubry, L., Insall, R., Gaskins, C., Devreotes, P. N., and Firtel, R. A. (1996). Seven helix chemoattractant receptors transiently stimulate mitogen activated protein kinase in *Dictyostelium*. Role of heterotrimeric G proteins. *J. Biol. Chem.*, **271**, 3351–4.

122. Jermyn, K., Berks, M., Kay, R., and Williams, J. (1987). Two distinct classes of prestalk-enriched messenger RNA sequences in *Dictyostelium discoideum*. *Development*, **100**, 745.

123. Williams, J., Ceccarelli, A., McRobbie, S., Mahbubani, H., Kay, R. R., Early, A., Berks, M., and Jermyn, K. A. (1987). Direct induction of *Dictyostelium* prestalk gene expression by DIF provides evidence that DIF is a morphogen. *Cell*, **49**, 185.

124. McRobbie, S. J., Jermyn, K. A., Duffy, K., Blight, K., and Williams, J. G. (1988). Two DIF-inducible, prestalk specific mRNAs of *Dictyostelium* encode extracellular matrix proteins of the slug. *Development*, **104**, 275.

125. Mann, S., Richardson, D., Lee, S., Kimmel, A., and Firtel, R. (1994). Expression of cAMP dependent protein kinase in prespore cells is sufficient to induce cell differentiation in *Dictyostelium*. *Proc. Natl Acad. Sci. USA*, **91**, 10561–5.

126. Zhukovskaya, N., Early, A., Kawata, T., Abe, T., and Williams, J. (1996). cAMP dependent protein kinase is required for the expression of a gene specifically expressed in *Dictyostelium* prestalk cells. *Dev. Biol.*, **179**, 27–40.

127. Aubry, L., Maeda, M., Insall, R., Devreotes, P. N., and Firtel, R. A. (1997). The *Dictyostelium* mitogen activated protein kinase ERK2 is regulated by Ras and cAMP-dependent protein kinase (PKA) and mediates PKA function. *J. Biol. Chem.*, **272**, 3883–6.

128. Anjard, C., Zeng, C., Loomis, W., and Nellen, W. (1998). Signal transduction pathways leading to spore differentiation in *Dictyostelium discoideum*. *Dev. Biol.*, **193**, 146–55.

129. Siegfried, E., Perkins, L. A., Capaci, T. M., and Perrimon, N. (1990). Putative protein kinase product of the *Drosophila* segment polarity gene, *zeste-white* 3. *Nature*, **354**, 825.

130. Bourouis, M., Moore, P., Ruel, L., Grau, Y., Heitzler, P., and Simpson, P. (1990). An early embryonic product of the gene *shaggy* encodes a serine/threonine protein kinase related to the CDC28/cdc2+ subfamily. *EMBO J.*, **9**, 2877.

131. Plyte, S. E., O'Donovan, E., Woodgett, J. R., and Harwood, A. J. (1999). Glycogen synthase kinase 3 (GSK-3) is regulated during *Dictyostelium* development via the serpentine receptor cAR-3. *Development*, **126**, 325–33.

132. Kuspa, A. and Loomis, W. F. (1992). Tagging developmental genes in *Dictyostelium* by restriction mediated integration of plasmin DNA. *Proc. Natl Acad. Sci. USA*, **89**, 8803–7.

133. Hoekstra, M. F., Demaggio, A. J., and Dhillon, N. (1991). Genetically, identified kinases in yeast. I: transcription, translation, transport and mating. *TIG*, **7**, 256.

134. Hoekstra, M. F., Demaggio, A. J., and Dhillon, N. (1991). Genetically, identified kinases in yeast. II: DNA metabolism and meiosis. *TIG*, **7**, 293–7.

135. Herskowitz, I. (1995). MAP kinase pathways in yeast: for mating and more. *Cell*, **80**, 187–97.

136. Sprague, G. F. (1991). Signal transduction in yeast mating. *TIG*, **7**, 393.

137. Marsh, L., Neiman, A. M., and Herkowitz, I. (1991). Signal transduction during pheromone response in yeast. *Annu. Rev. Cell. Biol.*, **7**, 699.

138. Hartwell, L. H. (1980). Mutants of *Saccharomyces cerevisiae* unresponsive to cell division control by polypeptide mating hormone. *J. Cell. Biol.*, **85**, 811.

139. Teague, M. A., Chaleff, D. T., and Errede, B. (1986). Nucleotide sequence of the yeast regulatory gene *STE7* predicts a protein homologous to protein kinases. *Proc. Natl Acad. Sci. USA*, **83**, 7371.

140. Elion, E. A., Grisafi, P. L., and Fink, G. R. (1990). *FUS3* encodes a *cdc2/CDC28*-related kinase required for the transition from mitosis to conjugation. *Cell*, **60**, 649.

141. Fujimura, H. (1990). Molecular cloning of the *DAC2/FUS3* gene essential for pheromone induced G1-arrest of the cell cycle in *Saccharomyces cerevisiae*. *Curr. Genet.*, **18**, 395.

142. Rhodes, N., Conell, L., and Errede, B. (1990). Ste11 is a protein kinase required for cell type specific transcription and signal transduction in yeast. *Genes Dev.*, **4**, 1862.

143. Courchesne, W. E., Kunisawa, R., and Thorner, J. (1989). A putative protein kinase overcomes pheromone-induced arrest of cell cycling in *S. cerevisiae*. *Cell*, **58**, 1107.

144. Leberer, E., Dignard, D., Harcus, D., Thomas, D. Y., and Whiteway, M. (1992). The protein kinase homologue Ste20 is required to link the yeast pheromone response G-protein bg subunits to downstream signalling components. *EMBO J.*, **11**, 4815.

145. Valtz, N., Peter, M., and Herskowitz, I. (1995). FAR 1 is required for orientated polarization of yeast cells in response to mating pheromones. *J. Cell Biol.*, **131**, 863–72.

146. Doo-Il Jeoung, D., Oehlen, W. M., and Cross, F. (1998). Cln 3 associated kinase activity in *Saccharomyces cerevisiae* is regulated by the mating factor pathway. *Mol. Cell. Biol.*, **18**, 433–41.

147. Peter, M., Neiman, A. M., Park, H-O, van Lohuizen, M., and Herskowitz, I. (1996). Functional analysis of the interaction between the small GTP binding protein Cdc 42 and the Ste 20 protein kinase in yeast. *EMBO J.*, **15**, 7046–59.

148. Oehlen, L. and Cross, F. (1998). The role of Cdc 42 in signal transduction and mating of the budding yeast *Saccharomyces cerevisiae*. *J. Biol. Chem.*, **273**, 8556–9.

149. Wu, C.., Whiteway, M., Thomas, D. Y., and Leberer, E. (1995). Molecular chracterisation of Ste 20, a potential mitogen activated protein or extracellular signal regulated kinase kinase (MEK) kinase kinase from *Saccharomyces cerevisiae*. *J. Biol. Chem.*, **270**, 15984–92.

150. Bardwell, L., Cook, J. G., Inouye, C. J., and Thorner, J. (1994). Signal propagation and regulation in the mating pheromone reponse pathway of the yeast *Saccharomyces cerevisiae*. *Dev. Biol.*, **166**, 363–79.

151. Stowers, L. D., Yelon, L. J., Berg, J., and Chant, J. (1995). Regulation of the polarization of T-cells towards antigen presenting cells by Ras related GTPase CDC 42. *Proc. Natl Acad. Sci. USA*, **92**, 5027–31.

152. Devreotes, P. N. and Zigmond, S. H. (1988). Chemotaxis in eukaryotic cells: a focus on leukocytes and *Dictyostelium*. *Annu. Rev. Cell Biol.*, **4**, 649–86.

153. Schrick, K., Garvik, B., and Hartwell, L. H. (1997). Mating in *Saccharomyces cerevesiae*: the role of the pheromone signal transduction pathway in the chemotropic response top pheromone. *Genetics*, **147**, 19–32.

154. Pryciak, P. M. and Huntress, F. A. (1998). Membrane recruitment of the kinase cascade scaffold protein ste 5 by the G bg complex underlies activation of the yeast pheromone response pathway. *Genes Dev.*, **12**, 2684–97.

155. Akada, R., Kallal, L., Johnson, D. I., and Kurjan, J. (1996). Genetic relationships between the G protein bg complex, Ste 5, ste 20p and Cdc 42p: investigation of effector roles in the yeast pheromone response pathway. *Genetics*, **143**, 103–17.

156. O'Rourke, S. M. and Herskowitz, I. (1998). The HOG 1 MAPK prevents cross-talk between the HOG and pheromone response MAPK pathways in *Saccharomyces cerevisiae*. *Genes Dev.*, **12**, 2875–86.

157. Lyons, D. M., Mahanty, S. K., Choi, K., Manandhar, M., and Elion, E. (1996). The SH3 domain protein Bem 1 coordinates mitogen activated protein kinase cascade activation with cell cycle control in *Saccharomyces cerevisiae*. *Mol. Cell. Biol.*, **16**, 4095–106.

158. Leeuw, T., Fourest-Lieuvin, A., Wu, C., Chenervert, J., Clark, K., Whiteway, M., Thomas, D. Y., and Leberer, E. (1995). Pheromone response in yeast: association of Bem1p with proteins of the MAP kinase cascade and actin. *Science*, **270**, 1210–13.

159. Simon, M., De Virgilio, C., Souza, B., Pringle, J. R., Abo, A., and Reed, S. I. (1995). Role for the Rho family GTPase Cdc 42 in yeast mating pheromone signal pathway. *Nature*, **376**, 702–5.

160. Choi, K.-Y., Satterberg, B., Lyons, D. M., and Elion, E. A. (1994). Ste 5 tethers multiple protein kinases in the MAP kinase cascade required for mating in *S. cerevisiae*. *Cell*, **78**, 499–512.

161. Printen, J. A. and Sprague, G. F. (1994). Protein protein interactions in the yeast pheromone response pathway: ste 5p interacts with all members of the MAP kinase cascade. *Genetics*, **138**, 609–19.

162. Peterson, J., Zheng, Y., Bender, L., Meyers, R, Cerione, R., and Bender, A. (1994). Interactions between the bud emergence protein Bem1p and Bem2p and Rho type GTPases in yeast. *J. Cell Biol.*, **127**, 1395–406.

163. Zheng, Y., Cerione R., and Bender, A. (1994). Control of the yeast bud site assembly GTPase Cdc 42. Catalysis of guanine nucelotide exchange by Cdc 24 and stimulation of GTPase by BEM3. *J. Biol. Chem.*, **269**, 2369–72.

164. Gartner, A., Nasmyth, K., and Ammerer, G. (1992). Signal transduction in *Saccharomyces cerevisiae* requires tyrosine and threonine phosphorylation of Fus3 and Kss1. *Genes Dev.*, **6**, 1280.

165. Errede, B., Gartner, A., Zhou, Z., Nasmyth, K., and Ammerer, G. (1993). MAP kinase-related Fus3 from *S. cerevisiae* is activated by Ste7 *in vitro*. *Nature*, **362**, 261.

166. Neiman, A. M. and Herskowitz, I. (1994). Reconstitution of a yeast protein kinase cascade in vitro: activation of the yeast MEK homologue STE 7 by STE 11. *Proc. Natl Acad. Sci. USA*, **91**, 3398–402.

167. Dolan, J. W., Kirkman, C., and Fields, S. (1989). The yeast Ste12 protein binds to the DNA sequence mediating pheromone induction. *Proc. Natl Acad. Sci. USA*, **86**, 5703.

168. Dolan, J. W. and Fields, S. (1990). Overproduction of the yeast Ste12 protein leads to constitutive transcriptional activation. *Genes Dev.*, **4**, 492–502.

169. Errede, B. and Ammerer, G. (1989). Ste12, a protein involved in cell type specific transcription and signal transduction in yeast is part of protein–DNA complexes. *Genes Dev.*, **3**, 1349.

170. Song, O. K., Dolan, J. W., Yuan, Y. I. O., and Fields, S. (1991). Pheromone-dependent phosphorylation of the yeast Ste12 protein correlates with transcriptional activation. *Genes Dev.*, **5**, 741.

171. Tyres, M., Tokiwa, G., Nash, R., and Futcher, B. (1992). The Cln3–Cdc28 kinase complex of *Saccharomyces cerevisiae* is regulated by proteolysis and phosphorylation. *EMBO J.*, **11**, 1773.

172. Tyres, M. and Futher, B. (1993). Far1 and Fus3 link the mating pheromone signal transduction pathway to three G_1-phase Cdc28 kinase complexes. *Mol. Cell Biol.*, **13**, 5659.

173. Zarzov, P., Mazzoni, C., and Mann, C. (1996). The SLY 2 (MPK1) MAP kinase is activated during periods of polarized cell growth in yeast. *EMBO J.*, **15**, 83–91.

174. Cairns, B. R., Ramer, S. W., and Kornberg, R. D. (1992). Order of action of components in the yeast pheromone response pathway revealed with a dominant allele of the Ste11 kinase and the multiple phosphorylation of the Ste7 kinase. *Genes Dev.*, **6**, 1305.

175. Stevenson, B. J., Rhodes, N., Errede, B., and Sprague, G. F. (1992). Constitutive mutants of the protein kinase Ste11 activate the yeast pheromone response pathway in the absence of the G protein. *Genes Dev.*, **6**, 1293.

176. Levin, D. E. and Errede, B. (1995). The proliferation of MAP kinase signalling pathways in yeast. *Curr. Opin. Cell Biol.*, **7**, 197–202.

177. Bardwell, L., Cook, J. G., Voora, D., Baggott, D. M., Martinez, A. R., and Thorner, J. (1998). Repression of yeast ste 12 transcription factor by direct binding of unphosphorylated Kss1 MAPK and its regulation by ste 7 MEK. *Genes Dev.*, **12**, 2887–98.

178. Madhani, H. D., Styles, C. A., and Fink, G. R. (1997). MAP kinases with distinct functions impart signalling specificity during yeast differentiation. *Cell*, **91**, 673–84.

179. Madhani, H. D. and Fink, G. R. (1997). Combinatorial control required for the specificity of yeast MAPK signalling. *Science*, **275**, 1314–17.

180. Peter, M. and Herskowitz, I. (1994). Direct inhibition of the yeast cyclin dependent kinase Cdc 28-Cln by Far 1. *Science*, **265**, 1228–31.

181. Fields, S. and Song O. (1989). A novel genetic system to detect protein–protein interactions. *Nature*, **340**, 245–6.

182. Furlanetto, R. W., Dey, B. R., Lopaczynski, W., and Nissley, S. P. (1997). 14–3–3 proteins interact with the insulin like growth factor receptor but not the insulin receptor. *Biochem. J.*, **327**, 765–71.

183. Fanger, G. R., Widmann, C., Porter, A. C., Sather, S., Johnson, G. L., and Vaillancourt, R. R. (1998). 14–3–3 proteins interact with specific MEK kinases. *J. Biol. Chem.*, **273**, 3476–83.

184. Ikeda, S., Kishida, S., Yamamoto, H., Murai, H., Koyama, S., and Kikuchi, A. (1998). Axin, a negative regulator of the wnt signalling pathway, forms a complex with GSK-3 b and b catenin and promotes GSK-3b -dependent phopshorylation of b catenin. *EMBO J.*, **17**, 1371–84.

185. Yamamoto, H., Kishida, S., Uochi, T., Koyama, S., Asashima, M., and Kikuchi, A. (1998). Axil, a member of the Axin family interacts with glycogen synthase kinase-3b and b catenin and inhibits axis formation of *Xenopus* embryos. *Mol. Cell. Biol.* **18**, 2867–75.

186. Hong, Y. R., Chen, C. H., Cheng, D. S., Howng, S. L., and Chow C. C. (1998). Human dynamin like protein interacts with glycogen synthase kinase 3b. *Biochem. Biophys. Res. Commun.*, **249**, 697–703.

187. Kim, Y. H., Choi, C. Y., Lee S. J., Conti, M. A., and Kim, Y. (1998). Homeodomain-interacting protein kinases, a novel family of co-repressors for homeodomain transcription factors. *J. Biol. Chem.*, **273**, 25875–9.

188. Moorthamer, M., Zumstein-Mecker, S., and Chaudhuri, B. (1999). DNA binding protein dbpA binds Cdh5 and inhibits its activity. *FEBS Lett.*, **446**, 343–50.

4 | Specificity in JAK–STAT signalling

ADRIENNE HALUPA and DWAYNE L. BARBER

1. Introduction

Since the original identification of *Janus* kinases by Wilks *et al.* (1, 2) and the isolation of STAT1 and STAT2 by the Darnell laboratory, an entire new field of discovery has emerged and come to dominate research in cytokine signalling during the last decade. This wealth of information now provides us with some clues as to the function of these novel tyrosine kinases and transcription factors. To date, four *Janus* kinases (Fig. 1) and seven STAT (Fig. 2) transcription factors have been identified in mammalian species. With the exception of TYK2 and STAT2, gene-targeting experiments have been reported for all members of the JAK–STAT family. *Drosophila* expresses a *Janus* kinase homologue, called *hopscotch*, indicating that the JAK–STAT pathway is conserved throughout evolution. The STAT transcription factors are more widespread, having been isolated in *Caenorhabditis elegans*, *Drosophila* spp., and *Dictyostelium* spp.

General aspects of JAK–STAT signalling have been elegantly described in several recent reviews (3–8). The major focus of this chapter will be a detailed discussion of

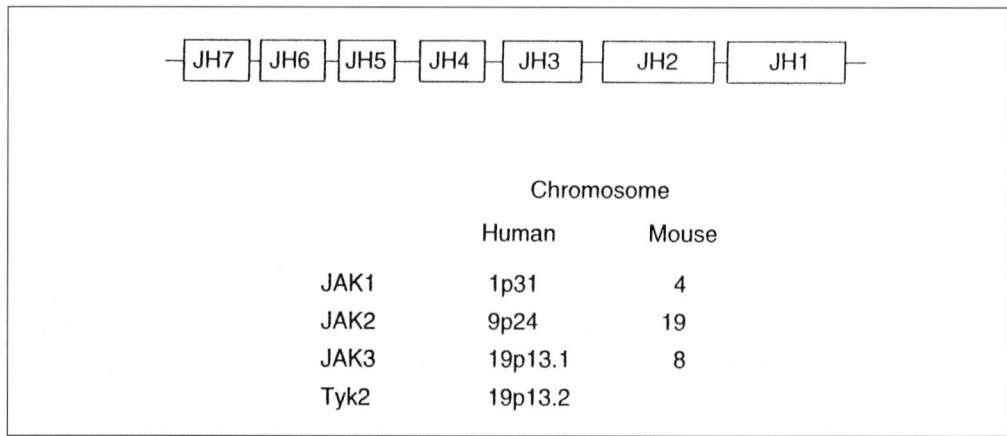

Fig. 1 Domain structure of JAK kinases. The conserved domains of JAK kinases are illustrated as well as the chromosomal localization in humans and mice. (Adapted from ref. 4.)

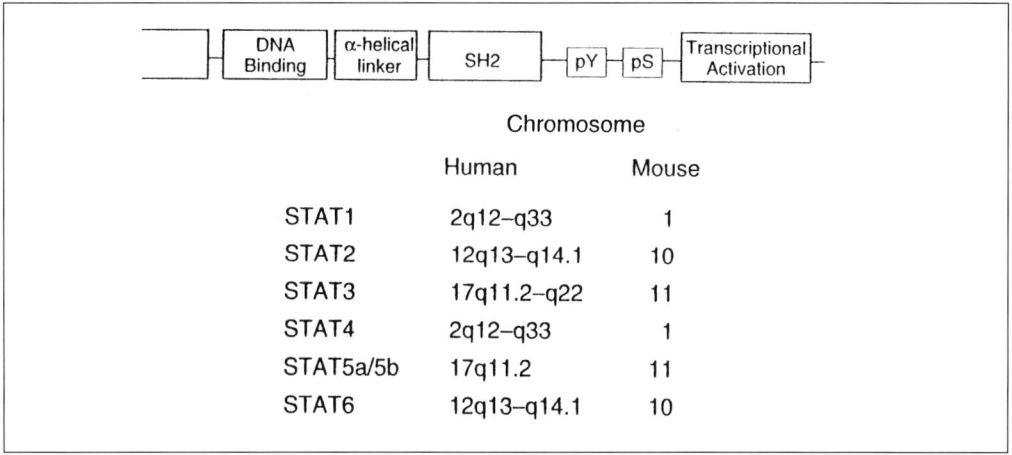

	Chromosome	
	Human	Mouse
STAT1	2q12–q33	1
STAT2	12q13–q14.1	10
STAT3	17q11.2–q22	11
STAT4	2q12–q33	1
STAT5a/5b	17q11.2	11
STAT6	12q13–q14.1	10

Fig. 2 Domain structure of STAT transcription factors. The conserved domains of STAT transcription factors are shown as well as the chromosomal localization in humans and mice. (Adapted from ref. 4.)

the gene-targeting experiments since this data is new and provides a definitive analysis of the function of each gene. A section describing the JAK–STAT pathway in development is followed by a discussion of the role of the JAK–STAT signalling pathway in disease pathogenesis. The final section of this chapter addresses future directions in this field.

2. JAK–STAT signalling—a general paradigm

Three pioneering approaches have led to the isolation of the various metazoan JAK and STAT proteins. Degenerate oligonucleotide PCR was initially utilized by Wilks and colleagues to identify orphan cytoplasmic tyrosine kinases, which they coined JAK1 and JAK2 (1, 2).

The Darnell laboratory utilized subtractive hybridization technology to identify a panel of interferon-α (IFN-α)- and interferon-γ (IFN-γ)-inducible genes (reviewed in ref. 9). Subsequent analysis of the promoter regions uncovered a conserved region in the promoters of IFN-α-inducible genes which they coined 'the interferon-stimulated response element' (ISRE). It was found that four proteins bind to the ISRE element. Isolation of the cDNA for each component revealed that the IFN-α-stimulated ISGF (interferon-stimulated gene factor)-3γ complex consists of STAT1α (91 kDa) (10), STAT1β (84 kDa) (10), STAT2 (113 kDa) (11), and the DNA binding component p48 (48 kDa) (12). IFN-γ-inducible genes have a distinctive novel element which was termed the 'gamma-activated sequence' (GAS). It was shown that IFN-γ activates tyrosine phosphorylation of STAT1, which undergoes dimerization via reciprocal SH2-mediated dimerization (13).

Mutagenesis experiments by Kerr and colleagues resulted in the isolation of a panel of cell lines that fail to respond to IFN-α and IFN-γ (reviewed in ref. 9). Complementation of each line with a cDNA library restored the IFN response. Subsequent

Table 1 Isolation of components of the JAK-STAT pathway through functional complementation

Complementation group	Response to ligand IFN-α	IFN-γ	Complementing protein
U1	−	+	TYK2
U2	+/−	+/−	p48
U3	−	−	STAT1
U4	−	−	JAK1
U5	−	+	?
U6	−	+	STAT2
γ1	+	−	JAK2
γ2	+	−	?

Adapted from reference 9.

isolation of the correcting cDNA resulted in the independent isolation of many of the components of the IFN-α and IFN-γ signalling pathway (refer to Table 1).

In general, JAK–STAT signalling can be represented by two distinct models (Fig. 3). IFN-α-stimulation leads to the activation of JAK1 and TYK2, which phosphorylates various tyrosine residues on the interferon-α receptors IFNAR1 and IFNAR2. STAT1 and STAT2 become tyrosine-phosphorylated and assemble with p48 into the ISGF3 heteromer. This complex regulates the transcription of IFN-α-inducible genes. IFN-γ activates JAK1 and JAK2 (Fig. 4). Tyrosine phosphorylation of the IFN-γ receptor

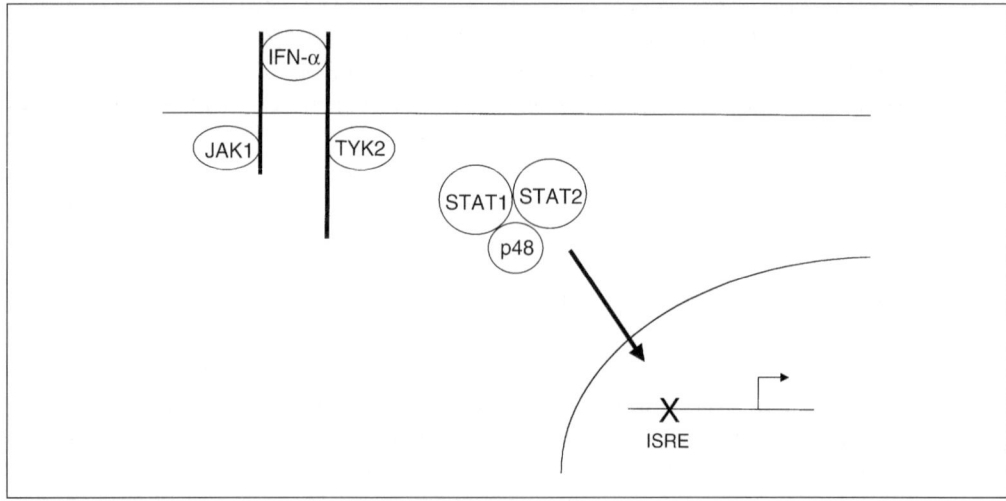

Fig. 3 IFN-α-dependent activation of the JAK–STAT pathway. Binding of IFN-α to the IFNAR1 and IFNAR2 chains activates the tyrosine kinase activity of JAK1 and TYK2. This results in the recruitment of STAT1 and STAT2 to the receptors and subsequent tyrosine phosphorylation of each STAT protein. Following tyrosine phosphorylation, the ISGF3 complex is formed with STAT1, STAT2, and p48. This regulates transcription from IFN-α-inducible genes that contain interferon-α stimulated response elements (ISRE).

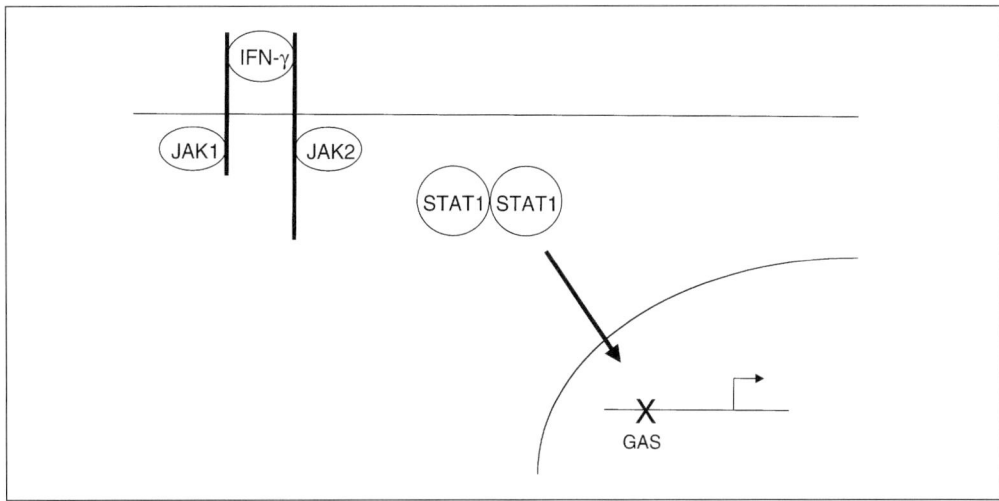

Fig. 4 IFN-γ-dependent activation of the JAK–STAT pathway. Binding of IFN-γ to the IFNGR activates the tyrosine kinase activity of JAK1 and JAK2. This results in the recruitment of STAT1 to the IFN-γ receptor and subsequent tyrosine phosphorylation of STAT1. Tyrosine-phosphorylated STAT1 dissociates from the receptor and dimerizes via reciprocal SH2-mediated dimerization. Once in the nucleus, STAT1 regulates transcription from IFN-γ inducible genes that contain interferon-γ activated sequences (GAS).

(IFNGR) results in the recruitment of STAT1 and its phosphorylation at Y701. This probably results in a conformational change which allows STAT1 to dissociate from the receptor, dimerize, enter the nucleus, and regulate the transcription of genes containing GAS elements (Fig. 1). All other cytokines also activate the homo- or heterodimerization of various combinations of STAT proteins.

Intense research by many investigators has resulted in the characterization of JAK–STAT activation downstream of many cytokines. What has emerged from this research is a rather cloudy view of the functional role of individual components of each family. Recent gene-targeting experiments have provided a clearer picture of distinct roles, especially for the individual STAT genes (summarized in Table 2).

3. Gene-targeting studies

3.1 JAK1

JAK1 was initially identified by Wilks and colleagues (1, 2,) utilizing degenerate oligonucleotide PCR with primers that correspond to conserved regions in all tyrosine kinases (15). This provided the first description of the domain structure of the *Janus* kinases; however, at that time JAK1 represented an orphan tyrosine kinase. Experiments utilizing an elegant genetic strategy indicated that JAK1 was downstream of IFN-α and IFN-γ signalling (16). Furthermore, complementation of a JAK1-deficient fibrosarcoma cell line with JAK1 cDNA restored responses to both IFN-α and IFN-γ. Subsequent investigation showed that the interleukin-2 (IL-2) family of cytokines—

Table 2 Gene-targeting experiments and resulting phenotypes

Gene	Phenotype	References
JAK1	Perinatal lethal Defects in IFN-α, IFN-γ, IL-10, gp130 signalling and B-cell development	28
JAK2	Embryonic lethal Similar to EPO and EPOR knockouts	42, 46
JAK3	Viable SCID-like phenotype	31–33, 54–55
TYK2	Viable	Unpublished
STAT1	Viable No response to viral challenge due to defective IFN-α signalling	64, 65
STAT2	Embryonic lethal	Unpublished
STAT3	Embryonic lethal Defects in visceral endoderm formation	87
STAT4	Viable Block in T_{H1} differentiation Defective IL-12 signalling	96, 97
STAT5A	Viable Defects in prolactin signalling	117
STAT5B	Viable Defects in growth hormone responses including sexual dimorphism	118
STAT5A/5B	Viable Defects in prolactin, growth hormone, and IL-2 signalling Anaemia in fetal erythropoiesis	48 119 47
STAT6	Viable Block in T_{H2}-mediated differentiation Defective IL-4 signalling	100–102
STAT4/6	Viable Default to incomplete T_{H1} differentiation	103

including IL-2 (17–21), IL-4 (18, 20, 22), IL-9 (22), IL-13 (23), and IL-15 (17, 20), IL-3 (24, 25), as well as the IL-6 family of cytokines including IL-6, IL-11, cardiotropin-1 (CT-1), ciliary neurotrophic factor (CNTF), leukaemia inhibitory factor (LIF), and oncostatin M (OSM) (26, 27)—all activate the tyrosine kinase activity of JAK1. Given that JAK1 appears to be a critical signalling intermediate downstream of many distinct families of cytokines, there was particular interest in the phenotype of JAK1 knockout animals.

Mice deficient in *JAK1* were viable and gross external examination revealed no developmental abnormalities, yet these mice failed to nurse and died within 24 hours of birth (28). *JAK1–/–* neonates tended to be smaller, weighing approximately 40% less than their heterozygote or wild-type littermates. Necropsy of newborns indicated no gross abnormalities in the heart, liver, lungs, kidney, or brain. However, the thymuses of *JAK1–/–* mice were much smaller than those of wild-type mice. Although the thymic epithelial structure appeared normal, the number of thymocytes was reduced 260-fold. Flow cytometry demonstrated that the distribution of CD4$^+$ and CD8$^+$ T cells in wild-type and *JAK1–/–* thymocyte populations was similar, indicating that the reduction of thymocytes reflects a primary defect in thymocyte production rather than a definitive block in thymocyte maturation.

Normal numbers of total splenocyte B cells were present in *JAK1–/–* mice, but analysis of B220 and IgM cell surface-marker expression revealed serious defects in the B-cell compartment. The B220$^+$IgM$^-$ cell population represents pro- and pre-B cells, natural killer (NK) cells, and a subpopulation of T cells. *JAK1*-deficient mice had 40-fold fewer B220$^+$IgM$^+$ B cells and 7- to 10-fold fewer B220$^+$IgM$^-$ B-cell precursors than wild-type mice. Pro-B cells, but not pre-B cells, express CD43. Whereas approximately 50% of the wild-type B220$^+$IgM$^-$ cell population was CD43$^+$, more than 90% of the *JAK1–/–* B220$^+$IgM$^-$ cell population expressed CD43. Therefore, B-lymphocyte differentiation in JAK1-deficient mice is blocked at the pro-B to pre-B cell transition, resulting in defective generation of mature B lymphocytes. Analysis of blood from *JAK1–/–* mice indicated normal numbers of erythrocytes, mononuclear phagocytes, neutrophils, and platelets. Mac-1$^+$ and/or Gr-1$^+$ myeloid cells in the fetal liver or newborn spleen were also normal. Since the haemopoietic defects of *JAK1*-deficient mice are limited to the lymphoid compartment, JAK1 must be crucial for the functioning of those cytokine receptors that are responsible for lymphoid cell development.

JAK1-deficient mice were, in fact, found to be defective in their responses to cytokines that bind to three distinct families of cytokine receptors: γ_c chain-containing receptors, class II receptors, and gp130 subunit-containing receptors. The γ_c chain is utilized by the receptors for IL-2, IL-4, IL-7, IL-9, and IL-15. When Rodig *et al.* tested fetal liver-cell responses to IL-2 and IL-4, they found that proliferation could only be induced in the presence of *JAK1* (28). The inability of *JAK1–/–* cells to respond to phorbol myristate acetate (PMA) and ionomycin suggests the presence of an inherent survival defect as well. In methylcellulose colony-forming assays using fetal liver cells from both wild-type and *JAK1–/–* mice, IL-7 stimulation induced colony formation in wild-type and heterozygous mice, but not in *JAK1–/–* mice. In addition, IL-7-stimulated, thymocyte-proliferation assays in liquid culture indicated a dependence on JAK1 for proliferation. Therefore, JAK1 appears to play an important role in lymphopoiesis induced by many, if not all, cytokine receptors that use the γ_c chain. It is likely, however, that the defective response to IL-7 is most significant in these mice, since IL-7 specifically acts early in lymphopoiesis to promote lymphocyte survival. This phenotype is similar to mice lacking expression of IL-7 (29), the IL-7 receptor (30), or JAK3 (31–33).

Activation of JAK1 also occurs following stimulation with IL-3, IL-5, GM-CSF, and G-CSF. IL-3 induced colony formation in *JAK1–/–* fetal liver cells, but colonies were smaller and fewer in number. In contrast, methylcellulose colony-forming assays using wild-type and *JAK1–/–* fetal liver cells revealed equivalent colony formation in response to M-CSF, G-CSF, GM-CSF, and IL-5.

The class-II receptor family includes receptors for IFN-α/β, IFN-γ, and IL-10. A defect in the ability of *JAK1*-deficient mice to respond to these cytokines was revealed in experiments with *JAK1–/–* embryonic macrophages. Whereas wild-type macrophages responded to lipopolysaccharide (LPS) and either IFN-α or IFN-γ by producing nitric oxide, *JAK1–/–* macrophages did not. *JAK1–/–* macrophages were also unable to respond to IFN-γ by upregulating intercellular adhesion molecule-1

(ICAM-1) expression or inducing expression of the third component of complement. However, *JAK1–/–* macrophages upregulated ICAM-1 expression in response to TNF-α, indicating that these cells were not unresponsive to cytokines in general, but rather that they were unable to respond to type-I and type-II IFNs.

The family of cytokine receptors that utilize the gp130 subunit include the IL-6, IL-11, CNTF, LIF, and OSM receptors. Electrophoretic mobility shift assays demonstrated diminished, but detectable, STAT3 activation in response to IL-6 and LIF in *JAK1–/–* macrophages, embryonic fibroblasts, and cardiomyocytes. Nevertheless, this decrease in STAT3 activation results in a biological response deficit, since *JAK1–/–* cardiomyocytes treated with IL-6, LIF, or CT-1 did not induce the hypertrophy observed in wild-type cells. Rodig *et al.* next tested the ability of *JAK1–/–* neurons to respond to LIF stimulation (28). Neurons from dorsal root ganglia of wild-type newborn mice remained viable when cultured in the presence of nerve growth factor (NGF) or LIF. Whereas the non-gp130-dependent cytokine NGF could support *JAK1–/–* neuron viability, LIF could not. Hence JAK1 is integral in signalling through gp130-containing receptors. Taken together, these results and those described above support the conclusion that, although JAK1 is ubiquitously expressed and activated by a wide variety of cytokines, it plays an essential and non-redundant role in signalling by only specific subsets of cytokine receptors.

3.2 JAK2

JAK2, in addition to JAK1, was initially identified by RT-PCR strategies by Wilks *et al.* (1, 2, 14). Functional studies revealed that JAK2 was activated downstream of several cytokine receptors including erythropoietin (EPO) (20, 34), the IL-3 family of cytokines including IL-3 (35), IL-5 (36), and GM-CSF (37), the IL-6 family of cytokines including IL-6, IL-11, CT-1, CNTF, and OSM (36), growth hormone (38), and prolactin (39, 40). Kerr and colleagues identified a mutant fibrosarcoma cell line which lacked JAK2 (41). Complementation of this cell line with the JAK2 cDNA restored IFN-γ responsiveness.

Mice in which the *JAK2* gene was deleted died around embryonic day 12.5, just prior to the switch from primitive to definitive erythropoiesis (42). Like *EPOR–/–* embryos (43–45) *JAK2–/–* embryos were severely anaemic. Analysis of *JAK2–/–* fetal livers, which were dramatically smaller than normal and were not red in colour, revealed a greater than 10-fold reduction in the number of nucleated cells and, similar to *EPOR–/–* fetal livers, the absence of erythroid cells. Wild-type fetal livers, in contrast, contained erythroid cells at all stages of differentiation. JAK2-deficient yolk sacs were of normal size, but yolk sac vessels contained greatly reduced numbers of circulating erythrocytes. Whereas wild-type embryos contained many yolk sac-derived primitive erythrocytes, both *JAK2–/–* and *EPOR–/–* embryos contained very few. Fetal liver-derived enucleated erythrocytes produced during definitive erythropoiesis were present in wild-type embryos, but could never be detected in either *JAK2–/–* or *EPOR–/–* embryos. Embryonic blood vessels contained reduced numbers of circulating nucleated erythrocytes. In contrast, peripheral blood vessels of *EPOR–/–* embryos retained substantial numbers of nucleated erythrocytes. To further confirm these findings, cytospin preparations of peripheral blood and the yolk sac were analysed after Giemsa staining. Heterozygotes contained cells at all stages of erythropoietic differentiation, but yolk sacs and peripheral blood from *JAK2–/–* embryos contained only proerythroblasts and very few nucleated erythrocytes. Enucleated erythrocytes were never observed. Therefore, EPOR signalling and definitive erythropoiesis in the fetal liver are dependent on JAK2.

Flow cytometry, using single-cell suspensions of fetal liver cells from day-12.5 embryos, was performed to determine the haemopoietic cell-population profiles in JAK2–/– mice. In wild-type mice, the majority of cells express low levels of CD44. However, CD44^{high} cells were predominant in *JAK2–/–* and *EPOR–/–* fetal livers. Normally, CD34^{low} cells are present in fetal livers at this stage of embryogenesis. CD34^{low}CD44^{high} cells were detected in *JAK2–/–* and *EPOR–/–* mice, but the proportion of these cells was somewhat increased in *EPOR–/–* fetal livers and considerably increased in *JAK2–/–* fetal livers. The pattern of c-kit expression was similar to that of CD44. Whereas c-kit^{neg} cells represent the majority of fetal liver cells in wild-type mice, most fetal liver cells of both *JAK2–/–* and *EPOR–/–* mice expressed c-kit. Ter-119 is expressed on erythroid precursors past the erythroid colony-forming unit (CFU-E) stage. In *JAK2–/–* fetal livers, the reduction of CD44^{low} cells appeared to be due to the absence of Ter-119^{pos} cells. A small number of Ter-119^{pos}CD44^{neg} primitive erythroid cells are present in both wild-type and *JAK2–/–* fetal livers. Therefore, although the absence of JAK2 disrupts definitive erythropoiesis, long-term reconstituting CD34^{low}c-kit^{pos} haemopoietic stem cells are present in JAK2-deficient mice.

In vitro colony assays to determine the numbers of erythroid blast-forming units (BFU-E) and CFU-Es were performed using fetal liver cells of *JAK2–/–*, *EPOR–/–*, and wild-type mice. Incubation of fetal liver cells in methylcellulose with IL-3, SCF, GM-CSF, and EPO resulted in readily detectable BFU-E and CFU-E colonies in *EPOR–/–* mice, but the number of cells was approximately 10-fold lower than in wild-type animals. Significantly, BFU-E and CFU-E could not be detected at all in JAK2–/– mice by these methods. When fetal liver cells were incubated in the presence of S17 stromal cells as well, benzidine staining revealed less than half as many haemoglobin-containing cells in *EPOR–/–* mice as compared to wild-type mice. *JAK2–/–* mice, however, contained almost no benzidine-positive cells. These experiments confirmed that JAK2 is crucial for definitive erythropoiesis, and suggest that the absence of JAK2 causes a more severe block in erythropoiesis than does the absence of the EPOR, probably because of the involvement of JAK2 in additional cellular processes.

JAK2 associates constitutively with the IFN-γ receptor β chain. To determine whether the IFN-receptor systems require JAK2, embryonic fibroblasts were stimulated with IFN-α or IFN-γ. Northern blotting after IFN-α stimulation revealed up-regulation of interferon-regulated factor-1 (IRF-1) in wild-type and *JAK2*-deficient cells, indicating that IFN-α signalling is unaffected by the loss of JAK2. In contrast, IFN-γ stimulation caused the induction of IRF-1 and guanylate binding protein-2 (GBP-2) in wild-type cells, but not in *JAK2–/–* cells. Therefore, JAK2 is critical for signalling through type-II, but not type-I, IFN receptors.

JAK2 is thought to associate with gp130, a subunit present in the LIF receptor. Stimulation with LIF, like IFN-α and IFN-γ, results in upregulation of IRF-1 in wild-type cells. Since *JAK2–/–* embryonic stem cells stimulated with LIF could induce IRF-1 mRNA transcription, JAK2 is not required for LIF signalling.

In another *JAK2* gene-targeting study by Parganas *et al.*, fetal liver cells from day-10 embryos were treated with various cytokines and analysed for their ability to form colonies (46). In contrast to wild-type cells, *JAK2–/–* cells failed to respond to IL-3 (in the presence or absence of EPO), thrombopoietin (TPO), GM-CSF, or IL-5, demonstrating that JAK2 plays a critical and non-redundant role in signalling through these receptors. G-CSF stimulation activates JAK2, but also activates JAK1 and TYK2. In response to G-CSF, wild-type and *JAK2–/–* fetal liver cells formed granulocytic colonies at roughly equivalent levels, indicating that JAK2 is either not required or is redundant in this receptor system. Additionally, these results indicated that the presence of haemopoietic precursor cells is not affected by JAK2 deficiency.

To confirm that the progenitor cells were responsive, the expression of several genes known to be active early in haemopoiesis was examined. Transcripts of EKLF, AML1, rhombotin-2, Gata-1, and Gata-2 were all present in fetal liver cells. Additionally, expression of the EPOR, c-myb, and Pu.1 was detected, although at reduced levels. Globin genes were also expressed, including the embryonic εy^2, ξ, and $\beta h1$ genes. The β^{maj} gene, which is specific to definitive erythropoiesis in the fetal liver, was expressed at reduced levels, supporting the conclusion that definitive erythropoiesis is impaired in *JAK2*-deficient mice. Furthermore, infection of day-10 to day-12 fetal liver cells with a JAK2-expressing retrovirus rescued erythroid colony formation.

Although CSF-1 and stem-cell factor (SCF), whose receptors have protein tyrosine kinase activity, utilize a different class of receptors from the cytokines studied above, both also activate JAK2. In response to CSF-1, wild-type and *JAK2–/–* fetal liver cells were able to form macrophage colonies, but fewer colonies were observed in the absence of JAK2. A similar pattern resulted upon stimulation with SCF. Although *JAK2*-deficient fetal liver cells formed mixed colonies in the presence of SCF, the number of colonies was lower than normal, again suggesting a requirement for JAK2 in the amplification of progenitor cells.

Cytokines of the IL-6 family and IFN-γ are known to activate JAK2 as well. In response to IL-6 stimulation, tyrosine phosphorylation of JAK1, TYK2, gp130, and STAT3, and induction of IRF-1 expression were detected in fibroblasts from both wild-type and *JAK2–/–* embryos. In contrast, tyrosine phosphorylation of JAK1 and STAT1, and induction of *IRF-1* and *SOCS-1*, were observed in IFN-γ-stimulated fibroblasts from wild-type but not *JAK2*-deficient embryos. JAK2 is essential for the antiviral response mediated by IFN-γ, since IFN-γ treatment protected wild-type cells from the encephalomyocarditis virus, yet failed to protect *JAK2–/–* cells.

To determine whether JAK2 is required for the generation of lymphoid progenitor cells, fetal liver cells from day-10 to day-12 embryos were used to reconstitute sublethally irradiated *JAK3*-deficient mice. As discussed below, JAK3 is required for amplification of early lymphoid progenitors. After 4 weeks, peripheral B220$^+$ B cells were detected in *JAK3–/–* mice reconstituted with fetal liver cells from either wild-type or *JAK2–/–* embryos. Thymuses of mice reconstituted with wild-type or *JAK2–/–* fetal liver cells contained similar numbers of CD4+CD8+ T cells, and these T cells proliferated in response to anti-CD3 or IL-2. These results demonstrated that JAK2 does not play a role in the development of lymphoid progenitors during embryonic development.

These two studies support a vital role for JAK2 in promoting erythropoiesis (42, 46). EPO specifically activates STAT5, yet the *STATa/b* gene-targeted mice display anaemia during embryogenesis (47) which does not appear to dramatically affect normal development (48). Given the critical role of JAK2 towards the proper establishment of all myeloid lineages, identification of critical downstream substrates will be important to our understanding of myeloid cell biology.

3.3 JAK3

JAK3 is the only *Janus* kinase with a restricted tissue expression, as it is primarily found in haemopoietic tissues. Several cytokines including IL-2 (18, 20, 49), IL-4 (20, 50), IL-7 (50), IL-9 (51), and IL-15 (20, 50) all activate the tyrosine kinase activity of this enzyme. JAK3 has been shown to specifically associate with the IL-2 receptor γ chain (γc) which is shared amongst the aforementioned cytokines. Considering that JAK3 has a restricted expression, there was considerable interest in developing an animal model to study JAK3 function. JAK3 was shown to be essential for B-cell maturation and competence of mature T cells (31–33).

Nosaka *et al.* demonstrated that the *JAK3* nullizygous animals had small thymuses and that mesenteric and peripheral lymph nodes were reduced in size (31). The absolute numbers of thymocytes were 1–10% of their heterozygous or wild-type littermates. At one week there were decreased numbers of CD4+CD8+ T cells; however, on ageing the CD4/CD8 ratio increased. *JAK3*-deficient thymocytes were non-responsive to concanavalin A (ConA) stimulation or PMA and CD3 cross-linking.

The spleens from *JAK3* gene-targeted animals were also reduced in size, and non-erythroid splenocytes were 10–25% of the controls. One-week-old mice displayed decreased numbers of B220+ cells. However, immature B220+sIgM (surface immuno-globulin M)- and mature B220+sIgM+ cells were detected in older mice. Splenocytes and lymph node cells from 4-week-old mice respond poorly to ConA, PMA plus CD3, or LPS. Single positive T cells were 5–10% of normal and B220+sIgM+ B cells were found at 1.6–3.3% of controls. The *IL-7* (29), *IL-7 R* (30) γc gene-targeted animals (52, 53) all display a profound block in B-cell maturation. Therefore, bone-marrow cells were analysed for their ability to respond to IL-7. As expected, there was no response to IL-7, confirming that the engagement of JAK3 by the IL-7 receptor plays a non-redundant role in B-cell differentiation.

Thomis *et al.* demonstrated that the block in B-cell differentiation occurred in the transition between pre-B and immature B cells, since the *JAK3*-deficient mice had normal levels of CD43+CD45R+ (pro-B), decreased CD43−CD45R+ (pre-B), but no CD45R+IgM+ (mature B) cells (32). Total thymic cell numbers were 0.5–10% of control animals and a similar increase in CD4+ cells was observed, similar to the

γc-deficient mice. The thymic architecture of the *JAK3–/–* mice was shown to be normal.

Spleen cellularity was observed to fluctuate wildly from 10% to 500% when compared to wild-type animals, and a similar block in B-cell differentiation was observed by an absence of CD45R$^+$IgM$^+$ cells (32). Peripheral lymph nodes had dramatically reduced numbers of cells (2%) and showed abnormal ratios of CD4$^+$ (60%) and CD8$^+$ (5%) cells. As described above, splenic T cells from *JAK3* knockout mice showed reduced proliferation in response to PMA plus ionomycin, ConA, anti-CD3, anti-CD3 plus anti-CD28. The level of IL-2 secretion from these cells after CD3 and CD28 crosslinking was reduced from 8.3 to 0.2 U/ml. The splenic T cells appeared to be activated, as upregulation of CD25, CD44, and CD69 were all observed and the cell size increased compared to controls.

To distinguish whether the T-cell phenotype resulted from an absence of JAK3 in peripheral cells or was a result of some other phenomenon such as aberrant thymic maturation, Berg and colleagues then utilized a reconstitution model in which they generated transgenic lines with *JAK3* under the regulatory control of the *lck* promoter (54). One transgenic line expressed JAK3 in thymus and spleen ($tg^{thy+spl}$), whereas another expressed JAK3 in the thymus alone (tg^{thy}). In addition, one transgenic line expressed a kinase inactive form of JAK3 (K851R) (tg^{kd}) in which expression was obtained in both the thymus and spleen. Neither *JAK3–/–* ($tg^{thy+spl}$) nor *JAK3–/–* (tg^{thy}) were able to restore B-cell function as no CD45$^+$IgM$^+$ cells were observed. However, the expression of wild-type JAK3 from either transgene restored thymus numbers and normal distribution of CD4$^+$ and CD8$^+$ cells. The reconstituted JAK3 mice have restored $\gamma\delta$ T cells in the thymus and normal numbers of natural killer cells in the spleen, but lymph node development is still impaired as in the *JAK3–/–* mice. Expression of a kinase-inactive form of JAK3 failed to reconstitute any of the B- or T-cell abnormalities of the JAK3-deficient animals.

Splenic T cells from *JAK3–/–* mice have an activated or memory phenotype, indicated by the presence of CD44^{hi}CD62L^{lo} cells (54), T cells from *JAK3–/–* ($tg^{thy+spl}$) mice are similar to the *JAK3+/–* T cells. Interestingly, the expression of CD44 and CD62L varies with the age of *JAK3–/–* (tg^{thy}) mice. Thymocytes from young mice expressed low levels of CD44 and high amounts of CD62L similar to the *JAK3+/–* thymocytes. However, this profile altered with maturation since by 42 days the expression of CD44 and CD62L paralleled that of the *JAK3–/–* animals. This suggests that constitutive expression of JAK3 is required for normal T-cell function.

Thymocytes from the thymus and spleen of *JAK3–/–* (tg^{thy+sp}) and *JAK3–/–* (tg^{thy}) animals have restored proliferative responses to CD3 and CD28 cross-linking (54). The *JAK3–/–* mice have a defect in IL-2 and IL-3 secretion. Similarly, expression of JAK3 from either transgenic line restores the levels of IL-2 and IL-3 production. Expression of the *JAK3–/–* (tg^{kd}) failed to restore proliferative responses or cytokine secretion. This study showed that JAK3 expression in a T-cell background could restore many of the abnormalities in the *JAK3–/–* mice, including the production of $\gamma\delta$ T cells and NK cells. However, B-cell function was still defective, indicating that the *JAK3* transgene failed to target an appropriate bone marrow-derived stem cell. Since

JAK3 functions in concert with JAK1 in lymphocyte signalling, it may have been expected that JAK1 activation would phosphorylate JAK3 in *trans*, resulting in the partial restoration of JAK3 function. However, a kinase-inactive JAK3 failed to restore any functions of the wild-type JAK3 enzyme.

Next, the Berg group examined T-cell function in the *JAK3–/–* animals (55). The expression of T-cell activation markers was monitored by incubating thymocytes on CD3-coated plates. Normal expression of CD62L and CD69 was observed in CD4$^+$CD8$^-$ thymocytes. In peripheral T cells, CD25, CD28, and CD44 were induced by 72 h anti-CD3 cross-linking in *JAK3 +/–* cells. Induction of these cell-surface markers was absent from *JAK3–/–* thymocytes. Calcium flux and bulk tyrosine phosphorylation appears to be normal in *JAK3–/–* thymocytes and splenic T cells.

The *JAK3–/–* mice were crossed into two different genetic backgrounds in order to determine if any of the defects in the *JAK3–/–* thymocytes could be attributed to inappropriate TCR engagement (55). The mice were crossed into the *OVA–TCR* background (specific for OVA and MHC class-I molecule, H-2K^b) and *2B4* (specific for cytochrome c plus MHC class-II molecule I-E^k). The expression of a functionally rearranged TCR did not restore normal thymocyte numbers in the *JAK3–/– 2B4 TCR* background, which is similar to normal *JAK3–/–* thymuses. Positive selection was observed to be normal as there was a large percentage of CD4$^+$CD8$^-$ and absence of CD4$^-$CD8$^+$ thymocytes.

The spleens were of normal size and displayed normal numbers of cells. The *JAK3–/–2B4* TCR contained very few CD4$^+$ T cells. Only 80% of CD4$^+$ T cells express the 2B4 TCR compared to 95% of *JAK3 +/–2B4* transgenic mice. The majority of the peripheral T cells were CD62L^{hi} and CD44^{lo}, which is indicative of a resting phenotype. This is similar to the resting T cells found in *JAK3 +/–* 2B4 TCR and not the activated T cells seen in the *JAK3–/–* mice.

Similarly, *JAK3–/– OVA–TCR-I* mice appeared to have normal positive selection in the thymus. Peripheral T cells appeared phenotypically naive. In the transgenic *JAK3–/–* background, thymocytes are not exposed to their TCR-specific antigen. This results in normal numbers of resting, peripheral T cells. This suggests that the non-transgenic *JAK3–/–* thymocytes are receiving an antigen-specific signal, which results in their activation and anergic state. The role of JAK3 in anergy was suggested from an earlier study (49).

Increased proliferation was observed in thymocytes isolated from *JAK3–/–* mice. Bromodeoxyuridine (BrdU) incorporation experiments indicate that 22–29% of T cells in the *JAK3–/–* compared to 7% of CD4$^+$ T cells from *JAK3 +/–*. There was also increased apoptosis in splenic T cells cultured in media alone or in the presence of CD3 and CD28 cross-linking as measured by propidium iodide staining and Annexin V incorporation. The mechanism is unknown.

These studies demonstrate a critical role for JAK3 in lymphoid development. Aberrant splice forms of JAK3 have also been described in breast epithelial-cell lines, suggesting a role for JAK3 in breast cancer (56). A fraction of patients with severe combined immune deficiency (SCID) suffer from mutations in *JAK3*, which is discussed in Section 5.

3.4 TYK2

TYK2 was originally isolated in a screen for novel tyrosine kinases. The first function for TYK2 was revealed when the *TYK2* cDNA complemented a cell line defective in IFN-α signalling (57). Muller and colleagues (H. Neubauer and M. Muller, unpublished data) are developing mice deficient in TYK2. Initial observations reveal that the mice are viable with normal expression of various cell-surface markers including CD3, CD4, CD5, CD8, CD14, CD19, CD21, CD28, and CD45 (H. Neubauer and M. Muller, unpublished data).

3.5 STAT1

Pioneering research by Darnell and colleagues resulted in the isolation, purification, and characterization of the ISGF complex assembled after IFN-α stimulation. Isolation of candidate cDNA clones revealed a novel transcriptional complex consisting of four components: STAT1, expressed as alternatively spliced 91- and 84-kDa proteins (10); STAT2, a 113-kDa protein (11); and p48, a member of the IRF-1 gene family (12). Both IFN-α and IFN-γ activate STAT1; however, IFN-γ stimulation activates a STAT1 homodimer and binds to a distinct GAS element. Other cytokines including IL-2 (58), IL-5 (59), IL-7 (60), IL-10 (61, 62), and EPO (63) all activate STAT1 in various cell lines, so it was of great interest to determine whether STAT1 responses were restricted to IFN-α and IFN-γ or mitigated signalling by several cytokines.

STAT1–/– mice had no gross developmental defects, but they tended to be runted and none have lived longer than 8 weeks (64, 65). All *STAT1*–/– mice have died within 48 hours of weaning, suggesting that loss of STAT1 renders mice susceptible to an infectious agent to which the mother provides milk-borne immunity. Mouse hepatitis virus (MHV) is a common pathogen to which only some immunocompromised animals succumb. Necropsy of *STAT1*–/– mice revealed signs of viral hepatitis, multifocal hepatic necrosis with syncytial cell formation, yet little inflammatory response. *STAT1*–/– mice generated under pathogen-free conditions did not develop any diseases, but were highly susceptible to infection with vesicular stomatitis virus (VSV), a pathogen that wild-type mice can clear.

To determine the basis for this susceptibility, Durbin *et al.* examined the immune systems of *STAT1*–/– mice for abnormalities (64). Flow cytometrical analysis indicated normal development of granulocytes, B lymphocytes, and monocytes in fetal livers, and of T lymphocytes in the neonatal thymus glands of *STAT1*–/– mice. Since IFN-α, β, and γ are important in the immune response to viral infection, the response of *STAT1*–/– splenocytes to these cytokines was examined. Northern blots revealed that whereas *ISG54*, *IRF-1*, and *ISGF3γp48* were induced by either IFN-α or IFN-γ in spleens of wild-type mice, no response to IFN occurred in *STAT1*-deficient spleens. Notably, the basal level of *p48* mRNA was lower in *STAT1*–/– mice, suggesting a possible role for STAT1 even in the absence of IFN.

To test for responses to other cytokines that also activate STAT1, macrophages were isolated from bone marrow and cultured for 7 days in CSF-1-containing medium. Although both wild-type and *STAT1*–/– mice produced morphologically normal macrophages, *STAT1*-deficient cells were unable to respond to IFN. In contrast, IL-6 induced the transcription of mRNA encoding FcγR1, IRF-1, and ICAM-1 in both wild-type and *STAT1–/–* mice.

The transcription factor IRF-1, which is induced by IFN through a STAT1-dependent mechanism, and ISGF3 may be responsible for IFN gene induction. Durbin *et al.* tested for the dependence of IFN-α and IFN-γ production on STAT1 activity when cells were infected with Newcastle disease virus (64). Northern blots revealed no differences in the induction of IFN mRNA in infected fibroblasts of either wild-type or *STAT1–/–* mice, indicating that normal levels of IFN expression were possible without the priming effects of IFN treatment.

Experiments by Meraz *et al.* revealed that wild-type and *STAT1–/–* mice had equivalent basal levels of IRF-1 mRNA, but whereas IFN-α or IFN-γ treatment enhanced IRF-1 mRNA production by 50-fold, levels of IRF-1 mRNA did not change in *STAT1–/–* tissues (65). The inability of IFN-α and IFN-γ to induce maximal levels of mRNA encoding guanylate-binding protein-1 (*GBP1*), MHC class-II *trans* activating protein (*CIITA*), *complement protein C3*, and *complement protein factor B* in *STAT1–/–* bone-marrow macrophages further confirmed that IFN-inducible genes could not be activated in the absence of STAT1.

The expression of other IFN-responsive genes was analysed in wild-type and *STAT1–/–* mice. In response to IFN-α or IFN-γ, peripheral blood T cells from wild-type mice upregulated cell-surface MHC class-I expression, whereas *STAT1*-deficient cells could not. Similarly, wild-type peritoneal and bone marrow-derived macrophages induced MHC class-II proteins upon stimulation with IFN-γ, whereas STAT1-deficient macrophages could not. IFN-γ treatment of bone marrow-derived macrophages resulted in expression of the B7–2 costimulatory molecule, but no response was observed in *STAT1–/–* cells.

A non-specific response to infection with microbial pathogens and viruses is the induction of inducible nitric oxide synthase (iNOS), which leads to the production of nitric oxide (NO). Treatment of peritoneal macrophages with IFN-α or IFN-γ and a stimulus such as LPS resulted in the accumulation of nitrite, a stable NO catabolic product, in the culture supernatant of wild-type cells, but not of *STAT1–/–* cells. In another test for antiviral activity, IFN was able to protect wild-type immortalized monkey kidney fibroblasts from the cytopathic effects of VSV in a dose-dependent manner, but no protective response was observed for *STAT1*-deficient cells.

To further examine the ability of *STAT1–/–* mice to defend themselves against bacterial infection, *STAT1–/–* mice were given a sublethal dose of *Listeria monocytogenes*. Since *L. monocytogenes* is an intracellular pathogen, responses to it are IFN-γ-dependent. The response was reminiscent of *IFN-α* (66) or *IFN-γ* (67) *receptor α* chain-deficient mice: wild-type and heterozygotes survived the challenge, whereas knockout mice failed to mount a response and died. As also demonstrated by Durbin *et al.*, wild-type mice survived infection with VSV (to which a response is dependent on the IFN-α receptor), but *STAT1–/–* mice succumbed (64). These results provide additional support to the conclusion that STAT1 is essential for IFN-α and IFN-γ-dependent responses.

GH, EGF, and IL-10 have also been reported to activate STAT1. However, GH-responsiveness of *STAT1–/–* mice appears to be intact, since *STAT1–/–* mice gained weight normally throughout development. Treatment of mice with EGF resulted in equivalent induction of c-*fos*, *junB*, and *Zif268* genes in both wild-type and *STAT1–/–* mice. Macrophages from both wild-type and *STAT1–/–* mice responded to IL-10 stimulation by inhibiting LPS-stimulated TNF production. Therefore, STAT1 is specifically required for signalling through the receptors for IFN-α and IFN-γ, but not through other cytokine receptors examined.

STAT1 may play an important role in bone development through FGF-dependent regulation of chondrocyte proliferation (68). Genetic evidence from *FGF* transgenic mice (69) and *FGFR3* gene-targeted mice (70, 71) suggest that FGFs act as negative regulators of bone growth. FGF stimulation of a rat chondrosarcoma cell line results in rapid tyrosine phosphorylation and nuclear localization of STAT1. FGF stimulates transcriptional activation of a GAS luciferase promoter containing four copies of the GAS sequence from the IRF-1 promoter.

To examine the effects of FGF *in vivo*, growth plate chondrocytes were isolated from 10-day-old wild-type and *STAT1–/–* mice (68). FGF1 treatment of wild-type chondrocytes resulted in a four- to fivefold reduction in proliferation, whereas no reduction was observed in the *STAT1* gene-targeted chondrocytes. Metatarsal cartilage was cultured from E15 wild-type and *STAT1* knockout embryos. FGF1 addition resulted in a dramatic decrease in DNA synthesis, whereas *STAT1–/–* cartilage displayed similar rates of DNA synthesis in the presence or absence of FGF1. Longer term culture indicated that FGF1 also influenced chondrocyte differentiation. FGF appeared to stimulate osteoblast proliferation, since FGF-treated metatarsals display a dramatic thickening of the periosteum, which was not observed in the *STAT1–/–* mice. It is unclear if this effect is directly or indirectly attributed to STAT1. Achondroplasia, thanatophoric dysplasia, and hypochondroplasia have been attributed to activating mutations within FGFR3. The mechanism by which FGFR3 influences chondrocyte function has not been elucidated, but these studies point towards a role for STAT1. Relevant transcriptional targets of STAT1 are unknown; however, the cyclin-dependent kinase inhibitor p21WAF1/CIP1 is one possibility.

3.6 STAT2

To date, the only cytokine which is known to activate the ISGF3 complex is IFN-α. No other cytokines have been documented to effect the tyrosine phosphorylation of STAT2. However, some evidence has suggested that STAT1:STAT2 (72, 73) and STAT2:STAT3 heterodimers can form (72). It was speculated that the phenotype of the STAT2 knockout mice would be similar to the *STAT1* nullizygous animals.

Identification of defects in *STAT2–/–* mice has proven to be difficult since these mice die during early embryogenesis (C. Schindler, unpublished data). In order to obtain some information on the role of STAT2 function, chimeric *RAG–/– STAT2–/–* mice were generated. *STAT2*-deficient ES cells are able to fully reconstitute both B- and T-lymphocyte development in *RAG–/–* mice, indicating that STAT2 does not play an essential role during lymphoid development (C. Schindler, unpublished data). Additional experiments are necessary to further elucidate the role of STAT2.

3.7 STAT3

STAT3 (also known as acute-phase reactive factor (APRF)) was identified as a STAT protein activated by the IL-6 family of cytokines (74). Subsequent studies revealed that many cytokines including IL-2 (75), IL-3 (76), IL-5 (77), IL-10 (61, 62, 78), G-CSF (79–81), TPO (82, 83), and EGF (74, 84–86) activate STAT3 in various backgrounds. Because STAT3 is a target downstream of many cytokines, it was felt that this STAT would play an important role in mouse development.

STAT3-deficient embryos develop normally up to embryonic day (E) 6, the egg cylinder stage, precluding a critical role for STAT3 during early postimplantation (87). However, *STAT3–/–* embryos tend to be smaller, suggesting that STAT3 does play a role in cell growth during this time. Mesoderm formation between the ecto-derm and endoderm fails to occur, and embryos begin to degenerate by E6.5. The embryos die around E7, the point at which gastrulation begins, and are completely resorbed by E7.5.

To identify earlier defects in *STAT3–/–* embryos, E3.5 blastocysts were cultured for 5 days. Both wild-type and *STAT3–/–* blastocysts produced trophoblast giant-cell out-growths, demonstrating that STAT3 is not required during the preimplantation period.

Niwa *et al.* have formulated a hypothesis based on the fact that STAT3 starts to be expressed exclusively in the visceral endoderm around E6 (88). The visceral endo-derm covers the upper side of egg cylinder-stage embryos and contributes to regulat-ing metabolic exchange with maternal blood. Hence, STAT3 may be an essential component of signalling pathways regulating visceral endoderm function, and lethality of *STAT3–/–* embryos may be due to the absence of sufficient nutrients for the embryo.

To examine the role of STAT3 in T-cell function, conditional targeting was per-formed using the *Cre–Lox* system. A targeting vector was designed in which *loxP* sites were introduced flanking exon 22 which contains the conserved site of tyrosine phosphorylation, Y705 (89). *STAT3$^{flox/flox}$* mice were crossed to mice which express Cre under the control of the *Lck* promoter in a heterozygous STAT3 background (*Lck–Cre/STAT3 +/–*). The resulting mice, *Lck–Cre/STAT3$^{flox/–}$*, produced two STAT3 proteins: the wild-type STAT3 and a truncated form. The *Lck–Cre/STAT3$^{flox/–}$* mice had normal ratios of CD4/CD8 in T cells and no differences in thymus, spleen, or lymph nodes was observed.

The proliferative capacity of thymocytes in response to IL-2, IL-6, and IL-7 was examined. Thymocytes from *Lck–Cre/STAT3$^{flox/–}$* mice showed diminished responses to ConA plus IL-2, PMA plus IL-2, and anti-CD3 plus IL-2 when compared to *Lck–Cre/STAT3$^{flox/+}$* mice (89). However, IL-7-dependent proliferation was observed to be normal. In addition, peripheral T cells showed decreased IL-6-dependent responses. Decreased [^{3}H]thymidine incorporation was observed in response to anti-CD3 plus IL-6 and ConA plus IL-6.

The investigators examined several IL-6-dependent responses in order to establish the mechanism for the IL-6 defect in T cells. Normal levels of the IL-6Rα and gp130 were expressed in lymphocytes from *Lck–Cre/STAT3*^{flox/–} mice (89). IL-6 protected peripheral T cells from apoptosis in the *Lck–Cre/STAT3*^{flox/+} mice; however, this protection was abolished from the conditionally targeted STAT3 lymphocytes. IL-6-dependent induction of Bcl2 was identical in both strains of mice.

Akira and colleagues proposed that STAT3 may be required for induction of the IL-2Rα chain, as has been implied for STAT5a (90). Peripheral T cells from *Lck–Cre/STAT3*^{flox/–} cells showed slightly decreased [^{3}H]thymidine incorporation in response to anti-CD3 plus IL-2, ConA plus IL-2. This effect was IL-2-dependent and was observed at lower concentrations of IL-2. The expression of the IL-2Rα chain when the *Lck–Cre/STAT3*^{flox/–} cells were stimulated with low concentrations of IL-2 and anti-CD3. However, cross-linking of the T cells with high concentrations of CD3 resulted in equivalent expression of the IL-2Rα chain, indicating that the decreased expression of IL-2Rα was IL-2-dependent.

The function of STAT3 in the myeloid lineage was examined by crossing the *STAT3*^{flox/flox} mice with mice in which the *cre* cDNA is inserted into the mouse *lysozyme M* gene by a knock-in approach (91). Lysozyme M has been shown to have expression restricted to the monocyte/macrophage and granulocyte lineages. The efficiency of Cre-mediated deletion of the floxed STAT3 allele was 97%. IL-6-mediated tyrosine phosphorylation was greatly reduced in peritoneal macrophages and neutrophils, whereas expression of STAT3 in thymocytes was not affected.

The *LysM–Cre/STAT3*^{flox/–} mice were exquisitely sensitive to LPS challenge; nine out of ten *LysM–Cre/STAT3*^{flox/+} mice survived a 20 µg challenge, whereas none of the mice targeted for STAT3 expression in the myeloid lineage survived (91). In addition, there was a dramatic elevation of the proinflammatory cytokines, IL-1β, IL-6, IL-10, IFN-γ, and TNF-α in the peripheral blood of *LysM–Cre/STAT3*^{flox/–} mice. STAT1 has been shown to be essential for macrophage activation. STAT1 is constitutively tyrosine-phosphorylated in peritoneal macrophages from the conditionally targeted STAT3 animals.

Peritoneal macrophages normally display a dose-dependent reduction of TNF-α production when stimulated with IL-10. However, macrophages from the *LysM–Cre/STAT3*^{flox/–} mice failed to elicit IL-10 functions, including decreased TNF-α and IL-6 secretion and lowered NO production. Bone marrow-derived macrophages proliferate in response to CSF-1, and pretreatment with IL-6 or IL-10 can inhibit this effect. Neither IL-6 nor IL-10 pretreatment of the STAT3-targeted macrophages inhibited their proliferation in CSF-1.

IL-10 has been shown to have biological functions in neutrophils in addition to macrophages. In a normal setting, IL-10 pretreatment reduces H_2O_2 and TNF-α production. However, IL-10 pretreatment did not affect H_2O_2 and TNF-α secretion in neutrophils isolated from *LysM–Cre/STAT3*^{flox/–} mice. Therefore, no functional responses to IL-10 were observed in neutrophils or macrophages from the *LysM–Cre/STAT3*^{flox/–} mice.

LysM–Cre/STAT3^{flox/–} mice appeared normal at birth; however, mortality in some animals was observed at 20–24 weeks which was attributed to enterocolitis (91). Persistent leucocytosis and anaemia was observed as well as infiltration of neutrophils, lymphocytes, and plasma cells into the lamina propria of the gut. Marked thickening of the colonic wall was observed as well as depletion of the mucin-producing goblet cells. Histochemical staining revealed expression of MHC class-II molecules on the colonic epithelium of the targeted mice but not on wild-type control mice. In addition, CD4-positive T cells were found in the lamina propria of *LysM–Cre/Stat3$^{flox/-}$* mice.

Previous studies have shown that enterocolitis is associated with a T-helper (T$_H$) type-1 phenotype. To examine the possibility that IL-10, a T$_{H2}$ cytokine, inhibits T$_{H1}$ cell development, splenic T cells were examined for cytokine production after CD3 cross-linking. Peripheral T cells from 12-week-old mice produced fourfold higher levels of IFN-γ. Increased levels of IFN-γ were observed after CD3 cross-linking, IL-12 or IL-18 stimulation in splenic T cells from 5-week-old mice which had not developed symptoms of enterocolitis. In addition, LPS stimulation of peritoneal macrophages resulted in elevated IL-12 production in the *LysM–Cre/STAT3$^{flox/-}$* mice when compared to the *LysM–Cre/STAT3$^{flox/+}$* mice. Therefore, splenic T cells display a T$_{H1}$ cell phenotype prior to the onset of chronic enterocolitis.

As described above, STAT3 is activated downstream of many different cytokines. Therefore, the phenotype of the STAT3 targeted animals is not too surprising. In addition, conditional gene-targeting experiments have revealed an important role of STAT3 in macrophage and neutrophil function as well as an unexpected role for IL-6 in mediating T-cell survival.

3.8 STAT4

STAT4 was initially isolated as an orphan STAT by the Darnell (92) and Ihle laboratories (93). Subsequent studies indicated that IL-12 activated STAT4 (94, 95). Since IL-12 has a specific role in T$_{H1}$ cell development, the phenotype of the STAT4 knockouts was eagerly anticipated by immunologists.

STAT4–/– mice were viable and appeared normal based on gross external examination. Flow cytometry indicated a normal distribution of major lymphoid populations expressing the cell-surface markers CD4, CD8, CD28, CD32, CD45, CD22, B220, CD34, CD2, and CD5 (96). Despite the fact that STAT4 is highly expressed in germ cells in the testes at the mid-round to early-elongating spermatid stage, *STAT4*-deficient males were fertile and spermatogenesis appeared to proceed normally (96).

Since STAT4 is activated by IL-12, responses to this cytokine were examined. Upon IL-12 stimulation, preactivated T and NK cells in wild-type mice proliferate. However, IL-12 restimulation of T and NK cells from *STAT4–/–* mice failed to induce proliferation (96), indicating that STAT4 is essential for both IL-12-induced IFN-γ secretion and T- and NK-cell proliferation. In addition, IL-12 stimulation of T cells from *STAT4–/–* mice does not result in the increase in IFN-γ secretion observed in wild-type T cells (97).

Since IL-12 is known to stimulate NK-cell cytotoxicity, Kaplan *et al.* assayed for NK activity in *STAT4–/–* cells (97). IL-12 stimulation of NK cells from wild-type mice resulted in enhanced cytotoxicity, as measured by lysis of the NK-sensitive YAC-1 target, whereas the cytotoxicity of STAT4-deficient NK cells remained at the un-induced, basal level.

Naive CD4$^+$ T cells differentiate into T$_{H1}$ cells in the presence of IL-12, and into T$_{H2}$ cells in the presence of IL-4 and IFN-γ (96, 97). However, STAT4-deficient T lymphocytes pretreated with IL-12 and activated by anti-CD3 produce significantly less IFN-γ and lymphotoxin, cytokines characteristically produced by T$_{H1}$ cells, than wild-type lymphocytes. STAT4-deficient T lymphocytes cultured in IL-4 and IFN-γ prior to activation with anti-CD3, on the other hand, produced significantly higher levels of the characteristic T$_{H2}$ cytokines IL-4, IL-5, and IL-10 than did wild-type lymphocytes. Significant amounts of IL-5 and IL-10 were produced by *STAT4–/–* cells even under conditions that favoured T$_{H1}$ differentiation. Therefore, STAT4 is required for the generation of T$_{H1}$ cells, and its absence results in a propensity towards the development of T$_{H2}$ cells.

3.9   STAT6

STAT6 was initially identified as a STAT protein downstream of IL-4 (98). A subsequent report suggested that IL-3 could also activate STAT6 (99). Given the importance of IL-4 in T$_{H2}$ cell development, there was considerable interest in the development of an animal model to study STAT6 function. Gene-targeting experiments were performed by three laboratories (100–102).

Like *STAT4–/–* mice, *STAT6–/–* mice were viable and grossly similar to wild-type mice. Normal numbers of lymphoid and myeloid cells were present, and flow cytometry for CD3, CD4, CD8, CD14, GR-1, IgM, and B220 failed to reveal any differences in the expression of these cell-surface markers (101). However, *STAT6–/–* mice were specifically defective in their responsiveness to IL-4, just as *STAT4–/–* mice are specifically defective in their response to IL-12.

IL-4 stimulation results in the proliferation of lymphocytes in wild-type mice. [^{3}H]Thymidine incorporation assays demonstrated that, whereas *STAT6–/–* lymphocytes proliferated normally in response to IL-2, anti-CD3, or LPS, the proliferative response of STAT6-deficient lymphocytes to IL-4 was almost completely abrogated (102).

In wild-type mice, IL-4 stimulation results in upregulation of MHC class-II antigens and the IL-4 receptor on both B and T cells. However, flow cytometry revealed that these genes were not upregulated on B or T lymphocytes of *STAT6–/–* mice. The basal levels of expression of these genes were similar in wild-type and mutant mice, and LPS stimulation of B cells of both genotypes led to the equivalent upregulation of MHC class-II antigens, but STAT6 is specifically required for the IL-4-induced increase in expression of IL-4-regulated genes (102). Similarly, IL-4 and LPS induced CD23 (FcεRII) and Thy-1 expression in wild-type B cells, but not in *STAT6–/–* B cells (101).

Another physiological response to IL-4 stimulation involves B-cell immunoglobulin class-switching to IgG1, and IgE in response to antigenic challenges depends on IL-4. Wild-type and *STAT6–/–* mice were injected with IgD-specific monoclonal antibody and sera were analysed by ELISA for the presence of all immunoglobulin isotypes. Prior to anti-IgD treatment, serum levels of IgG1, IgG2a, IgG2b, IgG3, IgA, and IgM were similar in both wild-type and *STAT6–/–* mice. Low levels of IgE were detected in wild-type mice, but none was detected in *STAT6–/–* mice. After anti-IgD treatment, dramatic increases in serum IgG1, but no changes in IgA and IgM serum levels, were observed in both wild-type and *STAT6–/–* mice. However, the increase in IgE levels observed in wild-type mice did not occur in *STAT6–/–* mice. In addition, serum levels of IgG2a, IgG2b, and IgG3 were increased in *STAT6–/–* mice as compared to wild-type mice. Therefore, STAT6 is required for $C\varepsilon$ gene transcription and IgE class-switching, but negatively regulates switching to IgG2a, IgG2b, and IgG3 (101).

To determine whether STAT6 is required for T_{H2} cell differentiation, CD4$^+$ spleen cells from both wild-type and *STAT6–/–* mice were cultured in an environment that favoured either T_{H1} cell development (IL-12) or T_{H2} cell development (IL-4). Culture supernatants were examined by ELISA for the presence of cytokines characteristic of T_{H1} cells (IFN-γ) or T_{H2} cells (IL-4, IL-5, and IL-10). Loss of STAT6 expression did not affect T_{H1} cell development, but T_{H2} cell development was impaired. The decrease in IFN-γ production observed in wild-type cells cultured under conditions that favour T_{H2} differentiation was not observed in *STAT6–/–* mice, indicating that loss of STAT6 expression also resulted in the loss of IL-4-mediated suppression of T_{H1} differentiation (101).

The defect in T_{H2} differentiation observed in *STAT6–/–* mice was more pronounced than that in IL-4-deficient mice, suggesting that a cytokine other than IL-4 has the ability to induce T_{H2} cell differentiation, but it also depends on a STAT6-containing signalling pathway. IL-13, whose high-affinity receptor employs the α-chain of the IL-4 receptor, is a candidate cytokine for this function. Support for this hypothesis came from experiments that showed that IL-13 stimulation resulted in T_{H2} cell development of wild-type spleen cells but not of *STAT6–/–* cells. Therefore, whereas naive lymphoid-cell development is normal in the absence of STAT6, IL-4- and IL-13-mediated T-lymphocyte development is specifically affected by the loss of STAT6 expression.

Since STAT6 is activated by IL-4 and IL-13, it was expected that functions mediated by these two cytokines would be affected by the loss of STAT6 expression. However, the lack of other defects in *STAT6–/–* mice was surprising because of the broad expression of STAT6 and its activation by IL-3 as well.

Given the similarity of *STAT6–/–* and *IL-4–/–* mice, it appears that STAT6 evolved to mediate IL-4-specific functions, just as STAT1 appears to have evolved to mediate IFN-specific functions. The specificity of these two STAT proteins and the fact that they are primarily responsible for the control of T_H cell differentiation can be exploited. The availability of mice which are specifically deficient in T_{H1} (*STAT4–/–*) or T_{H2} (*STAT6–/–*) differentiation will greatly facilitate study of the differences in the development of T_H cells.

3.10   STAT4/STAT6(–/–)

The discovery that the differentiation of T_H cells is controlled by only two STAT proteins prompted the development of *STAT4–/– STAT6–/–* mice in order to examine the ability of T_H cells to develop in the absence of both STAT proteins. Mice deficient in both STATs were grossly similar to wild-type mice, and lymphoid organs contained normal numbers of cells with normal expression of CD3, CD4, CD8, and B220 (103).

Restimulation of wild-type lymphocytes with anti-CD3 causes the secretion of IFN-γ and IL-4, whereas *STAT4–/–* lymphocytes fail to secrete IFN-γ and *STAT6–/–* lymphocytes fail to secrete IL-4. Interestingly, *STAT4–/– STAT6–/–* lymphocytes do not secrete IL-4 when restimulated with anti-CD3, but produce levels of IFN-γ that are similar to those produced by wild-type or *STAT6–/–* lymphocytes. It is known that, in terms of T_H cell differentiation, signalling by IL-4 is dominant over IL-12 signalling, and these results suggest that IL-4 signalling through STAT6 inhibits the ability of cells to develop into T_{H1} cells. Therefore, activated cells can produce IFN-γ even in the absence of STAT4, provided that STAT6 has not been activated by IL-4.

Since IFN-γ is thought to prime cells for additional IFN-γ production, anti-IFN-γ antibody was included in primary cultures of *STAT4–/– STAT6–/–* lymphocytes to determine if IFN-γ production would be affected by its removal. Anti-IFN-γ antibody did not affect the amount of IFN-γ produced by these cells, indicating that endogenous IFN-γ does not drive IFN-γ secretion in *STAT4–/– STAT6–/–* lymphocytes. This finding further supports the belief that IFN-γ plays an indirect role in T_{H1} cell differentiation by upregulating IL-12Rβ2 expression and hence causing cells to become responsive to IL-12.

Stimulation of activated T cells by IL-12 has been shown to prime cells to produce high levels of IFN-γ upon restimulation. *STAT4–/– STAT6–/–* lymphocytes cultured in IL-12 plus anti-IL-4 (to support T_{H1} cell development) produce four- to fivefold lower levels of IFN-γ. Furthermore, the number of IFN-γ-secreting cells in *STAT4–/– STAT6–/–* lymphocyte cultures is similar to that in STAT6–/– lymphocyte cultures and in T_{H1} cell-skewed cultures. Therefore, STAT4 activation by IL-12 results in an increase in IFN-γ production on a per cell basis, not in an increase in the number of T_{H1} cells.

Since IFN-γ-secreting cells can be produced in *STAT4–/– STAT6–/–* mice, Kaplan *et al.* examined the delayed-type hypersensitivity (DTH) reaction, a typical T_{H1} cell-mediated immune response, in these mice (103). Significantly, *STAT4–/– STAT6–/–* mice were found to contain functional T_{H1} cells which were able to elicit a DTH response.

These experiments have shown that the mechanisms which regulate the differentiation of T_{H1} and T_{H2} cells are distinct. STAT6 is essential for T_{H2} cell development, whereas both STAT4-dependent and -independent pathways apparently exist for T_{H1} cell development. STAT4 activation, in contrast to STAT6 activation, does not appear to provide a differentiative signal to T_H cells. Nevertheless, IL-12 and STAT4 are important in T_H cell development, with STAT4 playing a role in regulating the production of IFN-γ and perhaps other factors required for T_{H1} cell functions.

3.11 STAT5

STAT5 was originally identified as mammary gland factor (MGF), a prolactin target gene, by the Groner laboratory (104). Other investigations by several laboratories demonstrated that STAT5 was actually expressed as two co-localized genes, *STAT5a* and *STAT5b*, with high similarity (105, 106). Furthermore, a multitude of cytokines activate STAT5 including IL-2 (107), IL-3 (105, 106), IL-5 (105), IL-7 (60), IL-9 (108), IL-15 (17, 109), G-CSF (110), GM-CSF (105, 106), EPO (111–114), growth hormone (115, 116) and prolactin (104).

3.11.1 STAT5a

STAT5a–/– (117) and *STAT5b–/–* mice (118) were both found to have highly specific defects, despite the fact that they are expressed in most, if not all, tissues and are activated by many cytokines. Wild-type and *STAT5a*-deficient mice are similar in appearance, size, weight, and fertility, indicating that most cytokine responses are normal. Notably, the GH signalling pathway appears to be intact, as indicated by normal developmental weight gain of males and females.

However, the mammary glands of *STAT5a–/–* mice cannot produce and deliver milk to pups, even after extended periods of suckling (117). Defects in mammopoiesis were revealed by whole-mount analysis of mammary tissue from postpartum *STAT5a–/–* mice. Although ductal development was normal, *STAT5a–/–* mice had reduced lobuloalveolar outgrowth and secretory tissue was sparse. Alveoli were small and contained small lumina. The metabolic activity of *STAT5a–/–* tissue was confirmed by the detection of fat droplets in the cytoplasm of *STAT5a–/–* cells. Since terminal differentiation of mammary alveolar cells requires prolactin, the phenotype observed is probably due to the inability of *STAT5a–/–* mammary tissue cells to respond to prolactin.

Expression of certain milk-protein genes was found to be abnormal in *STAT5a–/–* mice, both 12 hours after parturition and after 3 days of suckling. Although normal levels of α-lactalbumin and WDNM1 RNA and near-normal levels of β-casein were present in *STAT5a–/–* mice, whey acidic protein (WAP) RNA and protein levels were greatly reduced. Nevertheless, WAP is synthesized, processed, and secreted into the alveolar lumina in *STAT5a–/–* mice. Hence it is unclear why *STAT5a*-deficient mice fail to lactate. The early stages of lactogenesis proceed normally since milk proteins accumulate and alveolar cells are metabolically active, but lumina fail to expand. The *WAP* gene, whose expression is tightly regulated during pregnancy, may be crucial for mammopoiesis, but its mechanism of action is not yet known.

Surprisingly, STAT5b protein (but not RNA) levels in *STAT5a*-deficient tissues were greatly reduced, and STAT5b phosphorylation was even more affected. This observation is consistent with the belief that docking of both STAT5a and STAT5b to the receptor is required for maximal activation, and that efficient phosphorylation of each of these two proteins depends on the presence of the other.

The *STAT5a–/–* mouse demonstrated remarkable specificity in its defects, and it was unexpected that STAT5b could not compensate for the STAT5a deficiency, despite the fact that these two proteins are only significantly different at the carboxyl termini. Liu *et al.* have hypothesized that since the mammary gland appeared late in organ evolution, redundant signalling pathways utilizing STAT1 or STAT3 have not yet developed. (117)

3.11.2 STAT5b

Loss of *STAT5b* expression caused the loss of the multiple responses that vary between males and females and are associated with the sexually dimorphic pattern of growth hormone (GH) secretion by the pituitary gland (118). Like GH-deficient *Little* mice, *STAT5b–/–* males (but not females) displayed decreased body weight gain, and grew at a lower rate which is characteristic of wild-type females. Skeletal analysis of *STAT5b–/–* males revealed no abnormalities other than a smaller size, and no histological defects or differences in weight of abdominal or thoracic organs could be detected. Livers of *STAT5b–/–* mice were, however, pale and enlarged. Plasma GH levels in *STAT5b*-deficient mice were normal or somewhat elevated in males, and normal in females. Since *STAT5b–/–* mice are not deficient in GH, they may be resistant to GH pulses. Pulsatile, but not continuous, GH production is known to activate STAT5b in the liver. This pattern of plasma GH is important in the regulation of sexual dimorphism of liver gene expression.

Major urinary proteins (MUP), a family of α2-macroglobulin-related proteins secreted by the liver, are excreted into the urine of male mice at levels that are approximately three times higher than in females. Since MUP are induced by pulsatile plasma GH, the excretion of these proteins in the urine of wild-type and *STAT5b–/–* mice was compared. The urine of *STAT5b–/–* male mice contained MUP levels that were similar to those found in the urine from wild-type females. *STAT5b–/–* females excreted less MUP than wild-type females. Thus STAT5b plays a role in the pulsatile GH induction of MUP in males, and may influence *MUP* gene expression in females as well.

Other sex-specific, liver-expressed gene products regulated by pulsatile GH include several cytochrome P450 CYP enzymes. Expression of $P450_{15\alpha}$ hydroxylase (CYP2A4), CYP3A, and the prolactin receptor occurs primarily in female mice. Expression of $P450_{15\alpha}$, CYP3A, and prolactin receptor in *STAT5b*-deficient males was increased to levels that are similar to those observed in wild-type females. In contrast, a decreased expression of the GH pulse-regulated, male-specific steroid 16α-hydroxylase protein CYP2D9 was observed in *STAT5b–/–* males. No differences in expression of the sex-independent, mouse-liver P450 protein, microsomal 2α-hydroxylase were observed between wild-type and *STAT5b–/–* males or females. Plasma levels of IGF-1, which is rapidly activated in the liver upon GH stimulation, were reduced in *STAT5b–/–* mice of both sexes.

As in *STAT5a–/–* females, *STAT5b–/–* females expressed milk-protein genes, yet had impaired mammary gland development and did not produce sufficient milk to feed their pups. In addition, loss of STAT5b expression affected female fertility, since they consistently aborted between day 8 and 17 of pregnancy. Pups from heterozygous females also had higher perinatal death rates. No maternal, placental, or fetal defects could be detected. Serum progesterone levels in wild-type females remain high until parturition, but decrease precipitously by day 12 of pregnancy in *STAT5b–/–* females. Pregnancies could be maintained upon the subcutaneous delivery of progesterone, which offsets this decrease.

Loss of STAT5b expression resulted in significantly less adipose tissue in males and especially females. However, some mice became obese starting at about 9 weeks of age. These mice had large testicular fat pads and increased amounts of abdominal fat. GH is important in the differentiation of preadipocytes to adipocytes, and since many *STAT5b–/–* cells examined did not contain lipid, adipocyte differentiation in *STAT5b–/–* mice may be impaired. In addition, loss of STAT5b expression in these mice caused a delay in hair growth.

Therefore, STAT5b appears to be responsible for mediating the sexually dimorphic effects of GH pulses in the liver and perhaps other tissues. STAT5a cannot compensate for STAT5b in this function, either due to intrinsic differences or variations in tissue expression patterns. STAT1 and STAT3, which are activated by GH along with STAT5, also cannot compensate for the loss of STAT5b function. This is perhaps to be expected, since neither of these STAT proteins respond to GH secretion in a pulsatile fashion.

3.11.3 STAT5a/b(–/–)

The apparent lack of redundancy of STAT5a and STAT5b revealed by gene targeting of the individual genes, despite their frequent redundancy in a variety of physiological responses, prompted considerable interest in the phenotype of mice deficient in both proteins. For reasons that are not yet known, approximately one-third of *STAT5a/b–/–* mice die within 48 hours of birth (48). Surprisingly, a critical, nonredundant role for both proteins in response to a number of cytokines that activate them, such as EPO and TPO, could not be shown. Normal levels of red blood cells, haemoglobin, haematocrits, platelets, neutrophils, and monocytes were observed. Colony assays failed to reveal abnormalities in the number of CFU-E or in responses to CSF-1 or SCF.

As expected, functions associated with GH were affected in *STAT5a/b–/–* mice (48). *STAT5a/b–/–* mice were considerably smaller than wild-type or *STAT5a–/–* mice, and even smaller than *STAT5b–/–* mice. Fat pads in these mice were also dramatically reduced. Absence of both STAT5a and STAT5b in male mice resulted in a reduction in *MUP* levels that was even greater than that observed in some *STAT5b–/–* males. *MUP* levels in *STAT5a/b–/–* females were reduced to undetectable levels. *IGF-1* levels were significantly reduced in *STAT5a/b–/–* males compared to females and wild-type mice. GH-dependent differences in CYP enzymes were also observed. As in *STAT5b–/–* males, levels of the male-suppressed CYP2A4 enzyme were elevated in *STAT5a/b–/–* males compared to that observed in wild-type and *STAT5a/b–/–* females. Similar observations were made with CYP3A and CYP4A enzymes. CYP2D9 expression, which is normally higher in males than in females, was lost in *STAT5b–/–* and *STAT5a/b–/–* males.

Like *STAT5b–/–* females, *STAT5a/b–/–* females do not produce sufficient progesterone to support pregnancy (48). Unlike *STAT5b–/–* females, however, *STAT5a/b–/–* females were infertile. Abnormalities were not detected upon examination of the oviducts and ovulation can occur normally. However, ovaries of wild-type females had large corpora lutea, whereas those from *STAT5a/b–/–* females did not. Follicle granulosa cells develop into corpora lutea in the presence of prolactin, or into corpora lutea atretica in its absence. The corpus luteum is the major site of progesterone production during pregnancy. Corpora lutea atretica, on the other hand, express 20α-hydroxysteroid dehydrogenase (20αSDH), which converts progesterone to an inactive derivative. *STAT5a/b–/–* mice had much more prevalent 20αSDH-expressing corpora lutea atretica in the ovaries. Therefore, the lack of responsiveness of *STAT5a/b*-deficient mice to prolactin may result in female infertility. Since mice deficient in either STAT5a or STAT5b were fertile, the two STAT5 proteins are functionally redundant in this respect.

In situ hybridization was used to examine the expression of cyclin D2 and p27, both of which are necessary for ovarian function (48). Cyclin D2 was only expressed in the granulosa cells of developing follicles of wild-type and *STAT5a/b–/–* mice, but not where STAT5 activation occurs. In contrast, p27 is highly expressed in corpora lutea of wild-type mice, but not in ovarian structures of *STAT5a/b–/–* mice. Hence, STAT5a and STAT5b play a role in the expression of p27, but not in the expression of cyclin D2.

Ihle and co-workers further characterized the lymphoid characteristics of these mice (119). There were normal numbers of single- and double-positive T cells in the thymus and periphery. With time, there was a decrease in the total numbers of T cells in the periphery. There was an increase in CD8$^+$ compared to CD8$^+$ single-positive cells over time. The expression of CD44 also increased in time, whereas there was a decrease in CD62L (L-Selectin). The mice were devoid of NK cells and displayed an absence of cytotoxic NK activity, similar to the *IL-2Rα* (120) and *IL-2Rβ* (121) knockouts. Thus, while STAT5 is not required for normal differentiation, it may be required for peripheral T-cell functions.

Peripheral T cells (splenic or lymph node-derived) from the *STAT5a/b–/–*, but not *STAT5a-* or *STAT5b*-deficient animals failed to respond to IL-2 in the presence of CD3 cross-linking. *STAT5a/b–/–* peripheral T cells also displayed no response to ConA, PMA plus ionomycin, or CD3 plus CD28 when compared to STAT5a/b cells. B cells were observed to be normal. Expression of IL-2 R-α, IL-2 R-β, IL-2 R-γc, and Fas was normal in all mice. There was fivefold lower IL-2 expression and increased IFN-γ expression, that is attributed to CD62L$^-$/CD44$^+$ T cells in the mutant mice which produce IFN-γ.

No upregulation of cyclin A, cyclin D2, cyclin D3, cyclin E, and cdk6 is observed in lymphocytes isolated from *STAT5a/b–/–* mice, but induction of p27, cdk2, and cdk4 is normal. The authors indicate that the promoters for cyclin D2 and cyclin D3 contain STAT5 binding elements. Cdk6 is the major cyclin D2- and D3-associated kinase in T cells.

Splenomegaly was observed with age in the *STAT5a/b–/–* mice. There was a four to tenfold elevation in splenic mass in 50% of the mice by 10–12 weeks of age. Hyperplasia of the red pulp occurred with extensive extramedullary haemopoiesis. Increased numbers of granulocytes and megakaryocytes and dramatically increased erythroid elements were observed. In addition, there was a dramatic increase in Ter-119 expression and in EPO-responsive colonies in colony assays. The nature of the splenomegaly was not discussed; however, several knockout models display similar erythroid hyperplasia including the *IL-2Rβ* knockout mice (121).

An unexpected role of STAT5 in fetal erythropoiesis was recently described (47). Significant anaemia was discovered in E13.5 fetal livers. This was accompanied by a decrease in the haematocrit from 27.5 in wild-type embryos to 11 in *STAT5a/b–/–* fetal livers. The blood smear from E13.5 embryos shows fewer non-nucleated erythrocytes and more adult-type nucleated erythroblasts, observations that are consistent with anaemia. No difference in BFU-E colony formation was observed: however, CFU-E formation was reduced by fourfold in the *STAT5* gene-targeted embryos. TdT-mediated dUTP-X nick end-labelling (TUNEL) assays indicated that fetal liver cells from the knockout embryos had an apoptotic rate that was 2.5-fold greater than the wild-type cells. Withdrawal of EPO resulted in high rates of apoptosis in both wild-type and knockout fetal liver cells. However, the addition of EPO only protected the wild-type erythroblasts and not the *STAT5a/b–/–* cells. To determine whether STAT5a/b directly regulate apoptosis in the erythroid lineage, dominant-negative STAT5a and STAT5b constructs were introduced into fetal liver cells via retroviral infection. Expression of dominant-negative STAT5a (83%) or STAT5b (71%) resulted in increased apoptosis of the transduced fetal liver cells (57%), 24 h after withdrawal of EPO.

Next Socolovsky *et al.* provided evidence that Bcl-X_L is a transcriptional target of STAT5, downstream of EPO activation (47). Bcl-X_L was shown to be a cycloheximide-independent, immediate early gene in HCD-57 cells. Analysis of the *Bcl-X_L* promoter revealed two GAS sites in the human and murine *Bcl-X* genes. Furthermore, EMSA analysis identified at least one STAT5 binding site in this region. When this sequence was analysed in transient assays, EPO stimulated an increase in transcription. Inclusion of a fragment encompassing (–588 to –90) resulted in a 2.5-fold increase in EPO-dependent luciferase activity. Dominant-negative STAT5a diminished EPO-dependent activation of Bcl-X_L and Cis, but did not affect c-myc or cyclin D2 induction when expressed in HCD-57 cells. Finally, constitutively active STAT5a was introduced into HCD-57 cells and was shown to protect cells from EPO-dependent apoptosis, at least in part through the maintenance of Bcl-X_L levels. Silva *et al.* also showed Bcl-XL to be a target of EPO-dependent STAT5 activation (122).

4. The JAK–STAT pathway and development

There appears to be remarkable conservation of the JAK–STAT signalling pathway in lower organisms suggesting, not surprisingly, that the complexity observed in higher organisms has evolved from primordial precursors (Fig. 5).

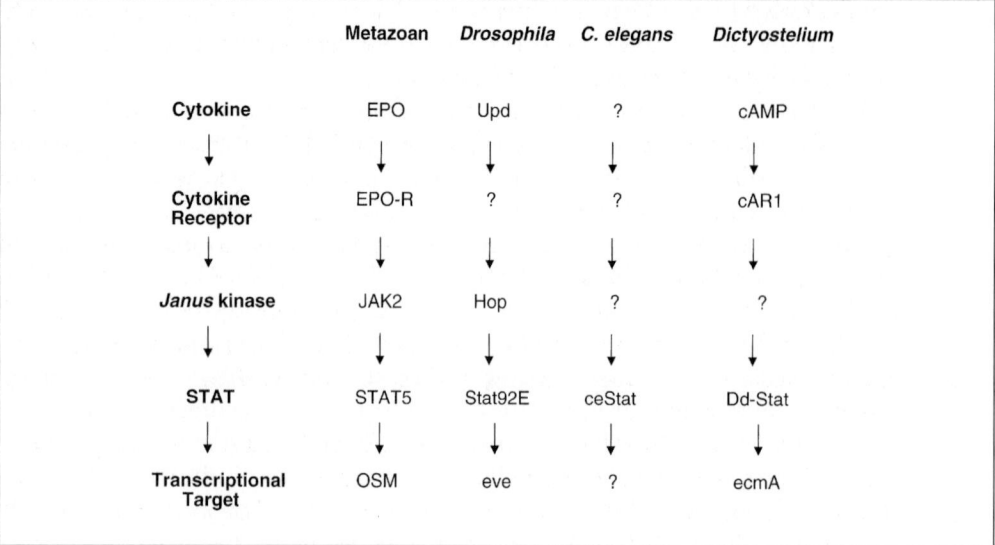

Fig. 5 Evolutionary conservation of the JAK–STAT pathway. The relationship of ligand–receptor–tyrosine kinase-transcription factor is shown for mammals, *Drosophila*, *C. elegans*, and *Dictyostelium*. In some cases, those individual components that have yet to be isolated are shown as question marks.

4.1 *Drosophila*

The JAK–STAT signalling pathway is conserved in *Drosophila* spp., although not all components in the signal transduction cascade have been elucidated. Flies mutant in an X-linked, zygotic-lethal mutation, *hopscotch* (*hop*), have segmentation defects (123). Analysis of the *hop* locus has revealed that the gene product is a *Janus* kinase which preserves the domain architecture of the mammalian enzymes, including the dual-kinase domain (124). Interestingly, overexpression of wild-type HOP mimics the *tumorless-lethal* (*tum-l*) phenotype originally described by Hanratty and colleagues (125). Indeed, *tum-l* corresponds to a single point mutation in *hop* and is a dominant gain-of-function allele (126, 127). Subsequently, a second *hop* activating point mutant, T42, was isolated within the JH2 domain of *hop* (128; see discussion below).

Drosophila also expresses a STAT protein, referred to as STAT92E (129) or marelle (130). Sequence alignments indicate that STAT92E retains all the functional domains of the mammalian STAT proteins and most closely resembles STAT5. Coexpression of HOP and STAT92E in *Drosophila* S2 cells results in tyrosine phosphorylation of STAT92E at a conserved tyrosine, Tyr-704 (129, 130). Importantly, a second site-suppressor screen of *hop^{tum-l}* resulted in the isolation of a loss-of-function *stat92e* allele, *stat^{Hi-jak}* (*STAT^{HJ}*) (131). This loss-of-function *stat92e* allele corresponds to a C to T mutation within the first intron of *stat92e*, which results in retention of intron 1. Site-selection experiments indicate that the optimal binding site of STAT92E is TTTCCCGGAAA (129), which highly resembles the mammalian GAS binding sites. STAT92E has been shown to regulate transcription of *even-skipped* (*eve*),which has two consensus STAT92E binding sites within its promoter (129, 130). Mutation of either site lowers Stripe 3 expression in *Drosophila* embryos, with the strongest effect mediated by the *eve* proximal binding site.

Upstream of *hop* and *stat92e*, it was recently reported that *unpaired* (*upd*) functions as a soluble factor capable of activating HOP (132). Loss of UPD expression results in identical defects in stripe-specific expression of pair-rule genes, as does the loss of HOP or STAT92E. *Upd* encodes a 2.2-kb message that corresponds to a basic propeptide of 46.8 kDa with a pI of 12. This secreted protein contains five sites for *N*-linked glycosylation and associates with the extracellular matrix. The association of UPD with the extracellular matrix may localize UPD expression to a localized area, similar to that shown for Wingless (WG) (133, 134). Coexpression of UPD and HOP in Clone 8 cells or co-culture of Clone 8–Upd and S2–Hop cells results in tyrosine phosphorylation of HOP.

This data suggests that UPD may be a diffusable factor which regulates HOP and STAT92E signalling. Whether UPD is a direct ligand or acts indirectly remains to be firmly established. UPD could also activate a signalling cascade that leads to the activation of HOP–STAT92E, or that UPD may regulate the expression of a ligand that stimulates the JAK–STAT signal transduction pathway in *Drosophila*. Unlike HOP and STAT92E which are coexpressed throughout the blastoderm embryo, UPD has zygotic expression. In addition, HOP and STAT92E are both maternally required, whereas UPD lacks a maternal phenotype. UPD has been shown to affect adult wing posture and has a role in oogenesis, but whether it contributes to the regulation of *Drosophila* haemopoiesis, spermatogenesis, and differentiation of adult structures remains to be determined. Although no metazoan UPD homologues have been identified to date, secondary structure analysis of UPD indicates that, like type-I cytokines, UPD contains α-helical regions. Identification of cell-surface receptors that couple to HOP and STAT92E remains elusive. In addition, it is unclear whether single JAK and STAT genes coordinate all JAK–STAT signalling in *Drosophila*.

4.2 *C. elegans*

With the sequence of the entire *C. elegans* genome reported in 1998, it is now possible to determine if the JAK–STAT pathway is also conserved in this species. *C. elegans* expresses an 80-kDa STAT homologue (ceSTAT) which contains all the domains conserved amongst STAT proteins with the exception of the amino-terminal domain. ceSTAT can be phosphorylated *in vitro* by mammalian kinases. However, expression of double-stranded RNA suggests that ceSTAT is not involved in developmental processes (Wang and Levy, unpublished data). Future experimentation in this excellent model system will undoubtedly reveal novel aspects of the JAK–STAT pathway in lower organisms.

4.3 *Dictyostelium*

Elements of the JAK–STAT pathway are also conserved in *Dictyostelium*. This unicellular slime mould undergoes a defined differentiation programme in response to nutrient deprivation (135). This process involves the aggregation of 10^5 cells into a mound and subsequent differentiation into stalk- and spore-forming cells. The chief regulator of this process is cAMP, which is released in successive waves after adenylyl cyclase activation. In turn, cAMP binds to a family of serpentine receptors which are expressed at specific developmental stages. Cyclic AMP receptor-1 (cAR1) is the major receptor expressed during early development and it also regulates postaggregative gene expression (136–140). Prespore differentiation is regulated in the majority of cells by extracellular cAMP (141) and differentiation inducing factor (DIF) (142–144), a chlorinated hexaphenone which induces the remaining cells to differentiate as prestalk cells. The prestalk cells form a tip that extends to generate a slug which falls over on to the matrix and is free to migrate.

The prestalk and stalk cells are characterized by the expression of two extracellular matrix genes, *ecmA* and *ecmB*, which are differentially regulated by DIF (144, 145). The *ecmA* gene is DIF-inducible and has a TTGA direct repeat in its promoter region. Conversely, *ecmB* has two inverted repeats in its promoter and is repressed by DIF.

A TTGA binding protein from the *ecmA* promoter was shown to be a STAT protein expressed in *Dictyostelium discoideum* (Dd-STAT) (146). This novel STAT has remarkable domain similarity with mammalian STATs in the amino terminus, DNA binding domain, and SH2 domain. Dd-STAT is expressed in all stages of *Dictyostelium* differentiation after tip aggregate formation. Induction of differentiation results in tyrosine phosphorylation, DNA binding, and nuclear localization of Dd-STAT. Interestingly, Dd-STAT can bind to both the *ecmA* and *ecmB* promoters, suggesting that it can serve as both an activator and repressor of transcription in this system. No Dd-STAT activation was observed in a *cAR1* deletion strain, indicating that Dd-STAT activation is coupled to the activation of a serpentine receptor (147). Previous studies in metazoan systems have provided evidence that serpentine receptors such as the angiotensin receptor (148) and serotonin 5-HT$_{2A}$ receptor (149) couple to STAT activation. It is unclear whether cAR1 recruits an as yet undiscovered *Dictyostelium* JAK kinase or if other *Dictyostelium* tyrosine kinases couple cAR1 engagement to Dd-STAT transcriptional activation.

5. Involvement of the JAK–STAT pathway in disease

Given the remarkable specificity of cytokine regulation and STAT gene function as revealed by gene-targeting experiments, it seemed rational that deregulation of the JAK–STAT signalling pathway in various lineages would have profound effects on proliferation, differentiation, and/or cell survival. Support for this hypothesis has been borne out of studies ranging from flies to humans.

5.1 *Hopscotch*$^{tumorous-lethal}$ (*hop*$^{tum-l}$)

Early experiments by Hanratty and colleagues had identified a mutation in *Drosophila* which gives rise to an overproliferation of the blood precursor cells, plasmatocytes, and their differentiated products, lamellocytes, which they referred to as *tumorless-*

lethal. This mutation results in the formation of melanotic masses and subsequent embryonic lethality in the larval or prepupal state. Once *hop* was identified, its over-expression was found to mimic the *tum-l* phenotype, raising the possibility that *tum-l* arose from a gain-of-function mutation in *hop.* Indeed, two groups have shown that *hoptum-l* arises from a G341E mutation within the JH4 domain of Hop (126, 127). In addition, a second gain-of-function mutation of *hop* was discovered by mutagenesis. Unlike G341E, which resides in a poorly conserved area of JAK kinases, E695K corresponds to a highly conserved glutamate residue found in the JH2 domain of all JAK kinases (128). Introduction of a similar mutation within the JH2 domain of JAK2 produced an enzyme with increased catalytic activity. However, overexpression of this mutant in Ba/F3 cells did not lead to increased proliferation.

5.2 v-src

JAK–STAT activation has been evaluated in several tumorigenesis models. v-src transformed fibroblast cell lines show constitutive tyrosine phosphorylation of JAK1 (150) and STAT3 (151). In addition, JAK1 associates with v-src in this setting. v-src transformation of the murine haemopoietic cell line 32D results in constitutive tyrosine phosphorylation of STAT1, STAT3, and STAT5 (152). STAT3, but not STAT1 nor STAT5, associates with v-src. Unlike v-src transformed fibroblasts, 32D v-src cells fail to activate JAK1, JAK2, or JAK3. The differences are probably due to the degree of v-src overexpression in the different cellular backgrounds.

5.3 v-abl

Constitutive tyrosine phosphorylation of JAK1, JAK3, STAT5, and STAT6 is observed in pre-B cell lines transformed with v-abl (153). A subsequent study demonstrated that JAK1 associates with a region within the DNA binding domain of v-abl (amino acids 858–1080), and inducible expression of a catalytically inactive version of JAK1 inhibits [^{3}H]thymidine incorporation when expressed in Ba/F3, a murine IL-3-dependent cell line (154). Ba/F3 cells expressing the v-abl Δ858–1080 mutant display a longer latency (16.5 days) as compared to Ba/F3-v-abl cells (9.5 days) in a nude mouse model. Although JAK1 catalytic activity correlates with this region of v-abl, it is unfortunate that the JAK1 binding site was not localized more precisely since the increased latency could reflect abrogation of the abl DNA binding function and not be a true reflection of the inability of v-abl to bind JAK1. A better approach would be to analyse the transformation ability of v-abl in the JAK1-deficient U4 cell line (16) or JAK1 knockout fibroblasts (28).

5.4 HTLV-1

Over time, T cells expressing HTLV-1 become factor-independent by an unknown mechanism. MT-2 cells, which are HLTV-1-transformed and IL-2-independent, show constitutive activation of JAK1, JAK3, STAT3, and STAT5 (155), all of which are relevant substrates downstream of IL-2 receptor engagement. Also, JAK3 associates with IL-2Rβ and IL-2Rγc. The murine T-cell lymphoma cell line LSTRA has a 40-fold elevation in Lck expression due to integration of the Moloney murine leukaemia virus adjacent to the *lck* gene (156). LSTRA cells display constitutive activation of JAK1, JAK2, STAT3, and STAT5. However, it is unclear whether JAK or STAT activation contributes to transformation, or is merely downstream of critical transformation events.

5.5   STAT activation in leukaemia

The molecular events for approximately 50% of human leukaemias continue to be poorly defined. Since leukaemias result from arrested differentiation in defined lineages, it is feasible to examine STAT activation from patients with myeloid and lymphoid leukaemia. In one study, constitutive activation of STAT1, STAT3, and STAT5 was observed in nuclear extracts isolated from five patients with primary acute myeloid leukaemia (AML). All patients showed constitutive activation of STAT3 and STAT5, and one patient also showed activation of STAT1 when peptide-specific STAT antibodies were used to supershift STAT proteins bound to a GAS oligonucleotide isolated from the IRF-1 gene in electrophoretic mobility shift assays (157). In contrast, another group showed that 10 out of 14 samples isolated from patients with primary AML showed constitutive activation of STAT1 and STAT3, and only one patient sample revealed deregulated STAT5 activation (158). For patients with acute lymphoblastic leukaemia (ALL), the former study demonstrated sporadic activation of STAT1 (1/3 patients) and constitutive activation of STAT5, but STAT3 activation was not observed in any of the patients (157). In the latter study, 3 out of 5 T-ALL and 12 out of 19 B-ALL patients demonstrated constitutive STAT5 tyrosine phosphorylation (158). All these studies examined STAT tyrosine phosphorylation in primary samples prior to chemotherapeutic intervention. Importantly, the molecular mechanism underlying each patient's disease may be dissimilar. As discussed above for STAT activation associated with retroviral oncogenesis, these data must be interpreted with caution since STAT activation may be a secondary or tertiary by-product of leukaemogenesis.

5.6   BCR–ABL

The causal agent in chronic myelogenous leukaemia (CML) is the BCR–ABL chromosomal translocation, which fuses the *breakpoint cluster region* (*Bcr*) gene (chromosome 22q11) with the Abelson tyrosine kinase (chromosome 9q34). The predominant BCR–ABL isoform is 210 kDa; 190-kDa and 230-kDa isoforms are also found. A coiled–coil domain in Bcr results in the oligomerization of Abl resulting in constitutive tyrosine kinase activity (159). Many intracellular substrates have been identified that bind to BCR–ABL or are activated downstream of BCR–ABL. Several studies have reported that cell lines established from CML patients, K562, or myeloid cell lines expressing p210-BCR–ABL predominantly activate STAT5 (24, 160, 162) as well as STAT1 (24, 161, 162) and STAT3 to a lesser extent (162). Interestingly, constitutive tyrosine phosphorylation of STAT6 is mediated by p190 BCR–ABL, but not by the p210 or p230 isoforms (162). Considering that p190 BCR–ABL participates in B-cell ALL, the activation of STAT6 may be physiologically relevant.

The mechanism by which activated tyrosine kinases such as BCR–ABL activate STAT proteins has not been definitively shown. One report indicated that JAK1 and JAK2 were constitutively activated in K562 cells (160); however, other studies have failed to substantiate that finding (24, 162). The overexpression of two distinct kinase-inactive forms of JAK2 has no effect on BCR–ABL- mediated STAT5 activation (162). Therefore, it is generally believed that STAT activation proceeds in the absence of endogenous JAK phosphorylation. It is presumed that BCR–ABL-mediated tyrosine kinase activity phosphorylates one or more STAT recruitment sites, which results in binding of the STAT protein and tyrosine phosphorylation by BCR–ABL. Muta-genesis studies have shown that deletions of the SH3 and SH2 domains of BCR–ABL abrogate STAT5 activation (163). Coexpression of BCR–ABL and a dominant inhib-itory form of STAT5B increases the latency of tumorigenesis in a SCID mouse model from 7 to 13 weeks (163). Expression of a dominant active form of STAT5B does not result in leukaemia when expressed in mice; however, coexpression of a BCR–ABL, SH3–SH2 deletion mutant and constitutively active STAT5B results in a latency period of 10 weeks. This data suggests that dominant active forms of STAT5 are in-sufficient to cause leukaemia alone; however, STAT5 can collaborate with one or more signals generated by a BCR–ABL construct deleted in the SH3 and SH2 domains. A definitive analysis of the role of STAT5 in BCR–ABL-mediated leukaemogenesis awaits bone-marrow transplants into a STAT5a/b-deficient background.

5.7 TEL–JAK2

The best evidence to date for the direct involvement of the JAK–STAT pathway in leukaemogenesis is the identification of the TEL–JAK2 chromosomal translocation. Three paediatric leukaemia patients have been found to express fusions of the helix–loop–helix or pointed (PNT) domain from the *ets* transcription factor TEL (chromosome 12p13) to three unique JAK2 fusion products (chromosome 9p24), all of which harbour various regions of the JAK2 JH1 kinase domain. Two fusions isolated from ALL patients result in an intact JAK2 JH1 domain and are primary transloca-tions (164, 165). The third specimen from a patient with atypical CML or chronic monomyelocytic leukaemia (CMML) was isolated from a complex t(9;15;12) which resulted in fusion of one TEL allele to the EVI1 transcription factor, whereas the other TEL allele fused to exon 12 of JAK2, resulting in a TEL–JAK2 fusion encompassing the JH2 and JH1 domains of JAK2 (164) (Fig. 6). Analysis of a Toronto leukaemia database has provided evidence for two additional patient samples which are both putative TEL–JAK3 translocations: one patient had non-Hodgkin's lymphoma with marrow involvement with t(12;19) (p13;p13) and a second patient had CML t(9;22) with a putative treatment-induced t(12;19) (p13;p13) translocation (J.A. Squire and D.L. Barber, unpublished observations). Recently, TEL translocations have been

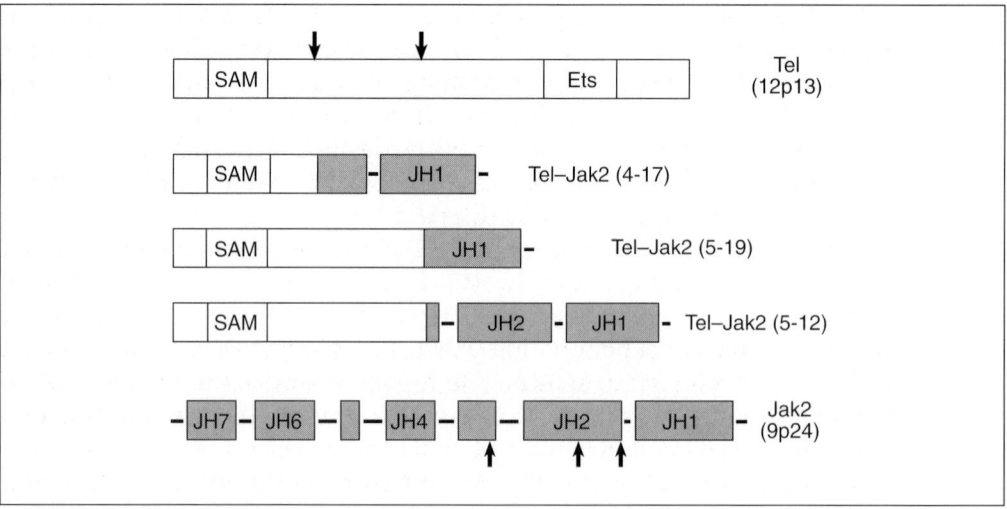

Fig. 6 TEL–JAK2 chromosomal translocations. Three patients have been identified harbouring t(9;12) (p24;p13) translocations corresponding to TEL–JAK2. Each fusion protein is shown. TEL exon 4 is fused to JAK2 exon 17 in TEL–JAK2 (4–17), which is derived from a patient with ALL. A second ALL patient was found with fusion of TEL exon 5 to JAK2 exon 19 (TEL–JAK2 (5–19)). In a patient with atypical CML, TEL–JAK2 (5–12) results in the fusion of TEL exon 5 to exon 12 of JAK2.

identified fused to PDGF-βR (166), Abl (167), Arg (168), TrkC (169, 170), as well as JAK2. Unlike TEL-AML1, which constitutes 20–30% of paediatric ALL, TEL tyrosine-kinase fusions are rather rare events.

Expression of TEL–JAK2 in cytokine-dependent haemopoietic cell lines results in factor-independence, and the resulting fusion proteins harbour constitutive tyrosine kinase activity. Constitutive tyrosine phosphorylation of STAT1 (171, 172), STAT3 (172), and STAT5 (165, 171, 172) have been observed downstream of TEL–JAK2 activation.

Importantly, an animal model has been developed to study TEL–JAK2 leukaemogenesis (171). TEL–JAK2(5–19), TEL–mJAK2 (TEL fused to murine JAK2), TELΔPNT–mJAK2 (deletion in the pointed domain), and TEL–mJAK2-KI (kinase-inactive murine JAK2) were subcloned into the retroviral expression vector pMSCV (98) and bone marrow was transduced with each construct. When lethally irradiated mice were reconstituted with the donor marrow, the animals developed a fatal mixed myeloproliferative and T-cell lymphoproliferative hyperplasia within 4–10 weeks (40 out of 40 animals). No evidence of disease was observed in animals transplanted with either TELΔPNT–mJAK2 (10 mice) or TEL–mJAK2-KI (10 mice) after 20 weeks. Examination of seven animals from the first transplant revealed an enlarged spleen and expansion of the red pulp by maturing myeloid elements and elevated levels of white blood cells. Extramedullary haemopoiesis was observed in the liver, lymph nodes, and adrenal medulla. Macrophage infiltration of the lung was also observed. In addition, enlarged lymph nodes were discovered in the abdomen and cervical regions in three out of seven animals.

Immunophenotypic analysis revealed that in transplanted animals, Gr-1$^+$ and Mac-1$^+$ granulocytes increased from 5 to 45% in the spleen. In some animals, a lymphoproliferative disorder predominated and resulted in normal numbers of Gr-1$^+$ and Mac-1$^+$ cells. Although there did not appear to be an expansion of mature B lymphocytes in the spleen, populations of single CD4$^+$, CD8$^+$ and double-positive T cells were found in the spleen. The presence of the CD4$^+$CD8$^+$ cells in the spleen is indicative of a T-cell proliferative disorder.

Proviral integration was observed in many tissues, including haemopoietic sources such as peripheral blood, bone marrow, spleen, thymus, lymph nodes, and sites of infiltration including liver and lung. A partial overlap in the integration pattern was observed in spleen and lymph nodes, which is suggestive of an oligoclonal disorder.

TEL–JAK2 represents an attractive model to study deregulated JAK2 signalling. Future studies in this area will undoubtedly involve the identification of additional signalling pathways that contribute to leukaemogenesis in addition to STAT activation as described above. It will also be important to study leukaemogenesis in an *in vivo* setting in order to identify the signalling events critical to transformation. Through these approaches it will be possible to distinguish between events shared with normal cytokine signalling and leukaemogenesis induced by other activated tyrosine kinases, including other TEL tyrosine-kinase fusions as well as BCR–ABL.

5.8   JAK3 and SCIDS

X-linked, severe combined immunodeficiency syndrome (XSCIDS) was originally described in a patient with a mutation in the IL-2 receptor γc chain (173). A subsequent report showed that a patient who exhibited a moderate phenotype with a γcL271Q mutation failed to associate with JAK3 (19). This implied that some of the XSCIDS phenotype could be attributed to mutations in JAK3. Several studies have shown that this is indeed the case. Several patients with various JAK3 mutations, which encompass frameshifts (174), deletions (175), missense (176), and nonsense mutations (174) have been described (summarized in Table 3). EBV-transformed B-cell lines were derived from the patient material. These cell lines failed to activate JAK3 in response to IL-2 or IL-4, and IL-2-dependent STAT5 activation was not observed. IL-4-dependent activation of STAT6 was reduced but consistently observed. One patient with a partial phenotype developed circulating CD4$^+$/CD45RO$^+$ T cells with an activated phenotype. As discussed above, the T42 mutant of Hop results from an E695K mutation within the JH2 domain of Hop and leads to constitutive Hop activity. A JAK3 C759R mutation also results in deregulated JAK3 tyrosine phosphorylation that is unaltered by serum starvation or cytokine induction. The Y100C mutation was characterized further by Johnston and colleagues (177). No IL-2-dependent tyrosine phosphorylation was observed in EBV-transformed B cell lines isolated from a JAK3 Y100C SCID patient, and both JAK3 Y100C and JAK3 Y100A failed to interact with IL-2 Rγc when coexpressed in 293T cells. Interestingly, JAK3 Y100F did interact with IL-2 Rγc, suggesting that Y100 must play an important functional role in contacting the cytoplasmic tail of γc.

Table 3 Summary of *JAK3* mutations associated with SCIDs

Patient	Genetic abnormality	Protein abnormality	Reference
A.P.1	nt 1172: Insertion of G	Truncated protein (aa 408)	174
A.P.2	nt 1695: C to A transition	Nonsense mutation (aa 565)	
G.M.	nt 394: A to G transition	Missense mutation Y100C	175
C.M.	Deletion of nt 2294–2444	Truncated protein (aa 876)	175
L.E.1	nt1537: A to G transition	Missense mutation (E481G)	176
L.E.2	Deletion of exons 10–12	Deletion of aa 482–596	
L.P.1	nt1428: C to T transition	C-terminal truncation R445Stop	176
L.P.2	nt 2370: T to C transition	Missense mutation C579R	
N.K.	Deletion of nt 1861–1881	Deletion of aa 586–592	176
V.L.	Deletion of nt 1537–1541 A to G transition forming new splice acceptor site	Missense mutation E481V and nonsense mutation P517Stop	176

aa, Amino acids.

   Human SCID patients are most frequently treated by allogeneic bone-marrow transplantation. Because of the low frequency of sibling donors, transplants are frequently performed with mismatched parental-donor cells. While the success of donor T-cell reconstitution is high, there is a coincident low rate of B-cell engraftment. This results in a requirement for intravenous gammaglobulin therapy for persistent hypogammaglobulinaemia. A recent study suggested that JAK3 represents an excellent candidate for gene therapy (178). Murine JAK3 was reconstituted into the JAK3−/− mouse and it was demonstrated to correct for murine SCID. A retroviral myeloproliferative sarcoma virus (MPSV)-based vector was selected for this study. Transduction of JAK3−/− bone-marrow cells demonstrated expression of JAK3 transcripts and protein. Peripheral B and T cells were analysed from the transplanted animals and B200$^+$IgM$^+$ T cells were observed. Some increase in CD4$^+$ and CD8$^+$ T cells was observed; however, the absolute numbers were not reported in this work. As mentioned above, JAK3−/− T cells are distinguished by an activated, anergic status with elevated CD44 and decreased CD62L expression. The JAK3 BMT expressed levels of CD44 and CD62L corresponding to the wild-type mouse. Of seven lines of JAK3 BMT T lymphocytes, four displayed normal levels of proliferation in response to ConA plus IL-2 stimulation, while three other lines showed lower but significant levels of [^{3}H]thymidine incorporation. Circulating IgA and IgG levels were found to be very low in JAK3−/− mice. Although reconstitution of JAK3 resulted in partial restoration of IgA and IgG, considerable fluctuation was observed.

   JAK3 bone-marrow transplant (BMT) mice were challenged with the influenza A virus (179). Wild-type bone marrow (BM) or JAK−/− BM was transduced with JAK3 as described above. The response of wild-type BMT and JAK3 BMT mice was tested by exposure to influenza virus 2 to 5 months after BMT. In two experiments, all the JAK3−/− mice (20) died by 20 days, whereas 14 out of 16 of the JAK3 BMT and 8 out of the 9 wild-type BMT mice survived to 11 months of age. Influenza-specific IgG was undetectable in the JAK3−/− plasma; however, JAK3 BMT animals elicited a normal virus-specific response, with the levels in the plasma remaining high until day 56. Virus-specific CD4$^+$ and CD8$^+$ T cells were found in the lymph nodes and spleens of JAK3 BMT and wild-type BMT at 7 and 13 days postinfection. Virus-specific CD8$^+$ T cells were found localized in the lungs of wild-type BMT and JAK3 BMT, whereas this population was absent in the lungs of JAK3–/– mice. Lung virus titres were determined at 7 and 13 days postexposure. At day 7, titres of the wild-type BMT and JAK3 BMT were lower than the JAK3–/– animals and by 13 days the wild-type BMT and JAK3 BMT had cleared the virus. These experiments suggest, as hypothesized, that JAK3 is an excellent candidate for gene therapy in a human setting.

5.9 Models of constitutive activation

TEL–JAK2 is a naturally occurring model of leukaemogenesis. However, constitutive activation has now been described for all components of the JAK–STAT signalling pathway. Activated cytokine receptors have been described as a product of retroviral oncogenes, as in the case of v-*mpl* (180, 181). A constitutive EPOR has been isolated by screening for factor-independence (182). Expression of EPOR R129C produces an erythroleukaemia in mice (183). Introduction of similar cysteines at the dimer interface of the thrombopoietin receptor (S368C, but not R117C, S120C, or S369C) results in constitutive activation (184). Activating mutations have also been reported in the extracellular domain of the IL-3 Rβc chain. These consist of point mutations (L356N, W358N, I374N, I374D, I374Q, I374F,V449E) (185, 186), duplication (37 amino acids) (187, 188), insertion (11 amino acids), domain truncation (188), and rearrangements (189). It is believed that the mechanism underlying the constitutive activation of IL-3Rβc involves a conformational switch in domain 4, since duplication and insertion events are localized to this domain.

The aforementioned TEL–JAK2 oncogene is the best model of constitutive JAK activation. Nevertheless, a recent study examined constitutive JAK2 activity by the construction of several Gyrase–JAK2 (Gyr–JAK2) constructs and by inducing dimerization with coumermycin (190), as was originally described for Raf1 (191). Coumermycin induced tyrosine phosphorylation of JAK2 when introduced as a GyrB–JAK2 JH2+JH1 fusion or when introduced between the amino-terminal (JH7–JH3 domains) and the carboxy-terminal (JH2–JH1) domains of JAK2. However, GyrB failed to activate JAK2 when fused at the amino terminus of full-length JAK2. Active GyrB–JAK2 constructs resulted in tyrosine phosphorylation of STAT5 but failed to elicit tyrosine phosphorylation of Shc and Shp2, two substrates downstream of GM-CSF stimulation. In addition, GyrB–JAK2 induced the transactivation of a GAS–luciferase promoter, but no activation of a c-*fos* promoter. Induction of *Cis*, a STAT5 target gene (192), and *myc* were observed. Activation of GyrB–JAK2 was shown to be dependent on the coumermycin concentration. However, coumermycin-induced JAK2 activation did not result in long-term protection of the Ba/F3–GyrB–JAK2 cells from apoptosis. This model of constitutive JAK activation would allow investigators to compare and contrast between coumermycin-induced activation and constitutive TEL–JAK2 transformation.

Constitutively active STAT5A mutants have been identified by Kitamura and colleagues using a retroviral mutagenesis strategy (193). One mutant harbours two mutations, H299R in the DNA binding domain, and S711F within the carboxy-terminal transactivation domain. This STAT5A mutant results in the conversion of Ba/F3 cells to factor-independence, constitutive tyrosine phosphorylation, DNA binding, and nuclear localization of STAT5A. In addition, expression of the STAT5A H229R, S711F mutant in Ba/F3 cells results in constitutive activation of a β-casein reporter in transient transfection experiments. STAT5A S771F with no accompanying H229 mutation results in constitutive tyrosine phosphorylation, DNA binding in EMSA experiments, and increases in transcriptional activation in reporter assays, but fails to enter the nucleus. STAT5A H299R results in transcriptional activation; however, no constitutive tyrosine phosphorylation, DNA binding, or nuclear localization was observed. Full transcriptional activation and factor-independence is only observed with the double point mutant. Recently, a second activated STAT5 mutant was isolated, containing a single point mutation (N643H) in the SH2 domain of STAT5A (Ariyoshi *et al.*, unpublished data). The mechanism by which these mutations activate STAT5 constitutively has not been addressed. Presumably in the former case, the H299R generates a higher affinity DNA binding domain and the S711F mutation binds more tightly or constitutively to an important transcriptional regulator. In the latter case, the SH2 mutation may result in a conformational change within the SH2 domain that influences the stability of the SH2 dimer.

6. Future directions

This chapter has only indirectly referenced the sheer volume of excellent biochemical approaches utilized to study JAK–STAT signalling in many unique systems. This final section discusses some of the outstanding issues.

The crystal structures of STAT1 (194), STAT3 (195), and the amino-terminal domain of STAT4 (196) have revealed many novel aspects of STAT function. For example, the DNA binding domain of STAT proteins has a similar fold as that found in p53 and NF-κB. An SH3 domain was suspected to exist in the STAT family (11). However, the crystal structure revealed that this was an extended helical domain with no elements reminiscent of SH3 domain structure. While all kinase domains have similar architecture, resolution of a *Janus* kinase structure would undoubtedly provide insight into the relationship of the pseudokinase domain to the functional JH1 domain and how the kinase domains are related to the remainder of the molecule. This may also provide clues as to the nature of mutations which give rise to constitutive activation of JAK3 (176) and Hopscotch (128).

Few studies have investigated the regulation of JAK kinases. Phosphorylation of critical regulatory loop tyrosines has been identified for JAK1 (197), JAK2 (198), and JAK3 (197, 199). All *Janus* kinases have a conserved H/K,D/E,YY motif. Phosphorylation of the amino-located tyrosine is required for catalytic activity of JAK1 and JAK2. However, phosphorylation of JAK3 occurs in the absence of tyrosines in the regulatory loop. No other phosphorylation sites have been mapped in JAK kinases despite an intensive effort by the O'Shea group (199).

Several studies have shown that STAT proteins are serine-phosphorylated after cytokine stimulation (200–209). A conserved proline-directed phosphorylation motif is found in STAT1, STAT3, and STAT4 (201). A recent report showed that p38^{hog1} mediated the phosphorylation of STAT1 and STAT3 after IL-2 and IL-12 stimulation of CD8^{+} T cells (209). Future experiments will undoubtedly determine whether p38^{hog1} activates serine phosphorylation of all STATs and determine the functional relevance of STAT serine phosphorylation.

While there is clear evidence that JAKs and other tyrosine kinases such as v-Src and BCR–ABL activate the tyrosine phosphorylation of STAT proteins, there are other critical substrates downstream of *Janus* kinase activation. For example, the JAK2 knockout mice have a phenotype similar to the EPO–/– and EPOR–/– mice, but STAT5a/b nullizygous animals display anaemia during fetal development; however, this is not manifested in detrimental adult erythropoietic defects. This indicates that other critical substrates must lie downstream of JAK2 in the erythroid pathway.

Proteins that reside in the nucleus express nuclear localization sequences that direct their entry into their correct subcellular compartment. STAT proteins are normally cytosolic, but upon tyrosine phosphorylation they undergo dimerization and are rapidly localized in the nucleus. The mechanism underlying this transport mechanism is unclear, as neither a linear nor bipartite nuclear localization motif is found in any of the STAT proteins.

As with any receptor signal transduction mechanism, there is a coordinate rise in activation of JAK kinase activity and tyrosine phosphorylation and DNA binding of the STAT proteins. However, mechanisms of densensitization are only now becoming clear. One of the mechanisms potentially involves the 'suppressor of cytokine signalling' family of proteins, discussed in Chapter 5 by Rottapel *et al*. A family of putative inhibitors referred to as protein inhibitor of activated STAT (PIAS) has been isolated by the Shuai laboratory (210, 211). One report has suggested that once STAT proteins are ubiquitinated, they undergo ubiquitin-mediated degradation (212). However, subsequent reports fail to support that initial finding, but suggest that ubiquitination, *per se*, does play an important role since proteasome inhibitors prolong *Janus* kinase activation and STAT tyrosine phosphorylation (213–215). STAT5 has been shown to be a target of protease regulation (216, 217). It is unclear if this is a common mode of regulation for all STAT proteins, and the protease that facilitates this conversion of STAT5 has not been identified.

The isolation of STAT1 and STAT2 by the Darnell laboratory was made possible through the availability of a panel of IFN-α and IFN-γ inducible genes and analysis of their distinct promoter elements. However, transcriptional targets downstream of other cytokines that may be targets of STAT regulation remain to be identified. The availability of genetically deficient cells for many of the STAT genes coupled with array technology will facilitate the isolation of critical substrates downstream of IL-2, IL-4, and IL-12, which provide functional activation of STAT5a/b, STAT6, and STAT4, respectively.

Given the high conservation of the JAK–STAT signalling pathway in lower organisms we can expect the identification of missing components involved in JAK–STAT signalling to be uncovered by genetic analysis. In mammals, examination of gene-targeted animals in the JAK–STAT signal transduction pathway has revealed that these genes are valuable models for haemopoietic development. Activated tyrosine kinases, including BCR–ABL and TEL–JAK2, activate various combinations of STAT proteins. Whether STAT activation is critical to transformation should be analysed in bone-marrow transplant models from animals deficient in the expression of the relevant STAT protein wherever possible. STAT activation may also play a role in leukaemogenesis outside of the realm of chromosomal translocations.

References

1. Wilks, A. F. (1989). Two putative protein-tyrosine kinases identified by application of the polymerase chain reaction. *Proc. Natl Acad. Sci. USA*, **86**, 1603.
2. Wilks, A. F., Harpur, A. G., Kurban, R. R., Ralph, S. J., Zurcher, G., and Ziemiecki, A. (1991). Two novel protein-tyrosine kinases, each with a second phosphotransferase-related catalytic domain, define a new class of protein kinase. *Mol. Cell. Biol.*, **11**, 2057.
3. Ihle, J. N. (1996). STATs: signal transducers and activators of transcription. *Cell*, **84**, 331.
4. O'Shea, J. J. (1997). Jaks, STATs, cytokine signal transduction, and immunoregulation: are we there yet? *Immunity*, **7**, 1.
5. Darnell, J. E., Jr. (1997). STATs and gene regulation. *Science*, **277**, 1630.
6. Pellegrini, S. and Dusanter-Fourt, I. (1997). The structure, regulation and function of the Janus kinases (JAKs). and the signal transducers and activators of transcription (STATs). *Eur. J. Biochem.*, **248**, 615.
7. Leonard, W. J. and O'Shea, J. J. (1998). Jaks and STATs: biological implications. *Annu. Rev. Immunol.*, **16**, 293.
8. Liu, K. D., Gaffen, S. L., and Goldsmith, M. A. (1998). JAK/STAT signaling by cytokine receptors. *Curr. Opin. Immunol.*, **10**, 271.
9. Darnell, J. E., Jr, Kerr, I. M. and Stark, G. R. (1994). JAK–STAT pathways and transcriptional activation in response to IFNs and other extracellular signaling proteins. *Science*, **264**, 1415.
10. Schindler, C., Fu, X. Y., Improta, T., Aebersold, R., and Darnell, J. E., Jr. (1992). Proteins of transcription factor ISGF-3: one gene encodes the 91- and 84-kDa ISGF-3 proteins that are activated by interferon alpha. *Proc. Natl Acad. Sci. USA*, **89**, 7836.
11. Fu, X.-Y. (1992). A transcription factor with SH2 and SH3 domains is directly activated by an interferon a-induced cytoplasmic protein tyrosine kinase(s). *Cell*, **70**, 323.
12. Veals, S. A., Schindler, C., Leonard, D., Fu, X. Y., Aebersold, R., Darnell, J. E., Jr, and Levy, D. E. (1992). Subunit of an alpha-interferon-responsive transcription factor is related to interferon regulatory factor and Myb families of DNA-binding proteins. *Mol. Cell. Biol.*, **12**, 3315.
13. Shuai, K., Horvath, C. M., Huang, L. H., Qureshi, S. A., Cowburn, D., and Darnell. J. E., Jr. (1994). Interferon activation of the transcription factor STAT91 involves dimerization through SH2-phosphotyrosyl peptide interactions. *Cell*, **76**, 821.
14. Harpur, A. G., Andres, A. C., Ziemiecki, A., Aston, R. R., and Wilks, A. F. (1992). JAK2, a third member of the JAK family of protein tyrosine kinases. *Oncogene*, **7**, 1347.

15. Hanks, S. K., Quinn, A. M., and Hunter, T. (1988). The protein kinase family: conserved features and deduced phylogeny of the catalytic domains. *Science*, **241**, 42.
16. Muller, M., Briscoe, J., Laxton, C., Guschin, D., Ziemiecki, A., Silvennoinen, O., Harpur, A. G., Barbieri, G., Witthuhn, B. A., Schindler, C., *et al.* (1993). The protein tyrosine kinase JAK1 complements defects in interferon-alpha/beta and -gamma signal transduction. *Nature*, **366**, 129.
17. Johnston, J. A., Bacon, C. M., Finbloom, D. S., Rees, R. C., Kaplan, D., Shibuya, K., Ortaldo, J. R., Gupta, S., Chen, Y. Q., Giri, J. D., *et al.* (1995). Tyrosine phosphorylation and activation of STAT5, STAT3, and Janus kinases by interleukins 2 and 15. *Proc. Natl Acad. Sci. USA*, **92**, 8705.
18. Witthuhn, B. A., Silvennoinen, O., Miura, O., Lai, K. S., Cwik, C., Liu, E. T., and Ihle, J. N. (1994). Involvement of the Jak3 Janus kinase in signalling by interleukin-2 and 4 in lymphoid and myeloid cells. *Nature*, **14**, 153.
19. Russell, S. M., Johnston, J. A., Noguchi, M., Kawamura, M., Bacon, C. M., Friedmann, M., Berg, M., McVicar, D. W., Witthuhn, B. A., Silvennoinen, O., *et al.* (1994). Interaction of IL-2R beta and gamma c chains with Jak1 and Jak3: implications for XSCID and XCID. *Science*, **266**, 1042.
20. Barber, D. L. and D'Andrea, A. D. (1994). Erythropoietin and interleukin-2 activate distinct JAK kinase family members. *Mol. Cell. Biol.*, **14**, 6506.
21. Miyazaki, T., Kawahara, A., Fujii, H., Nakagawa, Y., Minami, Y., Liu, Z. J., Oishi, I., Silvennoinen, O., Witthuhn, B. A., Ihle, J. N., *et al.* (1994). Functional activation of Jak1 and Jak3 by selective association with IL-2 receptor subunits. *Science*, **266**, 1045.
22. Yin, T., Tsang, M. L., and Yang, Y. C. (1994). JAK1 kinase forms complexes with interleukin-4 receptor and 4PS/insulin receptor substrate-1-like protein and is activated by interleukin-4 and interleukin-9 in T lymphocytes. *J. Biol. Chem.*, **269**, 26614.
23. Keegan, A. D., Johnston, J. A., Tortolani, P. J., McReynolds, L. J., Kinzer, C., O'Shea, J. J., and Paul, W. E. (1995). Similarities and differences in signal transduction by interleukin 4 and interleukin 13: analysis of Janus kinase activation. *Proc. Natl Acad. Sci. USA*, **92**, 7681.
24. Carlesso, N., Frank, D. A., and Griffin, J. D. (1996). Tyrosyl phosphorylation and DNA binding activity of signal transducers and activators of transcription (STAT) proteins in hematopoietic cell lines transformed by Bcr/Abl. *J. Exp. Med.*, **183**, 811.
25. Shikama, Y., Barber, D. L., D'Andrea, A. D., and Sieff, C. A. (1996). A constitutively activated chimeric cytokine receptor confers factor-independent growth in hematopoietic cell lines. *Blood*, **88**, 455.
26. Stahl, N., Boulton, T. G., Farruggella, T., Ip, N. Y., Davis, S., Witthuhn, B. A., Quelle, F. W., Silvennoinen, O., Barbieri, G, Pellegrini, S., Ihle, J. N., and Yancopoulos, G. D. (1994). Association and activation of Jak-Tyk kinases by CNTF-LIF-OSM-IL-6 beta receptor components. *Science*, **263**, 92.
27. Guschin, D., Rogers, N., Briscoe, J., Witthuhn, B., Watling, D., Horn, F., Pellegrini, S., Yasukawa, K., Heinrich, P., Stark, G. R., Ihle, J. N., and Kerr, I. M. (1995). A major role for the protein tyrosine kinase JAK1 in the JAK/STAT signal transduction pathway in response to interleukin-6. *EMBO J.*, **14**, 1421.
28. Rodig, S. J., Meraz, M. A., White, J. M., Lampe, P. A., Riley, J. K., Arthur, C. D., King, K. L., Sheehan, K. C., Yin, L., Pennica, D., Johnson, E. M., Jr, and Schreiber, R. D. (1998). Disruption of the Jak1 gene demonstrates obligatory and nonredundant roles of the Jaks in cytokine-induced biologic responses. *Cell*, **93**, 373.
29. von Freeden-Jeffry, U., Vieira, P., Lucian, L. A., McNeil, T., Burdach, S. E., and Murray, R. (1995). Lymphopenia in interleukin (IL)-7 gene-deleted mice identifies IL-7 as a nonredundant cytokine. *J. Exp. Med.*, **181**, 1519.

30. Peschon, J. J., Morrissey, P. J., Grabstein, K. H., Ramsdell, F. J., Maraskovsky, E., Gliniak, B. C., Park, L. S., Ziegler, S. F., Williams, D. E., and Ware, C. B. (1994). Early lymphocyte expansion is severely impaired in interleukin 7 receptor-deficient mice. *J. Exp. Med.*, **180**, 1955.

31. Nosaka, T. J. M., van Deursen, R. A., Tripp, W. E., Thierfelder, B. A., Witthuhn, A. P., McMickle, P. C., Doherty, G. C., Grosveld, and Ihle, J. N. (1995). Defective lymphoid development in mice lacking Jak3. *Science*, **270**, 800.

32. Thomis, D. C., Gurniak, C. B., Tivol, E., Sharpe, A. H., and Berg, L. J. (1995). Defects in B lymphocyte maturation and T lymphocyte activation in mice lacking Jak3. *Science*, **270**, 794.

33. Park, S. Y., Saijo, K., Takahashi, T., Osawa, M., Arase, H., Hirayama, N., Miyake, K., Nakauchi, H., Shirasawa, T., and Saito, T. (1995). Developmental defects of lymphoid cells in Jak3 kinase-deficient mice. *Immunity*, **3**, 771.

34. Witthuhn, B. A., Quelle, F. W., Silvennoinen, O., Yi, T., Tang, B., Miura, O., and Ihle, J. N. (1993). JAK2 associates with the erythropoietin receptor and is tyrosine phosphorylated and activated following stimulation with erythropoietin. *Cell*, **74**, 227.

35. Silvennoinen, O., Witthuhn, B. A., Quelle, F. W., Cleveland, J. L., Yi, T., and Ihle, J. N. (1993). Structure of the murine Jak2 protein-tyrosine kinase and its role in interleukin 3 signal transduction. *Proc. Natl Acad. Sci. USA*, **90**, 8429.

36. Ogata, N., Kouro, T., Yamada, A., Koike, M., Hanai, N., Ishikawa, T., and Takatsu, K. (1998). JAK2 and JAK1 constitutively associate with an interleukin-5 (IL-5) receptor alpha and beta c subunit, respectively, and are activated upon IL-5 stimulation. *Blood*, **91**, 2264.

37. Quelle, F. W., Sato, N., Witthuhn, B. A., Inhorn, R. C., Eder, M., Miyajima, A., Griffin, J. D., and Ihle, J. N. (1994). JAK2 associates with the beta c chain of the receptor for granulocyte–macrophage colony-stimulating factor, and its activation requires the membrane-proximal region. *Mol. Cell. Biol.*, **14**, 4335.

38. Argetsinger, L. S., Campbell, G. S., Yang, X., Witthuhn, B. A., Silvennoinen, O., Ihle, J. N., and Carter-Su, C. (1993). Identification of JAK2 as a growth hormone receptor-associated tyrosine kinase. *Cell*, **74**, 237.

39. Campbell, G. S., Argetsinger, L. S., Ihle, J. N., Kelly, P. A., Rillema, J. A., and Carter-Su, C. (1994). Activation of JAK2 tyrosine kinase by prolactin receptors in Nb2 cells and mouse mammary gland explants. *Proc. Natl Acad. Sci. USA*, **91**, 5232.

40. Rui, H., Kirken, R. A., and Farrar, W. L. (1994). Activation of receptor-associated tyrosine kinase JAK2 by prolactin. *J. Biol. Chem.*, **269**, 5364.

41. Watling, D., Guschin, D., Muller, M., Silvennoinen, O., Witthuhn, B. A., Quelle, F. W., Rogers, N. C., Schindler, C., Stark, G. R., Ihle, J. N., *et al.* (1993). Complementation by the protein tyrosine kinase JAK2 of a mutant cell line defective in the interferon-gamma signal transduction pathway. *Nature*, **366**, 166.

42. Neubauer, H., Cumano, A., Muller, M., Wu, H., Huffstadt, U., and Pfeffer, K. (1998). Jak2 deficiency defines an essential developmental checkpoint in definitive hematopoiesis. *Cell*, **93**, 397.

43. Lin, C.-S., Lim, S.-K., D'Agati, V., and Costantini, F. (1996). Differential effect of an erythropoietin receptor gene disruption on primitive and definitve erythropoiesis. *Genes Dev.*, **10**, 154.

44. Wu, H., Liu, X., Jaenisch, R., and Lodish, H. F. (1995). Generation of committed BFU-E and CFU-E progenitors does not require erythropoietin or the erythropoietin receptor. *Cell*, **83**, 59.

45. Kieran, M. W., Perkins, A. C., Orkin, S. H., and Zon, L. I. (1996). Thrombopoietin rescues *in*

vitro erythroid colony formation from mouse embryos lacking the erythropoietin receptor. *Proc. Natl Acad. Sci. USA*, **93**, 9126.

46. Parganas, E., Wang, D. Stravopodis, D., Topham, D. J., Marine, J. C., Teglund, S., Vanin, E. F., Bodner, S., Colamonici, O. R., van Deursen, J. M., Grosveld, G., and Ihle, J. N. (1998). Jak2 is essential for signaling through a variety of cytokine receptors. *Cell*, **93**, 385.

47. Socolovsky, M., Fallon, A. E. J., Wang, S., Brugnara, C., and Lodish, H. F. (1999). Fetal anemia and apoptosis of red cell progenitors in Stat5a–/–5b–/– mice: a direct role for Stat5 in Bcl-XL induction. *Cell*, **98**, 181.

48. Teglund, S., McKay, C., Schuetz, E., van Deursen, J. M., Stravopodis, D. Wang, D., Brown, M., Bodner, S., Grosveld, G., and Ihle, J. N. (1998). Stat5a and Stat5b proteins have essential and nonessential, or redundant, roles in cytokine responses. *Cell*, **93**, 841.

49. Boussiotis, V. A., Barber, D. L., Nakarai, T., Freeman, G. J., Gribben, J. G., Bernstein, G. M., D'Andrea, A. D., Ritz, J., and Nadler, L. M. (1994). Prevention of T cell anergy by signaling through the gamma c chain of the IL-2 receptor. *Science*, **266**, 1039.

50. Johnston, J. A., Wang, L. M., Hanson, E. P., Sun, X. J., White, M. F., Oakes, S. A., Pierce, J. H., and O'Shea, J. J. (1995). Interleukins 2, 4, 7, and 15 stimulate tyrosine phosphorylation of insulin receptor substrates 1 and 2 in T cells. Potential role of JAK kinases. *J. Biol. Chem.*, **270**, 28527.

51. Yin, T., Yang, L., and Yang, Y. C. (1995). Tyrosine phosphorylation and activation of JAK family tyrosine kinases by interleukin-9 in MO7E cells. *Blood*, **85**, 3101.

52. Cao, X., Shores, E. W., Hu-Li, J., Anver, M. R., Kelsall, B. L., Russell, S. M., Drago, J., Noguchi, M., Grinberg, A., Bloom, E. T., *et al.* (1995). Defective lymphoid development in mice lacking expression of the common cytokine receptor gamma chain. *Immunity*, **2**, 233.

53. DiSanto, J. P., Muller, W., Guy-Grand, D., Fischer, A., and Rajewsky, K. (1995). Lymphoid development in mice with a targeted deletion of the interleukin 2 receptor gamma chain. *Proc. Natl Acad. Sci. USA*, **92**, 377.

54. Thomis, D. C. and Berg, L. J. (1997). Peripheral expression of Jak3 is required to maintain T lymphocyte function. *J. Exp. Med.*, **185**, 197.

55. Thomis, D. C., Lee, W., and Berg, L. J. (1997). T cells from Jak3-deficient mice have intact TCR signaling, but increased apoptosis. *J. Immunol.*, **159**, 4708.

56. Lai, K. S., Jin, Y., Graham, D. K., Witthuhn, B. A., Ihle, J. N., and Liu, E. T. (1995). A kinase-deficient splice variant of the human JAK3 is expressed in hematopoietic and epithelial cancer cells. *J. Biol. Chem.*, **270**, 25028.

57. Velazquez, L., Fellous, M., Stark, G. R., and Pellegrini, S. (1992). A protein tyrosine kinase in the interferon alpha/beta signaling pathway. *Cell*, **70**, 313.

58. Frank, D. A., Robertson, M. J., Bonni, A., Ritz, J., and Greenberg, M. E. (1995). Interleukin 2 signaling involves the phosphorylation of Stat proteins. *Proc. Natl Acad. Sci. USA*, **92**, 7779.

59. van der Bruggen, T., Caldenhoven, E., Kanters, D., Coffer, P., Raaijmakers, J. A., Lammers, J. W., and Koenderman, L. (1995). Interleukin-5 signaling in human eosinophils involves JAK2 tyrosine kinase and Stat1 alpha. *Blood*, **85**, 1442.

60. van der Plas, D. C., Smiers, F., Pouwels, K., Hoefsloot, L. H., Lowenberg, B., and Touw, I. P. (1996). Interleukin-7 signaling in human B cell precursor acute lymphoblastic leukemia cells and murine BAF3 cells involves activation of STAT1 and STAT5 mediated via the interleukin-7 receptor alpha chain. *Leukemia*, **10**, 1317.

61. Finbloom, D. S. and Winestock, K. D. (1995). IL-10 induces the tyrosine phosphorylation of tyk2 and Jak1 and the differential assembly of STAT1 alpha and STAT3 complexes in human T cells and monocytes. *J. Immunol.*, **155**, 1079.

62. Wehinger, J., Gouilleux, F., Groner, B., Finke, J., Mertelsmann, R., and Weber-Nordt, R. M.

(1996). IL-10 induces DNA binding activity of three STAT proteins (Stat1, Stat3, and Stat5)., and their distinct combinatorial assembly in the promoters of selected genes. *FEBS Lett.*, **394**, 365.

63. Penta, K. and Sawyer, S. T. (1995). Erythropoietin induces the tyrosine phosphorylation, nuclear translocation, and DNA binding of STAT1 and STAT5 in erythroid cells. *J. Biol. Chem.*, **270**, 31282.

64. Durbin, J. E., Hackenmiller, R., Simon, M. C., and Levy, D. E. (1996). Targeted disruption of the mouse Stat1 gene results in compromised innate immunity to viral disease. *Cell*, **84**, 443.

65. Meraz, M. A., White, J. M., Sheehan, K. C., Bach, E. A., Rodig, S. J., Dighe, A. S., Kaplan, D. H., Riley, J. K., Greenlund, A. C., Campbell, D., Carver-Moore, K., DuBois, R. N., Clark, R., Aguet, M., and Schreiber, R. D. (1996). Targeted disruption of the Stat1 gene in mice reveals unexpected physiologic specificity in the JAK–STAT signaling pathway. *Cell*, **84**, 431.

66. Hwang, S. Y., Hertzog, P. J., Holland, K. A., Sumarsono, S. H., Tymms, M. H., Hamilton, J. A., Whitty, G., Bertoncello, I., and Kola, I. (1995). A null mutation in the gene encoding a type I interferon receptor component eliminates antiproliferative and antiviral responses to interferons alpha and beta and alters macrophage responses. *Proc. Natl Acad. Sci. USA*, **92**, 11284.

67. Huang, S., Hendriks, W., Althage, A., Hemmi, S., Bluethmann, H., Kamijo, R., Vilcek, J., Zinkernagel, R. M., and Aguet, M. (1993). Immune response in mice that lack the interferon-? receptor. *Science*, **259**, 1742.

68. Sahni, M., Amrosetti, D.-C., Mansukhani, A., Gertner, R., Levy, D., and Basilico, C. (1999). FGF signaling inhibits chondrocyte proliferation and regulates bone development through the STAT1 pathway. *Genes Dev.*, **13**, 1361.

69. Coffin, J. D., Florkiewicz, R. Z., Neumann, J., Mort-Hopkins, T., Dorn, G. W., Lightfoot, P., German, R., Howles, P. N., Kier, A., O'Toole, B. A., Sasse, J., Gonzalez, A. M., Baird, A., and Doetschman, T. (1995). Abnormal bone growth and selective translational regulation in basic fibroblast growth factor (FGF-2). transgenic mice. *Mol. Biol. Cell*, **6**, 1861.

70. Colvin, J. S., Bohne, B. A., Harding, G. W., McEwen, D. G., and Ornitz, D. M. (1996). Skeletal overgrowth and deafness in mice lacking fibroblast growth factor receptor 3. *Nature Genet.*, **12**, 390.

71. Deng, C. A., Wynshaw-Boris, A., Zhou, F., Kuo, A., and Leder, P. (1996). Fibroblast growth factor receptor 3 is a negative regulator of bone growth. *Cell*, **84**, 911.

72. Ghislain, J. J. and Fish, E. N. (1996). Application of genomic DNA affinity chromatography identifies multiple interferon-alpha-regulated Stat2 complexes. *J. Biol. Chem.*, **271**, 12408.

73. Li, X., Leung, S., Qureshi, S., Darnell, J. E., Jr, and Stark, G. R. (1996). Formation of STAT1–STAT2 heterodimers and their role in the activation of IRF-1 gene transcription by interferon-alpha. *J. Biol. Chem.*, **271**, 5790.

74. Zhong, Z., Wen, Z., and Darnell, J. E., Jr. (1994). Stat3: a STAT family member activated by tyrosine phosphorylation in response to epidermal growth factor and interleukin-6. *Science*, **264**, 95.

75. Nielsen, M., Svejgaard, A., Skov, S., and Odum, N. (1994). Interleukin-2 induces tyrosine phosphorylation and nuclear translocation of stat3 in human T lymphocytes. *Eur. J. Immunol.*, **24**, 3082.

76. Wang, Y., Morella, K. K., Ripperger, J., Lai, C. F., Gearing, D. P., Fey, G. H., Campos, S. P., and Baumann, H. (1995). Receptors for interleukin-3 (IL-3) and growth hormone mediate an IL-6-type transcriptional induction in the presence of JAK2 or STAT3. *Blood*, **86**, 1671.

77. Caldenhoven, E., van Dijk, T., Raaijmakers, J. A., Lammers, J. W., Koenderman, L., and De Groot, R. P. (1995). Activation of the STAT3/acute phase response factor transcription factor by interleukin-5. *J. Biol. Chem.*, **270**, 25778.

78. O'Farrell, A. M., Liu, Y., Moore, K. W., and Mui, A. L. (1998). IL-10 inhibits macrophage activation and proliferation by distinct signaling mechanisms: evidence for Stat3-dependent and -independent pathways. *EMBO J.*, **17**, 1006.

79. Tian, S. S., Lamb, P., Seidel, H. M., Stein, R. B., and Rosen, J. (1994). Rapid activation of the STAT3 transcription factor by granulocyte colony-stimulating factor. *Blood*, **84**, 1760.

80. Nicholson, S. E., Novak, U., Ziegler, S. F., and Layton, J. E. (1995). Distinct regions of the granulocyte colony-stimulating factor receptor are required for tyrosine phosphorylation of the signaling molecules JAK2, Stat3, and p42, p44MAPK. *Blood*, **86**, 3698.

81. Chakraborty, A., White, S. M., Schaefer, T. S., Ball, E. D., Dyer, K. F., and Tweardy, D. J. (1996). Granulocyte colony-stimulating factor activation of Stat3 alpha and Stat3 beta in immature normal and leukemic human myeloid cells. *Blood*, **88**, 2442.

82. Ezumi, Y., Takayama, H., and Okuma, M. (1995). Thrombopoietin, c-Mpl ligand, induces tyrosine phosphorylation of Tyk2, JAK2, and STAT3, and enhances agonists-induced aggregation in platelets *in vitro*. *FEBS Lett.*, **374**, 48.

83. Miyakawa, Y., Oda, A., Druker, B. J., Miyazaki, H., Handa, M., Ohashi, H., and Ikeda, Y. (1996). Thrombopoietin induces tyrosine phosphorylation of Stat3 and Stat5 in human blood platelets. *Blood*, **87**, 439.

84. Ruff-Jamison, S., Zhong, Z., Wen, Z., Chen, K., Darnell, J. E., Jr, and Cohen, S. (1994). Epidermal growth factor and lipopolysaccharide activate Stat3 transcription factor in mouse liver. *J. Biol. Chem.*, **269**, 21933.

85. Park, O. K., Schaefer, T. S., and Nathans, D. (1996). *In vitro* activation of Stat3 by epidermal growth factor receptor kinase. *Proc. Natl Acad. Sci. USA*, **93**, 13704.

86. Grandis, J. R., Drenning, S. D., Chakraborty, A., Zhou, M. Y., Zeng, Q., Pitt, A. S., and Tweardy, D. J. (1998). Requirement of Stat3 but not Stat1 activation for epidermal growth factor receptor-mediated cell growth *in vitro*. *J. Clin. Invest.*, **102**, 1385.

87. Takeda, K., Noguchi, K., Shi, W., Tanaka, T., Matsumoto, M., Yoshida, N., Kishimoto, T., and Akira, S. (1997). Targeted disruption of the mouse Stat3 gene leads to early embryonic lethality. *Proc. Natl Acad. Sci. USA*, **94**, 3801.

88. Niwa, H., Burdon, T., Chambers, I., and Smith, A. (1998). Self-renewal of pluripotent embryonic stem cells is mediated via activation of STAT3. *Genes Dev.*, **12**, 2048.

89. Takeda, K., Kaisho, T., Yoshida, N., Takeda, J., Kishimoto, T., and Akira, S. (1998). Stat3 activation is responsible for IL-6-dependent T cell proliferation through preventing apoptosis: generation and characterization of T cell-specific Stat3-deficient mice. *J. Immunol.*, **161**, 4652.

90. Nakajima, H., Liu, X. W., Boris-Wynshaw, A., Rosenthal, L. A., Imada, K., Finbloom, D. S., Hennighausen, L., and Leonard, W. J. (1997). An indirect effect of Stat5a in IL-2 induced proliferation: a critical role for Stat5a in IL-2 mediated IL-2 receptor alpha chain induction. *Immunity*, **7**, 691.

91. Takeda, K., Clausen, B. E., Kaisho, T., Tsujimura, T., Terada, N., Forster, I., and Akira, S. (1999). Enhanced Th1 activity and development of chronic enterocolitis in mice devoid of Stat3 in macrophages and neutrophils. *Immunity*, **10**, 39.

92. Zhong, Z., Wen, Z., and Darnell, J. E., Jr. (1994). Stat3 and Stat4: members of the family of signal transducers and activators of transcription. *Proc. Natl Acad. Sci. USA*, **91**, 4806.

93. Yamamoto, K., Quelle, F. W., Thierfelder, W. E., Kreider, B. L., Gilbert, D. J., Jenkins, N. A., Copeland, N. G., Silvennoinen, O., and Ihle, J. N. (1994). Stat4, a novel gamma interferon

activation site-binding protein expressed in early myeloid differentiation. *Mol. Cell. Biol.*, **14**, 4342.

94. Jacobson, N. G., Szabo, S. J., Weber-Nordt, R. M., Zhong, Z., Schreiber, R. D., Darnell, J. E., Jr, and Murphy, K. M. (1995). Interleukin 12 signaling in T helper type 1 (Th1) cells involves tyrosine phosphorylation of signal transducer and activator of transcription (Stat)3 and Stat4. *J. Exp. Med.*, **181**, 1755.

95. Bacon, C. M., Petricoin, E. F., 3rd, Ortaldo, J. R., Rees, R. C., Larner, A. C., Johnston, J. A., and O'Shea, J. J. (1995). Interleukin 12 induces tyrosine phosphorylation and activation of STAT4 in human lymphocytes. *Proc. Natl Acad. Sci. USA*, **92**, 7307.

96. Thierfelder, W. E., van Deursen, J. M., Yamamoto, K., Tripp, R. A., Sarawar, S. R., Carson, R. T., Sangster, M. Y., Vignali, D. A., Doherty, P. C., Grosveld, G. C., and Ihle, J. N. (1996). Requirement for Stat4 in interleukin-12-mediated responses of natural killer and T cells. *Nature*, **382**, 171.

97. Kaplan, M. H., Sun, Y. L., Hoey, T., and Grusby, M. J. (1996). Impaired IL-12 responses and enhanced development of Th2 cells in Stat4-deficient mice. *Nature*, **382**, 174.

98. Hou, J., Schindler, U., Henzel, W. J., Ho, T. C., Brasseur, M., and McKnight, S. L. (1994). An interleukin-4-induced transcription factor: IL-4 Stat. *Science*, **265**, 1701.

99. Quelle, F. W., Shimoda, K., Thierfelder, W., Fischer, C., Kim, A., Ruben, S. M., Cleveland, J. L., Pierce, J. H., Keegan, A. D., Nelms, K., *et al.* (1995). Cloning of murine Stat6 and human Stat6, Stat proteins that are tyrosine phosphorylated in responses to IL-4 and IL-3 but are not required for mitogenesis. *Mol. Cell. Biol.*, **15**, 3336.

100. Takeda, K., Tanaka, T., Shi, W., Matsumoto, M., Minami, M., Kashiwamura, S., Nakanishi, K., Yoshida, N., Kishimoto T., and Akira, S. (1996). Essential role of Stat6 in IL-4 signalling. *Nature*, **380**, 627.

101. Shimoda, K., van Deursen, J., Sangster, M.Y, Sarawar, S. R., Carson, R. T., Tripp, R. A., Chu, C., Quelle, F. W., Nosaka, T., Vignali, D. A., Doherty, P. C., Grosveld, G., Paul, W. E., and Ihle, J. N. (1996). Lack of IL-4-induced Th2 response and IgE class switching in mice with disrupted Stat6 gene. *Nature*, **380**, 630.

102. Kaplan, M. H., Schindler, U., Smiley, S. T., and Grusby, M. J. (1996). Stat6 is required for mediating responses to IL-4 and for development of Th2 cells. *Immunity*, **4**, 313.

103. Kaplan, M. H., Wurster, A. L., and Grusby, M. J. (1998). A signal transducer and activator of transcription (Stat)4-independent pathway for the development of T helper type 1 cells. *J. Exp. Med.*, **188**, 1191.

104. Wakao, H., Gouilleux, F., and Groner, B. (1994). Mammary gland factor (MGF) is a novel member of the cytokine regulated transcription factor gene family and confers the prolactin response. *EMBO J.*, **13**, 2182.

105. Mui, A. L., Wakao, H., O'Farrell, A. M., Harada, N., and Miyajima, A. (1995). Interleukin-3, granulocyte-macrophage colony stimulating factor and interleukin-5 transduce signals through two STAT5 homologs. *EMBO J.*, **14**, 1166.

106. Azam, M., Erdjument-Bromage, H., Kreider, B. L., Xia, M., Quelle, F., Basu, R., Saris, C., Tempst, P., Ihle, J. N., and Schindler, C. (1995). Interleukin-3 signals through multiple isoforms of Stat5. *EMBO J.*, **14**, 1402.

107. Lin, J. X., Mietz, J., Modi, W. S., John, S., and Leonard, W. J. (1996). Cloning of human Stat5B. Reconstitution of interleukin-2-induced Stat5A and Stat5B DNA binding activity in COS-7 cells. *J. Biol. Chem.*, **271**, 10738.

108. Demoulin, J. B., Uyttenhove, C., Van Roost, E., DeLestre, B., Donckers, D., Van Snick, J., and Renauld, J. C. (1996). A single tyrosine of the interleukin-9 (IL-9) receptor is required for STAT activation, antiapoptotic activity, and growth regulation by IL-9. *Mol. Cell. Biol.*, **16**, 4710.

109. Lin, J. X., Migone, T. S., Tsang, M., Friedmann, M., Weatherbee, J. A., Zhou, L., Yamauchi, A., Bloom, E. T., Mietz, J., John, S., and Leonard, W. J. (1995). The role of shared receptor motifs and common Stat proteins in the generation of cytokine pleiotropy and redundancy by IL-2, IL-4, IL-7, IL-13, and IL-15. *Immunity*, **2**, 331.

110. Nicholson, S. E., Starr, R., Novak, U., Hilton, D. J., and Layton, J. E. (1996). Tyrosine residues in the granulocyte colony-stimulating factor (G-CSF) receptor mediate G-CSF-induced differentiation of murine myeloid leukemic (M1). cells. *J. Biol. Chem.*, **271**, 26947.

111. Damen, J. E., Wakao, H., Miyajima, A., Krosl, J., Humphries, R. K., Cutler, R. L., and Krystal, G. (1995). Tyrosine 343 in the erythropoietin receptor positively regulates erythropoietin-induced cell proliferation and Stat5 activation. *EMBO J.*, **14**, 5557.

112. Quelle, F. W., Wang, D., Nosaka, T., Thierfelder, W. E., Stravopodis, D., Weinstein, Y., and Ihle, J. N. (1996). Erythropoietin induces activation of Stat5 through association with specific tyrosines on the receptor that are not required for a mitogenic response. *Mol. Cell. Biol.*, **16**, 1622.

113. Klingmuller, U., Bergelson, S., Hsiao, J. G., and Lodish, H. F. (1996). Multiple tyrosine residues in the cytosolic domain of the erythropoietin receptor promote activation of STAT5. *Proc. Natl Acad. Sci. USA*, **93**, 8324.

114. Gobert, S., Chretien, S., Gouilleux, F., Muller, O., Pallard, C., Dusanter-Fourt, I., Groner, B., Lacombe, C., Gisselbrecht, S., and Mayeux, P. (1996). Identification of tyrosine residues within the intracellular domain of the erythropoietin receptor crucial for STAT5 activation. *EMBO J.*, **15**, 2434.

115. Gouilleux, F., Pallard, C., Dusanter-Fourt, I., Wakao, H., Haldosen, L. A., Norstedt, G., Levy, D., and Groner, B. (1995). Prolactin, growth hormone, erythropoietin and granulocyte-macrophage colony stimulating factor induce MGF-Stat5 DNA binding activity. *EMBO J.*, **14**, 2005.

116. Smit, L. S., Vanderkuur, J. A., Stimage, A., Han, Y., Luo, G., Yu-Lee, L.Y, Schwartz, J., and Carter-Su, C. (1997). Growth hormone-induced tyrosyl phosphorylation and deoxyribonucleic acid binding activity of Stat5A and Stat5B. *Endocrinology*, **138**, 3426.

117. Liu, X., Robinson, G. W., Wagner, K. U., Garrett, L., Wynshaw-Boris, A., and Hennighausen, L. (1997). Stat5a is mandatory for adult mammary gland development and lactogenesis. *Genes Dev.*, **11**, 179.

118. Udy, G. B., Towers, R. P., Snell, R. G., Wilkins, R. J., Park, S. H., Ram, P. A., Waxman, D. J., and Davey, H. W. (1997). Requirement of STAT5b for sexual dimorphism of body growth rates and liver gene expression. *Proc. Natl Acad. Sci. USA*, **94**, 7239.

119. Moriggl, R., Topham, D. J., Teglund, S., Sexl, V., McKay, C., Wang, D., Hoffmeyer, A., van Deursen, J., Sangster, M. Y., Bunting, K. D., Grosveld, G. C., and Ihle, J. N. (1999). Stat5 is required for IL-2-induced cell cycle progression of peripheral T cells. *Immunity*, **10**, 249.

120. Willerford, D. M., Chen, J., Ferry, J. A., Davidson, L., Ma, A., and Alt, F. W. (1995). Interleukin-2 receptor ? chain regulates the size and content of the peripheral lymphoid compartment. *Immunity*, **3**, 521.

121. Suzuki, H., Duncan, G. S., Takimoto, H., and Mak, T. W. (1997). Abnormal development of intestinal intraepithelial lymphocytes and peripheral natural killer cells in mice lacking the IL-2 receptor beta chain. *J. Exp. Med.*, **185**, 499.

122. Silva, M., Benito, A., Sanz, C., Prosper, F., Ekhterae, D., Nunez, G., and Fernandez-Luna, J. L. (1999). Erythropoietin can induce the expression of Bcl-xL through Stat5 in erythropoietin-dependent progenitor cell lines. *J. Biol. Chem.*, **274**, 22165.

123. Perrimon, N. and Mahowald, A. P. (1986). l(1)hopscotch, a larval–pupal zygotic lethal with a specific maternal effect on segmentation in Drosophila. *Dev. Biol.*, **118**, 28.

124. Binari, R. and Perrimon, N. (1994). Stripe-specific regulation of pair-rule genes by hopscotch, a putative Jak family tyrosine kinase in Drosophila. *Genes Dev.*, **8**, 300.

125. Hanratty, W. P. and Ryerse, J. S. (1981). A genetic melanotic neoplasm of *Drosophila melanogaster*. *Dev. Biol.*, **83**, 238.

126. Luo, H., Hanratty, W. P., and Dearolf, C. R. (1995). An amino acid substitution in the Drosophila hopTum-l Jak kinase causes leukemia-like hematopoietic defects. *EMBO J.*, **14**, 1412.

127. Harrison, D. A., Binari, R., Nahreini, T. S., Gilman, M., and Perrimon, N. (1995). Activation of a Drosophila Janus kinase (JAK) causes hematopoietic neoplasia and developmental defects. *EMBO J.*, **14**, 2857.

128. Luo, H., Rose, P., Barber, D., Hanratty, W. P., Lee, S., Roberts, T. M., D'Andrea, A. D., and Dearolf, C. R. (1997). Mutation in the Jak kinase JH2 domain hyperactivates Drosophila and mammalian JAK–STAT pathways. *Mol. Cell. Biol.*, **17**, 1562.

129. Yan, R., Small, S., Desplan, C., Dearolf, C.R, and Darnell, J. E., Jr. (1996). Identification of a Stat gene that functions in Drosophila development. *Cell*, **84**, 421.

130. Hou, X. S., Melnick, M. B., and Perrimon, N. (1996). Marelle acts downstream of the Drosophila HOP/JAK kinase and encodes a protein similar to the mammalian STATs. *Cell*, **84**, 411.

131. Yan, R., Luo, H., Darnell, J. E., Jr, and Dearolf, C. R. (1996). A JAK–STAT pathway regulates wing vein formation in Drosophila. *Proc. Natl Acad. Sci. USA*, **93**, 5842.

132. Harrison, D. A., McCoon, P. E., Binari, R., Gilman, M., and Perrimon, N. (1998). Drosophila unpaired encodes a secreted protein that activates the JAK signaling pathway. *Genes Dev.*, **12**, 3252.

133. Binari, R. C., Staveley, B. E., Johnson, W. A., Godavarti, R., Sasisekharan, R., and Manoukian, A. S. (1997). Genetic evidence that heparin-like glycosaminoglycans are involved in wingless signaling. *Development*, **124**, 2623.

134. Hacker, U., Lin, X., and Perrimon, N. (1997). The Drosophila sugarless gene modulates Wingless signaling and encodes an enzyme involved in polysaccharide biosynthesis. *Development*, **124**, 3565.

135. Chen, M. Y., Insall, R. H., and Devreotes, P. N. (1996). Signaling through chemoattractant receptors in *Dictyostelium*. *Trends Genet.*, **12**, 52.

136. Mehdy, M. C. and Firtel, R. A. (1985). A secreted factor and cyclic AMP jointly regulate cell-type-specific gene expression in *Dictyostelium*. *Mol. Cell. Biol.*, **5**, 705.

137. Sun, T. J. and Devreotes, P. N. (1991). Gene targeting of the aggregation stage cAMP receptor cAR1 in *Dictyostelium*. *Genes Dev.*, **5**, 572.

138. Insall, R. H., Soede, R. D. M., Schaap, P., and Devreotes, P. N. (1994). Two cAMP receptors activate common signaling pathways in *Dictyostelium*. *Mol. Biol. Cell*, **5**, 703.

139. Soede, R. D. M., Insall, R. H., Devreotes, P. N., and Schaap, P. (1994). Extracellular cAMP can restore development in Dictyostelium cells lacking one, but not two subtypes of early cAMP receptors (cARs). Evidence for involvement of cAR1 in aggregative gene expression. *Development*, **120**, 1997.

140. Schnitzler, G. R., Briscoe, C., Brown, J. M., and Firtel, R. A. (1995). Serpentine cAMP receptors may act through a G protein-independent pathway in induce postaggregative development in *Dictyostelium*. *Cell*, **76**, 821.

141. Schaap, P. and vanDriel, R. (1985). Induction of post-aggregative differentiation in *Dictyostelium discoideum* by cAMP. Evidence for the involvement of the cell surface cAMP receptor. *Exp. Cell Res.*, **159**, 388.

142. Kay, R. R. and Jermyn, K. A. (1983). A possible morphogen controlling differentiation in *Dictyostelium*. *Nature*, **303**, 242.

143. Morris, H. R., Taylor, G. W., Masento, M. S., Jermyn, K. A., and Kay, R. R. (1987). Chemical structure of the morphogen differentiation inducing factor from *Dictyostelium discoideum*. *Nature*, **328**, 811.

144. Williams, J. G., Ceccarelli, A., McRobbie, S., Mahbubani, H., Kay, R. R., Early, A., Berks, M., and Jermyn, K. A. (1987). Direct induction of Dictyostelium prestalk gene expression by DIF provides evidence that DIF is a morphogen. *Cell*, **49**, 185.

145. Jermyn, K. A., Berks, M., Kay, R. R., and Williams, J. G. (1987). Two distinct classes of prestalk-enriched mRNA sequences in *Dictyostelium discoideum*. *Development*, **100**, 745.

146. Kawata, T., Shevchenko, A., Fukuzawa, M., Jermyn, K. A., Totty, N. F., Zhukovskaya, N. V., Sterling, A. E., Mann, M., and Williams, J. G. (1997). SH2 signaling in a lower eukaryote: a STAT protein that regulates stalk cell differentiation in *Dictyostelium*. *Cell*, **89**, 909.

147. Araki, T., Gamper, M., Early, A., Fukuzawa, M., Abe, T., Kawata, T., Kim, E., Firtel, R. A., and Williams, J. G. (1998). Developmentally and spatially regulated activation of a Dictyostelium STAT protein by a serpentine receptor. *EMBO J.*, **17**, 4018.

148. Marrero, M. B., Schieffer, B., Paxton, W. G., Heerdt, L., Berk, B. C., Delafontaine, P., and Bernstein, K. E. (1995). Direct stimulation of Jak/Stat pathway by the angiotensin II AT1 receptor. *Nature*, **375**, 247.

149. Guillet-Deniau, I., Burnoi, A. F., and Girard, J. (1997). Identification and localisation of a skeletal muscle serotonin 5-HT2A receptor coupled to the Jak/Stat pathway. *J. Biol. Chem.*, **272**, 14825.

150. Campbell, G. S., Yu, C. L. R., Jove, R., and Carter-Su, C. (1997). Constitutive activation of JAK1 in Src-transformed cells. *J. Biol. Chem.*, **272**, 2591.

151. Yu, C. L., Meyer, D. J., Campbell, G. S., Larner, A. C., Carter-Su, C., Schwartz, J., and Jove, R. (1995). Enhanced DNA-binding activity of a Stat3-related protein in cells transformed by the Src oncoprotein. *Science*, **269**, 81.

152. Chaturvedi, P., Sharma, S., and Reddy, E. P. (1997). Abrogation of interleukin-3 dependence of myeloid cells by the v-src oncogene requires SH2 and SH3 domains which specify activation of STATs. *Mol. Cell. Biol.*, **17**, 3295.

153. Danial, N. N., Pernis, A., and Rothman, P. B. (1995). JAK–STAT signaling induced by the v-abl oncogene. *Science*, **269**, 1875.

154. Danial, N. N., Losman, J. A., Lu, T., Yip, N., Krishnan, K., Krolewski, J., Goff, S. P., Wang, J. Y., and Rothman, P. B. (1998). Direct interaction of Jak1 and v-Abl is required for v-Abl-induced activation of STATs and proliferation. *Mol. Cell. Biol.*, **18**, 6795.

155. Migone, T. S., Lin, J. X., Cereseto, A., Mulloy, J. C., O'Shea, J. J., Franchini, G., and Leonard, W. J. (1995). Constitutively activated JAK–STAT pathway in T cells transformed with HTLV-I. *Science*, **269**, 79.

156. Yu, C. L., Jove, R., and Burakoff, S. J. (1997). Constitutive activation of the Janus kinase-STAT pathway in T lymphoma overexpressing the Lck protein tyrosine kinase. *J. Immunol.*, **159**, 5206.

157. Gouilleux-Gruart, V., Gouilleux, F., Desaint, C., Claisse, J. F., Capiod, J. C., Delobel, J., Weber-Nordt, R., Dusanter-Fourt, I., Dreyfus, F., Groner, B., and Prin, L. (1996). STAT-related transcription factors are constitutively activated in peripheral blood cells from acute leukemia patients. *Blood*, **87**, 1692.

158. Weber-Nordt, R. M., Egen, C., Wehinger, J., Ludwig, W., Gouilleux-Gruart, V., Mertelsmann, R., and Finke, J. (1996). Constitutive activation of STAT proteins in primary

lymphoid and myeloid leukemia cells and in Epstein–Barr virus (EBV)-related lymphoma cell lines. *Blood*, **88**, 809.

159. Tauchi, T., Miyazawa, K., Feng, G.-S., Broxmeyer, H. E., and Toyama, K. (1997). A coiled-coil tetramerization domain of BCR-ABL is essential for the interactions of SH2-containing signal transduction molecules. *J. Biol. Chem.*, **272**, 1389.

160. Shuai, K., Halpern, J., ten Hoeve, J., Rao, X., and Sawyers, C. L. (1996). Constitutive activation of STAT5 by the BCR-ABL oncogene in chronic myelogenous leukemia. *Oncogene*, **13**, 247.

161. Frank, D. A. and Varticovski, L. (1996). BCR/abl leads to the constitutive activation of Stat proteins, and shares an epitope with tyrosine phosphorylated Stats. *Leukemia*, **10**, 1724.

162. Ilaria, R. L., Jr and Van Etten, R. A. (1996). P210 and P190(BCR/ABL) induce the tyrosine phosphorylation and DNA binding activity of multiple specific STAT family members. *J. Biol. Chem.*, **271**, 31704.

163. Nieborowska-Skorska, M., Wasik, M. A., Slupianek, A., Salomoni, P., Kitamura, T., Calabretta, B., and Skorski, T. (1999). Signal transducer and activator of transcription (STAT)5 activation by BCR/Abl is dependent on intact Src homology (SH)3 and SH2 domains of BCR/ABL and is required for leukemogenesis. *J. Exp. Med.*, **189**, 1229.

164. Peeters, P., Raynaud, S. D., Cools, J., Wlodarska, I., Grosgeorge, J., Philip, P., Monpoux, F., Van Rompaey, L., Baens, M., Van den Berghe, H., and Marynen, P. (1997). Fusion of TEL, the ETS-variant gene 6 (ETV6), to the receptor-associated kinase JAK2 as a result of t(9;12) in a lymphoid and t(9;15;12) in a myeloid leukemia. *Blood*, **90**, 2535.

165. Lacronique, V., Boureux, A., Valle, V. D., Poirel, H., Quang, C. T., Mauchauffe, M., Berthou, C., Lessard, M., Berger, R., Ghysdael, J., and Bernard, O. A. (1997). A TEL-JAK2 fusion protein with constitutive kinase activity in human leukemia. *Science*, **278**, 1309.

166. Golub, T. R., Barker, G. F., Lovett, M., and Gilliland, D. G. (1994). Fusion of PDGF receptor beta to a novel ets-like gene, tel, in chronic myelomonocytic leukemia with t(5;12) chromosomal translocation. *Cell*, **77**, 307.

167. Golub, T. R., Goga, A., Barker, G. F., Afar, D. E., McLaughlin, J., Bohlander, S. K., Rowley, J. D., Witte, O. N., and Gilliland, D. G. (1996). Oligomerization of the ABL tyrosine kinase by the Ets protein TEL in human leukemia. *Mol. Cell. Biol.*, **16**, 4107.

168. Sato, Y., Iijima, Y., Ito, T., Oikawa, T., Eguchi, M., Eguchi-Ishimae, M., Kamada, N., Kishi, K., Asano, S., and Sakaki, Y. (1998). A new partner gene of the ETV/TEL, ARG (ABL related gene, or ABL2) cloned in a AML-M3 cell line with t(1;12) (q25;p13). *Blood*, **92**, 592a.

169. Knezevich, S. R., McFadden, D. E., Tao, W., Lim, J. F., and Sorensen, P. H. (1998). A novel ETV6-NTRK3 gene fusion in congenital fibrosarcoma. *Nature Genet.*, **18**, 184.

170. Eguchi, M., Eguchi-Ishimae, M., Tojo, A., Morishita, K., Suzuki, K., Sato, Y., Kudoh, S., Tanaka, K., Setoyama, M., Nagamura, F., Asano, S., and Kamada, N. (1999). Fusion of ETV6 to neurotrophin-3 receptor TRKC in acute myeloid leukemia with t(12;15) (p13;q25). *Blood*, **93**, 1355.

171. Schwaller, J., Frantsve, J., Aster, J., Williams, I. R., Tomasson, M. H., Ross, T. S., Peeters, P., Van Rompaey, L., Van Etten, R. A., Ilaria, R., Jr, Marynen, P., and Gilliland, D. G. (1998). Transformation of hematopoietic cell lines to growth-factor independence and induction of a fatal myelo- and lymphoproliferative disease in mice by retrovirally transduced TEL/JAK2 fusion genes. *EMBO J.*, **17**, 5321.

172. Ho, J. M.-Y., Beattie, B. K., Squire, J. A., Frank, D. A., and Barber, D. L. (1999). Fusion of the ets transcription factor TEL to Jak2 results in constitutive JAK–STAT signaling. *Blood*, **93**, 4354.

173. Noguchi, M., Yi, H., Rosenblatt, H. M., Filipovich, A. H., Adelstein, S., Modi, W. S., McBride, O. W., and Leonard, W. J. (1993). Interleukin-2 receptor gamma chain mutation results in X-linked severe combined immunodeficiency in humans. *Cell*, **73**, 147.

174. Russell, S. M., Tayebi, N., Nakajima, H., Riedy, M. C., Roberts, J. L., Aman, M. J., Migone, T. S., Noguchi, M., Markert, M. L., Buckley, R. H., O'Shea, J. J., and Leonard, W. J. (1995). Mutation of Jak3 in a patient with SCID: essential role of Jak3 in lymphoid development. *Science*, **270**, 797.

175. Macchi, P., Villa, A., Gillani, S., Sacco, M. G., Frattini, A., Porta, F., Ugazio, A. G., Johnston, J. A., Candotti, F., and O'Shea, J. J. (1995). Mutations of Jak-3 gene in patients with autosomal severe combined immune deficiency (SCID). *Nature*, **377**, 65.

176. Candotti, F., Oakes, S. A., Johnston, J. A., Giliani, S., Schumacher, R. F., Mella, P., Fiorini, M., Ugazio, A. G., Badolato R., Notarangelo, L. D., Bozzi, F., Macchi, P., Strina, D., Vezzoni, P., Blaese, R. M., O'Shea, J. J., and Villa, A. (1997). Structural and functional basis for JAK3-deficient severe combined immunodeficiency. *Blood*, **90**, 3996.

177. Cacalano, N. A., Migone, T. S., Bazan, F., Hanson, E. P., Chen, M., Candotti, F., O'Shea, J. J., and Johnston, J. A. (1999). Autosomal SCID caused by a point mutation in the N-terminus of Jak3: mapping of the Jak3-receptor interaction domain. *EMBO J.*, **18**, 1549.

178. Bunting, K. D., Sangster, M. Y., Ihle, J. N., and Sorrentino, B. P. (1998). Restoration of lymphocyte function in Janus kinase 3-deficient mice by retroviral-mediated gene transfer. *Nature Med.*, **4**, 58.

179. Bunting, K. D., Flynn, K. J., Riberdy, J. M., Doherty, P. C., and Sorrentino, B. P. (1999). Virus-specific immunity after gene therapy in a murine model of severe combined immunodeficiency. *Proc. Natl Acad. Sci. USA*, **96**, 232.

180. Souryi, M., Vigon, I., Penciolelli, J.-F., Heard, J.-M., Tambourin, P., and Wendling, F. (1990). A putative truncated cytokine receptor gene transduced by the myeloproliferative leukemia virus immortalizes hematopoietic progenitors. *Cell*, **63**, 1137.

181. Vigon, I., Mornon, J.-P., Cocault, L., Mitjavila, M.-T., Tambourin, P., Gisselbrecht, S., and Souryi, M. (1992). Molecular cloning and characterization of MPL, the human homolog of the v-mpl oncogene: identification of a member of the hematopoietic growth factor receptor superfamily. *Proc. Natl Acad. Sci. USA*, **89**, 5640.

182. Yoshimura, A., Longmore, G., and Lodish, H. F. (1990). Point mutation in the exoplasmic domain of the erythropoietin receptor resulting in hormone-independent activation and tumorigenicity. *Nature*, **348**, 647.

183. Longmore, G. D. and Lodish, H. F. (1991). An activating mutation in the murine erythropoietin receptor induces erythroleukemia in mice: a cytokine receptor superfamily oncogene. *Cell*, **67**, 1089.

184. Alexander, W. S., Metcalf, D., and Dunn, A. R. (1995). Point mutations within a dimer interface homology domain of c-mpl induce constitutive receptor activity and tumorigenicity. *EMBO J.*, **14**, 5569.

185. Jenkins, B. J., D'Andrea, R. J., and Gonda, T. J. (1995). Activating point mutations in the common b subunit of the human GM-CSF, IL-3 and IL-5 a receptors suggest the involvement of the b subunit dimerisation and cell type-specific molecules in signalling. *EMBO J.*, **14**, 4276.

186. Jenkins, B. J., Bagley, C. J., Woodcock, J., Lopez, A. F., and Gonda, T. J. (1996). Interacting residues in the extracellular region of the common b subunit of the human GM-CSF, IL-3 and IL-5 receptors involved in constitutive activation. *J. Biol. Chem.*, **271**, 29707.

187. D'Andrea, R., Rayner, J., Moretti, P., Lopez, A., Goodall, G. J., Gonda, T. J., and Vadas, M. (1994). A mutation of the common receptor subunit for interleukin-3 (IL-3), granulocyte–

macrophage colony-stimulating factor, and IL-5 that leads to ligand independence and tumorigenicity. *Blood*, **83**, 2802.

188. D'Andrea, R. J., Barry, S. C., Moretti, P. A. B., Jones, K., Ellis, S., Vadas, M. A., and Goodall, G. J. (1996). Extracellular truncations of hbc, the common signaling subunit for interleukin-3 (IL-3), granulocyte–macrophage colony-stimulating factor (GM-CSF) and IL-5, lead to ligand-independent activation. *Blood*, **87**, 2641.

189. Hannemann, J., Hara, T., Kawai, M., Miyajima, A., Ostertag, W., and Stocking, C. (1995). Sequential mutations in the interleukin-3 (IL3)/granulocyte-macrophage colony-stimulating factor/IL5 receptor beta-subunit genes are necessary for the complete conversion to growth autonomy mediated by a truncated beta C subunit. *Mol. Cell. Biol.*, **15**, 2402.

190. Mohi, M. G., Arai, K., and Watanabe, S. (1998). Activation and functional analysis of Janus kinase 2 in BA/F3 cells using the coumermycin/gyrase B system. *Mol. Biol. Cell*, **9**, 3299.

191. Farrar, M. A., Alberola-Ila, J., and Perlmutter, R. M. (1996). Activation of the raf-1 kinase cascade by coumermycin-induced dimerization. *Nature*, **383**, 178.

192. Matsumoto, A., Masuhara, M., Mitsui, K., Yokouchi, M., Ohtsubo, M., Misawa, H., Miyajima, A., and Yoshimura, A. (1997). CIS, a cytokine inducible SH2 protein, is a target of the JAK–STAT5 pathway and modulates STAT5 activation. *Blood*, **89**, 3148.

193. Onishi, M., Nosaka, T., Misawa, K., Mui, A. L., Gorman, D., McMahon, M., Miyajima, A., and Kitamura, T. (1998). Identification and characterization of a constitutively active STAT5 mutant that promotes cell proliferation. *Mol. Cell. Biol.*, **18**, 3871.

194. Chen, X., Vinkemeier, U., Zhao, Y., Jeruzalmi, D., Darnell, J. E., Jr, and Kuriyan, J. (1998). Crystal structure of a tyrosine phosphorylated STAT-1 dimer bound to DNA. *Cell*, **93**, 827.

195. Becker, S., Groner, B., and Muller, C. W. (1998). Three-dimensional structure of the Stat3beta homodimer bound to DNA. *Nature*, **294**, 145.

196. Vinkemeier, U., Moarefi, I., Darnell, J. E., Jr, and Kuriyan, J. (1998). Structure of the amino-terminal protein interaction domain of STAT-4. *Science*, **279**, 1048.

197. Liu, K. D., Gaffen, S. L., Goldsmith, M. A., and Greene, W. C. (1997). Janus kinases in interleukin-2-mediated signaling: JAK1 and JAK3 are differentially regulated by tyrosine phosphorylation. *Curr. Biol.*, **7**, 817.

198. Feng, J., Witthuhn, B. A., Matsuda, T., Kohlhuber, F., Kerr, I. M., and Ihle, J. N. (1997). Activation of Jak2 catalytic activity requires phosphorylation of Y1007 in the kinase activation loop. *Mol. Cell. Biol.*, **17**, 2497.

199. Zhou, Y. J., Hanson, E. P., Chen, Y. Q., Magnuson, K., Chen, M., Swann, P. G., Wange, R. L., Changelian, P. S., and O'Shea, J. J. (1997). Distinct tyrosine phosphorylation sites in JAK3 kinase domain positively and negatively regulate its enzymatic activity. *Proc. Natl Acad. Sci. USA*, **94**, 13850.

200. Zhang, X., Blenis, J., Li, H. C., Schindler, C., and Chen-Kiang, S. (1995). Requirement of serine phosphorylation for formation of STAT-promoter complexes. *Science*, **267**, 1990.

201. Wen, Z., Zhong, Z., and Darnell, J. E., Jr (1995). Maximal activation of transcription by Stat1 and Stat3 requires both tyrosine and serine phosphorylation. *Cell*, **82**, 241.

202. Eilers, A., Georgellis, D., Klose, B., Schindler, C., Ziemiecki, A., Harpur, A. G., Wilks, A. F., and Decker, T. (1995). Differentiation-regulated serine phosphorylation of STAT1 promotes GAF activation in macrophages. *Mol. Cell. Biol.*, **15**, 3579.

203. Ceresa, B. P. and Pessin, J. E. (1996). Insulin stimulates the serine phosphorylation of the signal transducer and activator of transcription (STAT3) isoform. *J. Biol. Chem.*, **271**, 12121.

204. Gotoh, A., Takahira, H., Mantel, C., Litz-Jackson, S., Boswell, H. S., and Broxmeyer, H. E. (1996). Steel factor induces serine phosphorylation of Stat3 in human growth factor-dependent myeloid cell lines. *Blood*, **88**, 138.

205. Zhu, X., Wen, Z., Xu, L. Z., and Darnell, J. E., Jr. (1997). Stat1 serine phosphorylation occurs independently of tyrosine phosphorylation and requires an activated Jak2 kinase. *Mol. Cell. Biol.*, **17**, 6618.

206. Chung, J., Uchida, E., Grammer, T. C., and Blenis, J. (1997). STAT3 serine phosphorylation by ERK-dependent and -independent pathways negatively modulates its tyrosine phosphorylation. *Mol. Cell. Biol.*, **17**, 6508.

207. Wen, Z. and Darnell, J. E., Jr. (1997). Mapping of Stat3 serine phosphorylation to a single residue (727) and evidence that serine phosphorylation has no influence on DNA binding of Stat1 and Stat3. *Nucl. Acids Res.*, **25**, 2062.

208. Cho, S. S., Bacon, C. M., Sudarshan, C., Rees, R. C., Finbloom, D., Pine, R., and O'Shea, J. J. (1996). Activation of STAT4 by IL-12 and IFN-alpha: evidence for the involvement of ligand-induced tyrosine and serine phosphorylation. *J. Immunol.*, **157**, 4781.

209. Gollob, J. A., Schnipper, C. P., Murphy, E. A., Ritz, J., and Frank, D. A. (1999). The functional synergy between IL-12 and IL-2 involves p38 mitogen-activated protein kinase and is associated with the augmentation of STAT serine phosphorylation. *J. Immunol.*, **162**, 4472.

210. Chung, C. D., Liao, J., Liu, B., Rao, X., Jay, P., Berta, P., and Shuai, K. (1997). Specific inhibition of Stat3 signal transduction by PIAS3. *Science*, **278**, 1803.

211. Liu, B., Liao, J., Rao, X., Kushner, S. A., Chung, C. D., Chang, D. D., and Shuai, K. (1998). Inhibition of Stat1-mediated gene activation by PIAS1. *Proc. Natl Acad. Sci. USA*, **95**, 10626.

212. Kim, T. K. and Maniatis, T. (1996). Regulation of interferon-gamma-activated STAT1 by the ubiquitin-proteasome pathway. *Science*, **273**, 1717.

213. Haspel, R. L., Salditt-Georgieff, M., and Darnell, J. E., Jr. (1996). The rapid inactivation of nuclear tyrosine phosphorylated Stat1 depends upon a protein tyrosine phosphatase. *EMBO J.*, **15**, 6262.

214. Yu, C. L. and Burakoff, S. J. (1997). Involvement of proteasomes in regulating JAK–STAT pathways upon interleukin-2 stimulation. *J. Biol. Chem.*, **272**, 14017.

215. Callus, B. A. and Mathey-Prevot, B. (1998). Interleukin-3-induced activation of the JAK/STAT pathway is prolonged by proteasome inhibitors. *Blood*, **91**, 3182.

216. Azam, M., Lee, C., Strehlow, I., and Schindler, C. (1997). Functionally distinct isoforms of STAT5 are generated by protein processing. *Immunity*, **6**, 691.

217. Meyer, J., Jucker, M., Ostertag, W., and Stocking, C. (1998). Carboxyl-truncated STAT5beta is generated by a nucleus-associated serine protease in early hematopoietic progenitors. *Blood*, **91**, 1901.

5 | On the road to destruction: suppression of protein tyrosine kinase signalling by SOCS family proteins

ROBERT ROTTAPEL, SUBBURAJ ILANGUMARAN, and PAULO DE SEPULVEDA

1. Introduction

The complexity of multicellular organisms established during embryogenesis and maintained during adult life relies on the continuous processing of diverse micro-environmental signals by various cell types. These environmental signals are often elaborated in the form of membrane-bound or soluble-peptide ligands that bind to cognate cell-surface receptors. One mechanism of transducing signals across the plasma membrane is to couple ligand-induced oligomerization of cell-surface receptors to the activation of protein tyrosine kinases (PTKs). Growth factor peptides such as insulin, epidermal growth factor, and steel factor bind to a class of transmembrane receptors with intrinsic protein tyrosine kinase activity (RTK). On the other hand, cytokines involved in inflammatory, antiviral, or hormonal growth responses bind to a class of receptors which are non-covalently linked to the cytoplasmic, non-receptor protein tyrosine kinases of the Janus kinase (JAK) family. PTK activation provokes intracellular signal amplification and diversification following covalent modification of a wide number of protein targets by tyrosine phosphorylation. Tyrosine phosphory-lation modifies enzymatic activity, induces multimeric protein assembly through the recruitment of SH2-containing proteins, and relocalizes signalling molecules to sub-cellular compartments where they exert their function. Cytokine receptors activate a complex network of downstream signalling pathways which mediate a diversity of biological responses including cell survival, proliferation, differentiation, and chemo-taxis. The initiation and termination of cytokine-induced signalling is tightly regulated. Many mechanisms have evolved to terminate signalling events following receptor activation. These include receptor internalization, receptor degradation, activation of antagonistic pathways including tyrosine phosphatases, GTPases, and upregulation of cyclin-dependent kinase inhibitors. This review focuses on the SOCS family of proteins, which attenuate PTK-mediated signalling by diverse mechanisms that include pseudosubstrate inhibition of JAK kinases and the targeting of downstream effector molecules to ubiquitin-mediated proteolytic destruction.

2. Discovery

The first member of the SOCS family, CIS, was identified during a screen to clone novel early-response genes induced by erythropoietin (EPO) using a cDNA subtraction technique (1). CIS mRNA was rapidly induced by EPO, IL-2, and IL-3, but not by G-CSF, IL-6, steel factor, or phorbol ester. The CIS cDNA encodes a 257-amino acid polypeptide containing a single SH2 domain with non-catalytic N-terminal and C-terminal flanking sequences. CIS thus refers to a 'cytokine-induced-SH2' containing protein. CIS was shown to physically bind to the phosphorylated form of the erythropoietin receptor (EPOR) and the β-chain of the IL-3 receptor in cotransfection experiments (1), and to reduce IL-3-dependent cell growth in FDC-P1 and Ba/F3 cells. CIS-mediated growth suppression is not a result of accelerated apoptosis or alteration in IL-3-dependent cell viability (1).

The second member of the CIS family was identified using three independent strategies. Starr *et al.* used a retroviral FDC-P1-derived cDNA library to identify genes that suppress IL-6-induced differentiation of the myelomonocytic leukaemic cell line M1 and cloned SOCS-1 (suppressor of cytokine signalling-1) (2). Endo *et al.* cloned the human form of SOCS-1 using the yeast two-hybrid approach to identify binding partners to the JH1 catalytic domain of JAK2 and named it JAB for JAK-binding protein (3). In an attempt to identify other members of the STAT family Naka *et al.* screened a murine thymus library using a monoclonal antibody directed against the phosphotyrosine binding motif of the STAT3 SH2 domain (GTFLLRFS) and cloned SSI (STAT-induced STAT inhibitor) which contained the similar motif (GTFLVRDS) (4). SSI is identical to both SOCS-1 and JAB. Subsequently, several other groups have cloned SOCS-1 by virtue of its capacity to bind to TEC (5) and KIT (3, 6).

3. Nomenclature

A search of the nucleic acid databank revealed that SOCS-1 shared homology with CIS and identified six EST (expressed sequence tag) sequences distinguished by a central SH2 domain and a conserved carboxy-terminal region designated the SOCS box (2, 7). The six putative additional members of this family were named SOCS-2 through SOCS-7. Several groups have identified members of the SOCS family and have assigned different names to the same SOCS family members (5, 8–10). Since all names to date have been based on a subset of functions associated with a limited number of SOCS family members, no single term best describes this family in its most general sense. For simplicity, we have chosen the term SOCS, as its name applies to

---

Table 1 SOCS family nomenclature

SOCS family members	Other names
SOCS-1	JAB, SSI-1, TIP3
SOCS-2	CIS-2, SSI-2
SOCS-3	CIS-3, SSI-3
SOCS-4	CIS-7
SOCS-5	CIS-6
SOCS-6	CIS-4
SOCS-7	CIS-5, NAP-4
CIS	CIS-1

---

the largest number of identified family members. The list of known SOCS family members is given in Table 1.

4. Genomic organization and domain topography of SOCS family members

The genomic organization of SOCS-1 is the only SOCS family member reported to date. A database search revealed that the mouse, rat, and human SOCS-1 genes are part of the protamine gene cluster (2). This group of four genes is localized on mouse chromosome 16 (11), rat chromosome 9 (12) and human chromosome 16 (13). The SOCS-1 gene lies 3' of the gene cluster, two kilobases downstream of the TPN2 gene. The SOCS-1 locus is composed of two exons separated by a 551 base-pair intron. The coding sequence is located in exon 2 (14). All five genes within this locus are transcribed in the same orientation. The genomic structure of SOCS-1 is illustrated in Fig. 1.

Phylogenetic relationships among the SOCS family members based upon analysis

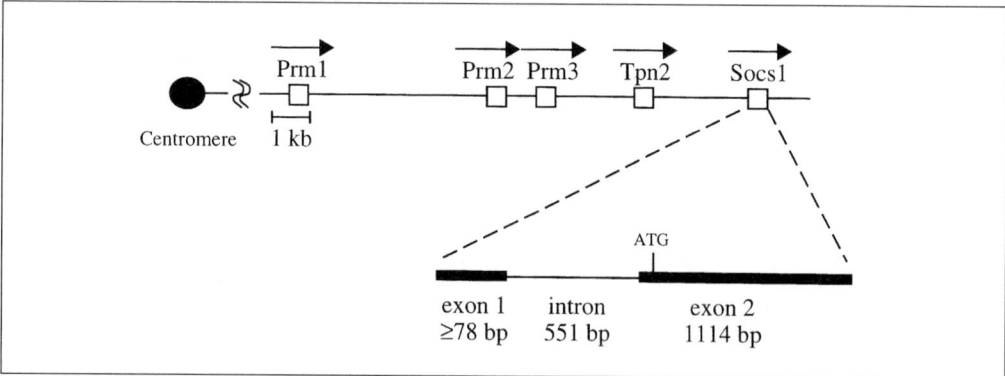

Fig. 1 The genomic structure of the SOCS-1 gene. The five genes of the protamine gene cluster on mouse chromosome 16 are shown. The organization and the size of SOCS-1 exons were deduced by comparing the sequence of several independent SOCS-1 cDNA clones with the genomic sequence of the protamine gene cluster present in the GenBank DNA database.

of the primary amino acid sequence and genomic structure suggest that they have evolved in pairs. CIS and SOCS-2, SOCS-1 and SOCS-3, SOCS-4 and SOCS-5, SOCS-6 and SOCS-7 are more closely related to each other than to other SOCS proteins (7). SOCS-4, SOCS-5, SOCS-6, and SOCS-7 can be distinguished from the first four members of the family by an extensive and poorly conserved N-terminal region (7).

The unifying feature of all SOCS family members is an SH2 domain followed by a conserved sequence of approximately 40 amino acids rich in arginine, leucine, and proline termed by various groups as the SOCS box (2), the CH (CIS homology) domain (8), or the SC (SSI C-terminal) motif (10). The N-terminal arms of the SOCS family members are variable in length and are not related to one another at the primary amino acid level.

The SOCS box homology is present in other protein contexts outside of the SOCS family and suggests that it subserves a conserved function. From homology searches with the SOCS box, 12 potential open reading frames present in the DNA databases have been identified (7). These other potential protein sequences characteristically contain protein–protein interacting domains, such as WD-40 repeats (WD-40 SOCS box; WSB-1 and –2), ankyrin repeats (ankyrin SOCS box; ASB-1, -2, -3), and SPRY domains (SPRY SOCS box; SSB-1, -2, -3). Two small GTPases RAR and Ras-like-GTPase were also identified as SOCS box-containing proteins (7).

The chicken WSB-1 homologue has recently been identified as a developmentally regulated protein known as SWiP-1 (SOCS box and WD-repeats in protein) (15). SWiP-1 is expressed in the somites and in the developing limbs. SWiP-1 expression is controlled by two opposing signals originating from structures adjacent to the segmental mesoderm: a positive signal from the notochord and a negative signal from the lateral mesoderm (15). SWiP-1 expression in the somites is upregulated by Sonic hedgehog (Shh) secreted from the notochord (15). Shh regulates, in part, the subdivision of the somite into dermomyotome and sclerotome and induction of the myotome (16–20). In the limb Shh emanates from the zone of polarizing activity (ZPA) likely patterns the anterior-posterior axis of the limb (21–24). cSWiP-1 expression correlates in time and space with regions being patterned by Shh (15). Shh binds to two potential cell-surface receptors and activates PKA, slimb (β-TRCP), and the transcription factor Cubitus interuptus (25). The function of SWiP-1 is not known but it may be a component of the Hedgehog signal transduction cascade. A human orthologue of SWiP-1/WSB-1 and a murine homologue SWiP-2/WSB-2 have been identified (7, 15).

5. Tissue expression and induction of SOCS family members

The expression pattern of SOCS-1, -2, -3, -5 and CIS proteins is complex and distinct for each family member (Table 2). SOCS-1 mRNA is highly expressed in thymus and spleen, SOCS-2 is principally expressed in liver, heart, kidney, placenta, prostate, and testis (10, 26), while SOCS-3 is expressed in peripheral leucocytes and at a low level in thymus, spleen, and lung (27). Both SOCS-5 and CIS are widely expressed (1, 7).

Table 2 mRNA Expression of SOCS family members in tissues

	Liver	Spleen	Thymus	Kidney	Lung	Heart	Brain	Stomach	Testis	PBL
SOCS-1							n.a.			n.a.
SOCS-2										
SOCS-2*										
SOCS-3										
SOCS-4	n.a.			n.a.	n.a.	n.a.	n.a.	n.a.		
SOCS-5								n.a.		n.a.
CIS										

mRNA levels of SOCS family members adapted from the following sources. SOCS-1 (2, 4); SOCS-2 (10); SOCS-2* (26); SOCS-3 (2,10); SOCS-4 (27); SOCS-5 (7); CIS (1, 2). Shaded boxes reflect the relative level of mRNA. white: low, grey: intermediate, black: high, n.a.: data not available.

Table 3 Induction of SOCS family proteins by cytokines

	EPO	TPO	G-CSF	GM-CSF	IL-2	IL-3	IL-4	IL-7	IL-13	IL-6	LIF	IL-12	IFN-γ	TNF-α
SOCS-1														
SOCS-2														
SOCS-3														
CIS														

SOCS family members mRNA levels are induced in response to different cytokines. mRNA levels are reflected by shaded boxes as follow: white: low, grey: intermediate, black: high, n.a.: data not available.

Transcriptional regulation of each SOCS family member demonstrates a discrete pattern in response to cytokine stimulation, and is summarized in Table 3. CIS was initially described as an EPO, IL-2, or IL-3 inducible gene (1), but its induction is responsive to a wide range of cytokines (2). Signalling through STAT5a/b is probably required for CIS transcription since STAT5a/b is a common signalling component for EPOR (28), IL-2R (29, 30), and IL-3R (31), STAT5 binding sites have been identified in the 5′ flanking region of the CIS gene (32), and CIS mRNA expression in the ovary is lost in the STAT5a/b double-knockout animal (33). IFN-γ is also a potent inducer of CIS but it activates STAT1 rather than STAT5, suggesting that other promoter elements regulating CIS are operative (34).

SOCS-1 transcripts are induced by the activation of cytokine- and growth-factor receptor families, including gp130 (2–4), common γ_c (2, 6, 35), common β_c (2, 6), INFγR (2, 34, 36), and the KIT RTK (6). Naka *et al.* showed that transfection of a dominant-negative form of STAT-3 into M1 cells blocked the ability of IL-6 to upregulate SOCS-1 mRNA, indicating, as had been observed for CIS, that signalling through the JAK–STAT pathway is necessary for the induction of SOCS-1 (4, 32).

Although SOCS-1 mRNA is abundant in thymus and spleen, the endogenous SOCS-1 protein has been difficult to detect by Western blot analysis. Synthesis of nascent SOCS-1 protein is controlled by initiation–elongation factors acting through 5′-untranslated region regulatory sequences (A. Veillette, personal communication). The SOCS-1 protein is unstable and is degraded by ubiquitin-mediated proteolysis (14). The signalling pathways which stabilize the SOCS-1 protein and the cellular factors that regulate the steady-state levels of SOCS-1, such as the cell-cycle state, have not been elucidated.

The *in vivo* response of SOCS family members has been studied by the intravenous administration of IL-6. Elevation of CIS, SOCS-1, SOCS-2, and SOCS-3 mRNA levels in the liver rapidly followed exposure to IL-6. CIS and SOCS-2 were rapidly induced and stayed elevated for up to 24 hours, whereas SOCS-1 and SOCS-3 expression were transient (2).

6. Structural aspects of SOCS-1 required for signal suppression

6.1 SOCS-1 is a pseudosubstrate inhibitor of JAK kinases

In order to address the mechanism by which SOCS-1 suppressed IL-6-induced differentiation of M1 cells, Starr *et al.* noted that both the gp130 chain of the IL-6 receptor and STAT3 were hypophosphorylated in cells expressing SOCS-1 when stimulated by IL-6 (2). Endo *et al.* and Naka *et al.* complemented these observations by demonstrating that SOCS-1 bound directly to JAK2 and TYK2 (3, 4). Coexpression of SOCS-1 diminished JAK1, JAK2, JAK3, and TYK2 tyrosine phosphorylation, suggesting that SOCS-1 may be acting as a general JAK pseudosubstrate inhibitor (3, 34, 35). This suppression of JAK family kinases by SOCS-1 was specific as it did not impair ligand-induced phosphorylation of the receptor tyrosine kinases FGFR, INSRβ, FLT3, or KIT (3, 4, 6). Together these data demonstrated that SOCS-1 was an inducible SH2-containing protein that suppressed JAK kinase activation by directly binding to the catalytic region of the enzyme. As SOCS-1 expression is induced by cytokine receptors, it is unlikely that SOCS-1 regulates the initiation of cytokine signalling but rather modulates the strength and/or duration of signalling.

6.1.1 Structural requirements for SOCS-1 pseudosubstrate inhibition

The observation that SOCS-1 was cloned by virtue of its capacity to bind to the kinase domain of JAK2 in a phosphotyrosine-dependent manner suggested that the SOCS-1 SH2 binding site was one of a limited subset of JAK2 autophosphorylation sites. Activation of protein tyrosine kinases characteristically require the phosphorylation of a conserved tyrosine present in the activation segment within the phosphotransferase domain. Solution of the Src family kinase structures demonstrated that, in the inactive state, Glu^{310} in helix αC of the N-lobe interacts with Arg^{385} in the activation segment creating a closed structure. Phosphorylation of the positive regulatory Tyr^{416} displaces the interaction of Glu^{310} from Arg^{385} and 'pulls' the activation loop out of the kinase active site allowing access of substrates to the catalytic core (37, 38). Tyrosine 1007 represents the conserved regulatory tyrosine in the activation segment of JAK2 whose phosphorylation is critical for kinase activity (39). Recombinant JAK2 JH1 domains mutated at this site no longer bound to SOCS-1 (40). Synthetic peptides representing the four major autophosphorylation sites in the JH1 domain were used to show that only peptides encompassing Y1007 coprecipitated SOCS-1 (40). The tyrosine motif at this site in JAK2 is EYYKV and is highly conserved among all JAK kinase family members. SOCS-1 also binds to the KIT and FLT3 receptors in a phosphotyrosine-dependent manner but probably at a different site (6). The activation segment tyrosine motifs of KIT and FLT3 are characterized by a positively charged residue at the +3 position (DSNYVVK and DSSYVVR, respectively) rather than a hydrophobic residue common to the JAK kinase family. Mutation of the tyrosine residue within this site does not impair the binding of SOCS-1 to KIT (14). The likelihood that SOCS-1 is binding to a site outside the activation segment of KIT and FLT3 is consistent with the fact that SOCS-1 does not impair the kinase activity of either of these kinases (6).

6.1.2 The pre-SH2 domain

The structural features of SOCS-1 that are necessary to suppress JAK kinase activity have been investigated by mutational analysis. Naka *et al.* observed that deletion of the SH2 domain and carboxy terminus of SOCS-1 disabled the protein's ability to interfere with IL-6- or LIF- induced apoptosis of M1 cells (4). In parallel with this observation, the expression of this truncated form of SOCS-1 protein was unable to suppress IL-6-mediated gp-130 or STAT3 phosphorylation (4). The SH2 domain of SOCS-1 is required to suppress JAK activation, as a loss-of-function point mutation within the phosphotyrosine binding groove failed to inhibit JAK kinase autophosphorylation or block transactivation of the IRF-1 IL-6 responsive element (41). Endo *et al.* noted that the isolated SH2 domain of SOCS-1 was not sufficient to inhibit JAK2 autophosphorylation or IL-2- and IL-3-induced c-*fos* activation (3). This suggested that the N-terminal and C-terminal flanking sequences were also important for JAK kinase suppression.

Truncation mutants of SOCS-1 were examined for their ability to inhibit IL-6-mediated M1 apoptosis, IL-6-responsive reporter assay, and JAK kinase binding and inhibition. Deletion of the first N-terminal 51 amino acids appeared to have no effect on SOCS-1 function (40–42), whereas deletion of the entire N-terminus up to the canonical boundary of the SH2 domain (amino acids 1–80) resulted in complete loss of SOCS-1 function. Thus, the minimal region within the N-terminus required for SOCS-1 function was contained within a stretch of 29 amino acids immediately proximal to the SH2 domain. This region was designated as the pre-SH2 domain and showed sequence homology within a similar region of other SOCS proteins (41). The N-terminal sequences of SOCS-1 and SOCS-3 appear to be functionally interchange-able. Chimeric SOCS-1 proteins containing the N-terminal sequences of CIS, SOCS-2, SOCS-3, SOCS-5, or SOCS-6, with the SH2 and SOCS box derived from SOCS-1 were tested for their ability to suppress LIF-mediated transactivation of a reporter con-struct. Only SOCS-3 sequences were able to substitute for the SOCS-1 sequences (42).

The pre-SH2 region lies in physical contiguity with the SOCS-1 SH2 domain to complete its canonical N-terminal boundary and participates in JAK kinase inhibition (Fig. 2). A minimum of 12 amino acids (Ile[68] through Gly[79]) immediately N-terminal to the SH2 domain are required for SOCS-1 binding to synthetic phosphopeptides representing the major autophosphorylation site of JAK2 (pY1007). These 12 amino acids have been termed the extended SH2 subdomain (40). The binding affinity of SOCS-1 to the catalytic domain of JAK2, however, was substantially increased by an additional 12 amino acids N-terminal to the extended SH2 subdomain. These residues contact JAK2 in a phosphotyrosine-independent manner, which may be important in increasing the binding avidity to JAK2 in a physiological setting. The extended SH2

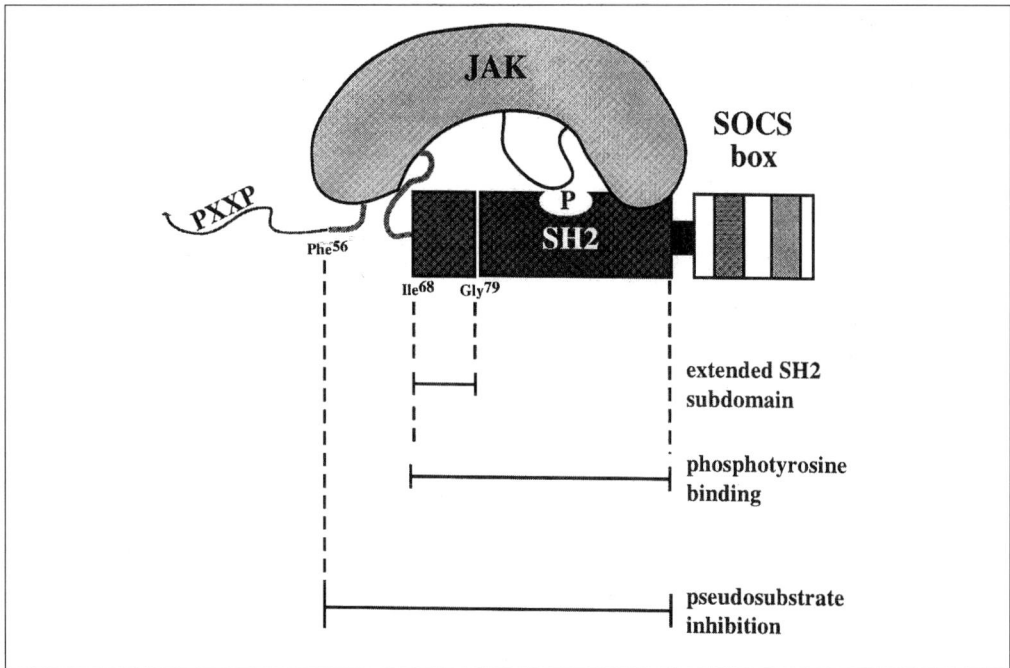

Fig. 2 Domain organization of SOCS-1. The N-terminus of SOCS-1 is composed of 56 amino acids that contain an SH3-binding diproline motif. Residues 56 through 68 are required for pseudosubstrate inhibition of JAK kinases and high avidity binding. Amino acids 68 through 79 represent an extended SH2 subdomain. These 12 amino acids lie in physical continuity with the canonical SH2 domain and are required for its phosphotyrosine binding function. The SH2 domain is shown. The C-terminus SOCS box is approximately 40 amino acids in length and has two blocks of homology corresponding to helix-1 of VHL (left) and helices 2 and 3 of VHL (right). JAK kinase binds to the SOCS-1 SH2 domain via its regulatory Tyr[1007] in the activation segment and to SOCS sequences involved in kinase inhibition.

subdomain of SOCS-1 is reminiscent of the STAT1 SH2 domain where the phosphate binding loop is buttressed by a helical structure within the linker domain just N-terminal to the SH2 domain (43). A series of hydrophobic side chains presented by helix α10 pack into the hydrophobic core of the SH2 domain underneath the phosphate binding loop. Ile[68] and Leu[75] in the SOCS-1 SH2 subdomain are conserved hydrophobic residues present in the STAT1 α10, which might perform a similar function in maintaining the integrity of phosphotyrosine binding. This idea is supported by the observation that mutation of either Ile[68] or Leu[75] to negatively charged residues destroyed binding to pY1007 phosphopeptides (40).

A second function of the extended SH2 subdomain is the suppression of JAK2 phosphorylation. Deletion of the first 60 N-terminal amino acids destroyed the inhibitory potential of SOCS-1 on JAK kinase activation, whereas truncation SOCS-1 mutants devoid of the first 51 amino acids were fully competent to inhibit JAK kinase autophosphorylation. A single point mutation at codon Phe[56], lost the capacity to inhibit cytokine signalling but was still capable of binding to JAK (40, 42). Therefore, phosphotyrosine binding *per se* requires a functional SH2 domain including residues 68–78 lying proximal to and in continuity with the classical boundary of the SOCS-1 SH2 domain, while amino acids 56–68 are required for maximal binding to the JAK kinase domain and inhibition of its autokinase activity (see Fig. 4) (40, 42). The manner in which residues 56–68 interfere with JAK kinase function is not known, but they might serve to block substrate accessibility by occluding the active site or competing with ATP binding within the N-terminal lobe of the kinase.

6.2 Receptor tyrosine kinase signalling is modulated by SOCS-1

KIT and FLT3 belong to the class-III family of RTKs which includes the PDGF-α and -β receptors and the CSF-1R. KIT and FLT3 have similar genomic organization and domain topology, suggesting that they arose from a common ancestral gene. The ligand-binding extracellular domain of KIT and FLT3 is composed of five immuno-globulin-type domains. Both receptors contain an intracellular kinase domain bisected by a non-catalytic region known as the kinase insert involved in substrate binding (44). KIT and FLT3 are expressed on early haemopoietic progenitor cells, but they have distinct function within the haemopoietic hierarchy. Mice which harbour mutations in the KIT ligand (encoded by *Sl*, for Steel locus) or within the KIT kinase (encoded by *W* for Dominant White Spotting locus) suffer from diminished haemopoietic stem-cell activity and defective erythroid- and mast-cell production (45–47). Mice in which the *Flt3* locus has been disrupted by homologous recombination have no stem-cell self-renewal defect, but are deficient in the steady-state numbers of CD43[+] pro-B cells and in their capacity to repopulate the B-lymphoid lineage during competition reconstitutive assays (48).

In an attempt to identify the signal transduction molecules that mediate KIT and FLT3 function in haemopoietic progenitor cells, the cytoplasmic domain of both receptors were employed as bait in a yeast two-hybrid screen against a murine, haemopoietic progenitor cell cDNA library. Using this approach SOCS-1 was cloned as a KIT-binding protein (6). SOCS-1 is an immediate-early gene which is induced following KIT activation in primary bone marrow-derived mast cells. SOCS-1 bound to the activated KIT receptor via its SH2 domain but, in contrast to its effects on JAK kinases, did not inhibit the KIT kinase activity. SOCS-1 could, nevertheless, suppress the proliferative response of Steel factor-dependent haemopoietic cell lines, which suggested that SOCS-1 may interfere with signalling pathways lying downstream of the KIT receptor (6).

7. SOCS-1 binds to multiple signal transduction molecules

7.1 SOCS-1 binds to SH3-containing proteins

The SOCS-1 N-terminus contains a type-I (residues 41–47) and a type-II (residues 34–39) diproline motif, the defining determinant for SH3-domain binding (49, 50). SOCS-1 binds via its N-terminal proline-rich arm to multiple SH3-containing proteins such as the adapter molecules GRB2, the p85 subunit of PI3K, NCK, and the T cell-specific tyrosine kinases FYN and ITK. Other proteins such as ABL, SRC, LCK, HS1, PLCγ, VAV, GAP, and α-spectrin do not bind to SOCS-1 via their SH3 domains (6). IL-6 or LIF signalling is not suppressed by SOCS-1 mutants deleted of the N-terminal sequences containing the diproline motifs (40–42). SOCS-1 also binds to TEC PTK and suppresses its activity in 293 cells. Deletion of the SH2 or the carboxy terminus did not abrogate the interactions of SOCS-1 with TEC, suggesting that the N-terminal proline-rich sequences of SOCS-1 were sufficient to mediate the interaction (5).

7.2 VAV is a SOCS-1 binding partner

A yeast two-hybrid screen using SOCS-1 as the bait identified the Dbl-family guanine nucleotide exchange factor (GEF) VAV as a potential binding partner (6). VAV is a haemopoietic-specific exchange factor for the small GTP-binding protein RAC (51). RAC activation induces lamellipodia formation and membrane ruffling (52). VAV is rapidly phosphorylated on tyrosine in response to the ligation of a wide variety of antigen and cytokine receptors (53–57) including KIT (58). The exchange activity of VAV is positively regulated by tyrosine phosphorylation, suggesting that VAV may be a positive effector of KIT function in haemopoietic cells (51).

SOCS-1 coprecipitates with either VAV or the VAV-related protein VAV-2 in mammalian cells (6). SOCS-1 binds to the acidic domain within the N-terminal region of VAV encompassing amino acids 1–199, a region that is highly conserved with VAV-2. Both VAV and VAV-2 are negatively regulated by their N-termini and can be rendered oncogenic by mutations within this region (59, 60). VAV binds to the SOCS-1 SH2 domain in a phosphotyrosine-independent manner (6). In comparison with other SH2 domains, the SOCS-1 SH2 domain is uncommonly basic in its amino acid content. Sequence-threading of the primary amino acid sequence of SOCS-1, using the atomic coordinates derived from the p85α structure, predicts the distribution of this positive charge to be localized to a discrete patch on the SOCS-1 SH2 surface (14). Therefore the interaction of the VAV acidic domain with the SOCS-1 SH2 domain may be electrostatic in nature.

7.3 SOCS-1 suppresses VAV function and targets VAV to the ubiquitin proteasome pathway

SOCS-1 not only binds to VAV and onco-VAV (the oncogenic form of VAV devoid of the first 60 amino acids), but can suppress the transforming activity of onco-VAV in NIH3T3 cells (133). This effect is specific to SOCS-1, as other SOCS family members have no effect in the foci formation assay. The suppression of onco-VAV by SOCS-1 does not require an intact phosphotyrosine-binding SOCS-1 SH2 domain. Cotransfection experiments in Cos7 cells demonstrated that the steady-state level of VAV was diminished in the presence of SOCS-1, but not SOCS-2 or SOCS-3. This effect was saturable by increasing the amount of transfected VAV. The half-life of nascent VAV protein was decreased by SOCS-1, and VAV protein levels were stabilized in the presence of proteasome inhibitors. The destruction of VAV protein is thus hastened in the presence of SOCS-1, which is likely to occur through a ubiquitin-mediated proteolytic pathway (133).

8. The SOCS box couples SOCS family proteins to the Elongin B/C ubiquitin ligase complex

Two groups have independently shown that the SOCS box targets SOCS-1 to the Elongin B/C complex, a component of a putative E3-ubiquitin ligase (61, 62). The connection of SOCS family proteins to the Elongin B/C complex has illuminated a second potential function of these adapter molecules in modulating signals. Elongin B/C has previously been shown to form a physical complex with VHL, the tumour suppressor gene product mutated in von Hippel–Lindau disease (63). The v-HL–Elongin B/C complex has been coined the VCB complex (VHL–Elongin C/Elongin B) (64). Other members of this protein complex have recently been identified to include Cullin-2 (Cul2) (65, 66), and Ring Box-1 (Rbx1) (67) also known as Regulator of Cullin-1 (Roc1) (68).

VHL is mutated in an autosomal dominant, familial cancer syndrome that predisposes individuals to a wide range of tumours including renal-cell carcinoma, phaeochromocytoma, cerebellar haemangioblastomas, and retinal angiomata (69). The gene product of VHL is also mutated in the majority of sporadic clear-cell renal carcinomas (70, 71). The mechanism by which mutations in VHL lead to tumour formation has been obscure until the recent realization that the VCB complex might function as a ubiquitin ligase (65, 66). With this realization, a potential model, which explains the oncogenic function of VHL, anticipates that mutations in VHL uncouple it from the VCB–ubiquitin ligase complex and lead to the accumulation of cellular proteins whose dysregulated expression is dangerous to the cell.

The possibility that VCB complexes carry ubiquitin ligase activity stems from the recognition that each component of the VCB complex shares sequence-similarity with subunits of the yeast SCF (Skp1-Cdc53/CUL1-F box protein)–E3 ubiquitin ligase complex which controls the expression of cell-cycle regulating proteins (72). The mammalian gene products Elongin C (73, 74), Cul2 (65), and Rbx/Roc1 (67) correspond to the yeast homologues Skp1, Cdc53, and Rbx, respectively. Elongin B contains a ubiquitin-like structure in the first N-terminal 80 amino acids (69, 75). The carboxy terminus of VHL contains a protein interaction motif which defines the interaction with Elongin C (76). This sequence is similar to a sequence in yeast proteins which binds to Skp1 (74), known as the F-box (73).

It is precisely this region of VHL that is also present within the SOCS box and has been termed the BC box (61). The presence of the conserved sequence (T/S/P)LXXX (C/S)XXX(L/I/V) common to both VHL and SOCS proteins suggested that SOCS box-containing proteins may also have the potential to bind to Elongin B/C (61). This was tested by Kamura *et al.* who demonstrated that the *Escherichia coli*-expressed SOCS box containing proteins SOCS-1, SOCS-3, RAR-1, ASB-2, or WSB-1 could all form a protein complex with Elongin B/C (61). Elongin B/C and SOCS-1 formed a physical complex *in vivo* in cotransfected 293T cells or in IFN-γ-treated fibroblasts. Point mutations within the SOCS box corresponding to mutations in VHL and Elongin A which disrupt binding to Elongin B/C similarly uncoupled SOCS-1 from Elongin B/C, suggesting that this region of the SOCS box was functionally similar to VHL and Elongin A (61). These observations were substantiated by Zhang *et al.* who used mass spectroscopy to identify proteins that bound to a GST–SOCS box affinity reagent from M1 cellular lysates (62). Two proteins of 18-kDa and 15-kDa molecular weight coprecipitated with the GST–SOCS fusion protein and were identified as Elongin B and Elongin C, respectively (62). The interaction of SOCS-1 with Elongin B and Elongin C *in vivo* was inducible, and depended on prior cytokine stimulation and was enhanced by proteasome inhibitors (62).

8.1 Elongin B/C is a putative ubiquitin ligase

The superfamily of E3 ubiquitin ligases share a common strategy in targeting proteins destined for proteolytic destruction. A linked chain of protein interactions recruit protein substrates into the ubiquitin-mediated proteasome pathway. The Skp1 subunit of the SCF complex binds to F-box-containing adapter proteins, which in turn bind to a limited subset of cellular proteins recruited for destruction (73). In this way, protein complexes nucleated by F-box-containing adapter proteins are funnelled into the ubiquitination machinery by virtue of the interaction between Skp1 and F-box-containing proteins. In the case of the yeast SCF E3 ligase, protein targets such as cyclins 1/2 bind to the specific F-box-containing adapter molecule, Grr1, which in turn couples to the Skp1 subunit of the SCF complex (77). Similarly, a second F-box protein Cdc4 binds to the CDK inhibitor SIC1, which then becomes targeted for

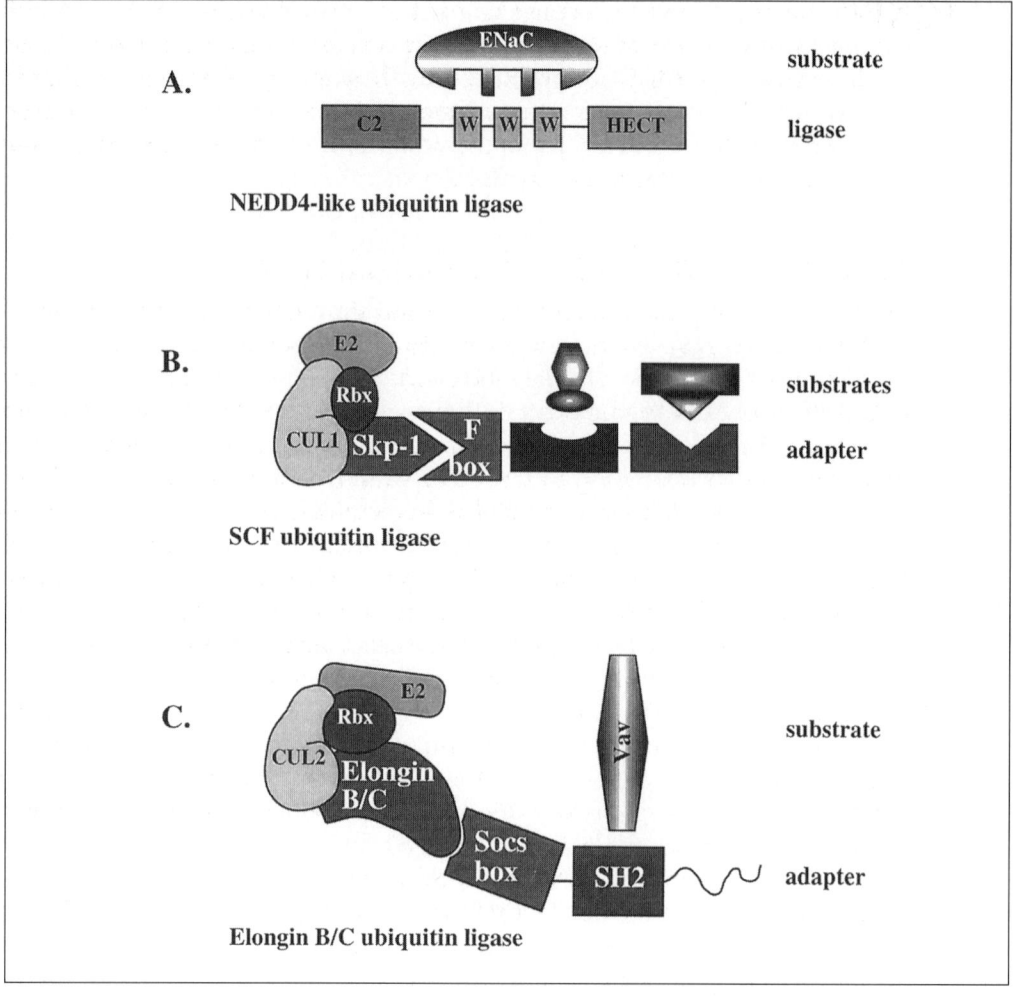

Fig. 3 E3 ubiquitin ligases. The NEDD4-like ubiquitin ligases (A) bind directly to their substrates (e.g. ENaC) via multiple copies of WW domains. The HECT E3 ubiquitin ligase is targeted to the plasma membrane in a Ca^{2+}-dependent manner through the C2 domain. Both the SCF (B) and the Elongin B/C (C) ubiquitin ligases are multiprotein complexes which use intermediate adapter proteins to couple to their substrates. The F-box and the SOCS box are the domains which link adapter proteins, the ubiquitin ligase machinery for SCF and Elongin BC, respectively. Substrates are shown as shaded grey shapes.

proteolytic degradation (73). Thus, this system affords specificity resulting from unique protein recognition events between substrates and adapter proteins, while generating diversity in the number of potential substrates targeted by the E3 ligase resulting from the large number of F-box-containing proteins.

The NEDD4 family of E3 ubiquitin ligases has a distinct modular organization. Family members are composed of an N-terminal C2 Ca^{2+} binding domain, multiple WW domains, and a C-terminal HECT (homologous to E6-AP C-terminus) domain which contains the ubiquitin ligase function. WW domains are hydrophobic structures composed of β-sheets, which bind proteins that have characteristic PPXY motifs. NEDD4 ubiquitin ligases bind directly to their proline-rich substrates through their WW domains without the requirement of interceding adapter proteins (78–80). A schematic comparison of protein components of the NEDD4, SCF, and Elongin B/C ubiquitin ligases is shown in Fig. 3.

8.2 Molecular structure of the VCB complex

Insights into the protein interactions within the VCB complex have come from the determination of its crystal structure (76). VHL has two domains: a 100-residue N-terminal domain, rich in β-sheet; and a smaller α-helical domain which corresponds to the BC box. The α-helical domain consists of three α-helices, which combine with an α-helix donated by Elongin C to form a stable four-helical bundle structure. The Elongin C structure consists of a three-stranded β-sheet packing against four α-helices, the last of which is separated by an extended loop. Helices H2, H3, H4 and the extended loop of Elongin C form a concave hydrophobic pocket into which the H1 helix of VHL binds. The H4 helix of Elongin C, which bulges out from the side of the concave surface, fits into an extended groove formed by the H1, H2, and H3 helices of the VHL α-domain and completes the intermolecular four-helix cluster. The complementing surfaces of Elongin C and VHL are hydrophobic (76).

Sequence-threading analysis of the 40 amino acid SOCS box over the three-dimensional structure of the VHL α-domain demonstrated an alignment of the hydrophobic residues within the SOCS box with the VHL residues critical for contact with Elongin C. Therefore, the SOCS box and the VHL α-domain may represent a common structural and functional motif (76).

Genetic studies of individuals suffering from the VHL disease have identified mutations within the BC box of VHL which uncouple it from Elongin C (72). Mutant forms of VHL that disrupt the VCB complex may lead to the accumulation of cellular proteins which potentiate tumour formation. One such VHL binding protein is the hypoxia-inducible factor-1 (HIF-1). HIF-1 plays a key role in cellular responses to hypoxia, including the regulation of genes involved in energy metabolism, angiogenesis, and apoptosis (81–84). The α-subunits of HIF are rapidly degraded by the proteasome under normal conditions, but are stabilized by hypoxia. HIF binds to VHL in an iron-dependent manner and accumulates in VHL-deficient cells. The degradation of HIF-1 is dependent on VHL (85). Constitutive expression of HIF-1 may underlie the angiogenic phenotype of VHL-associated tumours.

In a similar manner to VHL, SOCS-1 may function as an adapter protein recruiting target proteins such as VAV for destruction to the Elongin B/C ubiquitin ligase machinery, and in this way modulate intracellular signalling pathways (Fig. 4). SOCS-1 mutants lacking the C-terminal SOCS box bind to VAV but fail to suppress VAV-mediated transformation in fibroblasts (14). SOCS box mutants, however, do not impair binding to or inhibition of JAK2 and are fully capable of blocking the transactivation of an IL-6- or LIF-responsive reporter assay (41, 42). Inhibition of IL-6-

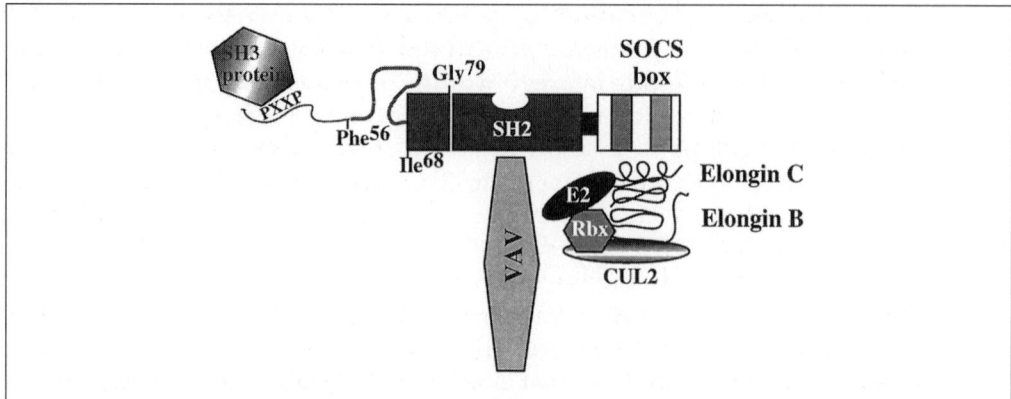

Fig. 4 SOCS-1 binds to VAV and is coupled to the Elongin B/C ubiquitin ligase. The SOCS box binds to the Elongin B/C ubiquitin ligase via α-helix 4 and loop 5 of Elongin C. Known components of the ubiquitin ligase are Elongin B, Cul2, Rbx, and a putative E2 ligase, thus far not identified. SOCS-1 binds to VAV via its SH2 domain in a phosphotyrosine-independent manner. SH3 proteins are recruited to the SOCS-1 complex via diproline motifs in the N-terminus.

mediated apoptosis was partially lost by truncation of the SOCS box (42). The SOCS box may therefore be targeting SOCS-1 to physically distinct and functionally separate signalling pathways.

An example of a SOCS-family protein participating in targeted protein degradation is CIS. CIS binds to the EPOR at a juxtamembrane tyrosine and suppresses EPO signalling. CIS, however, does not interfere with JAK kinase activation. CIS is a short-lived protein and is noted to be ubiquitinated in Cos cells (86). Constitutively expressed CIS protein in UT-7 cells can be stabilized by proteasome inhibitors such as LLnL or lactacystin, suggesting that CIS is degraded by the 26S proteasome (86). The EPOR, STAT5, and CIS–EPOR complexes are also stabilized by proteasome inhibitors. It is not known whether the activation-induced degradation of EPOR is dependent on CIS or if mutagenesis of the CIS binding site on EPOR at residue Tyr[401] would prolong the half-life of EPOR. Nor is it known if CIS deleted of its SOCS box can still target EPOR for degradation. Nevertheless, CIS-mediated degradation of the EPOR–STAT5 signalling complex, may be one mechanism by which CIS suppresses EPO signalling.

8.3   Problems with the SOCS–Elongin B/C hypothesis

The hypothesis that SOCS-1 functions as an adapter molecule, which targets a subset of cellular proteins for ubiquitination as a consequence of binding to the Elongin B/C ubiquitin ligase complex, has several predictions. In this model, deletion of the SOCS box should uncouple SOCS from Elongin C and thus stabilize SOCS-1 and its associated proteins. Indeed, two groups have observed the converse. Deletion of the SOCS box (SOCSΔCT) led to SOCS-1 protein destabilization and rendered it sensitive to ubiquitin-mediated degradation. The steady-state expression of SOCSΔCT protein was enhanced by proteasome inhibitors. These data led the authors to conclude that the SOCS box is important for protein stabilization, not destruction (41). Consistent with these observations, SOCS-1 protein's harbouring mutations in two conserved residues of the BC box required for VHL binding to Elongin C expressed at significantly lower levels than wild-type SOCS-1 (61). Pulse–chase analysis of transfected 293T cells demonstrated that the SOCS-1 proteins half-life was prolonged, rather than diminished, by the coexpression of Elongin B/C. These observations suggested that Elongin B/C stabilizes SOCS-1, a conclusion opposite to that predicted by the SOCS/ElonginB/C ubiquitin ligase model. An alternative model, which integrates these observations, states that SOCS-1 might function as a competitive inhibitor of the VCB complex and non-productively bind to Elongin B/C through its SOCS-box, precluding the interaction of other adapter molecules such as VHL from binding to Elongin B/C. This model would predict that overexpression of SOCS-1 would stabilize VHL-associated proteins from ubiquitin-mediated degradation. Another explanation rests on the realization that the VCB complex involves at least six proteins, all of which may be necessary for ubiquitin ligase activity. Massive overexpression of three components of this complex might disrupt this balanced stoichiometry and create partial protein complexes which are non-functional. In this way overexpression of Elongin B/C could dominantly interfere with the endogenous VCB complex and thus be observed to stabilize SOCS-1, when in fact its normal function is to degrade the SOCS-1 protein complex. Mutations in SOCS-1 which either delete the 40-amino acid carboxy terminus or generate disruptive point mutations may partially denature and thus destabilize the protein in a manner unrelated to its uncoupling from Elongin B/C.

9. Biology

9.1 CIS

CIS mRNA is abundantly expressed in the liver, kidney, heart, stomach, and lung, but is expressed at a low level in the brain and the spleen (1). Biochemical studies to date have demonstrated a function for CIS in negatively regulating EPO and IL-3 signalling. Its role in tissues where these cytokines are not active is currently unknown. CIS overexpression decreases the cell proliferation induced by EPO or IL-3 (1). CIS physically associates with the EPO and IL-3 receptors (1) but not with JAK kinases (8). CIS is upregulated by a subset of haemopoietic growth factors including IL-2, IL-3, GM-CSF, and EPO, but not Steel factor, G-CSF, and IL-6 (1). The transcriptional induction of CIS requires the activation of STAT5, which probably binds to four potential STAT5 binding sites within 200 bases upstream of the CIS transcription initiation site (32). Whereas SOCS-1 rapidly reduced the IL-6-induced gp130 tyrosine phosphorylation (2, 4), CIS did not decrease EPOR phosphorylation in UT-7 (1, 86). CIS associates with the second tyrosine residue (Tyr^{401}) of the intracellular domain of the EPOR (86). Tyr^{401} is phosphorylated *in vivo* and is one of two major binding sites for STAT5 on the EPOR (87, 88). CIS may potentially uncouple EPOR from STAT5 by

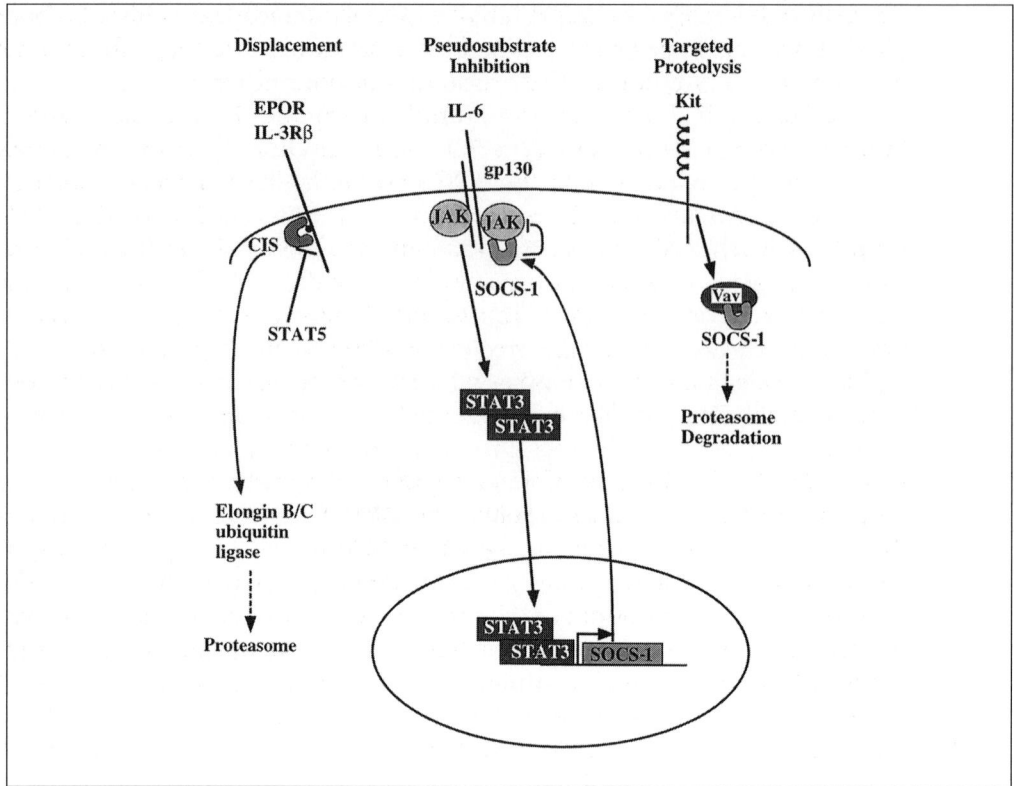

Fig. 5 SOCS family proteins use different mechanisms to attenuate signalling pathways. CIS suppresses EPO signalling by displacing STAT5 from binding to its juxtamembrane binding site on EPOR. CIS may also down-modulate EPOR by targeting it for ubiquitin-mediated proteolysis. SOCS-1 inhibits cytokine signalling by blocking JAK kinase activation as a pseudosubstrate. SOCS-1 is transcriptionally regulated by STAT3, whereas CIS is induced by STAT5. SOCS-1 binds to a putative ubiquitin ligase complex composed of Elongin B/C via its SOCS box. Proteins such as VAV which also bind to SOCS-1 may be targeted for degradation, thus modulating the signal transmitted by cell-surface receptors such as the Kit receptor tyrosine kinase.

competitive binding, and thus interfere with EPOR signalling. However, in the absence of Tyr^{401} STAT5 can still bind to EPOR at Tyr^{343}, which is not a CIS binding site (28, 87–90). Alternatively, CIS-mediated suppression of STAT5 activation might be by way of targeting EPOR to the proteasome degradation pathway as discussed above (Fig. 5).

9.2 SOCS-1 deficient mice

In order to clarify the functional and developmental role of SOCS-1 two groups used homologous recombination to disrupt the SOCS-1 locus (91, 92). The SOCS-1 deficient mice had no overt developmental defect. Although their birth weights were normal, significant weight discrepancies were noted within 1 week after birth and by 3 weeks of age there was universal lethality. Given the restricted expression of SOCS-

1 to cells of haemopoietic origin, an unexpected abnormality in the SOCS-1 deficient animals was fatty degeneration and necrosis of the liver as well as injury to the pancreas, heart, and lungs due to myelomonocytic infiltration.

Mice lacking SOCS-1 develop a complex pattern of haemopoietic abnormalities. Peripheral blood counts showed SOCS-1-deficient mice to be anaemic and lymphopenic. Blood granulocyte numbers are elevated and there is a deficiency in eosinophils. Lymphocyte development in the thymus and the spleen showed little difference at birth; however, by 10 days of age both thymocytes and splenocytes had decreased by 75–80% compared to wild-type controls. Splenic lymphoid follicles were poorly formed and often contained granulocytic infiltrates. Pre-B and B cells (B220+, CD43–) populations were significantly decreased, whereas pro-B cells appeared normal in number within the bone marrow. Lymph node cortices became depleted of lymphocytes over time. Splenic B and T cells were markedly decreased, though in equal proportion.

Thymic development appeared normal, with normal ratios of all subsets: $CD4^+CD8^+$, $CD4^+CD8^-$, $CD4^-CD8^+$ T cells. Thymocyte loss occurred in all subsets in equal proportion. Most of the cell death could be attributed to accelerated apoptosis in the thymus and secondary lymphoid organs of the SOCS-1-deficient mice. Immuno-histochemical staining suggested that in the absence of SOCS-1 the proapoptotic protein BAX accumulated without significant change in BCL2 levels (92). The observation that SOCS-1 can mediate the destruction of VAV and is a component of the putative Elongin B/C ubiquitin ligase complex suggests the possibility that SOCS-1 may similarly regulate the expression level of other proteins such as BAX. Disruption of the SOCS-1 gene could lead to the accumulation of BAX and the tendency of certain cell types to undergo apoptosis.

STAT6 phosphorylation was not properly downregulated in thymus-derived lymphocytes stimulated with IL-4 (92). The proliferation of SOCS-1-deficient thymocytes in response to αCD3 and IL-2 or IL-4 was augmented in comparison to wild-type controls.

The SOCS-1 knockout phenotype was a surprise given the anticipated function of SOCS-1 as a suppressor of the JAK–STAT pathway. The massive apoptosis evident in the thymus and spleen, as well as the accumulation of the BAX protein, suggested that SOCS-1 may regulate other signalling pathways. Granulocyte–macrophage progenitor cells exhibited heightened sensitivity to the inhibitory action of interferon-γ (IFN-γ) (93). SOCS-1 may thus be a negative regulator of IFN-γ in granulocyte progenitor cells. Dysregulation of IFN-γ activity in the absence of SOCS-1 could give rise to hepatocellular damage, as well as hyperresponsiveness of progenitor cells to granulocyte–macrophage colony-stimulating factor (GM-CSF) which could lead to the widespread tissue infiltration of granulocytes and monocytes (91).

9.2.1 SOCS-1 blocks interferon antiviral activity

Interferons (IFNs) provide pleiotropic functions to protect the host from viral infections. IFNs generate an early line of defence against viral infections by interfering with viral replication at stages of entry, uncoating, transcription, initiation of translation, and virion assembly. IFNs inhibit cell growth and initiate apoptosis in certain cell types, activities that aid in the suppression of infection and cancer. Lastly, IFNs have direct effects on the innate and adaptive immune response by modulating macrophage effector functions (94, 95), upregulating MHC expression (96), regulating antigen processing (97–100), facilitating the development of T_{H1} CD4+ T cells (101), and influencing Ig heavy-chain class-switching (102, 103).

Both type-I (α and β) and type-II (γ) IFNs signal through the JAK–STAT pathway and therefore may be regulated by SOCS family proteins. IFNs induce the expression of SOCS family-member mRNA in a cell type-dependent manner. IFN-γ induces the transcription of CIS, SOCS-1, SOCS-2, and SOCS-3 in murine bone marrow (2). IFN-γ, but not IFN-β, rapidly and potently upregulates the SOCS-1 and SOCS-3 mRNA message in NIH3T3 cells, whereas only SOCS-1 is induced in M1 cells (34, 104). Neither CIS nor SOCS-2 are induced in M1 cells by IFNs. SOCS-1 is more potently induced in M1 cells by IFN-γ than by IL-6, LIF, or GM-CSF (34), suggesting that a STAT1 responsive element is operative in the SOCS-1 5′-UTR. The antiproliferative effects of IFN-γ on M1 cells are blocked by the forced expression of SOCS-1 but not SOCS-3 (34, 104). This effect appears to be cell-type specific, because SOCS-3 is also able to block the antiproliferative effect of IFN-γ in HeLa cells (36). SOCS-2 cannot block the IFN effects in any cell-type tested. Vesicular stomatitis virus (VSV) is cytopathic for fibroblasts and HeLa cells. The antiviral effects of IFN protects fibroblasts from VSV-induced cell death. An NIH3T3 fibroblast cell line which constitutively expresses SOCS-1 is completely resistant to the anticytopathic effects of VSV (34, 36). IFN-γ signals through the STAT1 transcription factor which lies downstream of JAK1 and JAK2. As was observed with cytokines which signal through the gp130 receptor, SOCS-1 blocked the IFN-γ-induced phosphorylation and activation of JAK kinases (34), and STAT1 (34, 36) in adherent cells. In addition, the forced expression of SOCS-1 inhibited upregulation of the IFN-γ-inducible genes including IRF-1 and FcγR1 in M1 cells (34).

9.3 SOCS-2

The insulin growth factor-I receptor (IGF-IR) is a receptor tyrosine kinase highly related to the insulin receptor (INSR) and is involved in cellular growth, differentiation, and inhibition of apoptosis. IGF-IR binds to several adapter proteins including IRS1, Shc, and Grb-10 which couple it to downstream signalling events (105). Both SOCS-1 and SOCS-2 are capable of binding to IGF-IR, although neither suppress its autocatalytic function (26). The biological consequences of SOCS-2 binding to IGF-IR are not known. Mutagenesis analysis shows that tyrosine 950 and four tyrosines in the carboxy terminus of IGF-IR are not necessary for SOCS-2 binding, and suggests that the interaction occurs either through the kinase domain or the juxtamembrane region of the receptor (26). SOCS-2 is abundantly expressed in the fetal kidney and is a widely expressed in adult tissues (10, 26).

9.4 SOCS-3

SOCS-3 is another SOCS family member capable of inhibiting IL-6-mediated M1 cell differentiation. SOCS-1 and -3 were equally capable of suppressing IL-6 signalling. SOCS-3 can bind to JAK2 and inhibit its activity, but to a lesser extent than SOCS-1 (8, 104). Two-hybrid analysis demonstrated that SOCS-3 can bind to JAK2, LCK, PYK2, and FGFR, although the physiological relevance of these interactions remain to be determined (8). SOCS-3 can suppress EPO and growth hormone (GH) signalling both of which utilize JAK2, but not IFN-β, which utilizes JAK1 and TYK2 (8, 106). SOCS-3 interferes with IL-4-induced gene transcription in transient transfection assays, yet in stable cell lines SOCS-3 does not suppress IL-4 signalling (35).

9.4.1 SOCS-3 function in peripheral blood leucocytes

SOCS-3 mRNA is highly expressed in peripheral blood leucocytes (PBL). Transcription of SOCS-3 is rapidly induced in response to the IL-2 stimulation of several lympho-cyte cell lines and in human peripheral blood leucocyte-derived blasts. SOCS-3 is tyrosine-phosphorylated in response to IL-2 stimulation in Kit-225 cells, although the role tyrosine phosphorylation plays in the assembly of multiprotein complexes, JAK inhibition, or protein turnover is not known. IL-2 stimulation of lymphocytes acti-vates both JAK1 and JAK3 (103–105). SOCS-3 binds only to and is phosphorylated by JAK1 in transfected 293T cells. SOCS-3 can inhibit the phosphorylation of JAK1 in a manner reminiscent of SOCS-1, but it is a weak antagonist of JAK3 activation (27). The inhibitory effect of SOCS-3 on JAK1 can functionally block IL-2-mediated signal-ling (27). Ectopic expression of SOCS-3 in BaF3 cells harbouring the IL-2Rβ chain suppressed IL-2-mediated phosphorylation of STAT5b (27).

9.4.2 SOCS-3 and the regulation of IFN responsiveness

IL-10 has numerous effects on haemopoietic cells, but most notable is its role in the development of CD4$^+$ T$_{H2}$ lymphocytes. The polarization of T-cell development is thought to progress from a T$_{H0}$ phase to either a T$_{H1}$ (secreting IFN and IL-2) or T$_{H2}$ (secreting IL-4, IL-5 and IL-10) cell, depending on the presence or absence of various interleukins (110). IL-10 suppresses the T$_{H1}$ phenotype by blocking the expression of IL-12 required for inducing the expression and secretion of IFN-γ in T$_{H1}$ T cells (111). IL-10 can inhibit the induction of IFN-induced genes by preventing the assembly of ISGF3 and GRR/GAS transcription complexes (112). The suppression of gene expres-sion correlates with the IL-10-induced inhibition of the tyrosine phosphorylation of STAT1. One mechanism by which IL-10 suppresses IFN signalling is by virtue of its capacity to induce the rapid expression of SOCS-3 mRNA (112). Therefore SOCS-3 may not just be functioning as a negative feedback loop in IFN signalling but may play an important function in the mutual cross-talk between T$_{H1}$ and T$_{H2}$ cells.

9.4.3 SOCS-3 function in the pituitary

Leptin is an adipocyte-derived hormone that acts on specific regions of the brain to regulate food intake, energy expenditure, and neuroendocrine function (113–117).

Leptin binds to a receptor expressed in the hypothalamic nuclei that is related to the cytokine receptor superfamily and signals through the JAK–STAT pathway (118–122). Mice which lack either leptin (*ob/ob*) or its receptor (*db/db*) are severely obese (107, 121, 123). Peripheral administration of leptin leads to the upregulation of SOCS-3 mRNA —but not CIS, SOCS-1, or SOCS-2—in hypothalamic nuclei that express high levels of the leptin receptor (124). Signalling through the leptin receptor, as measured by a leptin-responsive element egr-1, is blocked by the forced expression of SOCS-3 (124). Phosphorylation of the leptin receptor is inhibited in those cells which constitutively express SOCS-3 (124). Defects at any point along the leptin signal transduction pathway could lead to leptin resistance and obesity. One such example is *lethal yellow* (A^y/a), which is a murine model of obesity whose locus is distinct from either *ob/ob* and *db/db* but is phenotypically similar to both. A^y/a mice develop obesity due to ectopic and unregulated overexpression of the agouti protein, a potent melanocortin receptor antagonist. These mice have central leptin resistance and express elevated levels of SOCS-3 mRNA in the dorsal medial hypothalamus (124). One possible cause of their obesity may be due to constitutive upregulation of SOCS-3 in the hypothalamus, which in turn may blunt leptin responsiveness.

The ciliary neurotrophic factor (CNTF) is another member of the IL-6 cytokine superfamily that regulates body weight (125). Leptin and CNTF induce SOCS-3 transcription in distinct but overlapping areas in the thalamus (126). Whereas peripheral administration of leptin to rats increased the SOCS-3 message in the dorsal medial nuclei and arcuate nuclei of the thalamus, CNTF potently upregulated SOCS-3 mRNA in the arcuate nucleus and the ependymal lining of the ventricles (126).

The hypothalamus–pituitary–adrenal (HPA) axis responds to inflammatory and stress stimuli. This response may be mediated by the IL-6 cytokine family members IL-11 and leukaemic inhibitory factor (LIF), which signal through the common gp130 receptor subunit to stimulate the induction of the pro-opiomelanocortin (POMC), a precursor form of ACTH. (127–129). LIF-induced ACTH secretion is dependent on the activation of STAT-3 and the induction of the POMC gene (130). SOCS-3 gene expression is rapidly induced by LIF in the pituitary *in vivo* and in corticotroph AtT-20 cells *in vitro* (131). Dominant-negative mutants of STAT-3 inhibit the LIF-dependent induction of SOCS-3 (132). Mutagenesis analysis of the SOCS-3 promoter reveals a STAT1/STAT-3 binding element, located at –72 to –64 (132). Forced expression of the SOCS-3 protein in AtT-20 cells inhibits LIF-induced transcription of the POMC gene and ACTH secretion (131). SOCS-3 expression also blocks further LIF-mediated induction of the SOCS-3 transcription, thereby limiting its accumulation by an autoregulatory negative feedback loop (132). The regulatory function of SOCS-3 may therefore serve to generate a 'pulse' of ACTH induction following the LIF stimulation of the HPA.

GH stimulated a rapid, transient induction of SOCS-3 mRNA, whereas SOCS-1, SOCS-2, and CIS were induced at a lower level with slower kinetics in the GH-responsive human fibroblast line 3T3-F442A (106). SOCS-3 mRNA was also induced in the liver of mice exposed to supraphysiological levels of GH administered intraperitoneally. Transactivation of the GH responsive serine protease inhibitor Spi 2.1

gene was blocked by the forced expression of SOCS-3 and SOCS-1, but not by SOCS-2 or CIS (106). SOCS-3 may thus regulate a family of neurotrophic cytokines including leptin, CNTF, LIF, and GH, all of which signal through JAK2 and modulate the hypothalamic–pituitary axis.

10. Conclusions and perspectives

The SOCS family represents a new set of adapter proteins that are defined by a central SH2 domain, a C-terminal SOCS box, and an N-terminus of variable length. Many SOCS family members are inducible proteins and create a negative feedback loop to modulate the proximal signalling pathways that stimulate their own synthesis. The SOCS proteins have evolved to target different components of signalling pathways. CIS may suppress EPO signalling by competing with STAT5 for a common binding site on the EPOR. SOCS-1 functions as a potent JAK kinase pseudosubstrate inhibitor. Both CIS and SOCS-1 may also function by targeting signalling proteins for degradation through the Elongin B/C ubiquitin ligase complex.

SOCS proteins integrate diverse signalling pathways as a consequence of forming multi-protein complexes. Identification of the protein partners for each of the family members will be an area of important future research. The SOCS family proteins are anticipated to bind to the Elongin B/C ubiquitin ligase complex through the SOCS box motif and thus recruit their respective binding partners into a degradation pathway. Like the F-Box proteins, the SOCS box-containing proteins create a new layer of complexity in our models of signal transmission within the cell. Signalling events can be attenuated or 'tuned' through the destruction of specific signalling molecules following the induction of SOCS family proteins.

VHL serves as an important example of a SOCS box protein which functions as a tumour suppressor gene. SOCS-1 is a potent suppressor of the JAK–STAT pathway and may be a candidate tumour suppressor for cytokine-driven haemopoietic malignancies such as multiple myeloma. Other clinical scenarios where SOCS-1 protein expression may be important are in those patients who are poor responders to IFN treatment for diseases such as hepatitis C or hairy-cell leukaemia. These individuals may express elevated SOCS-1 levels in these tissues leading to IFN resistance. In a similar manner, polymorphisms in the regulatory sequences of SOCS-3 which lead to constitutive elevated levels in the hypothalamus may account for some individuals who suffer from a leptin-resistance form of obesity.

The SOCS story has taught us that signal transduction is not simply the activation of preset pathways or circuits present in the cell. Rather, signalling is a dynamic process where inducible components can feed back upon the signalling network itself to 'sculpt' or modulate the quality and/or quantity of the incoming signal. SOCS proteins may serve to attenuate the amplitude or the duration of the 'signal', or may even alter the available pathways through which a signal can propagate (Fig. 5). One such example was observed when the expression of SOCS-1 in EML-C1 cells switched steel factor-dependent proliferation to a steel factor-dependent cytostasis (6). The induction of SOCS proteins may provide a potent mechanism by which a single receptor can generate a diversity of signalling outcomes.

References

1. Yoshimura, A., T. Ohkubo, T. Kiguchi, N. A. Jenkins, D. J. Gilbert, N. G. Copeland, T. Hara, and A. Miyajima. (1995). A novel cytokine-inducible gene CIS encodes an SH2-containing protein that binds to tyrosine-phosphorylated interleukin 3 and erythropoietin receptors. *EMBO J.*, **14**, 2816.

2. Starr, R., T. A. Willson, E. M. Viney, L. J. Murray, J. R. Rayner, B. J. Jenkins, T. J. Gonda, W. S. Alexander, D. Metcalf, N. A. Nicola, and D. J. Hilton. (1997). A family of cytokine-inducible inhibitors of signalling. *Nature*, **387**, 917.

3. Endo, T. A., M. Masuhara, M. Yokouchi, R. Suzuki, H. Sakamoto, K. Mitsui, A. Matsumoto, S. Tanimura, M. Ohtsubo, H. Misawa, T. Miyazaki, N. Leonor, T. Taniguchi, T. Fujita, Y. Kanakura, S. Komiya, and A. Yoshimura. (1997). A new protein containing an SH2 domain that inhibits JAK kinases. *Nature*, **387**, 921.

4. Naka, T., M. Narazaki, M. Hirata, T. Matsumoto, S. Minamoto, A. Aono, N. Nishimoto, T. Kajita, T. Taga, K. Yoshizaki, S. Akira, and T. Kishimoto. (1997). Structure and function of a new STAT-induced STAT inhibitor. *Nature*, **387**, 924.

5. Ohya, K., S. Kajigaya, Y. Yamashita, A. Miyazato, K. Hatake, Y. Miura, U. Ikeda, K. Shimada, K. Ozawa, and H. Mano. (1997). SOCS-1/JAB/SSI-1 can bind to and suppress Tec protein-tyrosine kinase. *J. Biol. Chem.*, **272**, 27178.

6. De Sepulveda, P., K. Okkenhaug, J. L. Rose, R. G. Hawley, P. Dubreuil, and R. Rottapel. (1999). Socs1 binds to multiple signalling proteins and suppresses steel factor-dependent proliferation. *EMBO J.*, **18**, 904.

7. Hilton, D. J., R. T. Richardson, W. S. Alexander, E. M. Viney, T. A. Willson, N. S. Sprigg, R. Starr, S. E. Nicholson, D. Metcalf, and N. A. Nicola. (1998). Twenty proteins containing a C-terminal SOCS box form five structural classes. *Proc. Natl Acad. Sci. USA*, **95**, 114.

8. Masuhara, M., H. Sakamoto, A. Matsumoto, R. Suzuki, H. Yasukawa, K. Mitsui, T. Wakioka, S. Tanimura, A. Sasaki, H. Misawa, M. Yokouchi, M. Ohtsubo, and A. Yoshimura. (1997). Cloning and characterization of novel CIS family genes. *Biochem. Biophys. Res. Commun.*, **239**, 439.

9. Matuoka, K., H. Miki, K. Takahashi, and T. Takenawa. (1997). A novel ligand for an SH3 domain of the adaptor protein Nck bears an SH2 domain and nuclear signaling motifs. *Biochem. Biophys. Res. Commun.*, **239**, 488.

10. Minamoto, S., K. Ikegame, K. Ueno, M. Narazaki, T. Naka, H. Yamamoto, T. Matsumoto, H. Saito, S. Hosoe, and T. Kishimoto. (1997). Cloning and functional analysis of new members of STAT induced STAT inhibitor (SSI) family: SSI-2 and SSI-3. *Biochem. Biophys. Res. Commun.*, **237**, 79.

11. Reeves, R. H., J. D. Gearhart, N. B. Hecht, P. Yelick, P. Johnson, and S. J. O'Brien. (1989). Mapping of PRM1 to human chromosome 16 and tight linkage of Prm-1 and Prm-2 on mouse chromosome 16. *J. Hered.*, **80**, 442.

12. Adham, I. M., C. Szpirer, H. Kremling, S. Keime, J. Szpirer, G. Levan, and W. Engel. (1991). Chromosomal assignment of four rat genes coding for the spermatid-specific proteins proacrosin (ACR), transition proteins 1 (TNP1) and 2 (TNP2), and protamine 1 (PRM1). *Cytogenet. Cell Genet.*, **57**, 47.

13. Viguie, F., L. Domenjoud, M. M. Rousseau, J. P. Dadoune, and P. Chevaillier. (1990).

Chromosomal localization of the human protamine genes, PRM1 and PRM2, to 16p13.3 by in situ hybridization. *Hum. Genet.*, **85**, 171.

14. De Sepulveda, P., S. Ilangumaran, and R. Rottapel. (Preliminary data.)

15. Vasiliauskas, D., S. Hancock, and C. D. Stern. (1999). SWiP-1: novel SOCS box containing WD-protein regulated by signalling centres and by shh during development. *Mech. Dev.*, **82**, 79.

16. Johnson, R. L., E. Laufer, R. D. Riddle, and C. Tabin. (1994). Ectopic expression of Sonic hedgehog alters dorsal-ventral patterning of somites. *Cell*, **79**, 1165.

17. Borycki, A. G., L. Mendham, and C. P. Emerson, Jr. (1998). Control of somite patterning by Sonic hedgehog and its downstream signal response genes. *Development*, **125**, 777.

18. Chiang, C., Y. Litingtung, E. Lee, K. E. Young, J. L. Corden, H. Westphal, and P. A. Beachy. (1996). Cyclopia and defective axial patterning in mice lacking Sonic hedgehog gene function. *Nature*, **383**, 407.

19. Fan, C. M., J. A. Porter, C. Chiang, D. T. Chang, P. A. Beachy, and M. Tessier-Lavigne. (1995). Long-range sclerotome induction by sonic hedgehog: direct role of the amino-terminal cleavage product and modulation by the cyclic AMP signaling pathway. *Cell*, **81**, 457.

20. Fan, C. M. and M. Tessier-Lavigne. (1994). Patterning of mammalian somites by surface ectoderm and notochord: evidence for sclerotome induction by a hedgehog homolog. *Cell*, **79**, 1175.

21. Riddle, R. D., R. L. Johnson, E. Laufer, and C. Tabin. (1993). Sonic hedgehog mediates the polarizing activity of the ZPA. *Cell*, **75**, 1401.

22. Chang, D. T., A. Lopez, D. P. von Kessler, C. Chiang, B. K. Simandl, R. Zhao, M. F. Seldin, J. F. Fallon, and P. A. Beachy. (1994). Products, genetic linkage and limb patterning activity of a murine hedgehog gene. *Development*, **120**, 3339.

23. Lopez-Martinez, A., D. T. Chang, C. Chiang, J. A. Porter, M. A. Ros, B. K. Simandl, P. A. Beachy, and J. F. Fallon. (1995). Limb-patterning activity and restricted posterior local-ization of the amino-terminal product of Sonic hedgehog cleavage. *Curr. Biol.*, **5**, 791.

24. Yang, Y., G. Drossopoulou, P. T. Chuang, D. Duprez, E. Marti, D. Bumcrot, N. Vargesson, J. Clarke, L. Niswander, A. McMahon, and C. Tickle. (1997). Relationship between dose, distance and time in Sonic Hedgehog-mediated regulation of anteroposterior polarity in the chick limb. *Development*, **124**, 4393.

25. Ingham, P. W. (1998). Transducing Hedgehog: the story so far. *EMBO J.*, **17**, 3505.

26. Dey, B. R., S. L. Spence, P. Nissley, and R. W. Furlanetto. (1998). Interaction of human suppressor of cytokine signaling (SOCS)-2 with the insulin-like growth factor-I receptor. *J. Biol. Chem.*, **273**, 24095.

27. Cohney, S. J., D. Sanden, N. A. Cacalano, A. Yoshimura, A. Mui, T. S. Migone, and J. A. Johnston. (1999). SOCS-3 is tyrosine phosphorylated in response to interleukin-2 and suppresses STAT5 phosphorylation and lymphocyte proliferation. *Mol. Cell. Biol.*, **19**, 4980.

28. Quelle, F. W., D. Wang, T. Nosaka, W. E. Thierfelder, D. Stravopodis, Y. Weinstein, and J. N. Ihle. (1996). Erythropoietin induces activation of Stat5 through association with specific tyrosines on the receptor that are not required for a mitogenic response. *Mol. Cell. Biol.*, **16**, 1622.

29. Hou, J., U. Schindler, W. J. Henzel, S. C. Wong, and S. L. McKnight. (1995). Identification and purification of human Stat proteins activated in response to interleukin-2. *Immunity*, **2**, 321.

30. Wakao, H., N. Harada, T. Kitamura, A. L. Mui, and A. Miyajima. (1995). Interleukin 2 and erythropoietin activate STAT5/MGF via distinct pathways. *EMBO J.*, **14**, 2527.

31. Mui, A. L., H. Wakao, T. Kinoshita, T. Kitamura, and A. Miyajima. (1996). Suppression of interleukin-3-induced gene expression by a C-terminal truncated Stat5: role of Stat5 in proliferation. *EMBO J.*, **15**, 2425.

32. Matsumoto, A., M. Masuhara, K. Mitsui, M. Yokouchi, M. Ohtsubo, H. Misawa, A. Miyajima, and A. Yoshimura. (1997). CIS, a cytokine inducible SH2 protein, is a target of the JAK-STAT5 pathway and modulates STAT5 activation. *Blood*, **89**, 3148.

33. Teglund, S., C. McKay, E. Schuetz, J. M. van Deursen, D. Stravopodis, D. Wang, M. Brown, S. Bodner, G. Grosveld, and J. N. Ihle. (1998). Stat5a and Stat5b proteins have essential and nonessential, or redundant, roles in cytokine responses. *Cell*, **93**, 841.

34. Sakamoto, H., H. Yasukawa, M. Masuhara, S. Tanimura, A. Sasaki, K. Yuge, M. Ohtsubo, A. Ohtsuka, T. Fujita, T. Ohta, Y. Furukawa, S. Iwase, H. Yamada, and A. Yoshimura. (1998). A Janus kinase inhibitor, JAB, is an interferon-gamma-inducible gene and confers resistance to interferons. *Blood*, **92**, 1668.

35. Losman, J. A., X. P. Chen, D. Hilton, and P. Rothman. (1999). Cutting edge: SOCS-1 is a potent inhibitor of IL-4 signal transduction. *J. Immunol.*, **162**, 3770.

36. Song, M. M. and K. Shuai. (1998). The suppressor of cytokine signaling, SOCS1 and SOCS3 but not SOCS2 proteins inhibit interferon-mediated antiviral and antiproliferative activities. *J. Biol. Chem.*, **273**, 35056.

37. Xu, W., S. C. Harrison, and M. J. Eck. (1997). Three-dimensional structure of the tyrosine kinase c-Src. *Nature*, **385**, 595.

38. Sicheri, F., I. Moarefi, and J. Kuriyan. (1997). Crystal structure of the Src family tyrosine kinase Hck. *Nature*, **385**, 602.

39. Feng, J., B. A. Witthuhn, T. Matsuda, F. Kohlhuber, I. M. Kerr, and J. N. Ihle. (1997). Activation of Jak2 catalytic activity requires phosphorylation of Y1007 in the kinase activation loop. *Mol. Cell. Biol.*, **17**, 2497.

40. Yasukawa, H., H. Misawa, H. Sakamoto, M. Masuhara, A. Sasaki, T. Wakioka, S. Ohtsuka, T. Imaizumi, T. Matsuda, J. N. Ihle, and A. Yoshimura. (1999). The JAK-binding protein JAB inhibits Janus tyrosine kinase activity through binding in the activation loop. *EMBO J.*, **18**, 1309.

41. Narazaki, M., M. Fujimoto, T. Matsumoto, Y. Morita, H. Saito, T. Kajita, K. Yoshizaki, T. Naka, and T. Kishimoto. (1998). Three distinct domains of SSI-1/SOCS-1/JAB protein are required for its suppression of interleukin 6 signaling. *Proc. Natl Acad. Sci. USA*, **95**, 13130.

42. Nicholson, S. E., T. A. Willson, A. Farley, R. Starr, J. G. Zhang, M. Baca, W. S. Alexander, D. Metcalf, D. J. Hilton, and N. A. Nicola. (1999). Mutational analyses of the SOCS proteins suggest a dual domain requirement but distinct mechanisms for inhibition of LIF and IL-6 signal transduction. *EMBO J.*, **18**, 375.

43. Chen, X., U. Vinkemeier, Y. Zhao, D. Jeruzalmi, J. E. Darnell, Jr, and J. Kuriyan. (1998). Crystal structure of a tyrosine phosphorylated STAT-1 dimer bound to DNA. *Cell*, **93**, 827.

44. Yarden, Y. and A. Ullrich. (1988). Growth factor receptor tyrosine kinases. *Annu. Rev. Biochem.*, **57**, 443.

45. Russel, E. S. (1979). Hereditary anemias of the mouse: a review for geneticists. *Adv. Genet.*, **20**, 357.

46. Bernstein, S. and E. Russell. (1959). Implantation of normal blood forming tissue in genetically anemic mice, without the irradiation of the host. *Proc. Exp. Biol. Med.*, **101**, 769.

47. Kitamura, Y., S. Go, and K. Hatanaka. (1978). Decrease of mast cells in W/Wv mice and their increase by bone marrow transplantation. *Blood*, **52**, 447.

48. Mackarehtschian, K., J. D. Hardin, K. A. Moore, S. Boast, S. P. Goff, and I. R. Lemischka.

(1995). Targeted disruption of the flk2/flt3 gene leads to deficiencies in primitive hemato-poietic progenitors. *Immunity*, **3**, 147.

49. Yu, H., J. K. Chen, S. Feng, D. C. Dalgarno, A. W. Brauer, and S. L. Schreiber. (1994). Structural basis for the binding of proline-rich peptides to SH3 domains. *Cell*, **76**, 933.

50. Sparks, A. B., J. E. Rider, N. G. Hoffman, D. M. Fowlkes, L. A. Quillam, and B. K. Kay. (1996). Distinct ligand preferences of Src homology 3 domains from Src, Yes, Abl, Cortactin, p53bp2, PLCgamma, Crk, and Grb2. *Proc. Natl Acad. Sci. USA*, **93**, 1540.

51. Crespo, P., K. E. Schuebel, A. A. Ostrom, J. S. Gutkind, and X. R. Bustelo. (1997). Phospho-tyrosine-dependent activation of Rac-1 GDP/GTP exchange by the vav proto-oncogene product. *Nature*, **385**, 169.

52. Ridley, A. J., H. F. Paterson, C. L. Johnston, D. Diekmann, and A. Hall. (1992). The small GTP-binding protein rac regulates growth factor-induced membrane ruffling. *Cell*, **70**, 401.

53. Bustelo, X. R., J. A. Ledbetter, and M. Barbacid. (1992). Product of vav proto-oncogene defines a new class of tyrosine protein kinase substrates. *Nature*, **356**, 68.

54. Bustelo, X. R. and M. Barbacid. (1992). Tyrosine phosphorylation of the vav proto-oncogene product in activated B cells. *Science*, **256**, 1196.

55. August, A., S. Gibson, Y. Kawakami, T. Kawakami, G. B. Mills, and B. Dupont. (1994). CD28 is associated with and induces the immediate tyrosine phosphorylation and activation of the Tec family kinase ITK/EMT in the human Jurkat leukemic T-cell line. *Proc. Natl Acad. Sci. USA*, **91**, 9347.

56. Margolis, B., P. Hu, S. Katzav, W. Li, J. M. Oliver, A. Ullrich, A. Weiss, and J. Schlessinger. (1992). Tyrosine phosphorylation of vav proto-oncogene product containing SH2 domain and transcription factor motifs. *Nature*, **356**, 71.

57. Weng, W. K., L. Jarvis, and T. W. LeBien. (1994). Signaling through CD19 activates Vav/mitogen-activated protein kinase pathway and induces formation of a CD19/Vav/phosphatidylinositol 3-kinase complex in human B cell precursors. *J. Biol. Chem.*, **269**, 32514.

58. Alai, M., A. L. Mui, R. L. Cutler, X. R. Bustelo, M. Barbacid, and G. Krystal. (1992). Steel factor stimulates the tyrosine phosphorylation of the proto-oncogene product, p95vav, in human hemopoietic cells. *J. Biol. Chem.*, **267**, 18021.

59. Katzav, S., D. Martin-Zanca, and M. Barbacid. (1989). vav, a novel human oncogene derived from a locus ubiquitously expressed in hematopoietic cells. *EMBO J.*, **8**, 2283.

60. Schuebel, K. E., X. R. Bustelo, D. A. Nielson, B. Song, M. Barbacid, D. Goldman, and I. L. Lee. (1996). Isolation and characterization of murine vav2, a member of the vav family of proto-oncogenes. *Oncogene*, **13**, 363.

61. Kamura, T., S. Sato, D. Haque, L. Liu, W. G. Kaelin, Jr, R. C. Conaway, and J. W. Conaway. (1998). The Elongin BC complex interacts with the conserved SOCS-box motif present in members of the SOCS, ras, WD-40 repeat, and ankyrin repeat families. *Genes Dev.*, **12**, 3872.

62. Zhang, J. G., A. Farley, S. E. Nicholson, T. A. Willson, L. M. Zugaro, R. J. Simpson, R. L. Moritz, D. Cary, R. Richardson, G. Hausmann, B. J. Kile, S. B. Kent, W. S. Alexander, D. Metcalf, D. J. Hilton, N. A. Nicola, and M. Baca. (1999). The conserved SOCS box motif in suppressors of cytokine signaling binds to elongins B and C and may couple bound proteins to proteasomal degradation. *Proc. Natl Acad. Sci. USA*, **96**, 2071.

63. Conaway, J. W., T. Kamura, and R. C. Conaway. (1998). The Elongin BC complex and the von Hippel-Lindau tumor suppressor protein. *Biochim. Biophys. Acta*, **1377**, M49.

64. Tyers, M. and A. R. Willems. (1999). One ring to rule a superfamily of E3 ubiquitin ligases. *Science*, **284**, 601.

65. Pause, A., S. Lee, R. A. Worrell, D. Y. Chen, W. H. Burgess, W. M. Linehan, and R. D. Klausner. (1997). The von Hippel-Lindau tumor-suppressor gene product forms a stable complex with human CUL-2, a member of the Cdc53 family of proteins. *Proc. Natl Acad. Sci. USA*, **94**, 2156.

66. Lonergan, K. M., O. Iliopoulos, M. Ohh, T. Kamura, R. C. Conaway, J. W. Conaway, and W. G. Kaelin, Jr. (1998). Regulation of hypoxia-inducible mRNAs by the von Hippel-Lindau tumor suppressor protein requires binding to complexes containing elongins B/C and Cul2. *Mol. Cell Biol.*, **18**, 732.

67. Kamura, T., D. M. Koepp, M. N. Conrad, D. Skowyra, R. J. Moreland, O. Iliopoulos, W. S. Lane, W. G. Kaelin, Jr, S. J. Elledge, R. C. Conaway, J. W. Harper, and J. W. Conaway. (1999). Rbx1, a component of the VHL tumor suppressor complex and SCF ubiquitin ligase. *Science*, **284**, 657.

68. Ohta, T., J. J. Michel, A. J. Schottelius, and Y. Xiong. (1999). ROC1, a homolog of APC11, represents a family of cullin partners with an associated ubiquitin ligase activity. *Mol Cell*, **3**, 535.

69. Latif, F., K. Tory, J. Gnarra, M. Yao, F. M. Duh, M. L. Orcutt, T. Stackhouse, I. Kuzmin, W. Modi, L. Geil, *et al.* (1993). Identification of the von Hippel-Lindau disease tumor suppressor gene. *Science*, **260**, 1317.

70. Gnarra, J. R., K. Tory, Y. Weng, L. Schmidt, M. H. Wei, H. Li, F. Latif, S. Liu, F. Chen, F. M. Duh, *et al.* (1994). Mutations of the VHL tumour suppressor gene in renal carcinoma. *Nature Genet.*, **7**, 85.

71. Foster, K., A. Prowse, A. van den Berg, S. Fleming, M. M. Hulsbeek, P. A. Crossey, F. M. Richards, P. Cairns, N. A. Affara, M. A. Ferguson-Smith, *et al.* (1994). Somatic mutations of the von Hippel-Lindau disease tumour suppressor gene in non-familial clear cell renal carcinoma. *Hum. Mol. Genet.*, **3**, 2169.

72. Kaelin, W. G., Jr and E. R. Maher. (1998). The VHL tumour-suppressor gene paradigm. *Trends Genet.*, **14**, 423.

73. Bai, C., P. Sen, K. Hofmann, L. Ma, M. Goebl, J. W. Harper, and S. J. Elledge. (1996). SKP1 connects cell cycle regulators to the ubiquitin proteolysis machinery through a novel motif, the F-box. *Cell*, **86**, 263.

74. Zhang, H., R. Kobayashi, K. Galaktionov, and D. Beach. (1995). p19Skp1 and p45Skp2 are essential elements of the cyclin A-CDK2 S phase kinase. *Cell*, **82**, 915.

75. Garrett, K. P., T. Aso, J. N. Bradsher, S. I. Foundling, W. S. Lane, R. C. Conaway, and J. W. Conaway. (1995). Positive regulation of general transcription factor SIII by a tailed ubiquitin homolog. *Proc. Natl Acad. Sci. USA*, **92**, 7172.

76. Stebbins, C. E., W. G. Kaelin, Jr, and N. P. Pavletich. (1999). Structure of the VHL-ElonginC-ElonginB complex: implications for VHL tumor suppressor function. *Science*, **284**, 455.

77. Skowyra, D., K. L. Craig, M. Tyers, S. J. Elledge, and J. W. Harper. (1997). F-box proteins are receptors that recruit phosphorylated substrates to the SCF ubiquitin-ligase complex. *Cell*, **91**, 209.

78. Rotin, D. (1998). WW (WWP) domains: from structure to function. *Curr. Top. Microbiol. Immunol.*, **228**, 115.

79. Sudol, M. (1998). From Src Homology domains to other signaling modules: proposal of the 'protein recognition code'. Oncogene, **17**, 1469.

80. Harvey, K. F., and S. Kumar. (1999). Nedd4-like proteins: an emerging family of ubiquitin-protein ligases implicated in diverse cellular functions. *Trends Cell Biol.*, **9**, 166.

81. An, W. G., M. Kanekal, M. C. Simon, E. Maltepe, M. V. Blagosklonny, and L. M. Neckers.

(1998). Stabilization of wild-type p53 by hypoxia-inducible factor 1alpha. *Nature*, **392**, 405.

82. Wang, G. L., B. H. Jiang, E. A. Rue, and G. L. Semenza. (1995). Hypoxia-inducible factor 1 is a basic-helix–loop–helix–PAS heterodimer regulated by cellular O_2 tension. *Proc. Natl Acad. Sci. USA*, **92**, 5510.

83. Bunn, H. F. and R. O. Poyton. (1996). Oxygen sensing and molecular adaptation to hypoxia. *Physiol. Rev.*, **76**, 839.

84. Carmeliet, P., Y. Dor, J. M. Herbert, D. Fukumura, K. Brusselmans, M. Dewerchin, M. Neeman, F. Bono, R. Abramovitch, P. Maxwell, C. J. Koch, P. Ratcliffe, L. Moons, R. K. Jain, D. Collen, E. Keshert, and E. Keshet. (1998). Role of HIF-1alpha in hypoxia-mediated apoptosis, cell proliferation and tumour angiogenesis. *Nature*, **394**, 485.

85. Maxwell, P. H., M. S. Wiesener, G. W. Chang, S. C. Clifford, E. C. Vaux, M. E. Cockman, C. C. Wykoff, C. W. Pugh, E. R. Maher, and P. J. Ratcliffe. (1999). The tumour suppressor protein VHL targets hypoxia-inducible factors for oxygen-dependent proteolysis. *Nature*, **399**, 271.

86. Verdier, F., S. Chretien, O. Muller, P. Varlet, A. Yoshimura, S. Gisselbrecht, C. Lacombe, and P. Mayeux. (1998). Proteasomes regulate erythropoietin receptor and signal trans-ducer and activator of transcription 5 (STAT5) activation. Possible involvement of the ubiquitinated Cis protein. *J. Biol. Chem.*, **273**, 28185.

87. Klingmuller, U., S. Bergelson, J. G. Hsiao, and H. F. Lodish. (1996). Multiple tyrosine residues in the cytosolic domain of the erythropoietin receptor promote activation of STAT5. *Proc. Natl Acad. Sci. USA*, **93**, 8324.

88. Gobert, S., S. Chretien, F. Gouilleux, O. Muller, C. Pallard, I. Dusanter-Fourt, B. Groner, C. Lacombe, S. Gisselbrecht, and P. Mayeux. (1996). Identification of tyrosine residues within the intracellular domain of the erythropoietin receptor crucial for STAT5 activation. *EMBO J.*, **15**, 2434.

89. Damen, J. E., H. Wakao, A. Miyajima, J. Krosl, R. K. Humphries, R. L. Cutler, and G. Krystal. (1995). Tyrosine 343 in the erythropoietin receptor positively regulates erythro-poietin-induced cell proliferation and Stat5 activation. *EMBO J.*, **14**, 5557.

90. Chretien, S., P. Varlet, F. Verdier, S. Gobert, J. P. Cartron, S. Gisselbrecht, P. Mayeux, and C. Lacombe. (1996). Erythropoietin-induced erythroid differentiation of the human erythro-leukemia cell line TF-1 correlates with impaired STAT5 activation. *EMBO J.*, **15**, 4174.

91. Starr, R., D. Metcalf, A. G. Elefanty, M. Brysha, T. A. Willson, N. A. Nicola, D. J. Hilton, and W. S. Alexander. (1998). Liver degeneration and lymphoid deficiencies in mice lacking suppressor of cytokine signaling-1. *Proc. Natl Acad. Sci. USA*, **95**, 14395.

92. Naka, T., T. Matsumoto, M. Narazaki, M. Fujimoto, Y. Morita, Y. Ohsawa, H. Saito, T. Nagasawa, Y. Uchiyama, and T. Kishimoto. (1998). Accelerated apoptosis of lymphocytes by augmented induction of Bax in SSI-1 (STAT-induced STAT inhibitor-1) deficient mice. *Proc. Natl Acad. Sci. USA*, **95**, 15577.

93. Metcalf, D., W. S. Alexander, A. G. Elefanty, N. A. Nicola, D. J. Hilton, R. Starr, S. Mifsud, and L. Di Rago. (1999). Aberrant hematopoiesis in mice with inactivation of the gene encoding SOCS-1. *Leukemia*, **13**, 926.

94. Huang, S., W. Hendriks, A. Althage, S. Hemmi, H. Bluethmann, R. Kamijo, J. Vilcek, R. M. Zinkernagel, and M. Aguet. (1993). Immune response in mice that lack the interferon-gamma receptor. *Science*, **259**, 1742.

95. Dalton, D. K., S. Pitts-Meek, S. Keshav, I. S. Figari, A. Bradley, and T. A. Stewart. (1993). Multiple defects of immune cell function in mice with disrupted interferon-gamma genes. *Science*, **259**, 1739.

96. Boehm, U., T. Klamp, M. Groot, and J. C. Howard. (1997). Cellular responses to interferon-gamma. *Annu. Rev. Immunol.*, **15**, 749.

97. Boes, B., H. Hengel, T. Ruppert, G. Multhaup, U. H. Koszinowski, and P. M. Kloetzel. (1994). Interferon gamma stimulation modulates the proteolytic activity and cleavage site preference of 20S mouse proteasomes. *J. Exp. Med.*, **179**, 901.

98. Groettrup, M., A. Soza, M. Eggers, L. Kuehn, T. P. Dick, H. Schild, H. G. Rammensee, U. H. Koszinowski, and P. M. Kloetzel. (1996). A role for the proteasome regulator PA28alpha in antigen presentation. *Nature*, **381**, 166.

99. Trowsdale, J., I. Hanson, I. Mockridge, S. Beck, A. Townsend, and A. Kelly. (1990). Sequences encoded in the class II region of the MHC related to the 'ABC' superfamily of transporters. *Nature*, **348**, 741.

100. Epperson, D. E., D. Arnold, T. Spies, P. Cresswell, J. S. Pober, and D. R. Johnson. (1992). Cytokines increase transporter in antigen processing-1 expression more rapidly than HLA class I expression in endothelial cells. *J. Immunol.*, **149**, 3297.

101. Hsieh, C. S., S. E. Macatonia, C. S. Tripp, S. F. Wolf, A. O'Garra, and K. M. Murphy. (1993). Development of TH1 CD4+ T cells through IL-12 produced by Listeria-induced macrophages. *Science*, **260**, 547.

102. Snapper, C. M., T. M. McIntyre, R. Mandler, L. M. Pecanha, F. D. Finkelman, A. Lees, and J. J. Mond. (1992). Induction of IgG3 secretion by interferon gamma: a model for T cell-independent class switching in response to T cell-independent type 2 antigens. *J. Exp. Med.*, **175**, 1367.

103. Snapper, C. M., C. Peschel, and W. E. Paul. (1988). IFN-gamma stimulates IgG2a secretion by murine B cells stimulated with bacterial lipopolysaccharide. *J. Immunol.*, **140**, 2121.

104. Suzuki, R., H. Sakamoto, H. Yasukawa, M. Masuhara, T. Wakioka, A. Sasaki, K. Yuge, S. Komiya, A. Inoue, and A. Yoshimura. (1998). CIS3 and JAB have different regulatory roles in interleukin-6 mediated differentiation and STAT3 activation in M1 leukemia cells. *Oncogene*, **17**, 2271.

105. Dey, B. R., K. Frick, W. Lopaczynski, S. P. Nissley, and R. W. Furlanetto. (1996). Evidence for the direct interaction of the insulin-like growth factor I receptor with IRS-1, Shc, and Grb10. *Mol Endocrinol.*, **10**, 631.

106. Adams, T. E., J. A. Hansen, R. Starr, N. A. Nicola, D. J. Hilton, and N. Billestrup. (1998). Growth hormone preferentially induces the rapid, transient expression of SOCS-3, a novel inhibitor of cytokine receptor signaling. *J. Biol. Chem.*, **273**, 1285.

107. Johnston, J. A., M. Kawamura, R. A. Kirken, Y. Q. Chen, T. B. Blake, K. Shibuya, J. R. Ortaldo, D. W. McVicar, and J. J. O'Shea. (1994). Phosphorylation and activation of the Jak-3 Janus kinase in response to interleukin-2. *Nature*, **370**, 151.

108. Barber, D. L. and A. D. D'Andrea. (1994). Erythropoietin and interleukin-2 activate distinct JAK kinase family members. *Mol. Cell. Biol.*, **14**, 6506.

109. Witthuhn, B. A., O. Silvennoinen, O. Miura, K. S. Lai, C. Cwik, E. T. Liu, and J. N. Ihle. (1994). Involvement of the Jak-3 Janus kinase in signalling by interleukins 2 and 4 in lymphoid and myeloid cells. *Nature*, **370**, 153.

110. O'Garra, A. and K. Murphy. (1994). Role of cytokines in determining T-lymphocyte function. *Curr. Opin. Immunol.*, **6**, 458.

111. Moore, K. W., A. O'Garra, R. de Waal Malefyt, P. Vieira, and T. R. Mosmann. (1993). Interleukin-10. *Annu. Rev. Immunol.*, **11**, 165.

112. Ito, S., P. Ansari, M. Sakatsume, H. Dickensheets, N. Vazquez, R. P. Donnelly, A. C. Larner, and D. S. Finbloom. (1999). Interleukin-10 inhibits expression of both interferon alpha- and interferon gamma- induced genes by suppressing tyrosine phosphorylation of STAT1. *Blood*, **93**, 1456.

113. Zhang, Y., R. Proenca, M. Maffei, M. Barone, L. Leopold, and J. M. Friedman. (1994). Positional cloning of the mouse obese gene and its human homologue. *Nature*, **372**, 425.

114. Halaas, J. L., K. S. Gajiwala, M. Maffei, S. L. Cohen, B. T. Chait, D. Rabinowitz, R. L. Lallone, S. K. Burley, and J. M. Friedman. (1995). Weight-reducing effects of the plasma protein encoded by the obese gene. *Science*, **269**, 543.

115. Pelleymounter, M. A., M. J. Cullen, M. B. Baker, R. Hecht, D. Winters, T. Boone, and F. Collins. (1995). Effects of the obese gene product on body weight regulation in ob/ob mice. *Science*, **269**, 540.

116. Ahima, R. S., D. Prabakaran, C. Mantzoros, D. Qu, B. Lowell, E. Maratos-Flier, and J. S. Flier. (1996). Role of leptin in the neuroendocrine response to fasting. *Nature*, **382**, 250.

117. Campfield, L. A., F. J. Smith, Y. Guisez, R. Devos, and P. Burn. (1995). Recombinant mouse OB protein: evidence for a peripheral signal linking adiposity and central neural networks. *Science*, **269**, 546.

118. Baumann, H., K. K. Morella, D. W. White, M. Dembski, P. S. Bailon, H. Kim, C. F. Lai, and L. A. Tartaglia. (1996). The full-length leptin receptor has signaling capabilities of interleukin 6-type cytokine receptors. *Proc. Natl Acad. Sci. USA*, **93**, 8374.

119. Bjorbaek, C., S. Uotani, B. da Silva, and J. S. Flier. (1997). Divergent signaling capacities of the long and short isoforms of the leptin receptor. *J. Biol. Chem.*, **272**, 32686.

120. Ghilardi, N., S. Ziegler, A. Wiestner, R. Stoffel, M. H. Heim, and R. C. Skoda. (1996). Defective STAT signaling by the leptin receptor in diabetic mice. *Proc. Natl Acad. Sci. USA*, **93**, 6231.

121. Lee, G. H., R. Proenca, J. M. Montez, K. M. Carroll, J. G. Darvishzadeh, J. I. Lee, and J. M. Friedman. (1996). Abnormal splicing of the leptin receptor in diabetic mice. *Nature*, **379**, 632.

122. Vaisse, C., J. L. Halaas, C. M. Horvath, J. E. Darnell, Jr, M. Stoffel, and J. M. Friedman. (1996). Leptin activation of Stat3 in the hypothalamus of wild-type and ob/ob mice but not db/db mice. *Nature Genet.*, **14**, 95.

123. Chen, H., O. Charlat, L. A. Tartaglia, E. A. Woolf, X. Weng, S. J. Ellis, N. D. Lakey, J. Culpepper, K. J. Moore, R. E. Breitbart, G. M. Duyk, R. I. Tepper, and J. P. Morgenstern. (1996). Evidence that the diabetes gene encodes the leptin receptor: identification of a mutation in the leptin receptor gene in db/db mice. *Cell*, **84**, 491.

124. Bjorbaek, C., J. K. Elmquist, J. D. Frantz, S. E. Shoelson, and J. S. Flier. (1998). Identification of SOCS-3 as a potential mediator of central leptin resistance. *Mol. Cell*, **1**, 619.

125. Gloaguen, I., P. Costa, A. Demartis, D. Lazzaro, A. Di Marco, R. Graziani, F. Paonessa, F. Chen, C. I. Rosenblum, L. H. Van der Ploeg, R. Cortese, G. Ciliberto, and R. Laufer. (1997). Ciliary neurotrophic factor corrects obesity and diabetes associated with leptin deficiency and resistance. *Proc. Natl Acad. Sci. USA*, **94**, 6456.

126. Bjorbaek, C., J. K. Elmquist, K. El-Haschimi, J. Kelly, R. S. Ahima, S. Hileman, and J. S. Flier. (1999). Activation of SOCS-3 messenger ribonucleic acid in the hypothalamus by ciliary neurotrophic factor. *Endocrinology*, **140**, 2035.

127. Akita, S., J. Malkin, and S. Melmed. (1996). Disrupted murine leukemia inhibitory factor (LIF) gene attenuates adrenocorticotropic hormone (ACTH) secretion. *Endocrinology*, **137**, 3140.

128. Akita, S., J. Webster, S. G. Ren, H. Takino, J. Said, O. Zand, and S. Melmed. (1995). Human and murine pituitary expression of leukemia inhibitory factor. Novel intrapituitary regulation of adrenocorticotropin hormone synthesis and secretion. *J. Clin. Invest.*, **95**, 1288.

129. Auernhammer, C. J. and S. Melmed. (1999). Interleukin-11 stimulates proopiomelanocortin gene expression and adrenocorticotropin secretion in corticotroph cells: evidence

for a redundant cytokine network in the hypothalamo-pituitary-adrenal axis. *Endocrinology*, **140**, 1559.

130. Bousquet, C., D. W. Ray, and S. Melmed. (1997). A common pro-opiomelanocortin-binding element mediates leukemia inhibitory factor and corticotropin-releasing hormone transcriptional synergy. *J. Biol. Chem.*, **272**, 10551.

131. Auernhammer, C. J., V. Chesnokova, C. Bousquet, and S. Melmed. (1998). Pituitary corticotroph SOCS-3: novel intracellular regulation of leukemia-inhibitory factor-mediated proopiomelanocortin gene expression and adrenocorticotropin secretion. *Mol. Endocrinol.*, **12**, 954.

132. Auernhammer, C. J., C. Bousquet, and S. Melmed. (1999). Autoregulation of pituitary corticotroph SOCS-3 expression: characterization of the murine SOCS-3 promoter. *Proc. Natl Acad. Sci. USA*, **96**, 6964.

133. De Sepulveda, P., S. Ilangumaran, and R. Rottapel. (2000). Suppressor of cytokine signaling-1 inhibits VAV function through protein degradation (In Process Citation). *J. Biol. Chem.*, **275**, 14005.

6 | Cyclin-dependent protein kinases

KATHLEEN L. GOULD

1. Introduction

Cyclin-dependent protein kinases (Cdks) regulate progression through the eukaryotic cell cycle. They are also involved in the regulation of other processes such as gene expression and phosphate metabolism. Cdks are the catalytic subunits of heterodimeric complexes and they approximate the size of a minimal protein-kinase catalytic domain ($\sim$ 34 kDa). Alone, they lack protein kinase activity. As their name implies, Cdks must associate with a member of the cyclin protein family to gain activity.

By far the best understood of the Cdks are those that have important roles in cell-cycle progression—Cdc2, Cdk2, and Cdk4. Cdks are activated transiently at particular stages of the cell cycle, and their activation or inactivation triggers the next events in the cycle. To ensure the coordination of these events, the activation of each Cdk is strictly regulated in a variety of ways including an elaborate cascade of protein phosphorylation events, interaction with subunits in addition to cyclins, and by regulated proteolysis of the cyclin and other subunits. Due to the wealth of knowledge concerning Cdks and the limitations of space, it is impossible to include all pertinent information about Cdks in this chapter or to cite all the original references. Numerous recent reviews emphasize various aspects of Cdk function and regulation, and readers will be referred to these where appropriate.

2. Discovery of Cdks

Cdks were first identified in the budding yeast *Saccharomyces cerevisiae*, and the fission yeast *Schizosaccharomyces pombe*, in genetic screens for conditional lethal mutants unable to progress through the cell cycle at an elevated temperature, although they can continue to grow (reviewed in ref. 1). Such mutants are called CDC, for cell division cycle. CDC mutants arrest at the particular point in the cell cycle where the mutant gene product normally functions. There are nearly 50 CDC mutants from each yeast whose arrest points are scattered through every stage of the cell cycle (G_1, S, G_2, or M), and some of these correspond to Cdks.

2.1 Nomenclature

Several CDC mutants identify orthologous gene functions in the two yeasts. However, since mutants were numbered in the order in which they were isolated, genes with the same number in both yeasts (i.e. *S. cerevisiae CDC2* and *S. pombe cdc2*$^+$) are not related and, conversely, related genes have different numbers (i.e. *CDC28* and *cdc2*$^+$). A further confusion in the terminology of cell-cycle genes arises because of the different conventions for naming genes in the two yeast systems (discussed in ref. 1). Wild-type genes in *S. pombe* are written in lower case italicized letters (i.e. *cdc2*$^+$), whereas wild-type genes in *S. cerevisiae* are written in upper case italicized letters (i.e. *CDC28*). The protein products of the genes are indicated by non-italicized letters followed by a 'p', with the initial letter capitalized (i.e. Cdc2p and Cdc28p). There are yet more names for similar gene functions when higher eukaryotic proteins are considered, and these proteins will not be indicated with a 'p'.

2.2 A brief history of Cdc2

The study of CDC mutants in yeast helped to identify two major control points in the eukaryotic cell cycle (reviewed in ref. 1). The first operates prior to S phase, the time during which DNA is replicated, and is called 'START'. The second controls the onset of M phase, the stage of nuclear division. The *cdc2*$^+$ gene attracted considerable interest when it was found to be essential for passage through both major cell-cycle transition points in *S. pombe*. When it was sequenced, it was found to be related to the *S. cerevisiae CDC28* gene. *CDC28* was first found to be essential for progress past START, and subsequently for entry into mitosis as well. The predicted protein sequences of both *CDC28* and *cdc2*$^+$ indicated that they functioned as protein kinases. By expressing *CDC28* in *S. pombe* strains deficient for *cdc2*$^+$ function, it was found that Cdc28p and Cdc2p are functional homologues. That is to say, the *CDC28* gene product could rescue temperature-sensitive mutations in *cdc2*$^+$. As *S. cerevisiae* and *S. pombe* are evolutionary quite divergent, this result was the first indication that cell-cycle control proteins might be similar in all eukaryotes (1). The pivotal role of Cdc2p in regulating the cell cycle of all eukaryotes was fully recognized when a functional homologue of *CDC28/cdc2*$^+$ (hereafter referred to as Cdc2) was isolated from a human cDNA library through a functional complementation approach (2). Functional homologues have now been identified in many other eukaryotes (3, 4).

Cdc2 was also identified as a component of maturation promoting factor (MPF). MPF was originally defined as an activity present in the cytosol of mature *Xenopus laevis* eggs which, when injected into immature *Xenopus* oocytes, would promote the process of maturation (cell-cycle transit from prophase of meiosis I through to metaphase of meiosis II) in the absence of protein synthesis (reviewed in ref. 5). Hence, MPF activity is all that is required to induce the resumption of meiosis. Highly purified preparations of MPF possess protein-serine/threonine kinase activity and are composed of two proteins. As mentioned above, one is Cdc2; the other is a member of the cyclin family of proteins (5). Just as all eukaryotes possess a Cdc2p homologue, all eukaryotic cells in mitosis or meiosis possess MPF activity. For this reason, MPF is now more broadly defined as M-phase promoting factor since it will induce mitosis as well as meiosis.

2.3 A family of Cdks

The protein kinase activity associated with Cdc2p is sharply periodic through the cell cycle, peaking as cells enter mitosis and falling away again as cells exit mitosis. This activity profile matches the timing of Cdc2p function in yeast, determined genetically, as well as the appearance and disappearance of MPF activity in higher eukaryotes. Thus, the role of Cdc2p as a general eukaryotic M-phase promoting factor is well established (reviewed in ref. 6). However, in higher eukaryotic cells, Cdc2p does not appear to have a major role in promoting the G_1/S transition as it does in yeast. Injection of neutralizing antibodies raised against Cdc2p into fibroblasts does not prevent DNA replication, although it does prevent entry into mitosis (7). Similarly, depleting Cdc2p from frog egg extracts inhibits M phase but not DNA replication (8). Also, a temperature-sensitive *CDC2* mutation identified in a hamster cell line arrests cells in G_2, as expected, but does not prevent DNA replication (9).

The above results are now fully appreciated since other Cdc2-related protein kinases have been identified in higher eukaryotic cells which function at other points in the cell cycle (4). These family members also require a cyclin subunit for activity, and thus have been designated Cdks (Table 1). In fact, the term Cdk1 is now often used for Cdc2. Besides Cdc2, the only other Cdks able to complement Cdc2p/Cdc28p function in yeast are Cdk2 and Cdk3 and, not surprisingly, they are the most related to Cdc2 at the amino acid sequence level (10). Cdk3 has received relatively little attention, most likely since it is difficult to detect in mammalian cells (11). In contrast, early studies of Cdk2 function were the first to demonstrate that the functions performed by a single enzyme in yeast (Cdc2p/Cdc28p) have been delegated to at least two protein kinases in higher eukaryotic cells. As examples, when Cdk2 is depleted from frog cell-free extracts using antibodies specific for it, the extracts can no longer replicate DNA but mitosis in the extracts is not impaired (8). Further, microinjection of anti-Cdk2 antibodies into fibroblasts inhibits DNA replication (12, 13). Thus, Cdk2 function is required for the initiation of S phase but Cdk2 is not required for the G_2/M transition; that function is performed by Cdc2.

Based on sequence similarity and the requirement for a cyclin-like subunit, there are now 10 protein kinases termed Cdks, and even more protein kinases related to these enzymes have been identified in a variety of organisms (4). In addition to the already mentioned Cdks (Cdc2, Cdk2, and Cdk3), Cdk4, Cdk6, and Cdk7 are important for cell-cycle progression (4, 14). However, other Cdks do not function in cell-cycle control, or perform a dual function in cell-cycle control and other processes (Table 1). For example, Cdk5 does not function in cell-cycle control, but it is essential for normal brain development and might be important for neuronal cytoskeleton structure (15, 16). Cdk7 functions in one capacity as CAK (see Section 4.2.1) and also in transcriptional regulation (17). The sections below will discuss Cdc2, Cdk2, or

Table 1 Cdks and cyclins

Cdks	Cyclins	Role
Multicellular organisms		
Cdc2 (Cdk1)	Cyclin B1, B2	Cell cycle
Cdk2	Cyclin A	Cell cycle
	Cyclin E	Cell cycle
Cdk3	Cyclin E2	Cell cycle
Cdk4	Cyclin D1, D2, D3	Cell cycle
Cdk5	p35	Neuronal development
Cdk6	Cyclin D1, D2, D3	Cell cycle
Cdk7 (CAK)	Cyclin H	Transcription
Cdk8	Cyclin C	Transcription
Cdc2/cyclin B	Cyclin F	Localization
?	Cyclin G	?
S. cerevisiae		
Cdc28p	Cln1–3	Cell cycle
	Clb1-6	Cell cycle
Pho85p	Pho80	Transcription
	Pcl1/Hcs26	Cell cycle
	Pcl2/OrfD	Cell cycle
	Clg1, Pcl5-10	?
Kin28p	Ccl1	Transcription
Srb10p	Srb11	Transcription
Ctk1p	Ctk11	Transcription
S. pombe		
Cdc2p	Cdc13p	Cell cycle
	Cig1p	Cell cycle
	Cig2p	Cell cycle
	Puc1p	Cell cycle
Mop1p/Crk1p (CAK)	Mcs2p	Cell cycle/transcription?

Cdk4 to best illustrate paradigms of Cdk regulation, structure, or function. Other reviews discuss the functions of individual Cdks in more depth (16–18).

3. Cyclins

Cyclins were first identified in sea urchin and clam embryos as proteins which accumulated during interphase and were specifically degraded during mitosis (reviewed in refs 5, 19, 20). Two subclasses of cyclins (A and B) were initially recognized based on their slightly different kinetics of accumulation and destruction. Cyclin B accumulates in somatic cells during G_2 and is degraded during mitosis, whereas cyclin A begins to accumulate during S phase and is degraded slightly before cyclin B (21). Cyclin A, in fact, is required for S phase in addition to its requirement for mitosis (22–24). Cyclin B, like Cdc2p, has been conserved throughout eukaryotic evolution and, as mentioned above, is the second recognized component of MPF (5). In general, however, there is considerably less conservation in sequence between the cyclins of yeast and higher eukaryotes (and between the yeast cyclins) as there is amongst the Cdks.

3.1 A proliferation of cyclins

The cyclin family now extends well beyond cyclins A and B, both in terms of differential expression during the cell cycle and in sequence diversity (Table 1). In contrast to the 'mitotic cyclins' A and B, other cyclins accumulate in G_1 and persist for different periods of the cell cycle (reviewed in refs 5, 19, 20, 25). G_1 cyclins were first identified in *S. cerevisiae* and called Clns1–3, although others have subsequently been added to the list (25). In metazoans, G_1 cyclins have been classified into C, D, and E types because their sequences are significantly different from those of the Clns (reviewed in ref. 4). Only one G_1 cyclin gene has been reported in *S. pombe* (*Puc1p*) and it falls into a distinct category based on its sequence and apparent function (26).

In addition to the numerous categories of cyclins, there are multiple members of most cyclin types. For example, there are several different cyclin Bs, Ds, and Clns and they are distinguished from one another with numbers following their designations, i.e. B1, B2. In *S. pombe*, there are three known cyclin B-related proteins encoded by the $cdc13^+$, $cig1^+$, and $cig2^+$ genes (reviewed in ref. 27). In *S. cerevisiae*, there are six cyclins related to cyclin B; these are termed Clbs1–6 (25).

All cyclins contain a conserved stretch of approx. 150 amino acids, which has been termed the 'cyclin box' (19). New members are added to the cyclin family based on the presence of a cyclin box. Not surprisingly, the cyclin box appears to be the primary region through which cyclins bind Cdks (28). Beyond the cyclin box, the primary sequences of the various cyclins diverge considerably. Despite this, there does appear to be a degree of functional conservation between all cyclins. *S. cerevisiae* cells are viable in the absence of any two, but not three, of *CLNs* 1–3, suggesting that some aspect of *CLN* function is redundant (29, 30). Moreover, cyclins C, D, and E can functionally substitute for the absence of *CLNs* 1–3 in yeast when overexpressed (31–34). That is, in fact, one method by which they were identified. Even the mitotic cyclins, when overexpressed, can substitute for the G_1 cyclins in yeast. On the other hand, there is also evidence to indicate that individual members of the cyclin classes, both G_1 and mitotic, are functionally distinct. As examples, each cyclin D has it own pattern of cell-cycle and tissue expression (reviewed in ref. 14). Of the four mitotic Clbs, the loss of Clb1p function has the most profound effect on the process of meiosis (35).

3.2 Specificities of interactions

Given the existence of so many cyclins and Cdks, the possibilities for combinatorial control appear endless. However, the number of complexes that actually form is limited (reviewed in refs 3, 4). In metazoans, cyclin B is complexed primarily with Cdc2p. Cdk2 will form complexes with cyclin A or cyclin E. However, these complexes appear to be temporally distinct, with the activity of the Cdk2–cyclin E complex peaking at

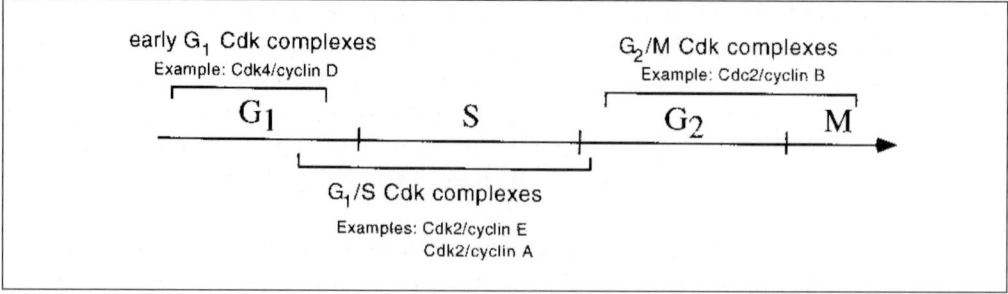

Fig. 1 Model of cell-cycle progression driven by various Cdk–cyclin complexes.

the onset of S phase and generally preceding that of the Cdk2–cyclin A complex. Cdk4 and Cdk6 bind cyclin Ds exclusively and their activities peak earlier in G_1. The theme to have emerged over the last decade is that certain complexes are the predominant activators of different transition points in the cell cycle (Fig. 1). The cyclin B–Cdc2 complex acts as the primary promoter of mitosis; Cdk2 complexes are primarily responsible for triggering S phase, and Cdk4 complexes serve to integrate extracellular signals and direct the cell-cycle engine according to the cell's environment.

4. Regulation

The activities of the Cdk–cyclin complexes are periodic throughout the cell cycle. This periodicity is achieved by regulation at many levels and is necessary for the proper coordination of cell-cycle events.

4.1 Transcriptional

In yeast, the amount of Cdc2 mRNA and protein are constant throughout the cell cycle and during exit from the cell cycle. In vertebrate cells, however, Cdc2 mRNA and protein levels can vary depending upon the cell's growth state. In general, Cdc2 levels are very low in non-proliferating tissues such as brain (36), and are undetectable in senescent or serum-deprived cells (37–39). Cdk2 (40, 41) and Cdk4 (42) mRNA and protein levels also increase as quiescent cells re-enter the cycle. However, transcriptional regulation of Cdks is not considered to be an important mode of cell-cycle regulation in cycling cells. Transcriptional regulation of cyclin genes does, however, contribute to cell-cycle control.

In *S. cerevisiae*, there are nine cyclins which bind and activate Cdc28p. All nine cyclins are transcriptionally regulated in successive waves throughout the cell cycle (reviewed in ref. 43). Transcription of *CLN3* occurs at the M/G_1 transition and directs transcription of *CLN1* and *CLN2* in G_1 (44–46). Transcription of *CLB5* and *CLB6* occur at the G_1/S transition, and transcription of *CLBs* 1–4 occurs still later in the cell cycle (43). A defined promoter element termed the ECB (early cell-cycle box), to which the Mcm1p transcription factor binds, is responsible for M/G_1-specific transcription of

CLN3 (44). Transcription of the *CLB1* and *CLB2* genes also depends on Mcm1p in collaboration with an unknown factor (47). How the Mcm1p transcription factor is regulated is sure to be a subject of future investigation. Cell-cycle periodicity of *CLN1* and *CLN2* transcription depend upon the transcription factors, Swi4p and Swi6p, which are regulated by Cln3p–Cdc28p activity (43). A Swi4–Swi6 complex is activated by Cln–Cdc28p complexes and displaced from DNA by Clb–Cdc28p complexes in G_2 (48).

The transcription of cyclins in multicellular organisms is also periodic in the cell cycle. Transcription of cyclin Ds is induced by mitogens early in G_1 and transcription of cyclin E follows later in G_1 (14). The next cyclin mRNA to accumulate is cyclin A, and the cyclin B mRNAs accumulate still later (4). Since overexpression of cyclin D genes correlates with loss of growth control (4), the mechanism(s) which control cyclin D mRNA levels are of interest. However, factors responsible for regulating cyclin D mRNA levels have not yet been identified.

4.2 Phosphorylation

As described above, binding to a cyclin subunit is essential for Cdk activity. However, Cdks must also be in the appropriate phosphorylation state in order to display activity. There are two sites within the protein subject to phosphorylation and these phosphorylation events have different effects on the activity of the complexes.

4.2.1 Activatory phosphorylation

For activity, Cdc2, Cdk2, and Cdk4 must be phosphorylated on a conserved threonine residue corresponding to position 161 in human Cdc2 within the so-called T loop of the kinase (reviewed in ref. 17). In all organisms studied, with the exception of *S. cerevisiae*, phosphorylation at this site is mediated by a protein kinase termed CAK (for Cdk activating kinase) (17) (Fig. 2). CAK is itself a Cdk, Cdk7, and its activity depends upon cyclin H. In addition to phosphorylating Cdks, Cdk7 is a component of the basal transcription complex TFIIH and is involved in transcriptional initiation by virtue of its ability to phosphorylate the RNA polymerase-II carboxy-terminal domain. In *S. cerevisiae*, CAK activity resides in a novel monomeric enzyme

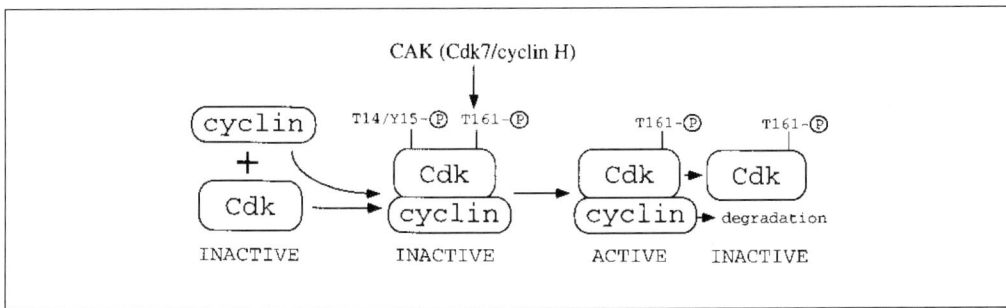

Fig. 2 Model of Cdk activation by cyclin binding and CAK phosphorylation.

termed Cak1 (49–51). While essential for activity, CAK phosphorylation is not rate-limiting and does not appear to participate in the cell-cycle regulation of Cdc2, Cdk2, or Cdk4 activities (17).

4.2.2 Inhibitory phosphorylation

The activity of *S. pombe* Cdc2p is inhibited by phosphorylation at Tyr15 (52). In vertebrate cells, Cdc2 and Cdk2 are inhibited by phosphorylation at Tyr15 and also Thr14 (reviewed in ref. 53). These residues lie within a domain involved in binding ATP known as the glycine loop. While these phosphorylation events inhibit the activity of the enzyme, the mechanism of inhibition remains unknown since phosphorylation at these sites does not block ATP binding *in vitro* (54). In most organisms, except *S. cerevisiae*, dephosphorylation of these residues is instrumental in triggering activation of the Cdc2–cyclin B complex and consequent entry into mitosis (reviewed in ref. 55). Phosphorylation of these residues is also used in many organisms to restrain Cdc2–cyclin B activity when DNA replication is incomplete or when DNA has been damaged (reviewed in refs 56–58). Phosphorylation of Cdk4 at the Tyr15 equivalent has also been reported to occur in the presence of DNA damage (59), and in response to growth inhibition by TGFβ (60). The inactivation of these Cdks allows the cell extra time to either complete DNA replication or repair damaged DNA before proceeding through the cell cycle (61). In *S. cerevisiae*, Cdc28p activity is inhibited through phosphorylation of these sites, not in response to DNA damage, but in response to a morphogenetic checkpoint (62). This checkpoint delays entry into mitosis in the absence of actin polarization and budding (63). Genes responsible for regulating Cdc2 activation were first identified genetically in *S. pombe* as $cdc25^+$ and $wee1^+$ (64, 65). Their protein products, in concert with homologues, were subsequently shown to regulate Cdc2 activity by controlling the phosphorylation state of Tyr15/Thr14 (Fig. 3).

Consistent with its role as a dose-dependent activator of mitosis (64), Cdc25 is a dual-specificity protein phosphatase capable of removing phosphate from both Tyr15 and Thr14 (reviewed in ref. 66). Cdc25 activity, in turn, is regulated at the level of its transient accumulation and also by its phosphorylation state (reviewed in ref. 66). In *Drosophila* and *S. pombe*, Cdc25 levels accumulate prior to entry into M phase (67–69) and increases in Cdc25 mRNA levels have been detected at this time, at least in *Drosophila* and human (70, 71). The reduction in Cdc25 levels following M phase is at least partly due to ubiquitin-mediated proteolysis (72). In vertebrates, three Cdc25-related enzymes have been identified: Cdc25A, B, and C (66). These enzymes are not redundant, displaying distinct cell-cycle and developmental roles. Whereas Cdc25C is the form active at the G_2/M transition, Cdc25A and Cdc25B act at earlier times in the cell cycle (66).

Phosphorylation of Cdc25 influences its activity both positively and negatively. Both Cdc2 and the POLO family of kinases are able to phosphorylate Cdc25 and activate its catalytic activity (73–76). This is speculated to be part of a feedback loop that leads to maximal activation of Cdc25 and Cdc2 activities at the onset of mitosis (66). Several protein kinases also have the ability to phosphorylate Cdc25 leading to

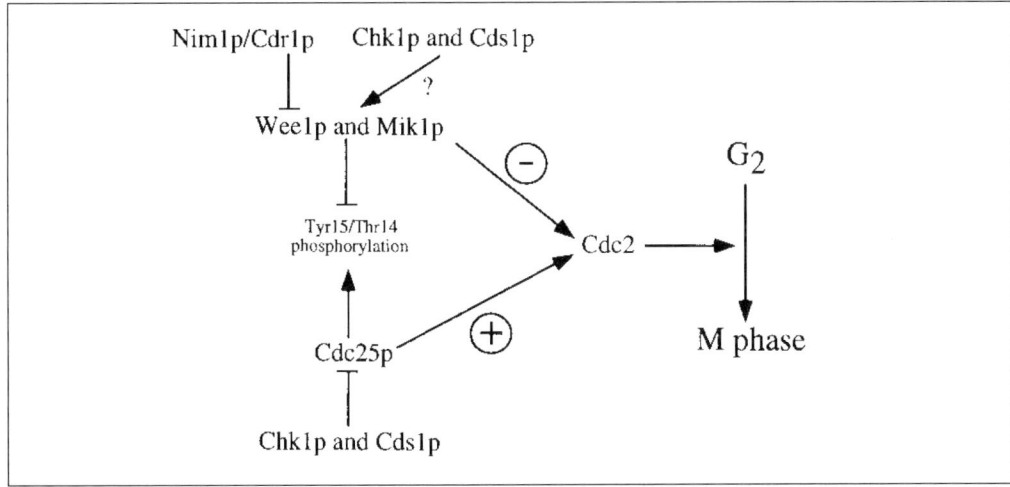

Fig. 3 Model of Cdc2 regulation at the G_2/M transition by protein kinases and phosphatases.

its inhibition *in vivo* (77–83) (Fig. 3). Two of these kinases, Cds1p and Chk1p, are components of the DNA damage or DNA replication checkpoints (61). Phosphorylation by these kinases does not inhibit the specific activity of Cdc25 but creates binding sites for 14–3–3 proteins (77, 80, 84). Binding of 14–3–3 proteins correlates with the re-localization of Cdc25, which is normally localized in the nucleus, to the cytoplasm (reviewed in 85). In this manner, damaged or unreplicated DNA signals through protein kinases to isolate Cdc25 from its target, and thereby contribute to preventing entry into mitosis (reviewed in 85).

The primary protein kinase responsible for phosphorylating Tyr15 is Wee1 (reviewed in refs 53, 55), first identified as a dosage-dependent inhibitor of mitosis (65). In *S. pombe*, a second protein kinase which can also phosphorylate Tyr15 is Mik1 (86, 87) (Fig. 3). In vertebrates also, a Wee1-related protein kinase has been identified, termed Myt1 (88–90). Interestingly, unlike Wee1 and Mik1, Myt1 phosphorylates Thr14 exclusively (88). Like other components in this regulatory network, Wee1 and Mik1 are subject to regulation via phosphorylation. Wee1 is phosphorylated and inhibited by Nim1/Cdr1 as well as unidentified protein kinases (reviewed in refs 53, 55) (Fig. 3); Myt1 can be phosphorylated and inhibited by p90[rsk] (91). Although a direct kinase–substrate relationship has not been established, the Cdr2p protein kinase also acts upstream of Wee1p in *S. pombe* (92, 93). Recently, it has been shown that *S. cerevisiae* Wee1, termed Swe1 (94), and *Xenopus* Wee1 (95) are subject to ubiquitin-mediated proteolysis which is apparently triggered by their phosphorylation (96). The protein kinase Chk1, which is activated by DNA damage, can phosphorylate Wee1 directly *in vitro* (97). Perturbations of the actin cytoskeleton or inhibition of DNA replication can delay Wee1/Swe1 destruction, indicating that increasing the stability of Wee1/Swe1 is one means of preventing Cdc2 activation when the cell cycle is perturbed (95, 96).

Still other protein kinases and phosphatases have been identified genetically as upstream regulators of Cdc2 activation, but their exact biochemical function has yet to be delineated (reviewed in refs 53, 55, 58). The complexity of this network reflects the roles of Cdks as focal points in signal transduction cascades which coordinate transitions in the cell cycle.

4.3 Inhibitors

The activity of Cdk–cyclin complexes is also regulated by inhibitor proteins (CKIs) (Table 2). CKIs fall into two broad categories based upon whether they bind solely to the catalytic subunit or whether they bind to Cdk complexes (Fig. 4). The former category has been identified only in higher eukaryotic cells and is termed the Ink family. There are four known Inks (p15, p16, p18, and p19); they bind to Cdk4/6 and prevent cyclin D binding (98). The second category of inhibitor is found in all eukaryotes; these bind to the Cdk complexes. In mammalian cells, this family includes p21, p27, and p57 (98). At the G_2/M transition, a major mechanism of regulating Cdk activity appears to be the modulation of Tyr15/Thr14 phosphorylation. In contrast, during G_1 and at the G_1/S transition, CKI inhibition appears to be a preferred mechanism of regulating Cdk activity. In fact, the levels of many inhibitors are sensitive to extracellular stimuli (reviewed in ref. 98). Levels of p21 are upregulated in response to DNA damage and contribute to delaying cell-cycle progression under these circumstances. In *S. cerevisiae*, there are two CKIs: Far1p and Sic1p. Far1p accumulates in response to mating pheromone; it inhibits Cdc28p–Clnp complexes and contributes to pheromone-induced cell-cycle arrest in G_1. Sic1p inhibits Cdc28p–Clbp complexes during G_1 and, in so doing, contributes to the coordination of events at START. In *S. pombe*, Rum1p plays a similar role to Sic1p. It inhibits Cdc2–cyclin B complexes in G_1 so that the cells do not initiate mitosis prematurely (reviewed in ref. 99).

It has not been clear until recently whether one molecule of inhibitor bound to a Cdk complex is sufficient to inactivate it, since there have been reports that antibodies to Cdks immunoprecipitate active kinases (reviewed in ref. 98). Careful kinetic studies of the interaction between p21 and Cdk2–cyclin A (100) and the crystal structure of

Table 2 Cdk inhibitors

Mammalian cells

INK family	Binds only CdK4/Cdk6 and inhibits them by preventing cyclin binding
p21 CIP/WAF	Binds and inhibits all Cdk complexes
p27 KIPI	Binds and inhibits all Cdk complexes
p57 KIP2	Binds and inhibits all Cdk complexes

S. cerevisiae

Far1p	Binds and inhibits Cln–Cdc28p complexes
Sic1p	Binds Cln–Cdc28p and Clb– Cdc28p complexes, but inhibits only Clb–Cdc28p complexes

S. pombe

Rum1p	Binds and inhibits Cdc2p– Cdc13p complexes

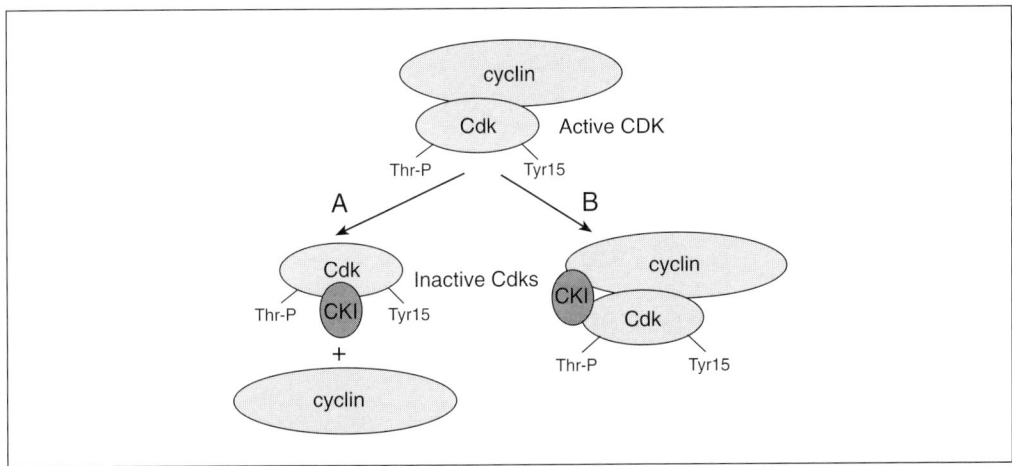

Fig. 4 Models of Cdk complex inhibition by CKIs. In (A) CKI binds Cdk and prevents cyclin binding. In (B) CKI binds Cdk–cyclin complex and prevents it from interacting with ATP and substrate.

Cdk2–cyclin A complex bound to p27 (101), however, indicate that one molecule of inhibitor is able to completely block Cdk kinase activity. The confusion probably arose from the dynamic interaction of the inhibitor with the Cdk complex; as it is transiently released during immunocomplex kinase assays, the kinase is free to phosphorylate its substrates (100).

4.4 Regulated cyclin proteolysis

Inactivation of Cdks is accomplished by regulated destruction of the cyclin subunit via ubiquitin-mediated proteolysis (reviewed in refs 102, 103) (Fig. 5). This is a multi-step process in which a ubiquitin-activating enzyme (E1) first activates ubiquitin by formation of a high-energy thiolester bond. E1 then transfers ubiquitin to an E2 or UBC (ubiquitin-conjugating) enzyme. The E2 may transfer ubiquitin directly to the substrate, targeting the substrate for destruction by the 26S proteasome. Alternatively, the specificity of this reaction may be determined by association of the E2 with a ubiquitin protein ligase (E3) to catalyse ubiquitination of a target protein (reviewed in refs 104, 105).

An E3 ubiquitin ligase that recognizes A- and B-type cyclins was identified biochemically in clam and frog extracts, and genetically in fission and budding yeasts, and is known as the anaphase promoting complex or cyclosome (APC/C) (reviewed in refs 102, 103). The APC/C is a multi-subunit complex of approx. 20S that has been conserved throughout evolution; most or all of its components have been identified in several species. The APC/C targets proteins containing a destruction box motif for ubiquitin-dependent proteolysis during mitosis and G_1 phases (Fig. 6). In mitotic cyclins, this motif is nine amino acids in length. While G_2/M-phase cyclins were the first targets known, other APC/C target proteins have subsequently been identified

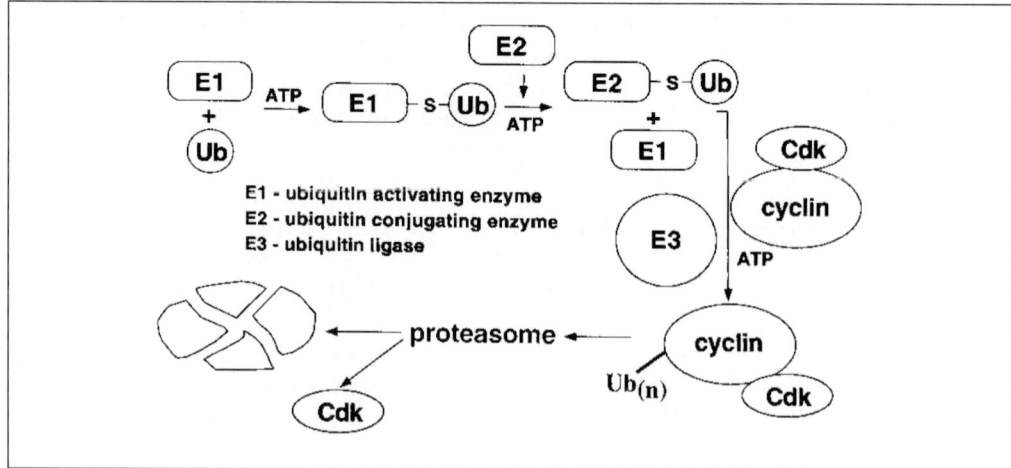

Fig. 5 Pathway for ubiquitin-mediated degradation of cyclins.

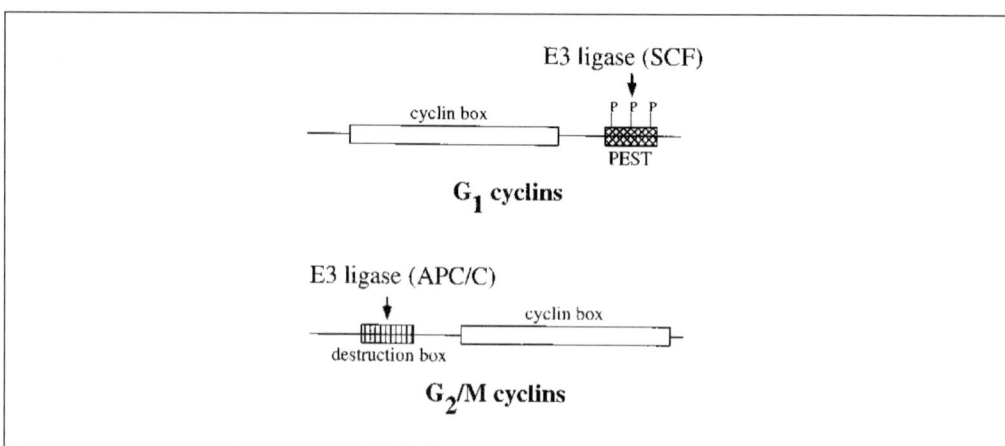

Fig. 6 Structures of G_1 and G_2/M cyclins and recognition by different E3 ligases for ubiquitin-mediated proteolysis.

whose destruction is also important for proper progression through mitosis and into the next cell cycle (102, 103).

An E3 ubiquitin ligase that recognizes other types of cyclins has also been identified and characterized (reviewed in refs 103, 106, 107). It has been termed SCF for the Skp1, Cullin, F-box protein complex (Fig. 6). SCF is heterogeneous in composition, containing Skp1, a member of the cullin family, and one of many different F-box containing proteins; the identity of the F-box component provides substrate specificity to the complex. To be recognized efficiently by an SCF complex, cyclins must first be phosphorylated. In the cases of cyclin E and G_1 cyclins in *S. cerevisiae*, there is considerable evidence that the kinases responsible are the Cdks themselves, providing a built-in mechanism to ensure the timely inactivation of the Cdk. In the case of cyclin

D1, the protein kinase responsible for triggering its destruction has yet to be identified. In addition to recognizing G_1 cyclins, SCF complexes target the Cdk inhibitors Sic1, Far1, and p27[Kip1] for ubiquitin-mediated proteolysis (103, 106, 107). Again, their recognition by SCF complexes is triggered by their phosphorylation; the phosphorylation can be carried out by Cdks.

5. Interaction with Suc1p/Csk1p

A protein which has long been known to bind certain Cdks was first identified in *S. pombe* as the *suc1*[+] gene product, p13 (reviewed in ref. 108). Overexpression of *suc1*[+] complemented certain temperature-sensitive mutations in *cdc2* and mutations in the *suc1*[+] gene also suppressed certain *cdc2* mutations. The *S. cerevisiae* homologue of *suc1*[+], *CSK1*, was also identified as a high copy suppressor of *CDC28* mutations, and homologues in humans (CskHs1 and CskHs2) were identified based on sequence homology. Despite many studies investigating its role in the cell cycle, the function of Suc1p/Csk1p is not yet understood. Purified Suc1p/Csk1p exhibits a strong and selective affinity for Cdc2 and Cdk2 and will bind to these protein kinases in the absence of other subunits. Suc1p/Csk1p does not bind Cdk4/6 or other Cdks, with the possible exception of Cdk3. Thus, Suc1p/Csk1p has been used extensively in cell-cycle studies as an affinity reagent to isolate Cdk complexes. Suc1p/Csk1p binding does not interfere with nor does it stimulate Cdk activity.

The *suc1*[+] and *CSK1* genes are essential, and there is evidence from studies in both yeasts and in *Xenopus* oocyte cell-free systems that Suc1p/Csk1p functions in mitosis and also at other stages of the cell cycle such as the G_1/S transition in *S. cerevisiae* and in G_2 in *S. pombe* and *Xenopus* (108). In *Xenopus* and clam oocyte cell-free systems, a connection has been made between Suc1p/Csk1p function and phosphorylation of the APC/C (109, 110). Suc1p/Cskp1 enhances the ability of Cdc2–cyclinB to phosphorylate subunits of the APC/C, which presumably helps in APC/C activation (110). Interestingly, in *Xenopus* extracts, it has been shown that cyclin binding stimulates Suc1p/Csk1p binding to Cdks (111). Thus, the role of Suc1p/Csk1p might be to direct active Cdk complexes to important targets during mitosis and other stages of the cell cycle.

6. Structure

The crystal structures of various forms of Cdk2 have allowed an understanding of the structural basis for Cdk regulation by various interacting proteins. The structures that have been solved are unmodified and uncomplexed Cdk2 (112), Cdk2 complexed with cyclin A (113), Cdk2 complexed with cyclin A and phosphorylated by CAK (114), Cdk2 complexed with the Suc1p homologue, CKSHs1 (115), and Cdk2 complexed with both cyclin A and the inhibitor, p27[Kip1] (101). Cdk2, like other protein kinases, consists of N- and C-terminal lobes with the active site formed between the two. What has been learned from the crystal structures is that cyclin binding and CAK phosphorylation cause significant conformational changes that allow for substrate entry into the catalytic site and proper alignment of the phosphate groups of bound ATP to allow for the phosphotransfer reaction (reviewed in ref. 116). The inhibitor, p27[Kip1], binds to both Cdk2 and its cyclin subunit with the effect of preventing both substrate and ATP binding (101). The structure of Cdk2 bound to CskHs1 revealed that the binding site of Csk1Hs1 is within the C-terminal lobe and does not affect the structure of the active site (115). This was expected since Suc1p beads are able to bind active Cdk complexes. However, as CskHs1 binds close enough to the active site, it has been speculated that it might contribute to a substrate recognition interface (116).

7. Localization

Far more is understood about the localization of Cdc2 and its regulators than any of the other Cdks, and this section will focus on these data. The question of Cdc2 localization has been addressed in several systems by indirect immunofluorescent methods as well as cell-fractionation experiments. Unfortunately, many of the early studies utilized antibodies against the conserved PSTAIRE domain rather than antibodies specific for individual Cdks. Nevertheless, the available evidence suggests that Cdc2 is distributed throughout the cell, with a significant fraction in the nucleus and also associated with the centrosomes (spindle pole bodies (SPBs) in yeast) (7, 117–119). These structures are likely to contain substrates relevant to an enzyme responsible for orchestrating entry into mitosis.

Mitotic cyclins have distinct localization patterns in a variety of organisms. The relative specificity of cyclin localization compared with that of Cdc2 has led to the suggestion that cyclins target Cdks to particular intracellular sites and substrates. Cyclin A is detected in the nucleus throughout the cell cycle and it partially colocalizes with chromatin (120, 121). In *S. pombe*, cdc13–cyclin B accumulates during interphase in the nucleus and is also present at SPBs (117). In vertebrate cells too, cyclin Bs are present at the centrosome, accumulate in the nucleus at the G_2/M transition, and localize to the mitotic spindle (122–126), although human cyclin B2 is associated with the Golgi (125). In *Xenopus*, the localization of cyclin B to the nucleus is critical for its function (127), and the amount of nuclear cyclin B1 influences the timing of mitosis in human cells as well (126). The accumulation of cyclin B in the nucleus depends upon its phosphorylation state (127). When phosphorylated, it is exported from the nucleus in a manner dependent upon the nuclear export factor, CRM1 (128, 129). Nuclear exclusion of cyclin B1 may play a role in the response to DNA damage (126, 130). Keeping cyclin B1–Cdc2 out of the nucleus and isolated from its substrates might be one mechanism that cells utilize to delay or prevent the onset of mitosis (reviewed in ref. 85).

What about the regulators of Cdc2? Many of these also appear to be nuclear proteins. CAK is found in the nucleus (131–133). As mentioned previously, mitotic forms of Cdc25 accumulate in the nucleus (134, 135) and there is evidence that keeping Cdc25 from the nucleus is another means of preventing mitosis in the presence of damaged DNA (85). Wee1 has also been shown to be a nuclear protein in *S. pombe* and human cells (136–138). In contrast, other regulators of Cdc2 activity are found in the cytoplasm. One form of Cdc25, Cdc25B, is exclusively cytoplasmic (139). The Wee1 homologue, Myt1, which contains a transmembrane domain, is associated with membranes (88–90, 140) and is localized to the endoplasmic reticulum and Golgi complex (88). The localization of Nim1 also appears to be exclusively cytoplasmic (137). Thus, there appears to be a subset of Cdk complexes and regulators present in the cytoplasm to coordinate the cytoplasmic changes that occur during M phase.

The theme emerging from these studies is that the localization of these cell-cycle regulators is dynamic and is likely to be a very important aspect in the regulation of the cell cycle. The isolation of various components from one another can be an efficient regulatory mechanism. In the instances in which the localization question has been addressed, the same theme appears to apply to other Cdks. For example, cyclin D1 is found in the nucleus during G_1 but moves out of the nucleus when S phase begins (reviewed in ref. 14).

8. Substrates

Since it has been well established that the activities of the various Cdk complexes drive progression through the cell cycle, an important avenue of investigation is to identify relevant substrates. This task is always a difficult one. The substrate specificities of Cdks are similar and, like most protein kinases, Cdks can phosphorylate a wide assortment of proteins *in vitro* (reviewed in ref. 141). To determine whether substrates are functionally relevant, three minimal criteria must be met. First, the sites phosphorylated by Cdks *in vitro* should be identical to those phosphorylated in cells. Second, the level of phosphorylation through the cell cycle should parallel the activity of the specific Cdk. Third, phosphorylation should alter the protein's function in a manner consistent with its cell-cycle function.

A complication in the Cdk field is that the substrate specificity of Cdks overlaps with that of at least one other protein kinase family, the MAP kinase family. Both classes of protein kinase are considered to be 'proline-directed', and can phosphorylate similar or identical sites in substrate proteins. An example would be the phosphorylation of lamins by Cdc2 and MAP kinases at the same sites (142). Overlapping substrate specificity is an especially important consideration in instances in which these other protein kinases are activated at the same time as the Cdks. A second complication is that although a common recognition sequence for Cdks in substrates is S/T–P–X–K/R, not all proteins considered to be substrates actually contain this sequence (141, 143). For example, the regulatory myosin-II light chain is phosphorylated *in vivo* and *in vitro* by Cdc2 on serines that lack an adjacent proline (144). Similarly, some target sites lack adjacent basic residues (141). Thus, the absence of this consensus site should not necessarily rule out a protein as a potential target.

Despite these difficulties, several probable *in vivo* substrates have been identified for Cdc2, Cdk2, and Cdk4/6 (Table 3). In the case of Cdc2, most of the substrates are involved in the structural reorganization of the cell during mitosis. As examples, phosphorylation of the nuclear lamins triggers dissociation of the lamina (141),

Table 3 Proposed Cdk substrates

Substrate	Reference
Cytoskeletal proteins	
Kinesin, Eg5	148
Nuclear lamins	141[a]
Golgi matrix protein GM140	146
Caldesmon	141, 155[a]
Myosin regulatory LC	144
Desmin	155[a]
Vimentin	141, 155[a]
MAP4	156
Chromatin associated proteins	
13S condensin complex	145
HMGI/Y, P1	155[a]
Histone H1	141, 155[a]
Cell-cycle regulators	
Cdc25	66[a]
Cyclins (E, C1n2p, B)	157, 158, 141[a]
CKIs (Far1p, Sic1p, p27^{KIP1})	103, 106, 107[a]
APC/C	147[a]
Cdc6/Cdc18p	150, 151[a]
DNA binding proteins	
MCMs	150, 151, 159[a]
pRb and related proteins	141, 155[a]
p53	155[a]
Swi5p	155[a]
c-Myb	155[a]
c-Fos/jun	141[a]
Replication protein A	155[a]
SV40 large T antigen	155[a]
B-Myb	160[a]
Protein kinases and phosphatases	
c-Src	155[a]
c-Abl	155[a]
PP1	161
Others	
NPAT	162
Stathmin/Op18	163
Poly(A) polymerase	164
Nucleolin, No38	141[a]
RNA polymerase II	17[a]
Rab1/Rab4	141[a]

[a]Original references provided within.

myosin light-chain phosphorylation is correlated with the onset of cytokinesis (144), phosphorylation of the 13S condensin complex activates its supercoiling activity and presumably aids in chromosome condensation (145), phosphorylation of the Golgi matrix protein GM130 is required for Golgi fragmentation at mitosis (146), phosphorylation of several subunits of the APC/C is correlated with increased ubiquitin ligase activity (reviewed in ref. 147), and phosphorylation of the Eg5 kinesin is important for its proper intracellular localization and spindle formation (148). Cdc2 is also thought to contribute to its own activation, by phosphorylating and activating Cdc25 family members and perhaps inhibiting Wee1 family members (see Section 4.2.2).

The activity of Cdc2 and Cdk2 during all phases of the cell cycle is generally measured using histone H1 as an exogenous substrate (141). Whether histone H1 is a physiologically relevant substrate of Cdks, however, remains unclear. It has long been proposed that histone H1 phosphorylation promotes chromosome condensation, but unequivocal results have been elusive. Also, there is evidence that chromosome condensation can occur in the absence of histone H1 phosphorylation (reviewed in ref. 149). Other chromatin-associated and nucleolar proteins are substrates for Cdc2 *in vitro* and are hyperphosphorylated during mitosis, possibly contributing to the process of chromosome condensation and nucleolar breakdown (141). Several transcription factors also are hyperphosphorylated during mitosis, and can be phosphorylated by Cdc2 *in vitro* (141). It has long been proposed that this might be a general mechanism of reducing their affinity for DNA in preparation for chromosome condensation (143).

It is certain that several proteins required for initiation of replication are targets of Cdks, and that phosphorylation by Cdks regulates initiation both positively and negatively (reviewed in refs 150, 151). One protein that has received a good deal of attention in this regard is the *S. cerevisiae* Cdc6 replication protein and its homologues, termed Cdc6 in higher eukaryotes and Cdc18 in *S. pombe*. In this case, Cdk phosphorylation causes loss of Cdc6/Cdc18 function. In *S. pombe*, phosphorylation by the mitotic Cdk leads to ubiquitin-mediated degradation; in other organisms, phosphorylation by mitotic Cdks is correlated with its relocalization (reviewed in ref. 151). Other initiator proteins are also phosphorylated by Cdks, most likely Cdk2 in higher eukaryotes, but the details of how Cdk complexes regulate replication initiation must still be worked out (150, 151). Cdk inhibitors (Sic1p, Far1p, and p27) and G_1 cyclins (Cln2p and cyclin E) are substrates of Cdk complexes. In these cases, phosphorylation targets the proteins for ubiquitin-mediated degradation (reviewed in refs 103, 106, 107). Thus, Cdk phosphorylation helps to ensure the irreversibility of cell-cycle progression.

Cdk4/6–cyclin D does not phosphorylate histone H1 and has a very distinct substrate specificity from that of other Cdk complexes (152). The only known targets of Cdk4/6–cyclin D *in vivo* are the retinoblastoma protein, pRb, and its relatives (14). pRb is a tumour suppressor protein which, in its hypophosphorylated form, inhibits S phase. It does so, at least in part, by binding to the transcription factor, E2F, and repressing its transactivation potential (reviewed in ref. 153). Upon phosphorylation of Rb by Cdk4/cyclin D and then Cdk2 complexes, E2F is released and is able to induce the transcription of target genes (14, 154). As mentioned earlier, the activity of Cdk4/6–cyclin D complexes is responsive to mitogens. The actions of these complexes are likely to be responsible for enabling cells to pass a point in G_1 beyond which cells no longer require mitogenic stimulation for another round of cell division.

9. Conclusions

In this chapter, I have briefly described some of the salient features of the Cdk protein kinase family. The most well-known members of this family regulate important transitions in the eukaryotic cell cycle. Protein kinases with similar structure and modes of regulation continue to be discovered and have proven to have roles in other aspects of cellular metabolism. Cdks are small protein kinases and their activities are regulated at many levels. The activities of cell-cycle Cdks oscillate during the cell cycle and this oscillation is controlled not only by their association with a member of the cyclin family of proteins, but also by a network of protein kinases and protein phosphatases. Cdks target a wide variety of proteins to coordinate the major events of the cell cycle. These substrates are found throughout the cell and are involved in structural reorganization during mitosis, the transition from G_1 into S phase, and the initiation and maintenance of S phase. The combined efforts to understand Cdk function in a variety of organisms should continue to yield rapid advances in our understanding of eukaryotic cell-cycle regulation.

References

1. Forsburg, S. L. and Nurse, P. (1991). Cell cycle regulation in the yeasts *Saccharomyces cerevisiae* and *Schizosaccharomyces pombe*. *Annu. Rev. Cell Biol.*, **7**, 227.
2. Lee, M. G. and Nurse, P. (1987). Complementation used to clone a human homologue of the fission yeast cell cycle control gene cdc2. *Nature*, **327**, 31.
3. Lees, E. (1995). Cyclin dependent kinase regulation. *Curr. Opin. Cell Biol.*, **7**, 773.
4. Pines, J. (1995). Cyclins and cyclin-dependent kinases: a biochemical view. *Biochem. J.*, **308**, 697.
5. Minshull, J. (1993). Cyclin synthesis: who needs it? *Bioessays*, **15**, 149.
6. Nurse, P. (1990). Universal control mechanism regulating onset of M-phase. *Nature*, **344**, 503.
7. Riabowol, K., Draetta, G., Brizuela, L., Vandre, D., and Beach, D. (1989). The cdc2 kinase is a nuclear protein that is essential for mitosis in mammalian cells. *Cell*, **57**, 393.
8. Fang, F. and Newport, J. W. (1991). Evidence that the G1–S and G2–M transitions are controlled by different cdc2 proteins in higher eukaryotes. *Cell*, **66**, 731.
9. Th'ng, J. P., Wright, P. S., Hamaguchi, J., Lee, M. G., Norbury, C. J., Nurse, P., and Bradbury, E. M. (1990). The FT210 cell line is a mouse G2 phase mutant with a temperature-sensitive CDC2 gene product. *Cell*, **63**, 313.
10. Meyerson, M., Enders, G. H., Wu, C. L., Su, L. K., Gorka, C., Nelson, C., Harlow, E., and Tsai, L. H. (1992). A family of human cdc2-related protein kinases. *EMBO J.*, **11**, 2909.
11. van den Heuvel, S. and Harlow, E. (1993). Distinct roles for cyclin-dependent kinases in cell cycle control. *Science*, **262**, 2050.
12. Pagano, M., Pepperkok, R., Lukas, J., Baldin, V., Ansorge, W., Bartek, J., and Draetta, G. (1993). Regulation of the cell cycle by the cdk2 protein kinase in cultured human fibroblasts. *J. Cell Biol.*, **121**, 101.
13. Tsai, L. H., Lees, E., Faha, B., Harlow, E., and Riabowol, K. (1993). The cdk2 kinase is required for the G1-to-S transition in mammalian cells. *Oncogene*, **8**, 1593.
14. Sherr, C. J. (1995). D-type cyclins. *Trends Biochem. Sci.*, **20**, 187.

15. Ohshima, T., Ward, J. M., Huh, C. G., Longenecker, G., Veeranna, Pant, H. C., Brady, R. O., Martin, L. J., and Kulkarni, A. B. (1996). Targeted disruption of the cyclin-dependent kinase 5 gene results in abnormal corticogenesis, neuronal pathology and perinatal death. *Proc. Natl Acad. Sci. USA*, **93**, 11173.

16. Tang, D. and Wang, J. H. (1996). Cyclin-dependent kinase 5 (Cdk5) and neuron-specific Cdk5 activators. *Progr. Cell Cycle Res.*, **2**, 205.

17. Harper, J. W. and Elledge, S. J. (1998). The role of Cdk7 in CAK function, a retro-retrospective. *Genes Dev.*, **12**, 285.

18. Lenburg, M. E. and O'Shea, E. K. (1996). Signaling phosphate starvation. *Trends Biochem. Sci.*, **21**, 383.

19. Hunt, T. (1991). Cyclins and their partners: from a simple idea to complicated reality. *Semin. Cell Biol.*, **2**, 213.

20. Pines, J. (1991). Cyclins: wheels within wheels. *Cell Growth Differ.*, **2**, 305.

21. Pines, J. and Hunter, T. (1990). Human cyclin A is adenovirus E1A-associated protein p60 and behaves differently from cyclin B. *Nature*, **346**, 760.

22. Girard, F., Strausfeld, U., Fernandez, A., and Lamb, N. J. (1991). Cyclin A is required for the onset of DNA replication in mammalian fibroblasts. *Cell*, **67**, 1169.

23. Pagano, M., Pepperkok, R., Verde, F., Ansorge, W., and Draetta, G. (1992). Cyclin A is required at two points in the human cell cycle. *EMBO J.*, **11**, 961.

24. Zindy, F., Lamas, E., Chenivesse, X., Sobczak, J., Wang, J., Fesquet, D., Henglein, B., and Brechot, C. (1992). Cyclin A is required in S phase in normal epithelial cells. *Biochem. Biophys. Res. Commun.*, **182**, 1144.

25. Andrews, B. and Measday, V. (1998). The cyclin family of budding yeast: abundant use of a good idea. *Trends Genet.*, **14**, 66.

26. Forsburg, S. L. and Nurse, P. (1994). Analysis of the *Schizosaccharomyces pombe* cyclin puc1: evidence for a role in cell cycle exit. *J. Cell Sci.*, **107**, 601.

27. Fisher, D. and Nurse, P. (1995). Cyclins of the fission yeast *Schizosaccharomyces pombe*. *Semin. Cell Biol.*, **6**, 73.

28. Lees, E. M. and Harlow, E. (1993). Sequences within the conserved cyclin box of human cyclin A are sufficient for binding to and activation of cdc2 kinase. *Mol. Cell. Biol.*, **13**, 1194.

29. Cross, F. R. (1990). Cell cycle arrest caused by CLN gene deficiency in *Saccharomyces cerevisiae* resembles START-I arrest and is independent of the mating-pheromone signal-ling pathway. *Mol. Cell. Biol.*, **10**, 6482.

30. Richardson, H. E., Wittenberg, C., Cross, F., and Reed, S. I. (1989). An essential G1 function for cyclin-like proteins in yeast. *Cell*, **59**, 1127.

31. Koff, A., Cross, F., Fisher, A., Schumacher, J., Leguellec, K., Philippe, M., and Roberts, J. M. (1991). Human cyclin E, a new cyclin that interacts with two members of the CDC2 gene family. *Cell*, **66**, 1217.

32. Lahue, E. E., Smith, A. V., and Orr-Weaver, T. L. (1991). A novel cyclin gene from Drosophila complements CLN function in yeast. *Genes Dev.*, **5**, 2166.

33. Leopold, P. and O'Farrell, P. H. (1991). An evolutionarily conserved cyclin homolog from Drosophila rescues yeast deficient in G1 cyclins. *Cell*, **66**, 1207.

34. Lew, D. J., Dulic, V., and Reed, S. I. (1991). Isolation of three novel human cyclins by rescue of G1 cyclin (Cln) function in yeast. *Cell*, **66**, 1197.

35. Grandin, N. and Reed, S. I. (1993). Differential function and expression of *Saccharomyces cerevisiae* B-type cyclins in mitosis and meiosis. *Mol. Cell. Biol.*, **13**, 2113.

36. Draetta, G., Beach, D., and Moran, E. (1988). Synthesis of p34, the mammalian homolog of the yeast cdc2$^+$/CDC28 protein kinase, is stimulated during adenovirus-induced pro-liferation of primary baby rat kidney cells. *Oncogene*, **2**, 553.

37. Stein, G. H., Drullinger, L. F., Robetorye, R. S., Pereira-Smith, O. M., and Smith, J. R. (1991). Senescent cells fail to express cdc2, cycA, and cycB in response to mitogen stimulation. *Proc. Natl Acad. Sci. USA*, **88**, 11012.

38. Dalton, S. (1992). Cell cycle regulation of the human cdc2 gene. *EMBO J.*, **11**, 1797.

39. Lee, M. G., Norbury, C. J., Spurr, N. K., and Nurse, P. (1988). Regulated expression and phosphorylation of a possible mammalian cell-cycle control protein. *Nature*, **333**, 676.

40. Elledge, S. J., Richman, R., Hall, F. L., Williams, R. T., Lodgson, N., and Harper, J. W. (1992). CDK2 encodes a 33-kDa cyclin A-associated protein kinase and is expressed before CDC2 in the cell cycle. *Proc. Natl Acad. Sci. USA*, **89**, 2907.

41. Rosenblatt, J., Gu, Y., and Morgan, D. O. (1992). Human cyclin-dependent kinase 2 is activated during the S and G2 phases of the cell cycle and associates with cyclin A. *Proc. Natl Acad. Sci. USA*, **89**, 2824.

42. Matsushime, H., Ewen, M. E., Strom, D. K., Kato, J. Y., Hanks, S. K., Roussel, M. F., and Sherr, C. J. (1992). Identification and properties of an atypical catalytic subunit (p34PSK–J3/cdk4) for mammalian D type G1 cyclins. *Cell*, **71**, 323.

43. Nasmyth, K. (1996). At the heart of the budding yeast cell cycle. *Trends Genet.*, **12**, 405.

44. McInerny, C. J., Partridge, J. F., Mikesell, G. E., Creemer, D. P., and Breeden, L. L. (1997). A novel Mcm1-dependent element in the SWI4, CLN3, CDC6, and CDC47 promoters activates M/G1-specific transcription. *Genes Dev.*, **11**, 1277.

45. Stuart, D. and Wittenberg, C. (1995). CLN3, not positive feedback, determines the timing of CLN2 transcription in cycling cells. *Genes Dev.*, **9**, 2780.

46. Dirick, L., Bohm, T., and Nasmyth, K. (1995). Roles and regulation of Cln–Cdc28 kinases at the start of the cell cycle of *Saccharomyces cerevisiae*. *EMBO J.*, **14**, 4803.

47. Althoefer, H., Schleiffer, A., Wassmann, K., Nordheim, A., and Ammerer, G. (1995). Mcm1 is required to coordinate G2-specific transcription in *Saccharomyces cerevisiae*. *Mol. Cell. Biol.*, **15**, 5917.

48. Koch, C., Schleiffer, A., Ammerer, G., and Nasmyth, K. (1996). Switching transcription on and off during the yeast cell cycle: Cln/Cdc28 kinases activate bound transcription factor SBF (Swi4/Swi6) at start, whereas Clb/Cdc28 kinases displace it from the promoter in G2. *Genes Dev.*, **10**, 129.

49. Espinoza, F. H., Farrell, A., Erdjument-Bromage, H., Tempst, P., and Morgan, D. O. (1996). A cyclin-dependent kinase-activating kinase (CAK) in budding yeast unrelated to vertebrate CAK. *Science*, **273**, 1714.

50. Thuret, J. Y., Valay, J. G., Faye, G., and Mann, C. (1996). Civ1 (CAK in vivo), a novel Cdk-activating kinase. *Cell*, **86**, 565.

51. Kaldis, P., Sutton, A., and Solomon, M. J. (1996). The Cdk-activating kinase (CAK) from budding yeast. *Cell*, **86**, 553.

52. Gould, K. L. and Nurse, P. (1989). Tyrosine phosphorylation of the fission yeast *cdc2*⁺ protein kinase regulates entry into mitosis. *Nature*, **342**, 39.

53. Lew, D. J. and Kornbluth, S. (1996). Regulatory roles of cyclin dependent kinase phosphorylation in cell cycle control. *Curr. Opin. Cell Biol.*, **8**, 795.

54. Atherton-Fessler, S., Parker, L. L., Geahlen, R. L., and Piwnica-Worms, H. (1993). Mechanisms of p34cdc2 regulation. *Mol. Cell. Biol.*, **13**, 1675.

55. Berry, L. D. and Gould, K. L. (1996). Regulation of Cdc2 activity by phosphorylation at T14/Y15. *Progr. Cell Cycle Res.*, **2**, 99.

56. Nurse, P. (1997). Checkpoint pathways come of age. *Cell*, **91**, 865.

57. Osmani, S. and Ye, X. (1997). Targets of checkpoints controlling mitosis: lessons from lower eukaryotes. *Trends Cell Biol.*, **7**, 283.

58. Russell, P. (1998). Checkpoints on the road to mitosis. *Trends Biochem. Sci.*, **23**, 399.

59. Terada, Y., Tatsuka, M., Jinno, S., and Okayama, H. (1995). Requirement for tyrosine phosphorylation of Cdk4 in G1 arrest induced by ultraviolet irradiation. *Nature*, **376**, 358.

60. Iavarone, A. and Massague, J. (1997). Repression of the CDK activator Cdc25A and cell-cycle arrest by cytokine TGFbeta in cells lacking the CDK inhibitor p15. *Nature*, **387**, 417.

61. Elledge, S. J. (1996). Cell cycle checkpoints: preventing an identity crisis. *Science*, **274**, 1664.

62. Sia, R. A., Herald, H. A., and Lew, D. J. (1996). Cdc28 tyrosine phosphorylation and the morphogenesis checkpoint in budding yeast. *Mol. Biol. Cell*, **7**, 1657.

63. Lew, D. J. and Reed, S. I. (1995). A cell cycle checkpoint monitors cell morphogenesis in budding yeast. *J. Cell Biol.*, **129**, 739.

64. Russell, P. and Nurse, P. (1986). *cdc25*⁺ functions as an inducer in the mitotic control of fission yeast. *Cell*, **45**, 145.

65. Russell, P. and Nurse, P. (1987). Negative regulation of mitosis by *wee1*⁺, a gene encoding a protein kinase homolog. *Cell*, **49**, 559.

66. Draetta, G. and Eckstein, J. (1997). Cdc25 protein phosphatases in cell proliferation. *Biochim. Biophys. Acta*, **1332**, M53.

67. Edgar, B. A., Sprenger, F., Duronio, R. J., Leopold, P., and O'Farrell, P. H. (1994). Distinct molecular mechanism regulate cell cycle timing at successive stages of Drosophila embryogenesis. *Genes Dev.*, **8**, 440.

68. Ducommun, B., Draetta, G., Young, P., and Beach, D. (1990). Fission yeast cdc25 is a cell-cycle regulated protein. *Biochem. Biophys. Res. Commun.*, **167**, 301.

69. Moreno, S., Nurse, P., and Russell, P. (1990). Regulation of mitosis by cyclic accumulation of p80cdc25 mitotic inducer in fission yeast. *Nature*, **344**, 549.

70. Lucibello, F. C., Truss, M., Zwicker, J., Ehlert, F., Beato, M., and Muller, R. (1995). Periodic cdc25C transcription is mediated by a novel cell cycle-regulated repressor element (CDE). *EMBO J.*, **14**, 132.

71. Edgar, B. A., Lehman, D. A., and O'Farrell, P. H. (1994). Transcriptional regulation of string (cdc25): a link between developmental programming and the cell cycle. *Development*, **120**, 3131.

72. Nefsky, B. and Beach, D. (1996). Pub1 acts as an E6-AP-like protein ubiquitin ligase in the degradation of cdc25. *EMBO J.*, **15**, 1301.

73. Hoffmann, I., Draetta, G., and Karsenti, E. (1994). Activation of the phosphatase activity of human cdc25A by a cdk2–cyclin E dependent phosphorylation at the G1/S transition. *EMBO J.*, **13**, 4302.

74. Hoffmann, I., Clarke, P. R., Marcote, M. J., Karsenti, E., and Draetta, G. (1993). Phosphorylation and activation of human cdc25-C by cdc2–cyclin B and its involvement in the self-amplification of MPF at mitosis. *EMBO J.*, **12**, 53.

75. Izumi, T. and Maller, J. L. (1995). Phosphorylation and activation of the Xenopus Cdc25 phosphatase in the absence of Cdc2 and Cdk2 kinase activity. *Mol. Biol. Cell*, **6**, 215.

76. Kumagai, A. and Dunphy, W. G. (1996). Purification and molecular cloning of Plx1, a Cdc25-regulatory kinase from Xenopus egg extracts. *Science*, **273**, 1377.

77. Peng, C. Y., Graves, P. R., Thoma, R. S., Wu, Z., Shaw, A. S., and Piwnica-Worms, H. (1997). Mitotic and G2 checkpoint control: regulation of 14–3–3 protein binding by phosphorylation of Cdc25C on serine-216. *Science*, **277**, 1501.

78. Sanchez, Y., Wong, C., Thoma, R. S., Richman, R., Wu, Z., Piwnica-Worms, H., and Elledge, S. J. (1997). Conservation of the Chk1 checkpoint pathway in mammals: linkage of DNA damage to Cdk regulation through Cdc25. *Science*, **277**, 1497.

79. Furnari, B., Rhind, N., and Russell, P. (1997). Cdc25 mitotic inducer targeted by chk1 DNA damage checkpoint kinase. *Science*, **277**, 1495.

80. Peng, C. Y., Graves, P. R., Ogg, S., Thoma, R. S., Byrnes, M. J., 3rd, Wu, Z., Stephenson, M. T., and Piwnica-Worms, H. (1998). C-TAK1 protein kinase phosphorylates human Cdc25C on serine 216 and promotes 14–3–3 protein binding. *Cell Growth Differ.*, **9**, 197.

81. Kumagai, A., Guo, Z., Emami, K. H., Wang, S. X., and Dunphy, W. G. (1998). The Xenopus Chk1 protein kinase mediates a caffeine-sensitive pathway of checkpoint control in cell-free extracts. *J. Cell Biol.*, **142**, 1559.

82. Zeng, Y., Forbes, K. C., Wu, Z., Moreno, S., Piwnica-Worms, H., and Enoch, T. (1998). Replication checkpoint requires phosphorylation of the phosphatase Cdc25 by Cds1 or Chk1. *Nature*, **395**, 507.

83. Blasina, A., de Weyer, I. V., Laus, M. C., Luyten, W., Parker, A. E., and McGowan, C. H. (1999). A human homologue of the checkpoint kinase Cds1 directly inhibits Cdc25 phosphatase. *Curr. Biol.*, **9**, 1.

84. Kumagai, A., Yakowec, P. S., and Dunphy, W. G. (1998). 14–3–3 proteins act as negative regulators of the mitotic inducer Cdc25 in Xenopus egg extracts. *Mol. Biol. Cell*, **9**, 345.

85. Pines, J. (1999). Four dimensional control of the cell cycle. *Nature Cell Biology*, **1**, E73.

86. Lundgren, K., Walworth, N., Booher, R., Dembski, M., Kirschner, M., and Beach, D. (1991). mik1 and wee1 cooperate in the inhibitory tyrosine phosphorylation of cdc2. *Cell*, **64**, 1111.

87. Lee, M. S., Enoch, T., and Piwnica-Worms, H. (1994). mik1$^+$ encodes a tyrosine kinase that phosphorylates p34cdc2 on tyrosine 15. *J. Biol. Chem.*, **269**, 30530.

88. Liu, F., Stanton, J. J., Wu, Z., and Piwnica-Worms, H. (1997). The human Myt1 kinase preferentially phosphorylates Cdc2 on threonine 14 and localizes to the endoplasmic reticulum and Golgi complex. *Mol. Cell. Biol.*, **17**, 571.

89. Booher, R. N., Holman, P. S., and Fattaey, A. (1997). Human Myt1 is a cell cycle-regulated kinase that inhibits Cdc2 but not Cdk2 activity. *J. Biol. Chem.*, **272**, 22300.

90. Mueller, P. R., Coleman, T. R., Kumagai, A., and Dunphy, W. G. (1995). Myt1: a membrane-associated inhibitory kinase that phosphorylates Cdc2 on both threonine-14 and tyrosine-15. *Science*, **270**, 86.

91. Palmer, A., Gavin, A. C., and Nebreda, A. R. (1998). A link between MAP kinase and p34(cdc2)/cyclin B during oocyte maturation: p90(rsk) phosphorylates and inactivates the p34(cdc2). inhibitory kinase Myt1. *EMBO J.*, **17**, 5037.

92. Breeding, C. S., Hudson, J., Balasubramanian, M. K., Hemmingsen, S. M., Young, P. G., and Gould, K. L. (1998). The *cdr2($^+$)* gene encodes a regulator of G2/M progression and cytokinesis in *Schizosaccharomyces pombe*. *Mol. Biol. Cell*, **9**, 3399.

93. Kanoh, J. and Russell, P. (1998). The protein kinase cdr2, related to Nim1/Cdr1 mitotic inducer, regulates the onset of mitosis in fission yeast. *Mol. Biol. Cell*, **9**, 3321.

94. Kaiser, P., Sia, R. A., Bardes, E. G., Lew, D. J., and Reed, S. I. (1998). Cdc34 and the F-box protein Met30 are required for degradation of the Cdk-inhibitory kinase Swe1. *Genes Dev.*, **12**, 2587.

95. Michael, W. M. and Newport, J. (1998). Coupling of mitosis to the completion of S phase through Cdc34-mediated degradation of Wee1. *Science*, **282**, 1886.

96. Sia, R. A., Bardes, E. S., and Lew, D. J. (1998). Control of Swe1p degradation by the morphogenesis checkpoint. *EMBO J.*, **17**, 6678.

97. O'Connell, M. J., Raleigh, J. M., Verkade, H. M., and Nurse, P. (1997). Chk1 is a wee1 kinase in the G2 DNA damage checkpoint inhibiting cdc2 by Y15 phosphorylation. *EMBO J.*, **16**, 545.

98. Peter, M. (1997). The regulation of cyclin-dependent kinase inhibitors (CKIs). *Progr. Cell Cycle Res.*, **3**, 99.

99. Labib, K. and Moreno, S. (1996). rum1: a CDK inhibitor regulating G1 progression in fission yeast. *Trends Cell Biol.*, **6**, 62.

100. Hengst, L., Gopfert, U., Lashuel, H. A., and Reed, S. I. (1998). Complete inhibition of Cdk/cyclin by one molecule of p21(Cip1). *Genes Dev.*, **12**, 3882.

101. Russo, A. A., Jeffrey, P. D., Patten, A. K., Massague, J., and Pavletich, N. P. (1996). Crystal structure of the p27Kip1 cyclin-dependent-kinase inhibitor bound to the cyclin A–Cdk2 complex. *Nature*, **382**, 325.

102. Townsley, F. M. and Ruderman, J. V. (1998). Proteolytic ratchets that control progression through mitosis. *Trends Cell Biol.*, **8**, 238.

103. Peters, J.-M. (1998). SCF and APC: the Yin and Yang of cell cycle regulated proteolysis. *Curr. Opin. Cell Biol.*, **10**, 759.

104. Ciechanover, A. (1994). The ubiquitin–proteasome proteolytic pathway. *Cell*, **79**, 13.

105. Hochstrasser, M. (1995). Ubiquitin, proteasomes, and the regulation of intracellular protein degradation. *Curr. Opin. Cell Biol.*, **7**, 215.

106. Krek, W. (1998). Proteolysis and the G1–S transition: the SCF connection. *Curr. Opin. Genet. Dev.*, **8**, 36.

107. Patton, E. E., Willems, A. R., and Tyers, M. (1998). Combinatorial control in ubiquitin-dependent proteolysis: don't Skp the F-box hypothesis. *Trends Genet.*, **14**, 236.

108. Pines, J. (1996). Cell cycle: reaching for a role for the Cks proteins. *Curr. Biol.*, **6**, 1399.

109. Sudakin, V., Shteinberg, M., Ganoth, D., Hershko, J., and Hershko, A. (1997). Binding of activated cyclosome to p13(suc1). Use for affinity purification. *J. Biol. Chem.*, **272**, 18051.

110. Patra, D. and Dunphy, W. G. (1998). Xe-p9, a Xenopus Suc1/Cks protein, is essential for the Cdc2-dependent phosphorylation of the anaphase-promoting complex at mitosis. *Genes Dev.*, **12**, 2549.

111. Egan, E. A. and Solomon, M. J. (1998). Cyclin-stimulated binding of Cks proteins to cyclin-dependent kinases. *Mol. Cell. Biol.*, **18**, 3659.

112. De Bondt, H. L., Rosenblatt, J., Jancarik, J., Jones, H. D., Morgan, D. O., and Kim, S. H. (1993). Crystal structure of cyclin-dependent kinase 2. *Nature*, **363**, 595.

113. Jeffrey, P. D., Russo, A. A., Polyak, K., Gibbs, E., Hurwitz, J., Massague, J., and Pavletich, N. P. (1995). Mechanism of CDK activation revealed by the structure of a cyclinA–CDK2 complex. *Nature*, **376**, 313.

114. Russo, A. A., Jeffrey, P. D., and Pavletich, N. P. (1996). Structural basis of cyclin-dependent kinase activation by phosphorylation. *Nat. Struct. Biol.*, **3**, 696.

115. Bourne, Y., Watson, M. H., Hickey, M. J., Holmes, W., Rocque, W., Reed, S. I., and Tainer, J. A. (1996). Crystal structure and mutational analysis of the human CDK2 kinase complex with cell cycle-regulatory protein CksHs1. *Cell*, **84**, 863.

116. Morgan, D. O. (1996). The dynamics of cyclin dependent kinase structure. *Curr. Opin. Cell Biol.*, **8**, 767.

117. Alfa, C. E., Ducommun, B., Beach, D., and Hyams, J. S. (1990). Distinct nuclear and spindle pole body population of cyclin–cdc2 in fission yeast. *Nature*, **347**, 680.

118. Bailly, E., Doree, M., Nurse, P., and Bornens, M. (1989). p34cdc2 is located in both nucleus and cytoplasm; part is centrosomally associated at G2/M and enters vesicles at anaphase. *EMBO J.*, **8**, 3985.

119. Pockwinse, S. M., Krockmalnic, G., Doxsey, S. J., Nickerson, J., Lian, J. B., van Wijnen, A. J., Stein, J. L., Stein, G. S., and Penman, S. (1997). Cell cycle independent interaction of CDC2 with the centrosome, which is associated with the nuclear matrix-intermediate filament scaffold. *Proc. Natl Acad. Sci. USA*, **94**, 3022.

120. Maldonado-Codina, G. and Glover, D. M. (1992). Cyclins A and B associate with chromatin and the polar regions of spindles, respectively, and do not undergo complete degradation at anaphase in syncytial Drosophila embryos. *J. Cell Biol.*, **116**, 967.

121. Pines, J. and Hunter, T. (1991). Human cyclins A and B1 are differentially located in the cell and undergo cell cycle-dependent nuclear transport. *J. Cell Biol.*, **115**, 1.

122. Ookata, K., Hisanaga, S., Okano, T., Tachibana, K., and Kishimoto, T. (1992). Relocation and distinct subcellular localization of p34cdc2–cyclin B complex at meiosis reinitiation in starfish oocytes. *EMBO J.*, **11**, 1763.

123. Gallant, P. and Nigg, E. A. (1992). Cyclin B2 undergoes cell cycle-dependent nuclear translocation and, when expressed as a non-destructible mutant, causes mitotic arrest in HeLa cells. *J. Cell Biol.*, **117**, 213.

124. Bailly, E., Pines, J., Hunter, T., and Bornens, M. (1992). Cytoplasmic accumulation of cyclin B1 in human cells: association with a detergent-resistant compartment and with the centrosome. *J. Cell Sci.*, **101**, 529.

125. Jackman, M., Firth, M., and Pines, J. (1995). Human cyclins B1 and B2 are localized to strikingly different structures: B1 to microtubules, B2 primarily to the Golgi apparatus. *EMBO J.*, **14**, 1646.

126. Jin, P., Hardy, S., and Morgan, D. O. (1998). Nuclear localization of cyclin B1 controls mitotic entry after DNA damage. *J. Cell Biol.*, **141**, 875.

127. Li, J., Meyer, A. N., and Donoghue, D. J. (1997). Nuclear localization of cyclin B1 mediates its biological activity and is regulated by phosphorylation. *Proc. Natl Acad. Sci. USA*, **94**, 502.

128. Yang, J., Bardes, E. S., Moore, J. D., Brennan, J., Powers, M. A., and Kornbluth, S. (1998). Control of cyclin B1 localization through regulated binding of the nuclear export factor CRM1. *Genes Dev.*, **12**, 2131.

129. Hagting, A., Karlsson, C., Clute, P., Jackman, M., and Pines, J. (1998). MPF localization is controlled by nuclear export. *EMBO J.*, **17**, 4127.

130. Toyoshima, F., Moriguchi, T., Wada, A., Fukuda, M., and Nishida, E. (1998). Nuclear export of cyclin B1 and its possible role in the DNA damage-induced G2 checkpoint. *EMBO J.*, **17**, 2728.

131. Tassan, J. P., Schultz, S. J., Bartek, J., and Nigg, E. A. (1994). Cell cycle analysis of the activity, subcellular localization, and subunit composition of human CAK (CDK-activating kinase). *J. Cell Biol.*, **127**, 467.

132. Darbon, J. M., Devault, A., Taviaux, S., Fesquet, D., Martinez, A. M., Galas, S., Cavadore, J. C., Doree, M., and Blanchard, J. M. (1994). Cloning, expression and subcellular localization of the human homolog of p40MO15 catalytic subunit of cdk-activating kinase. *Oncogene*, **9**, 3127.

133. Labbe, J. C., Martinez, A. M., Fesquet, D., Capony, J. P., Darbon, J. M., Derancourt, J., Devault, A., Morin, N., Cavadore, J. C., and Doree, M. (1994). p40MO15 associates with a p36 subunit and requires both nuclear translocation and Thr176 phosphorylation to generate cdk-activating kinase activity in Xenopus oocytes. *EMBO J.*, **13**, 5155.

134. Girard, F., Strausfeld, U., Cavadore, J. C., Russell, P., Fernandez, A., and Lamb, N. J. (1992). cdc25 is a nuclear protein expressed constitutively throughout the cell cycle in nontransformed mammalian cells. *J. Cell Biol.*, **118**, 785.

135. Millar, J. B., Blevitt, J., Gerace, L., Sadhu, K., Featherstone, C., and Russell, P. (1991). p55CDC25 is a nuclear protein required for the initiation of mitosis in human cells. *Proc. Natl Acad. Sci. USA*, **88**, 10500.

136. Baldin, V. and Ducommun, B. (1995). Subcellular localisation of human wee1 kinase is regulated during the cell cycle. *J. Cell Sci.*, **108**, 2425.

137. Wu, L., Shiozaki, K., Aligue, R., and Russell, P. (1996). Spatial organization of the Nim1–Wee1–Cdc2 mitotic control network in *Schizosaccharomyces pombe*. *Mol. Biol. Cell*, **7**, 1749.

138. Heald, R., McLoughlin, M., and McKeon, F. (1993). Human wee1 maintains mitotic timing by protecting the nucleus from cytoplasmically activated Cdc2 kinase. *Cell*, **74**, 463.

139. Gabrielli, B. G., De Souza, C. P., Tonks, I. D., Clark, J. M., Hayward, N. K., and Ellem, K. A. (1996). Cytoplasmic accumulation of cdc25B phosphatase in mitosis triggers centrosomal microtubule nucleation in HeLa cells. *J. Cell Sci.*, **109**, 1081.

140. Kornbluth, S., Sebastian, B., Hunter, T., and Newport, J. (1994). Membrane localization of the kinase which phosphorylates p34cdc2 on threonine 14. *Mol. Biol. Cell*, **5**, 273.

141. Nigg, E. A. (1993). Targets of cyclin-dependent protein kinases. *Curr. Opin. Cell Biol.*, **5**, 187.

142. Peter, M., Sanghera, J. S., Pelech, S. L., and Nigg, E. A. (1992). Mitogen-activated protein kinases phosphorylate nuclear lamins and display sequence specificity overlapping that of mitotic protein kinase p34cdc2. *Eur. J. Biochem.*, **205**, 287.

143. Moreno, S. and Nurse, P. (1990). Substrates for p34cdc2: in vivo veritas? *Cell*, **61**, 549.

144. Satterwhite, L. L., Lohka, M. J., Wilson, K. L., Scherson, T. Y., Cisek, L. J., Corden, J. L., and Pollard, T. D. (1992). Phosphorylation of myosin-II regulatory light chain by cyclin–p34cdc2: a mechanism for the timing of cytokinesis. *J. Cell Biol.*, **118**, 595.

145. Kimura, K., Hirano, M., Kobayashi, R., and Hirano, T. (1998). Phosphorylation and activation of 13S condensin by Cdc2 in vitro. *Science*, **282**, 487.

146. Lowe, M., Rabouille, C., Nakamura, N., Watson, R., Jackman, M., Jamsa, E., Rahman, D., Pappin, D. J., and Warren, G. (1998). Cdc2 kinase directly phosphorylates the cis-Golgi matrix protein GM130 and is required for Golgi fragmentation in mitosis. *Cell*, **94**, 783.

147. Hershko, A. (1997). Roles of ubiquitin-mediated proteolysis in cell cycle control. *Curr. Opin. Cell Biol.*, **9**, 788.

148. Blangy, A., Lane, H. A., d'Herin, P., Harper, M., Kress, M., and Nigg, E. A. (1995). Phosphorylation by p34cdc2 regulates spindle association of human Eg5, a kinesin-related motor essential for bipolar spindle formation *in vivo*. *Cell*, **83**, 1159.

149. Roth, S. Y. and Allis, C. D. (1992). Chromatin condensation: does histone H1 dephosphorylation play a role? *Trends Biochem. Sci.*, **17**, 93.

150. Jallepalli, P. V. and Kelly, T. J. (1997). Cyclin-dependent kinase and initiation at eukaryotic origins: a replication switch? *Curr. Opin. Cell Biol.*, **9**, 358.

151. Leatherwood, J. (1998). Emerging mechanisms of eukaryotic DNA replication initiation. *Curr. Opin. Cell Biol.*, **10**, 742.

152. Kitagawa, M., Higashi, H., Jung, H. K., Suzuki-Takahashi, I., Ikeda, M., Tamai, K., Kato, J., Segawa, K., Yoshida, E., Nishimura, S., and Taya, Y. (1996). The consensus motif for phosphorylation by cyclin D1–Cdk4 is different from that for phosphorylation by cyclin A/E–Cdk2. *EMBO J.*, **15**, 7060.

153. Dynlacht, B. D. (1997). Regulation of transcription by proteins that control the cell cycle. *Nature*, **389**, 149.

154. Bartek, J., Bartkova, J., and Lukas, J. (1996). The retinoblastoma protein pathway and the restriction point. *Curr. Opin. Cell Biol.*, **8**, 805.

155. Gould, K. L. (1994). Cyclin-dependent protein kinases. In *Protein kinases* (ed. J. R. Woodgett), pp. 149. IRL Press, Oxford.

156. Ookata, K., Hisanaga, S., Sugita, M., Okuyama, A., Murofushi, H., Kitazawa, H., Chari, S., Bulinski, J. C., and Kishimoto, T. (1997). MAP4 is the *in vivo* substrate for CDC2 kinase

in HeLa cells: identification of an M-phase specific and a cell cycle-independent phosphorylation site in MAP4. *Biochemistry*, **36**, 15873.

157. Won, K. A. and Reed, S. I. (1996). Activation of cyclin E/CDK2 is coupled to site-specific autophosphorylation and ubiquitin-dependent degradation of cyclin E. *EMBO J.*, **15**, 4182.

158. Lanker, S., Valdivieso, M. H., and Wittenberg, C. (1996). Rapid degradation of the G1 cyclin Cln2 induced by CDK-dependent phosphorylation. *Science*, **271**, 1597.

159. Hendrickson, M., Madine, M., Dalton, S., and Gautier, J. (1996). Phosphorylation of MCM4 by cdc2 protein kinase inhibits the activity of the minichromosome maintenance complex. *Proc. Natl Acad. Sci. USA*, **93**, 12223.

160. Saville, M. K. and Watson, R. J. (1998). The cell-cycle regulated transcription factor B-Myb is phosphorylated by cyclin A/Cdk2 at sites that enhance its transactivation properties. *Oncogene*, **17**, 2679.

161. Dohadwala, M., da Cruz e Silva, E. F., Hall, F. L., Williams, R. T., Carbonaro-Hall, D. A., Nairn, A. C., Greengard, P., and Berndt, N. (1994). Phosphorylation and inactivation of protein phosphatase 1 by cyclin- dependent kinases. *Proc. Natl Acad. Sci. USA*, **91**, 6408.

162. Zhao, J., Dynlacht, B., Imai, T., Hori, T., and Harlow, E. (1998). Expression of NPAT, a novel substrate of cyclin E–CDK2, promotes S-phase entry. *Genes Dev.*, **12**, 456.

163. Andersen, S. S., Ashford, A. J., Tournebize, R., Gavet, O., Sobel, A., Hyman, A. A., and Karsenti, E. (1997). Mitotic chromatin regulates phosphorylation of Stathmin/Op18. *Nature*, **389**, 640.

164. Colgan, D. F., Murthy, K. G., Prives, C., and Manley, J. L. (1996). Cell-cycle related regulation of poly(A). polymerase by phosphorylation. *Nature*, **384**, 282.

7 | Mechanisms and biology of signalling by serine/threonine kinase receptors for the TGFβ superfamily

MICHAEL KLÜPPEL, PAMELA A. HOODLESS, JEFFREY L. WRANA, and LILIANA ATTISANO

1. Overview

In a multicellular organism, communication between cells and their environment is of fundamental importance for all developmental and physiological processes. This communication relies on the transfer of information from one cell to another through specific signalling molecules. From worms to fruit flies to mammals, the transforming growth factor β (TGFβ) superfamily of growth factors represents a family of signalling molecules that is of critical importance for a diverse array of biological processes. Here, we will review the current state of knowledge of TGFβ signalling, with special emphasis on the recent advances in our understanding of the intracellular mechanisms that transmit TGFβ-like signals from the cell surface to the nucleus. In addition, we provide an overview of the involvement of TGFβ signalling pathways in development and disease.

2. The ligands

2.1 The TGFβ superfamily of ligands

The TGFβ superfamily comprises the prototypic member TGFβ, as well as a host of TGFβ-related factors. Over 40 family members from animals as diverse as *Drosophila* spp., *Caenorhabditis elegans*, zebrafish, and humans have been described. Historically, these factors have been subdivided into three groups on the basis of their structural and functional characteristics: the TGFβs, the activins/inhibins, and the bone morphogenic proteins (BMPs). However, identification of new family members that appear to bridge these subdivisions, demonstrates the existence of a broad continuum of related factors.

TGFβ was first identified as a factor that induced the anchorage-independent growth of normal rat-kidney fibroblasts. However, it is now known that TGFβ is a multifunctional growth factor with wide-ranging and often opposite effects on many cellular processes (reviewed in refs 1, 2). For example, TGFβ is involved in the regulation of proliferation and the control of differentiation of a variety of cell lineages including preadipocytes, myoblasts, and osteogenic precursors. TGFβ is also involved in the regulation of the growth and cellular processes of many cells of the immune system, and, of relevance, transgenic mice lacking TGFβ die of an acute inflammatory response (3). The activins and inhibins, originally identified by their ability to regulate hormone secretion in pituitary cells, are also involved in mammalian erythroid differentiation and mesoderm development (reviewed in ref. 2). The BMP family is the most heterogeneous group, with many of the first members (BMP2 to BMP7) originally identified by their effects on bone morphogenesis (4). It is now known that these and other BMP-related proteins are also important for inductive interactions during early development (5). For example, the nodals, which include mammalian nodal, *Xenopus* nodal-related (Xnr1, -2, -3, and -4), and the zebrafish *squint* and *cyclops* genes appear to play crucial roles in dorsal mesoderm patterning, organizer development, and determination of left–right asymmetry (reviewed in ref. 6).

In *Drosophila*, the BMP homologues, decapentaplegic (DPP), 60A, and screw are also implicated in the regulation of cell fate during development (reviewed in ref. 7). Further, three BMP-related ligands have been identified in *C. elegans*: DAF-7, which is important for dauer larvae formation (8), UNC-129, which directs axon guidance (9), and DBL-1/CET-1 which regulates overall body size and male tail patterning (10, 11). Thus, TGFβ superfamily members are important regulators of homeostatic processes as well as multiple and diverse developmental events.

2.2 Structure of TGFβ superfamily members

The mature biologically active forms of TGFβ superfamily members are typically composed of dimers of two 12–15-kDa subunits, which, in most cases, appear to be linked by a single disulfide bond (1, 2). Members of the lefty subgroup lack the cystine residue involved in dimerization, suggesting that this group of ligands either function as monomers or that dimers are maintained through non-covalent interactions. Although TGFβ family members exist primarily as homodimers, heterodimers such as TGFβ1, -2, and BMP2/7 have been described (1, 2, 12, 13). The active form of the ligand is expressed as the C-terminal portion of a larger precursor form. Cleavage of the mature ligand from the proregion occurs after a conserved RXXR sequence and is thought to be mediated by furin (also known as subtilisin-like proprotein convertase, SPC1) and furin-related proteases. For example, efficient cleavage of TGFβ, activin A, Müllerian-inhibiting substance, BMP4, and nodal by furins has been demonstrated (14; and references therein). There is considerable overlap in the expression pattern of various TGFβ-related ligands and furin during early mouse development (15). Furthermore, consistent with a role for furin in the activation/processing of TGFβ super-family members, a furin/SPC1-deficient mouse displays numerous defects suggestive of a loss of TGFβ superfamily signals (15). Interestingly, Vg1, a *Xenopus* BMP-like ligand can only induce axial mesoderm when fused to the BMP4 prodomain (16, 17). Presumably, this chimeric protein bypasses the usual requirement for the endogenous Vg1 protease. Thus, it appears that the proteolytic processing of TGFβ superfamily ligands is an important regulatory event.

TGFβ isoforms (β1, β2, and β3) are usually secreted in an inactive form, with the mature peptide remaining non-covalently associated with its N-terminal propeptide, known as the latency-associated protein or LAP (reviewed in ref. 18). In most cases, the complex of LAP and TGFβ (known as the small latent complex, SLC) also associates with latent TGFβ binding proteins (LTBPs) to form what is known as the large latent complexes. Activation of these complexes is an important event, and a variety of proteins have been implicated in this process (reviewed in ref. 18). For example, plasmin can activate latent TGFβ1 both *in vitro* and in cell culture by cleaving LAP. Thrombospondin activates both the SLC and the LLC *in vitro* by binding to the N-terminal region of LAP and inducing a conformational change in the latent complex. Further, similar patterns of inflammation have been observed in TGFβ1 and thrombospondin knockout mice, consistent with a role for thrombospondin in TGFβ activation (19). LAP can also bind specific integrins through an arginine–glycine–aspartic acid (RGD) sequence, and a recent study demonstrated that TGFβ1 LAP binding to integrin vβ6 activated the latent complex in a spatially restricted manner (20). Together, these diverse studies suggest that activation of the TGFβ latent complex *in vivo* is likely to occur through a number of distinct mechanisms.

Our first insights into the structure of TGFβ superfamily ligands came from the analysis of TGFβ1 and -2. X-ray crystallographic and solution nuclear magnetic resonance (NMR) studies revealed that TGFβ has an extended structure with six cystines forming an unusual structure known as the 'cystine knot' (21–23). Interestingly, a similar structure is found in other growth factors such as nerve growth factor (NGF) and platelet-derived growth factor (PDGF) (24). Since these six cystines are conserved in all TGFβ superfamily members, it has been postulated that other TGFβ-related molecules would also form similar structures. Indeed, solving of the crystal structure of BMP7/OP-1 which is only 36% identical to TGFβ2 has confirmed this prediction (25). Comparisons of these ligand structures, along with mutational analysis of TGFβ2, have also identified a region that is thought to be important for determining the specificity of receptor interaction (25, 26). This ridge shows a high degree of amino acid variation in different ligands and comprises an α-helix and exposed 'finger-tip' loops that would be consistent with a role in receptor contact. Currently, the region of the receptor that binds the ligand is not known.

2.3 Interaction of TGFβ superfamily ligands with soluble proteins

BMPs play important roles in early developmental events; however, the selective inactivation of BMP signalling pathways is also critical in determining cell fate particularly during gastrulation (reviewed in ref. 5). One mechanism whereby this antagonism of BMP signalling occurs is through the actions of a group of structurally unrelated, secreted proteins that includes noggin, chordin, follistatin, and the DAN family of factors.

Many of these proteins were first identified as dorsalizing factors. For example, *Xenopus* noggin and chordin are secreted from the Spemann's organizer, can dorsalize ventral mesoderm, act as neural inducers, and antagonize BMP4 signals in micro-injection assays (reviewed in refs 27, 28). Similarly, in *Drosophila*, the short-gastrulation gene (*sog*), which is the structural and functional homologue of the *Xenopus* chordin gene, functions as an antagonist of DPP, a BMP2/4 orthologue (29). Furthermore, the *dino* (*or chordino*) mutant of zebrafish, which lacks chordin, displays a partially ventralized phenotype and, like chordin, *dino* appears to antagonize BMP signals (30). An understanding of the biochemical basis for the effects of these proteins in developmental processes came with the demonstration that noggin and chordin can directly bind BMPs with affinities similar to that of the ligands for their receptors (31, 32). Thus, these antagonists block BMP signalling by preventing BMP interaction with its receptors.

Recently, the DAN family of secreted factors, including DAN, cerberus, and grem-lin have also been shown to antagonize BMP signals. DAN is a mouse zinc-finger protein that is downregulated in transformed cells and when injected in *Xenopus* can dorsalize embryos (33). Cerberus appears to function in *Xenopus'* head induction and the expression pattern of a cerberus-like protein in mouse suggests it is involved in anterior neural induction (34–37). Gremlin can induce axis duplication in *Xenopus*, though it is normally expressed in neural crest cells and their derivatives (33). Im-portantly, all DAN family members have been shown to directly interact with BMP ligands (33). However, unlike DAN and gremlin, cerberus, can also antagonize activin and nodal-related signals (33, 38). Furthermore, cerberus can disrupt non-TGFβ super-family signals such as wingless/Wnt signalling through direct binding of Wnts to cerberus (38). Interestingly, some of the BMP antagonists are also subject to regula-tion. *Drosophila* Tolloid and its *Xenopus* homologue Xolloid are metalloproteases that can cleave SOG or chordin, respectively, and thereby enhance DPP/BMP activity (39, 40). Thus, it appears that early developmental events occurring in response to members of the TGFβ superfamily are tightly controlled by a group of structurally unrelated extracellular proteins that function to enhance or antagonize BMP activity.

Follistatin is another soluble protein secreted by the Spemann's organizer that can block the potent mesoderm-inducing activity of activin (41) through direct protein interactions (42). However, follistatin can also interact with BMPs, albeit with a lower affinity (43), suggesting that it might also be involved in regulating BMP-mediated events.

Several studies have shown that TGFβ can also bind to a number of soluble proteins including α_2-macroglobulin, soluble betaglycan, decorin and biglycan, thrombospon-din, β-amyloid precursor protein, α-fetoprotein, fetuin, and type-II collagen (44, 45; and reviewed in refs 2, 46). These proteins appear to have diverse biological roles—some acting as antagonists and thereby altering the availability of ligand for the signalling receptors, while others appear to provide an extracellular storage site for the ligands.

3. The receptors

3.1 The TGFβ superfamily signals through a heteromeric serine/threonine kinase receptor complex

TGFβ superfamily members signal through a family of transmembrane kinase receptors, classified as type-I or type-II receptors (reviewed in refs 47–49). Both receptor types comprise a short, cysteine-rich extracellular domain, a single transmembrane domain, and an intracellular kinase domain with specificity towards serine and threonine residues. Although they share these common structural features, the type-I and type-II receptors perform distinct functions within a heteromeric receptor complex.

Numerous studies have shown that a heteromeric complex comprising both type-I and type-II receptors forms in the presence of ligand and that this complex is required for signal transduction (reviewed in refs 47–49). Importantly, formation of this complex is a cooperative event. Thus, for TGFβ and activin, ligand binds to the type-II receptor which then leads to the recruitment of the type-I receptor. A variation of this model exists for some BMP receptor complexes, in that cooperation of type-II receptors with type-I receptors is required to bind ligand. Interestingly, diverse biological responses can be generated through the formation of heteromeric complexes composed of various combinations of type-I and type-II receptors. The receptor complex is likely to be at least a heterotetramer and is formed from pre-existing homomeric complexes of type-I and type-II receptors (50, 51). Type-II receptor homodimers are essential for signalling and are formed through interactions at multiple contact points throughout the receptors (50–53). The type-II receptor is a constitutive kinase, and once type-I is recruited into the complex it becomes phosphorylated on Ser and Thr residues by the receptor-II kinase. This phosphorylation event occurs in the 'GS domain' located just upstream of the kinase domain of all type-I receptors (54–56). Once phosphorylated, receptor-I is activated to signal to downstream targets. Thus, a requirement for both type-II and type-I receptors is a general feature for signalling by the TGFβ superfamily.

3.2 Structural requirements in type-II receptors

To date, five vertebrate type-II receptors have been identified: TβRII, BMPRII, and AMHRII, which bind TGFβ, BMPs, and Müllerian-inhibiting substance (MIS), respectively; and ActRII and ActRIIB, which can bind both activins and BMPs (Table 1). In *Drosophila*, the one identified type-II receptor, *punt* (previously known as AtrII) binds DPP and is postulated to mediate signals for *60A* and *screw*. In *C. elegans*, the

Table 1 Ligands, receptors, and Smads in vertebrate signal transduction

Ligand	Type-II receptor	Type-I receptor	Receptor-regulated Smads	Common Smad	Inhibitory Smads
TGFβ	TβRII	ALK5(TβRI)	Smad2 Smad3		
Activin	ActRII ActRIIB	ALK4(ActRIB)			
BMPs	ActRII ActRIIB BMPRII	ALK2(ActRI) ALK3 (BMPRIA) ALK6(BMPRIB)	Smad1 Smad5 Smad8	Smad4 (Smad10?)	Smad6 Smad7
?	?	ALK1 (TSRI)	Smad1		
MIS	AMHR	?	?		
?	?	ALK7	?		

A summary of the known components of TGFβ superfamily signalling pathways in vertebrates is shown. Ligands interact with heteromeric complexes of type-I and type-II receptors. Receptor activation leads to phosphorylation of receptor-regulated Smads (Smad1, 2, 3, 5, 8) which promotes their interaction with the common Smad, Smad4. The inhibitory Smads, Smad6, and 7 block TGFβ superfamily signalling.

one identified type-II receptor, DAF-4, is required in the dauer pathway and for body size determination and male tail patterning (reviewed in refs 47–49).

The type-II receptors are constitutively active and are subject to autophosphorylation (53, 54, 56, 57). Importantly, addition of ligand does not alter this kinase activity. In the TGFβ type-II receptor, three major autophosphorylation sites have been identified: Ser213 in the juxtamembrane region, and Ser409 and Ser416 in the substrate recognition T-loop region of the kinase (53). These serines are differentially required such that phosphorylation of Ser213 and Ser409 is needed for activation of kinase activity and propagation of signals, whereas phosphorylation of Ser416 appears to inhibit receptor function (53). Currently, it is not known whether autophosphorylation in the other type-II receptors is similarly required for both positive and negative regulation of receptor kinase activity. Further insights into the structural requirements for the type-II receptors have been derived from a number of mutational studies. The intracellular kinase domain of the type-II receptors contains two inserts and a C-terminal serine/threonine-rich tail. However, unlike similar regions in tyrosine kinase receptors, the kinase insert 1 and the C-terminal extension in TβRII are not required for signalling, although kinase insert 2 is required to maintain catalytic activity (58). Furthermore, analysis of chemically mutagenized, TGFβ-resistant, Mv1Lu cell lines has revealed that a mutation in TβRII at Pro525 to leucine yields a receptor that can autophosphorylate but that cannot recognize its substrate, the type-I receptor (59).

3.3   Structural requirements in type-I receptors

Type-I receptors possess Ser/Thr kinase domains that are closely related to type-II receptors, but which are distinguishable from type-IIs since they lack a Ser/Thr-rich tail and contain a highly conserved sequence motif at the amino terminus of the kinase known as the 'GS domain' (reviewed in refs 47–49). The type-I receptors can be further subdivided into three groups based on sequence similarity within the kinase domain. The first group includes the TGFβ and activin type-I receptors, TβRI and ActRIB, the orphan receptors ALK7 and *Xenopus* XTrR-I, and the closely related *Drosophila* type-I receptor, BABOON (BABO, previously known as ATRI; (60)). The second group comprises the vertebrate BMP receptors, ALK3 and ALK6 (also known as BMPRIA and BMPRIB), and the *Drosophila* DPP receptor, thick veins (TKV). The third group includes the vertebrate BMP7 receptor ALK2 (61), the orphan receptor ALK1, and the *Drosophila* DPP receptor Saxophone (SAX). Two type-I receptors have been described in *C. elegans*, *daf-1* and *sma-6*, though due to sequence divergence, they have not been assigned to the above described subgroups (62, 63).

Initiation of signalling occurs when the type-II receptor phosphorylates the type-I receptor within the 30-amino acid glycine- and serine-rich GS domain located in the juxtamembrane region. Within the GS domain of the TGFβ type-I receptor, TβRI, there are five clustered serine and threonine residues (amino acids 185–192) that have been shown to be target phosphorylation sites (55). While mutation of any one site does not disrupt signalling, substitution of two or three of these residues reduces or abrogates signalling, respectively. Interestingly, mutation of a nearby threonine residue at amino acid position 204 (or a similarly positioned glutamine residue in certain type-I receptors) to an aspartic acid yields constitutively active type-I receptors. These activated receptors can signal in the absence of ligand and type-II receptor (55), thereby exemplifying the critical role of the type-I receptor as the downstream signalling component of the receptor complex. Recently, the crystal structure of an inactive version of the TGFβ type-I receptor kinase domain was solved (64). The conformation of the inactive kinase domain is maintained by interactions between the GS domain and the amino-terminal lobe of the kinase, and it is presumed that phosphorylation of the GS domain disrupts this inhibitory interaction. Other mutational analyses of the type-I receptor have revealed that alterations of specific residues (serine 172 and threonine 176) in the juxtamembrane domain of TβRI can differentially block growth inhibition and extracellular matrix protein production (65). Further, serine 165, which is also a receptor-II target phosphorylation site, appears to be involved in the modulation of TGFβ signalling since its mutation leads to increased growth inhibition and extracellular matrix formation as well as decreased TGFβ-induced apoptosis (56). Besides the juxtamembrane regions, a particularly important domain in type-I receptors is the L45 loop. This region located within subdomains IV and V of the kinase is crucial for specifying interactions with the Smad downstream targets (66) (and see below).

3.4   Betaglycan and endoglin

In the TGFβ system, TβRI and TβRII are essential for signal transduction; however, two other proteins, betaglycan and endoglin, also appear to bind ligand. Betaglycan is a 200–300-kDa transmembrane proteoglycan with a short cytoplasmic domain that lacks any known signalling motifs (67, 68). Betaglycan binds all three isoforms of TGFβ (β1, -2, and -3) and plays a crucial role in presenting TGFβ2 to the signalling receptors (69). Interestingly, the cytoplasmic domain of betaglycan is highly related to the cytoplasmic tail of endoglin. Endoglin is a 180-kDa protein that is highly expressed in endothelial and mesangial cells (70, 71). Unlike betaglycan, endoglin binds TGFβ1 and -β3 (71) and can also interact with activin A, BMP7, and BMP2 (72). However, endoglin cannot bind ligand on its own, but rather requires coexpression of the appropriate receptor complex (72, 73). Currently, the function of endoglin in the TGFβ receptor system is unclear, but it is thought that endoglin may modulate TGFβ signalling.

4.   The signalling pathway

In the last 3 years, the identification of components of the signalling pathway has produced tremendous insights into the mechanism through which TGFβ family ligands generate responses in cells. The principal signal mediators are a family of novel intracellular proteins known as Smads. However, a number of other receptor-interacting proteins, including FKBP12, farnesyl transferase-α, TRIP-1, phosphatase 2A, TRAP, STRAP, and IAPs have been isolated. In addition, a MAP kinase pathway has also been implicated in TGFβ signalling through the MAP kinase kinase kinase (MAPKKK) protein, TAK1 and its regulator TAB1. While it is clear that Smads are critical and essential mediators of TGFβ superfamily signalling, analysis of these other interacting proteins suggest that they may also play a role in mediating or modifying signals.

4.1   Receptor-interacting proteins

4.1.1   FK506/rapamycin binding protein (FKBP12)

Numerous groups, using a yeast two-hybrid approach, have isolated the FK506/ rapamycin binding protein, FKBP12, as a receptor-I interacting protein (74–76). FKBP12, an abundantly expressed protein, interacts with FK506 and rapamycin and thereby mediates their immunosuppressive activities (77). It appears that FKBP12 can interact with all tested type-I receptors but does not bind type-II receptors (74–76). Mutational studies with TβRI have shown that it is possible to disrupt FKBP12 binding to the receptor without altering its signalling capacity, suggesting that FKBP12 is not directly involved in transducing the TGFβ signal (74). Furthermore, it has been shown that constitutively active TβRI is a more potent activator when mutations in the GS domain (L193A with P194A), which disrupt FKBP12 binding to the receptor, are introduced (75). Thus, it appears that FKBP12 binding is actually inhibitory to the signalling pathway. The ability of FKBP12 to act as a negative modulator of TGFβ signalling appears to be the result of FKBP12-mediated prevention of receptor-II transphosphorylation of receptor I (78). Interestingly, solving the crystal structure of the TGFβ type-I receptor bound to FKBP12 demonstrates that FKBP12 binds to the GS domain, thereby protecting the type-II receptor phosphorylation sites and stabilizing the inactive conformation of the receptor (64). Thus, FKBP12 might play a role in preventing spurious ligand-independent activation of TGFβ receptors, thereby providing a mechanism for tight control of TGFβ signalling.

4.1.2 Farnesyl transferase-α (FT-α)

Using a yeast two-hybrid screen, several groups identified farnesyl transferase (FT)-α as a TβRI-interacting protein (79–81). Farnesyl transferase is responsible for the farnesylation of RAS, a modification that permits RAS membrane association and its biological activity (82). The α-subunit of FT, which is shared with other isoprenyl transferases such as geranylgeranyl transferase, appears to be important for regulating and stabilizing the catalytic β-subunit. Analysis of FT-α association with TGFβ receptors indicates that the interaction is specific for TβRI or the highly related ActRIB and does not occur with the TGFβ type-II receptor (79–81). Current evidence, however, does not support a role for farnesylation in TGFβ signalling since inhibitors of FT do not disrupt typical TGFβ responses, nor does TGFβ treatment alter FT activity or the overall level of protein isoprenylation (81). Interestingly, it has been reported that an inhibitor of geranylgeranyl transferase-I, but not of farnesyl transferase-I, prevents TGFβ-dependent stabilization of elastin mRNA (83), although it is unclear whether this is a direct effect. Further study will be required to determine the relevance of FT-α association with TβRI.

4.1.3 TGFβ receptor-interacting protein-1 (TRIP-1)

TRIP-1 was first identified in a yeast-two hybrid screen by its interaction with the cytoplasmic domain of the TGFβ type-II receptor, TβRII (84). TRIP-1, which contains a WD domain, associates specifically with the TGFβ type-II receptor and becomes phosphorylated by receptor-II alone or in a complex with receptor-I. Recent work suggests that TRIP-1 can selectively modulate TGFβ signals by specifically repressing transcription from the plasminogen activator inhibitor-1 promoter, without effecting TGFβ inhibition of cyclin-A promoter transcription (85). However, a direct role for TRIP-1 in signalling is uncertain, since receptor association and TRIP-1 phosphorylation occur independently of ligand binding and the inhibitory effect of TRIP-1 on PAI-1 transcription occurs through a receptor-dependent as well as -independent mechanism.

4.1.4 Phosphatase 2A

Phosphatase 2A is a serine/threonine phosphatase which can regulate signalling and cell-cycle progression. A trimeric enzyme, phosphatase 2A comprises a catalytic (C), a structural (A), and one of several regulatory (B) subunits. The Bα regulatory subunit contains WD repeats, has been shown to interact with the cytoplasmic domain of TGFβ type-I receptors, and is a target of their kinase activity (86). Evidence suggests that this interaction may play a role in the antiproliferative effects of TGFβ, though further work to define how it might function in the pathway is required.

4.1.5 TGFβ receptor-associated protein (TRAP)

A novel protein known as TRAP-1, was recently isolated in a yeast two-hybrid screen (87). TRAP-1 interacts, in yeast and mammalian cells, with a constitutively activated but not wild-type version of the type-I receptor. Interestingly, expression of a fragment of TRAP-1 that interacts with the receptor can block the induction of TGFβ-dependent transcription, possibly by functioning as a dominant-negative. Further studies are required to define what role TRAP may play in receptor signalling.

4.1.6 Serine/threonine kinase receptor-associated protein (STRAP)

A yeast-two hybrid screen with the TGFβ type-I receptor led to the identification of a novel WD-containing protein called STRAP (88). STRAP can associate with both type-I and type-II TGFβ receptors and can inhibit TGFβ-mediated transcriptional activation. Interestingly, STRAP enhances Smad7- but not Smad6-mediated inhibition of transcription. Thus, STRAP appears to play a role in the negative regulation of gene expression, possibly by promoting Smad7 interactions with the receptor.

4.1.7 TAK1, TAB1, and IAP

TAK1 (TGFβ activated kinase) was identified in a novel yeast-based screen for mammalian mitogen-activated, protein kinase-kinase-kinase (MAPKKK) proteins, a component of the MAP kinase (MAPK) cascade (89); however, several lines of evidence have implicated TAK1 as a mediator of TGFβ signalling. An activated form of TAK1, in which the N-terminal 22 amino acids are deleted, can induce the expression of a TGFβ-responsive promoter and a kinase-deficient version of TAK1 partially inhibits TGFβ-dependent responses. Furthermore, *in vitro* kinase assays demonstrate that TAK1 can be activated by both TGFβ and BMP4 but not EGF, supporting the idea that TAK1 has a role in signal transduction by multiple TGFβ ligands. TAK1 is activated by TAB1, a protein identified in a yeast two-hybrid screen for TAK1 interacting proteins (90). Interestingly, experiments in *Xenopus* embryos (see below, Section 5.2.1.3) have suggested that TAK1 and TAB1 may actually function as mediators of BMP signalling rather than activin or TGFβ.

Recently, a protein, XIAP (X-chromosome-linked inhibitor of apoptosis protein) was cloned in a screen for TAB1 interacting factors (91). This new protein, previously implicated in controlling apoptosis, interacts with both TAB1 and the intracellular domain of the BMP type-I receptor, ALK3 (BMPRIA) in a ligand-independent manner. Thus XIAP may provide a link between the receptor complex and downstream TAB1/TAK1 signalling molecules. Interestingly, two *Drosophila* homologues of IAP (DIAP1 and DIAP2) were cloned in a yeast two-hybrid screen for proteins that interact with the DPP type-I receptor, TKV; however, TAK1- and TAB1-related proteins have yet to be identified in *Drosophila* (92). Currently, it is unknown whether XIAP interacts with any other type-I receptors, and the mechanism by which ALK3–XIAP–TAB1 complexes activate TAK1 in response to a ligand stimulus remains unresolved.

TAK1 and TAB1 have also been implicated in the activation of stress-activated protein kinase/c-Jun N-terminal kinase (SAPK/JNK) through the induction of TAK1 activity by both haemopoietic progenitor kinase-1 (HPK1), and the second messenger signal, ceramide (93, 94). SAPK/JNK is a subgroup of MAP kinase proteins that play critical roles in cellular responses to proinflammatory cytokines, environmental stresses, and apoptotic agents. In addition, TAK1/TAB1 was found to activate NF-κB, inducing the degradation of IκB and translocation of the NF-κB p50/p65 heterodimer (95). However, there is little evidence that TGFβ family members play a direct role in the activation of SAPK/JNK or NF-κB. Thus, the precise role of the TAK/TAB pathway in TGFβ signalling remains to be determined, since a direct and specific response to TGFβ family ligands through TAK1 is yet to be discovered. Moreover, it is unclear how and if the TAK1/TAB1 pathway integrates with the major TGFβ intracellular signalling pathway mediated by the Smad proteins (see below). Regardless, it appears that TAK1/TAB1 has the potential to function in multiple intracellular signalling pathways, possibly providing an important link permitting cross-regulation between pathways and modulating cellular responses to the TGFβ family of ligands.

4.2 The Smad family

A number of independent studies have demonstrated that the critical intracellular mediators of TGFβ superfamily signals from the cell surface to the nucleus are the Smad proteins (reviewed in refs 7, 49, 96–100). Family members have now been identified in species as diverse as *Drosophila*, *C. elegans*, *Xenopus*, zebrafish, mouse, and humans and, in all cases, biochemical and developmental studies have supported the observation that Smads play an essential role in mediating TGFβ superfamily signals.

The first member of this family of signal transducers, known as MAD, for Mothers against dpp, was identified in a *Drosophila* screen for maternal enhancers of weak *dpp* alleles (101, 102). Importantly, flies expressing decreased levels of MAD protein suppressed the mutant wing and eye phenotypes generated by the expression of a constitutively active version of the DPP type-I receptor, TKV (103, 104). Furthermore, MAD was shown to be required for DPP function specifically in cells which respond to DPP (105). Together, these studies place MAD as a downstream component of the signalling pathway. Another *Drosophila* Smad, known as MEDEA, was originally described in the same screen that lead to the identification of MAD (101), and has also been isolated in a second maternal-effect enhancer screen (106) and by degenerate polymerase chain reaction (PCR) as a MAD-related protein (107). Like MAD, MEDEA also functions downstream of TKV in the DPP pathway (106–108). Other *Drosophila* Smads include DAD, which modulates DPP signalling (109) and dSmad2 which mediates signals downstream of the type-I receptor, BABO (60).

Parallel to the earlier studies in *Drosophila*, three related homologues of MAD, *sma-2*, *sma-3*, and *sma-4*, were identified in *C. elegans* (110). Mutations in these three genes yields worms with reduced body size and abnormal male tail rays, a subset of the phenotypes observed in mutants of the *C. elegans* type-II receptor gene, *daf-4*. Recent work suggests that DAF-4 along with the type-I receptor, SMA-6, regulate body size and male tail ray formation through SMA-2, SMA-3, and SMA-4 (63). Furthermore, it is thought that DAF-4 association with the type-I receptor, DAF-1 controls dauer larvae formation through the Smad family proteins, DAF-3, DAF-14, and DAF-8 (63, 111).

Several vertebrate homologues of MAD have now been identified in human, mouse, rat, zebrafish, and *Xenopus* (reviewed in refs 7, 49, 96–100). These homologues are termed Smads to symbolize the genes originally identified, *sma* and *Mad*. The first human Smad protein, Smad4 (first named DPC4), was identified as a candidate tumour suppressor gene involved in pancreatic cancer (112). Other members have been identified by homology screening of libraries, expressed sequence tag (EST) databases, and by polymerase chain reaction (PCR) techniques. In addition, mouse Smad2 was cloned from a functional screen in *Xenopus* to identify molecules involved in mesoderm induction (113). At present, at least nine distinct Smad proteins (Smad1 through 10; Table 1) have been reported in vertebrates (114; and reviewed in refs 7, 49, 96–100). Sequence comparisons of all family members indicate the presence of three distinct domains, two highly conserved N-terminal (termed MH1 for MAD homology-1) and C-terminal (termed MH2) domains separated by a divergent or non-conserved central region. These domains contain no known structural motifs, indicating that Smad proteins form a novel family of signalling molecules. However, all mutations identified in *Drosophila*, *C. elegans*, and human cancers map to highly conserved residues within the MH domains, suggesting that these regions are critical to the function and regulation of the protein (reviewed in refs 7, 49, 96–100).

The Smad family can be subdivided into three classes based on their function in the signalling cascade: the receptor-regulated Smads (R-Smads), which propagate the specificity of the signal in response to ligands; the common Smads (Smad4 in vertebrates, *Medea* in *Drosophila*, and possibly *sma-4* in *C. elegans*), which are thought to function as mediators for signalling by all TGFβ family members; and the inhibitory Smads that function to regulate the ligand response by blocking signal transduction.

4.2.1 The receptor-regulated Smads

Receptor-regulated Smads are specifically phosphorylated in response to ligand

The primary defining feature of the R-Smads is that they are rapidly and specifically phosphorylated in response to ligand. This was first demonstrated with Smad1 (originally named MADR1), which was shown to be specifically phosphorylated within minutes in response to BMP signalling through ALK3, ALK6, or ALK2 but not by TGFβ or activin signalling through TβRI or ActRIB (61, 103, 115, 116). Smad1 is highly related to *Drosophila* MAD and, like Smad1, MAD is phosphorylated in response to BMP signalling (117). This phosphorylation is necessary for signal transduction since a point mutation in Smad1 that corresponds to a null allele of *Mad* in *Drosophila* is not phosphorylated in response to BMP2 (103). Comparisons between Smad protein sequences has identified two other closely related vertebrate Smad proteins, Smad5 and Smad8, and these Smads also mediate BMP signalling (118, 119).

In contrast to the BMP R-Smads, Smad2 and Smad3 are phosphorylated by TGFβ and activin signalling through TβRI and ActRIB, respectively (120–122). Similarly, a new *Drosophila* Smad, dSmad2, which is highly related to vertebrate Smad2 and -3, is a target of the *Drosophila* type-I receptor, BABO (60). Although the endogenous ligand for BABO is not known, this receptor can induce TGFβ/activin-responsive promoters in mammalian cells and thus is likely to mediate signals for a *Drosophila* activin-like ligand.

The distinct roles of BMP and TGFβ/activin R-Smads are supported by numerous functional studies. In *Xenopus*, overexpression of different Smad proteins is sufficient to initiate developmental signals that recapitulate the responses observed for their respective ligands. For example, ectopic expression of the BMP R-Smads -1, -5, or -8 in *Xenopus* embryos stimulates ventral mesoderm induction, inhibits neural differentiation, and induces epidermis in ectodermal cells, similar to the effects observed with ligands BMP2 and 4 (reviewed in refs 7, 49, 96–100). In contrast, expression of Smad2 induces dorsal mesoderm, similar to that observed in response to activin, Vg1, or nodal (reviewed in refs 7, 49, 96–100). Interestingly, mutants of Smad2 identified in colorectal carcinomas that fail to be phosphorylated can inhibit the induction of dorsal mesodermal cell fates induced by activin or Vg1 (120, 123).

In mammalian cells, differential regulation of transcription by Smad proteins has also been observed. The most commonly used assay to measure TGFβ and activin responsiveness is the 3TP promoter, which combines portions of the plasminogen activator inhibitor (PAI-1) promoter with three TPA-responsive elements. This promoter is highly responsive to TGFβ and activin, but responds poorly to BMPs. Consistent with this, 3TP is strongly activated by Smad2 or Smad3 but not Smad1 (124–126). Similarly, BMP-dependent induction of the promoter from the homeobox gene *Tlx-2* can be enhanced by expression of Smad1 and inhibited by expression of a truncated version of Smad1, whereas it is not induced by activin or TGFβ (61, 127). Collectively, these results indicate that the R-Smads can be separated into two groups, the BMP and the activin/TGFβ R-Smads, and that specific phosphorylation of these Smads in response to ligand is critical for protein function and signal transduction.

Phosphorylation of receptor-regulated Smads by the activated type-I receptor

Ligand-dependent phosphorylation of R-Smads occurs on two serine residues in the motif SSXS located at the immediate C-terminus of the protein (122, 128–131). Although three serines are present, only the last two (Ser[465] and Ser[467] in Smad2) are phosphorylated in response to signalling (130, 131). The first serine residue, however, is required for efficient phosphorylation of the other two. Interestingly, phosphorylation of both terminal serines is required for all downstream functions of R-Smad since mutation of either Ser[465] or Ser[467] of Smad2 results in a non-functional protein (130, 131). Other phosphorylation sites are present in R-Smads; however, these sites are not induced by TGFβ-ligand signalling. In particular, Smad1 has been shown to be phosphorylated in the non-conserved region of the protein by the Erk family of MAP kinases in response to EGF or HGF (132). Similarly, HGF has been shown to induce phosphorylation of Smad2 (133). Together these results suggest that these additional sites may provide mechanisms for cross-regulation between signalling pathways.

The tight regulation of R-Smad phosphorylation by BMP and TGFβ/activin signalling pathways can now be attributed to the direct and specific interaction of Smad proteins with the Ser/Thr kinase receptor complexes (124, 128, 129, 131). In mammalian cells, Smad2 and Smad3 have been shown to interact directly with activated TGFβ and activin-receptor complexes (123, 124, 128, 131). Similarly, dSmad2 has been shown to specifically interact with *Drosophila* BABO (60). These interactions are transient since Smad proteins are released immediately upon phosphorylation, but this association can be stabilized by expression of a kinase-deficient version of the type-I receptor which permits trapping of the Smad substrate (128). Alternatively, mutations of the phosphorylation site or the addition of a 3′ tag, which disrupts phosphorylation of the Smad protein, also stabilizes the interaction between Smads and the receptor (124, 128). *In vitro* kinase assays using purified receptors have further demonstrated that Smad proteins are direct substrates of the receptor complex and that phosphorylation is mediated by the activated type-I receptor (124, 128, 129). As described above, the region of the receptor with which the Smads interact is known as the L45 loop. Substitution of this αC–β4–β5 region in subdomains III to V of the kinase region in ActRI with the comparable region in TβRI, conferred TGFβ signalling on the chimeric receptor (66) and specified interaction with Smad2 and not Smad1 (134).

R-Smads function in the nucleus

Several studies have demonstrated that R-Smads are cytosolic proteins that accumulate in the nucleus in response to signalling proteins (reviewed in refs 7, 49, 96–100). *In vivo* studies in *Drosophila* indicate that in the absence of signalling, MAD is localized to the cytoplasm but translocates to the nucleus after activation of DPP signalling (105). Further, in *Xenopus* a lacZ/Smad2 fusion protein has been localized to the nucleus at the anterior portion of the axis, in a region in which activin, Vg1, and/or nodal are thought to signal dorsal mesoderm induction (113). Receptor-dependent phosphorylation on the SSXS motif is required for this nuclear localization, since a phosphorylation site mutant of Smad2 fails to translocate in response to TGFβ signalling (128). Moreover, an N-terminal truncation of Smad2 that contains the MH2 domain and a small portion of the non-conserved domain, is more active than full-length Smad2 and is constitutively nuclear (113). Numerous studies now indicate that Smad proteins function in the nucleus to activate transcription (see below).

4.2.2 The common Smads

In contrast to R-Smads, Smad4, which lacks the SSXS motif, is not phosphorylated by any of the receptor complexes (121, 124, 128). However, Smad4, is able to cooperate with R-Smads in both BMP and TGFβ/activin signalling. For example, Smad4 synergizes with Smad2 or Smad3 to activate TGFβ/activin-responsive promoters such as 3TP, PAI-1, or *goosecoid* (124, 125, 135). Furthermore, expression in *Xenopus* of a truncated form of Smad4, which has a small C-terminal deletion, can antagonize dorsal mesoderm induction by Smad2 and ventral mesoderm induction by Smad1, indicating that Smad4 can interact functionally with R-Smads in both BMP and TGFβ/activin signalling pathways (125).

The molecular basis for this functional cooperativity lies in the ability of Smad4 to form heteromeric complexes with R-Smads upon stimulation by BMP4 or TGFβ/activin (125). Mutation of the phosphorylation site in R-Smads abrogates heteromeric complex formation, demonstrating that this interaction is dependent on phosphorylation. These observations have led to the model that Smad4 is a central mediator of TGFβ superfamily signalling. Interestingly, in the absence of signalling, Smad4, and its *Drosophila* counterpart MEDEA, are completely localized in the cytoplasm of the cells (108, 123, 136). Upon ligand activation, Smad4/MEDEA interacts with a R-Smad which recruits Smad4/MEDEA into the nucleus (108, 123, 136). The heteromeric complex can then function to activate specific patterns of gene expression. Thus, the common Smad plays a more general role in signal transduction, while specificity is provided by the R-Smad protein with which it interacts. The idea that Smad4 is critical for all signalling by the TGFβ superfamily is being challenged by evidence that TGFβ superfamily signalling can occur independently of MEDEA/Smad4, although the existence of another Smad4-like protein cannot be completely excluded (108, 137).

4.2.3 The inhibitory Smads

Two inhibitory Smads have been described to date, Smad6 and Smad7 (in *Xenopus* Smad7 was originally called Smad8). The inhibitory Smads, as the name implies, act to downregulate signalling by TGFβ family proteins (reviewed in refs 7, 49, 96–99). Both Smad6 and Smad7 contain the highly conserved MH2 domain found in all Smad proteins, though the SSXS motif that serves as the phosphorylation site in R-Smads is absent. In contrast, both proteins have divergent MH1 domains, suggesting that these proteins play a novel role in signal transduction. Furthermore, two forms of Smad6 exist, a long form and a truncated version which lacks an MH1 domain completely (138–141).

For Smad7, the mechanism for the inhibitory activity lies in the ability of Smad7 to strongly interact with the TGFβ, activin, and BMP receptor complexes (142–144). Since Smad7 lacks the SSXS motif, it cannot be phosphorylated by the type-I receptor and results in the formation of a stable receptor/Smad complex. Similar to R-Smads, this association requires that the type-II receptor activates the type-I receptor by phosphorylation of the GS domain. The formation of a stable complex blocks interactions of the receptor with the R-Smads, thereby preventing phosphorylation of these Smads, association with Smad4, and activation of downstream responses (142, 143). In support of this mechanism, Smad7 has been shown in *Xenopus* embryos to inhibit both BMP and activin responses (145–147). Interestingly, in contrast to R-Smads, Smad7 has been shown to be predominantly localized to the nucleus in the absence of signalling and to translocate to the cytoplasm upon TGFβ receptor activation (148).

Smad6 has also been shown to form ligand-dependent stable associations with TGFβ, activin, and BMP type-I receptors (140, 141). Further, Smad6 was shown to block ALK3 (but not ALK6)-dependent phosphorylation of Smad1 as well as blocking

TβRI-dependent phosphorylation of Smad2, but not Smad3 (140). In contrast to the observed effect of Smad6 on Smad1 and Smad2 phosphorylation, experiments in *Xenopus* embryos have shown that Smad6 blocks BMP signalling without disrupting activin activity. In agreement with this, studies using GAL4 fusion proteins of Smad1 and Smad2 with a GAL4-dependent promoter, demonstrate that Smad6 blocks Smad1 but not Smad2 activity (141). Furthermore, mink lung-epithelial cells (Mv1Lu) stably expressing Smad6 did not inhibit endogenous TGFβ responses such as growth inhibition, while similar cells expressing Smad7 did (148). Interestingly, Smad6 was shown to inhibit the formation of Smad1/4 complexes by forming heteromers with Smad1 (141). Thus, unlike Smad7, Smad6 may inhibit BMP signalling by interfering with Smad1/4 complexes rather than preventing Smad1 association with the receptor.

Interestingly, while Smad6 and Smad7 function to block signalling by TGFβ super-family ligands, the mRNA for both proteins can be induced in cultured cells by TGFβ, activin, and BMPs (148–151). This suggests that a feedback loop acts to regulate signalling by TGFβ family members. Furthermore, the mRNA for Smad6 and Smad7 can be induced by EGF (150), and that of Smad7 by interferon-γ (152), suggesting that the inhibitory Smads may provide a mechanism for other signalling pathways to inhibit TGFβ signal transduction.

4.2.4 New Smads

Recently, a new Smad, Smad10, was identified in *Xenopus* which is most closely related to Smad4 (83 and 91% identity in the MH1 and MH2 domains, respectively, and 63% identity overall (114). However, *Xenopus* injection studies have shown that, unlike Smad4, Smad10 does not inhibit BMP signalling through Smad1. This would imply that Smad10 does not function as a common Smad. Interestingly, Smad10 contains the motif SSVN at the carboxy terminus similar to the SSXS motif found in R-Smads. However, the lack of one of the critical serine residues suggests that Smad10 may not be phosphorylated in response to ligand and may not act like a R-Smad. Smad10 may have a different role from the other identified Smads, although what this function might be remains to be determined.

4.2.5 Smad proteins interact with nuclear factors

In the last 2 years, much research has focused on how Smad proteins function in the nucleus to activate gene transcription (reviewed in ref. 97). Several studies have focused on deciphering Smad interactions with TGFβ-responsive elements in a variety of gene promoters such as PAI-1, type VII collagen, *goosecoid*, and JunB. Early studies in *Drosophila* with the DPP-responsive gene *vestigial* established that the MH1 domain of MAD can bind directly to DNA, although the full-length protein was unable to bind (153). It is now firmly established that the vertebrate Smads 3 and 4 (reviewed in ref. 97) and *C. elegans* DAF-3 (154) can also bind directly to DNA through the MH1 domain. In contrast, Smads 1, 2, and 5 have not been shown to interact directly with DNA despite attempts to demonstrate binding on numerous DNA fragments. Interestingly, Smad2 is nearly identical to Smad3 except for two small insertions in the MH1 domain. An alternatively spliced variant of Smad2 that is expressed in certain cell types deletes one of these regions and yields a Smad that is capable of DNA binding (155). Therefore, this region appears to be important for association with DNA. Binding-site selection methods have yielded an optimized binding site for Smad3 and Smad4 containing two inverted repeats of a core GTCT sequence (156). At least one copy of this sequence has been found in the TGFβ-responsive regions of the type-II collagen, PAI-1, and JunB genes (157–162). However, Smad proteins also interact with sequences that do not contain this element but are GC-rich, such as in the *goosecoid* promoter and the *vestigial* quadrant enhancer, indicating that the specificity of DNA binding is not stringent (135, 153).

Although Smad proteins can bind DNA directly, it is generally believed that Smad proteins interact with other sequence-specific DNA-binding proteins to form higher order complexes that activate transcription (reviewed in ref. 97). Several transcription factor targets have now been identified, which encompass a wide range of nuclear factors that reflect the diversity of responses regulated by TGFβ family members. These factors include the FAST family of forkhead proteins, the helix–loop–helix factor TFE3, the leucine-zipper factor AP-1 containing c-Fos and c-Jun, the zinc-finger proteins Evi-1 and Gli3, and the vitamin D receptor (VDR) (135, 157, 162–168).

The most extensively characterized nuclear factor downstream of TGFβ/activin is the FAST (forkhead activin signal transducer) family. FAST-1 is a winged-helix DNA binding factor (forkhead) that was originally identified in *Xenopus* by its ability to bind to a 50 base-pair activin responsive element (ARE) found in the promoter region of the gene *Mix.2* (169). Human and mouse homologues of FAST-1 have now been identified which can bind to the *Mix.2* ARE (135, 164, 170, 171). These vertebrate FASTs are 40% and 39% identical, respectively, to *Xenopus* FAST-1 at the amino acid level, with the highest region of homology corresponding to the forkhead domain. Mouse FAST, termed FAST-2, has been shown to bind a TGFβ/activin-responsive region (TARE) of the *goosecoid* promoter (135). In the absence of signalling, FAST binds constitutively to the response element of both the *Mix.2* and *goosecoid* genes. However, in the presence of activin or TGFβ, a higher order complex, termed ARF (activin-responsive factor) or TRF (TGFβ/activin-responsive factor), assembles on the DNA, and in addition to FAST contains both Smad2 and Smad4 (123, 135, 136, 163, 164, 170). Formation of this complex is required for transcriptional activation of these elements in response to ligand. Analysis of the complex has indicated that FAST binds a sequence-specific DNA element and that Smad2 interacts directly with FAST to recruit Smad4 into the nuclear complex. In the *goosecoid* promoter, Smad4 has been shown to form contacts with DNA in a region adjacent to the FAST binding site and this contact is essential for efficient formation of TRF. Interestingly, work with FAST-2 has demonstrated that Smad3 can replace Smad2 in the TRF complex (135, 170). However, this FAST-2/Smad3/Smad4 complex inhibits rather than activates transcriptional responses to ligand from the *goosecoid* promoter (135). This inhibition is possibly due to Smad3 competing with Smad4 for binding to the Smad DNA element. This work suggests that Smad3 can function as a transcriptional activator or repressor depending on the context within the promoter.

The basic helix–loop–helix transcription factor, TFE3, was isolated in a novel screen to identify proteins involved in TGFβ-induced transcription from the PAI-1 promoter (162). In the TGFβ response element, TFE3 binds to an E-box adjacent to a Smad binding element. Importantly, both sites appear to be required for TGFβ induction. The TGFβ-responsive PAI-1 promoter also has an AP-1 site, to which a complex of c-fos and c-jun bind. It has been shown that Smad3 interacts with both c-jun and c-fos forming a multiprotein complex containing Smad4 that leads to an increase in TGFβ-induced transcription (157, 165). Evi-1 is another transcriptional regulator that appears to be involved in TGFβ signalling. The oncoprotein Evi-1 is thought to promote growth and inhibit differentiation. Expression of Evi-1 has been demonstrated to block growth-inhibitory responses to TGFβ and this inhibition occurs through the interaction of Evi-1 with Smad3 (166). In the presence of Evi-1, Smad3 is phosphory-lated and associates with Smad4 in response to ligand. However, formation of Smad3/4 complexes on the PAI-1 promoter are inhibited, suggesting that Evi-1 func-tions in the nucleus to block Smad3. Another zinc-finger protein, Gli3, may also play a role in transcriptional regulation by Smad proteins. Gli3 is an antagonist for hedge-hog signalling and Gli3 proteins truncated at the carboxy terminus can interact with Smads (167). The significance of this observation is yet to be determined since Gli3 truncations have only been associated with some human syndromes and the full-length protein does not bind Smads. Previous observations of a cooperative interaction between TGFβ and vitamin D prompted an examination of the interaction of Smads with the nuclear Vitamin D receptor (VDR; 168). Indeed, Smad3 was found to act as a coactivator for VDR through its ability to form a complex with the steroid receptor coactivator-1 protein in a vitamin D- and TGFβ-dependent manner. Thus, it appears that Smad3 may mediate cross-talk between Vitamin D and TGFβ signalling pathways.

In addition to transcriptional regulators, Smads have also been implicated in inter-actions with the coactivators CBP/p300 and MSG-1. CBP/p300 act as coactivators for a number of distinct transcription factors, i.e. by bridging transcription factors and the basal transcriptional machinery and through their intrinsic histone acetyltrans-ferase activity. Enhancement of Smad-mediated transcription occurs through the direct interaction of CBP/p300 with the C-terminus of Smad2 and Smad3, an inter-action that is enhanced by TGFβ-induced phosphorylation (172–176). Studies in *Drosophila* have similarly demonstrated that CBP can interact with the MH2 domain of MAD (177). Another transcriptional coactivator, MSG-1, has also been implicated in transcriptional activation by Smads. It this case, MSG-1 interacts with the C-terminus of Smad4 and thereby enhances TGFβ-dependent transcription (178). Recently, a homeodomain protein, TGIF, has been shown to repress TGFβ-induced transcription by recruiting histone deacetylases to a Smad target promoter (179). This suggests that induction of TGFβ target genes may be a balance between the recruit-ment of activators such as CBP/p300 and interactions with repressors such as TGIF/deacetylates.

Taken together these studies indicate that Smad proteins can interact with numer-ous nuclear factors. In contrast to the relatively direct pathway of Smad proteins from the receptor to the nucleus, this diversity may provide the molecular basis for the wide array of tissue-specific, growth and differentiation responses observed *in vivo* to

TGFβ family ligands. The next few years are bound to produce many other factors important in gene regulation by the TGFβ family of ligands and will hopefully lead to a better understanding of these processes.

4.2.6 The structure of Smads

Smad proteins contain three major regions: the highly conserved MH1 and MH2 domains and the intervening non-conserved linker region. The MH2 domain is thought to contain the elements necessary to interact with nuclear factors and activate transcription. Since removal of the MH1 and linker domains of R-Smads yields a protein that can constitutively activate transcription (113, 115), it is thought that the MH1 domain functions to inhibit the MH2 domain and that activation of signalling pathways relieves this repression through a conformational change. Support for this hypothesis has been provided by experiments demonstrating that the MH1 interacts with the MH2 domain, and that mutations of the MH1 domain of Smad2 identified in colorectal cancer may stabilize this interaction and prevent association with Smad4 and signalling (180).

Determination of the crystal structure of the MH2 domain of Smad4 has revealed the presence of five α-helices (termed H1 to H5) and three large loops (L1, L2, and L3) (181). These elements enclose a β-sandwich with twisted anti-parallel β-sheets. In the crystal, the Smad4 MH2 domain forms a trimer—many of the mutations in Smad proteins identified in cancers map to the protein–protein interface predicted by this homotrimer, suggesting that disruption of the homomeric interface interferes with protein function (181). Currently, the stoichiometric composition of a heteromeric R-Smad and Smad4 complex is uncertain. Examination of the crystallographic data suggests that the heteromeric R-Smad/Smad4 complex is composed of two Smad trimers which form a hexamer (181). However, biochemical data indicates that Smads can exist as monomers and that the ligand-activated heteromeric complex is actually a trimer (182).

Functional analysis of R-Smads has identified the L3 loop and α-helix 1 of the MH2 domain as the determinants controlling specificity of the interaction between Smads and the receptor (116, 183). The L3 loop is a 17-amino acid region that is invariant in R-Smads except for two amino acids. Exchange of these two amino acids between Smad1 and Smad2 causes a switch in receptor interaction and phosphorylation by BMP (ALK3/ALK6) and TGFβ (TβRI) receptors, respectively (183). In addition to the L3 loop, the nearby α-helix-1 is essential for the differential interaction of Smad1 with the ALK3/ALK6 versus the ALK1/ALK2 receptors (116). The L3 loop of Smad4, which does not interact with the receptors, is essential for the interaction with the R-Smads (181). In addition to the L3 loop, the α-helix-2 (H2) of Smad2 has been identified as an important determinant for interaction with the nuclear factor FAST-1. Substitution of the Smad1 H2 with that of Smad2 yielded a chimeric molecule which could interact with FAST-1 and activate the ARE promoter in response to BMP signalling (134). Thus, H2 is postulated to mediate the specific interactions of Smad proteins with nuclear factors. Together, this work suggests that the L3, H1, and H2 regions encode the determinants in Smad which maintain specificity in the signalling pathway.

The MH1 domain of Smad3 has been crystallized in the presence of a 16 base-pair DNA fragment containing two Smad binding elements (SBE) with the core sequence 5′-GTCT-3′ (184). In the crystal, two MH1 domains interact with one DNA molecule. However, the two peptides do not physically interact and there does not appear to be any cooperativity in DNA binding. The MH1 domain itself forms a compact globular fold with four α-helices, six short β-strands, and five loops. DNA binding is accomplished by a novel DNA-binding structure composed of a β-hairpin, formed by two of the β-strands, which contacts the DNA in the major groove (184). This β-hairpin falls into one of the most highly conserved regions of Smad proteins, and, since different Smads show varying abilities to bind DNA, determinants outside this region must be required for establishing specificity of binding. In the case of Smad2, inserts in the MH1 domain appear to displace the β hairpin, providing a structural explanation for the inability of Smad2 to bind DNA.

4.2.7   A Smad/receptor anchor protein

Receptor-mediated phosphorylation of Smads is a critical step in initiating the intracellular signalling cascade. Thus, controlling this event is likely be an important regulatory point in the pathway. Recently, a novel protein named SARA (for Smad anchor for receptor activation) was identified in a screen for Smad2-interacting proteins and was shown to play an important role in this process (185). SARA comprises a FYVE domain, a Smad2-binding domain, and a TGFβ receptor-binding domain. The FYVE domain is a motif that in other proteins has been shown to bind phosphatidyl-inositol-3-phosphate and can thereby anchor proteins to membranes. In SARA, this FYVE domain was shown to be required for its subcellular localization into punctate spots that also contain the TGFβ-receptors. Furthermore, the interaction of SARA with Smad2 was not observed in the presence of Smad4, consistent with a role for SARA upstream of Smad2/Smad4 heteromeric complex formation. Importantly, disruption of the FYVE domain, without altering the Smad2-binding domain, caused mislocalization of both SARA and Smad2 and inhibited TGFβ-dependent transcription. Thus, SARA appears to play an important role in the initiation of ligand-dependent signalling by recruiting Smad2 to the TGFβ receptor complex (185).

4.2.8   A model for TGFβ signalling

Extensive biochemical and biological studies on various aspects of the TGFβ signalling pathway have been carried out. These accumulated investigations have led to the formulation of a model for TGFβ signalling that serves as the prototype for signalling by all serine/threonine kinase receptors (Fig. 1). Signalling is initiated when TGFβ binds to receptor-II. Ligand-bound receptor-II phosphorylates receptor-I, thereby activating its kinase activity. Membrane-bound SARA recruits Smad2 to specific cellular regions that also contain TGFβ receptors. This leads to the formation of a complex containing SARA, Smad2, and the activated receptor. Smad2 is then phosphorylated on the C-terminal serines by the type-I receptor kinase. This leads to dissociation of Smad2 from SARA and the receptor and induces the formation of a Smad2/Smad4 complex. This complex then translocates into the nucleus and associates with DNA

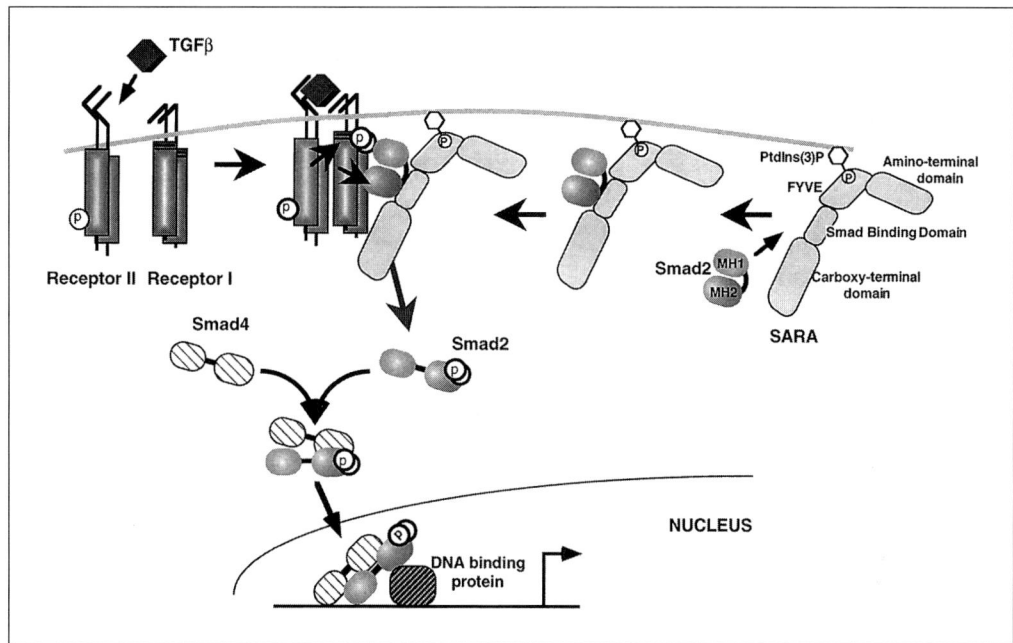

Fig. 1 A model for TGFβ signal transduction. TGFβ binding to receptor-II leads to the recruitment of receptor-I into a heteromeric complex. Receptor-II phosphorylates and thereby activates receptor-I. Smad2 bound to SARA interacts with the receptor complex thereby promoting receptor-I-mediated phosphorylation of Smad2. Smad2 dissociates from SARA and the receptor complex, forms a heteromeric complex with Smad4, and then translocates into the nucleus. In the nucleus, the Smad2/4 complex interacts with DNA binding proteins to regulate the transcription of specific target genes.

binding proteins to activate expression of target genes. In the meantime, SARA is freed and can recruit additional Smad2 to the receptor complex. Currently, a SARA-like protein that might recruit Smad1 to the BMP receptors has not been identified, nor have any nuclear targets for the BMP-regulated Smads. Nevertheless, it seems likely that the same general mechanism for the generation of biological signals through BMP-regulated Smad proteins will apply.

5. The biology of TGFβ signalling

Members of the TGFβ superfamily of polypeptide growth factors regulate a wide range of cellular functions, depending on cell type, its state of differentiation, and its environment. These functions include proliferation, cell-cycle inhibition, apoptosis, terminal differentiation, extracellular matrix production, and specification of developmental fate (1, 2). From a developmental point of view, a remarkable characteristic of TGFβ-like factors is their ability to specify multiple cell and tissue types in a complex spatial and temporal fashion. These factors play crucial roles in the establishment of the basic body plan during gastrulation as well as in the development of nearly all organs and tissues at later stages of embryogenesis (reviewed in refs 5, 98, 100).

Advances in our knowledge of the roles of TGFβ signalling during development have come from studies in several model organisms including genetic loss-of-function experiments in *Drosophila* and mice, ectopic expression experiments in *Xenopus* and chick, and genetic screens in the nematode *C. elegans* and in the zebrafish, *Danio rerio*. It is beyond the scope of this review to discuss the advances of TGFβ research in all model organisms mentioned. However, recent investigations in the areas of gastrulation and axis formation in *Xenopus* and mouse, as well as patterning of the paraxial somitic mesoderm in chick and mouse, have led to major advances in our understanding of the *in vivo* roles of members of the TGFβ signalling machinery and will be the focus of this section.

5.1 Mutations in TGFβ signalling components in mice

In recent years, the mutational analysis of components of the TGFβ signalling pathways in the mouse has resulted in dramatic advances in our understanding of the pleiotropic functions of these pathways in vertebrate development. Table 2 lists mutations in TGFβ signalling components generated by homologous recombination in mouse embryonic stem (ES) cells. Some of the results clearly define the importance of a distinct ligand/receptor pathway for a specific developmental process. For instance, the TGFβ superfamily ligand, Müllerian-inhibiting substance (MIS), is expressed in the fetal testis and is responsible for the regression of the Müllerian ducts, the precursor of the female reproductive organs. Male mice deficient in MIS develop both male and female reproductive organs (205). Moreover, mice deficient in the MIS type-II receptor are phenotypically identical to the MIS ligand-deficient mice, and the MIS-ligand/MIS-receptor double mutants are indistinguishable from the single mutants (206). These results suggest that the MIS receptor may be the only functional type-II receptor for MIS *in vivo* and that this ligand/receptor pair functions in a pathway required for Müllerian duct regression.

Other knockout experiments have provided some rather unexpected results, which were not necessarily predictable from examination of expression patterns, function in tissue-culture cells, or from overexpression studies in *Xenopus*. For example, BMP7 is expressed during the early stages of gastrulation in the node and notochord (194). However, mice mutant in BMP7 have no defects in gastrulation but display defects in eye development and die as a result of abnormal kidney development (199, 200). This disparity between expression pattern and mutant phenotypes is commonly observed in the TGFβ superfamily and is possibly due to the redundancy of many TGFβ-like factors. In addition, the phenotypes of ligands often do not have the same phenotype as mutants in the corresponding receptor. This is most likely due to promiscuous interactions between ligands and receptors. For example, the type-II receptors ActRII and ActRIIB are able to bind activin, BMP2, BMP7, GDF5, and other members of the TGFβ superfamily (100). Moreover, recent experiments unravelled the importance of TGFβ signalling pathways in extraembryonic tissues in the control of epiblast patterning, a notion that had not been addressed in the frog, simply because frogs lack extraembryonic tissues.

5.2 Gastrulation

5.2.1 TGFβ signalling pathways in *Xenopus* axis formation

Ligands, receptors, and modulators

The analysis of mesoderm induction in the *Xenopus* embryo has provided ample information that activins, Vg1, *Xenopus* nodal-related factors (Xnrs), and BMPs have the capacity to induce and pattern mesoderm to generate the complete range of ventral and dorsal cell fates that are typically assumed during embryogenesis (reviewed in ref. 98, 100, 225). Overexpression and misexpression of various ligands and receptors of TGFβ signalling pathways have suggested that maternal Vg1 or another TGFβ-like molecule located in the endoderm of the vegetal hemisphere is important for endoderm development and induces mesodermal fates in adjacent ectodermal equatorial cells. This mesoderm inducer is speculated to be under the transcriptional control of VegT, a T-box transcription factor expressed in the vegetal hemisphere (226). The mesoderm is then patterned through the ventralizing activity of BMP4 and the dorsalizing activities of activin, Vg1, and Xnrs. TGFβ superfamily signalling also controls patterning of the ectoderm: under the influence of BMP4 the ectoderm commits to an epidermal rather than a neural fate. However, in anterior regions, the activity of ectodermal BMP4 is blocked by noggin and chordin, two BMP-binding factors secreted by the Spemann organizer. As described above (Section 2.3), binding of noggin and chordin to BMP4 prevents the interaction of BMP4 with its receptors and leads to neural induction and the formation of neuroectoderm (reviewed in ref. 28). Interestingly, a knockout of the mouse *noggin* gene has revealed that although a normal neural tube appears to develop at the early stages, severe neural defects do occur later in development (227). In addition, the activin-binding protein follistatin functions in a similar way to antagonize the activin-dependent induction of anterior mesoderm in prospective neuroectoderm. The expansion of Spemann's organizer field is limited by the ubiquitously expressed Xolloid metalloprotease (a BMP1 homologue), which cleaves chordin, but not noggin, at two specific sites. This cleavage of chordin releases BMP4 from the chordin–BMP4 complex and reinstates its ventralizing biological activity (40). In an analogous fashion, the tolloid metalloprotease in *Drosophila* has been shown to cleave Sog (the structural and functional homologue of chordin) and thereby activate DPP signalling (39). Thus, a double inhibition mechanism appears to promote the ventralizing activity of BMP/DPP and spatially limits the dorsalizing activity of chordin/Sog.

Smads in *Xenopus* axis formation

Analysis of the function of Smads in *Xenopus* development established their role as intracellular mediators of TGFβ-like signals *in vivo*. Most of the Smads analysed in *Xenopus* thus far can be grouped into three categories: ventralizing Smads, dorsalizing Smads, and Smads that mediate both ventralizing and dorsalizing signals. Smad1 and Smad5 have been shown to mediate ventralizing BMP signals. Thus, misexpression of these Smads recapitulates the phenotypes of ectopic BMP2, 4, and 7 expression: ventralization and induction of ventral mesoderm in embryos and induction of epidermis

Table 2 The expression patterns and phenotypes of mouse mutations in TGFβ superfamily member ligands and in the signalling components

Gene	Function	Normal embryonic expression	Mutant phenotype	Gastrulation	Ref.
activin A	Ligand	Present in oocyte and preimplantation embryo E5.5–E8.5: decidual expression only surrounding the whole embryo E8.5: weak expression in the heart Later expression ini mesenchyme of developing face, whiskers, hair follicles, tooth bud, heart and digestive tract	Die within 24 h of birth. Lack whiskers and incisors and have abnormalities in the mouth.	Normal	186–89
activin B	Ligand	Present in oocyte and preimplantation embryo E5.5–E8.5: weak expression around ectoplacental cone E8.5: weak expression in the heart	Viable with defective eyelid development and female are unable to rear offspring.	Normal	186–88, 190
α-inhibitin	Ligand	Present in oocyte E7.5: no embryonic or decidual expression E10.5: in developing gonads and placenta	Both males and females are normal and fertile but develop gonadal stromal tumours	Normal	186–88, 191
nodal	Ligand	E5.5: primitive ectoderm E6.5: proximal posterior ectoderm and primitive endoderm overlying streak E7.5: restricted to node	Arrests shortly after gastrulatioin with no evidence of primitive streak formation. Chimeric analysis demonstrates the requirement of nodal in the most caudal aspect of embryonic ectoderm for primitive streak formation and gastrulation as well as the importance of nodal expression in the visceral endoderm for AP patterning of the embryo	Mesoderm evident in 25% 10% have expression of Embryonic ectoderm proliferates but then degenerates.	192, 193, 235
BPM2	Ligand	E6.5: detected by PCR analysis in embryo: present in decidua E7.5: mesoderm of amnion/chorion visceral endoderm and presumptive precardiac mesoderm E9.5: limb buds and myocardial layer of the heart and edges of neural folds later stages in tooth buds, whisker follicles, and craniofacial mesenchyme	Arrests between E7.5 and E9.5. Malformation of amnion and chorion. Abnormal cardiac development. BMP2/BMP7 double-heterodygote mutants have no defects.	Normal except for persistence of open proamniotic canal in 69% of null mutants.	194–96
BMP4	Ligand	E6.5: low levels in posterior primitive streak E7.5: amnion, allantois, and posterior primitive streak	Two primary phenotypes observed. Early phenotype: arrests at gastrulation with no primitive streak formation. Later	Early phenotype shows extra embryonic mesoderm but no *brachyury* expression. Possible	195, 214, 215, 220

Name	Type	Expression		Phenotype	Reference
BMP5 (short ear)	Ligand	E9.5: posterior primitive streak, posterior ventral mesoderm, foregut, branchial arches, heart and limb buds	Normal	phenotype: truncated or disorganized posterior structures as well as absence of lens induction. BMP4/BMP7 double-heterozygote mutants develop minor defects in rib cage and distal parts of limbs. defects in cell survival and proliferation in epiblast layer. Late phenotype lacks posterior mesoderm.	197, 198, 220
BMP6	Ligand	Expressed at early stages of skeletal development. Also in soft tissues including lungs, liver, ureter, bladder, and intestines	Normal	Viable and fertile, but have effects in cartilage and bone including altered sternum and the loss of one pair of ribs. Bones are smaller and weaker. Also have reduced size of external ear. BMP5/BMP6 double-homozygote mutants display exacerbated sternal defects. BMP5/BMP7 double-heterozygote mutants have no defects.	222
BMP7	Ligand	Expressed throughout skeletal development. Also in developing brain and diverse sites from preimplantation on	Normal	Viable and fertile, but display delay in ossification of sternum. BMP5/BMP6 double-homozygote mutants display exacerbated sternal defects.	199, 200, 220
BMP8B	Ligand	E6.5: anterior primitive streak and node E7.5: allantois, axial mesoderm, low level in visceral endoderm E8.5: notochord and definitive endoderm, non-neural ectoderm, myocardia cells, and somatopleure E9.5: developing brain limb bud E10.5: developing kidney, limb buds	Normal	Severe defects in kidney and eye development. Die at birth. Polydactyly in hind limbs, minor skeletal defects in ribs and skull. BMP2/BMP7 and BMP5/BMP7 double-heterozygote mutants have no defects. BMP4/BMP7 double-heterozygote mutants develop minor defects in rib cage and distal parts of limbs.	201
GDF-5 (branchipodism)	Ligand	Male germ cells of testis and trophoblast cells of the placenta	Normal	Viable. Females are normal while males display germ-cell deficiencies or infertility.	207
GDF-8	Ligand	Expressed in embryonic and adult skeletal muscle	Normal	Viable with defects in the number and length of bones in the limb	221
GDF-9	Ligand	Expression is restricted to oocyte from primary one-layer follicle stage until ovulation	Normal	Viable. Mice display increased body size due to increased skeletal muscle growth.	208
Lefty-1	Ligand	Expressed on left side of ventral neural tube	Normal	Females are viable but infertile. Follicular developments arrests at one-layer stage.	250

Table 2 Continued

Gene	Function	Normal embryonic expression	Mutant phenotype	Gastrulation	Ref.
TGFβ1	Ligand	Extraembryonic mesoderm, cardiac mesoderm, endothelial cells of blood vessels, fetal liver, and surrounding decidua	Appear normal until approximately 3 weeks after birth, then exhibit 'wasting' and develop multifocal inflammatory disease.	Normal	3, 186, 268
TGFβ2	Ligand	During embryogenesis expressed in chondrocytes, osteocytes; precadiac mesoderm, cardiac myocytes, kidney, gut, sensory epithelia in eye and ear, CNS and PNS	Die after birth. Defects include cardiac, lung, craniofacial, limb, spinal column, eye, inner ear, and urogenital tract. Non-overlapping with TGFβ2.	Normal	217
TGFβ3	Ligand	Expressed in epithelial tissues during embryogenesis	Viable with cleft palate and delayed lung morphogenesis.	Normal	223, 224
MIS (Müllerian-inhibiting substance)	Ligand	E12.5: in fetal testis. Expression persists after birth in testis but at lower levels in Sertoli cells. Also in ovaries 6 days after birth	Mice are viable, but males are infertile since they develop both male and female reproductive organs.	Normal	205
MIS type II receptor (AMHR)	Type-II receptor	In the mesenchymal cells adjacent to the Müllerian duct epithelium. Also in Sertoli cells and granulosa cells of fetal and adult testes and ovaries	Identical to above MIS deficient mice.	Normal	206
ALK3 (BMPRI)	Type-I receptor	E7.0: epiblast and mesoderm Later stages: ubiquitous but absent from liver	Die between E7.0 and E9.5 with no primitive streak formation.	No mesoderm induction occurs and no *brachyury* expressioin. Defects in epiblast cell proliferation	202
ActRII	Type-II receptor	Present in oocytes E5.5-E8.5: weak expression in embryonic ectoderm. Also in decidua	Viable, but females are sterile. A few had skeletal and facial abnormalities.	Normal	189, 203
ActRIB	Type-I receptor	E5.5-6.0: low in extraembryonic ectoderm and epiblast E6.5-7.0: ectoderm and epiblast and weak in ectoplacental cone and proximal visceral endoderm E7.5: all three embryonic germ layers	Disorganized epiblast and extraembryonic ectoderm. Chimeric analysis indicates that −/− cells are able to contribute to mesoderm but cannot form a primitive streak. ActRIB is needed in extraembryonic tissues for proper gastrulation.	Arrest prior to gastrulation at egg cylinder. No mesoderm induction. −/− Extraembryonic tissues look normal but result in severe gastrulation defects in chimeras.	209
ActRIIB	Type-II receptor	During embryogenesis expressed in CNS and PNS, male and female reproductive systems, and epithelial and endothelial structures	Mice die after birth with skeletal, cardiac, pulmonary, and splenic abnormalities.	Normal	236

Gene	Function	Expression	Phenotype	Refs
Smad2	Intracellular signal transducer	E6.0: weakly but ubiquitously expressed Ubiquitous expression still present at late gastrulation	Variable phenotype depending on allele. Die between E8 and E9. Disruption of proximal–distal axis. Extraembryonic ectoderm and epiblast disorganized. Chimeric analysis indicates that Smad2 is required in the visceral endoderm to promote epiblast patterning.	Smad2^{Robm1}: entire epiblast adopts posterior mesoderm fate, giving rise to yolk sac and fetal blood cells. Smad2^{mh1}: and Smad2$^{mh2/lacZ}$: no mesoderm formation. Small embryos. Smad2^{ΔC}: delayed extraembryonic tissues at early stages. No mesoderm	113, 210–12 216, 213
Smad3	Intracellular signal transducer	Widespread expression	Variable phenotype depending on allele, Smad3$^{ex2/ex2}$ develop metastatic colorectal cancer; Smad3$^{ex8/ex8}$ and Smad3$^{3ex1/ex1}$ lead to diminished T-cell responsiveness to TGFβ.	Normal	276, 284
Smad4	Intracellular signal transducer	E6.0: weakly but ubiquitously expressed Ubiquitous expression still present at late gastrulation	Die before E7.5. Decreased proliferation in epiblast.	No mesoderm formation. Chimeric analysis indicates that abnormal visceral endoderm development is primary defect, causing the gastrulation defect in the epiblast.	214, 215
Smad5	Intracellular signal transducer	Widespread expression during embryogenesis with highest levels in mesenchyme and somites. Ubiquitously expressed in adults	Embryos die between E9.5 and E11.5 with multiple defects including disorganization of blood vessels.	Normal	218, 219

in animal caps which are derived from the ectodermal portion of the *Xenopus* embryo (118, 228, 229). Smad2, on the other hand, mediates the dorsalizing effects of activin, Vg1, and Xnrs. Overexpression of full-length Smad2 or its MH2 domain in animal caps is sufficient to induce ventral mesoderm markers at low doses and dorsal mesoderm markers at high doses (113, 120, 228). These pathway-specific Smads promote distinct dorsal and ventral fates even when the upstream receptors are blocked by overexpression of dominant-negative receptor constructs, suggesting that no additional receptor-mediated signals are required for the biological activity of the ligands during axis formation (120, 228).

In contrast to R-Smads, misexpression of the common Smad, Smad4, in animal caps leads to the induction of both ventral and dorsal mesoderm (230), consistent with a role of Smad4 in both BMP and activin signalling. Through its common role in both BMP and activin signalling, Smad4 might also be a key regulator of cross-regulation and integration of multiple TGFβ-like signals (125, 231).

Recently, the roles of several inhibitory Smads in *Xenopus* axis formation have been analysed. Smad6, when overexpressed in embryos, phenocopies the effect of blocking BMP4 signalling, thus causing dorsalization of mesoderm and neuralization of ectoderm (141, 145). In addition, Smad6 might be able to partially block the activity of activin. Smad6 therefore functions as an intracellular antagonist of BMP and possibly activin signalling. Smad7 (and the independently named *Xenopus* homologue Smad8) can also block both activin and BMP signalling, albeit at different thresholds. Low concentrations of Smad7 dorsalize ventral mesoderm, whereas higher concentrations inhibit mesoderm induction completely (145, 146). In animal caps, this inhibition of both BMP and activin signalling leads to neural induction. These results suggest that the inhibitory Smads have important modulatory functions in BMP and activin signalling during *Xenopus* embryogenesis.

Very recently, a *Xenopus* Smad10 has been cloned and its biological activity during embryogenesis has been analysed (114). As described above, neural induction by noggin, chordin, or follistatin functions by blocking BMP or activin signalling in the ectoderm, thus preventing the ectoderm from assuming epidermal or mesodermal fates, respectively. However, these inhibitory molecules can only induce anterior, but not posterior, neural fates. The endogenous Spemann organizer, however, induces both anterior and posterior neural fates. This implies that an organizer signal(s) other than noggin, chordin, and follistatin are important for posterior neural fate induction. Ectopic expression of Smad10 leads to dorsalization of ventral mesoderm; moreover, Smad10 is able to induce both anterior and posterior neural fates in animal caps. Surprisingly, Smad10 does not interfere with BMP-dependent induction of mesodermal markers; therefore, it appears not to function in a homologous fashion to noggin or chordin. These results suggest that Smad10 might play a role in mediating signals from the Spemann organizer in the ectoderm to promote neural fate.

BMP signalling through alternative pathways

Recently, an alternative intracellular signalling cascade downstream of BMP receptors has been described (89, 90). In this cascade, BMP type-I receptors activate the

MAP kinase kinase kinase, TAK1, through two intermediate molecules, named TAB1, an activator of TAK1, and XIAP, which appears to link receptor complexes with the TAB1/TAK1 complex (see Section 4.1.7, and refs 89, 90). Overexpression studies in *Xenopus* embryos confirmed the involvement of these molecules in BMP signalling. When expressed ectopically in *Xenopus* embryos, TAB1 and TAK1 induced cell death (90). When this was circumvented by coexpressing bcl-2, the cells were not only rescued, but also ventralized. Moreover, a kinase-deficient form of TAK1 could partially rescue the ventralization caused by a constitutively active BMP type-IA receptor and also block the expression of ventral mesoderm markers induced by Smad1 or Smad5 (232). Similarly, XIAP1 could enhance the ventralizing effect of TAB1 and TAK1 (91). These results suggest that some aspects of BMP signalling might be mediated by an intracellular signalling pathway utilizing XIAP, TAB1, and TAK1 to elicit BMP-like responses during *Xenopus* embryogenesis.

5.2.2 TGFβ signalling and mammalian axis determination

Overexpression experiments in the frog support the notion that BMP and activin-like signals are sufficient to induce ectodermal and mesodermal patterning during gastrulation. This appears to be achieved by a complex balance between ventralizing signals such as BMPs, dorsalizing signals such as activins and Vg1, and inhibitors of TGFβ ligands, such as noggin, chordin, and follistatin. While the models of gastrulation appear to be better understood in *Xenopus*, it is important to determine whether similar models hold true for mammalian development. Indeed, recent genetic loss-of-function experiments of components of the TGFβ signalling pathways in mice have not only demonstrated the necessity for components of these pathways during embryonic patterning, they also shed light on the spatial requirements of these molecules during gastrulation (210, 214, 235).

Ligands and receptors

Gene-targeting approaches in the mouse demonstrated that the BMP pathway is important for mesoderm induction and gastrulation. Embryos deficient in the BMP2/4 receptor, ALK3 (BMPRIA), exhibit a complete absence of mesoderm (202). This lack of mesoderm could be secondary to an observed proliferation defect in the epiblast, since the onset of gastrulation is thought to require 1400–1500 cells in the epiblast, indicating that mesoderm induction and cell proliferation are highly coupled (233). However, teratomas derived from mutant embryos were also impaired in mesoderm formation, suggesting that some BMP signalling is important for mesoderm induction.

A similar phenotype was observed in mice deficient in BMP4, a known ligand for ALK3. BMP4 mutant embryos could be separated into two main phenotypes, suggesting two critical phases in development during which BMP4 is required (195). The early phenotype was similar to the ALK3 mutants in that they arrested at gastrulation and produced no embryonic mesoderm. Surviving embryos arrested later in development with disorganized or truncated posterior structures as well as reduction of extraembryonic mesoderm. This latter phenotype is possibly due to a rescue of some embryos by another ligand. A potential candidate is BMP2, a second ALK3 ligand, since it is expressed at low levels during gastrulation (195). BMP2 might be able to compensate for BMP4 during gastrulation; however, it is not required for gastrulation. BMP2 mutant embryos display defects in heart formation and in closure of the exocoelomic cavity later in development (196). Taken together, these results suggest a role of BMPs in cell proliferation and survival of the ectoderm and in mesoderm formation. Moreover, they accentuate the notion that the highly related BMP2 and BMP4, which interact with common receptors, nevertheless carry out unique functions during cardiac development and mesoderm formation, respectively.

Interestingly, recent experiments in mouse embryoid bodies suggest that BMP2 and BMP4 are capable of promoting, and are required for differentiation of the visceral endoderm, a part of the extraembryonic tissues (234). Therefore, one has to consider the possibility that the gastrulation defects seen in the epiblasts of BMP4 and ALK3 mutant embryos are secondary to visceral endoderm defects. This causal relationship has been demonstrated for several other components of TGFβ signalling pathways (discussed below). The generation and analysis of chimeric embryos consisting of wild-type epiblasts and BMP4 or ALK3 mutant extraembryonic tissues (and vice versa) would address this important question.

Another factor of the TGFβ superfamily implicated in mesoderm induction and gastrulation is nodal (192, 193). The development of nodal mutant embryos is arrested shortly after gastrulation with no evidence of primitive streak formation. Mesoderm is observed in 25% of the embryos with 10% exhibiting brachyury expression, suggesting that the requirement for nodal in mesoderm formation is not absolute. Indeed, the generation of chimeric embryos derived from nodal mutant and wild-type cells demonstrated that nodal is absolutely required for primitive streak formation, but that nodal−/− cells are able to contribute to mesoderm (235). Moreover, nodal expression in the visceral endoderm is required for normal head development. Embryos lacking nodal in the visceral endoderm lack forebrain structures, thus suggesting that the anterior visceral endoderm is able to function as a head-inducing organizer, with nodal being one of the signals that mediate this activity.

ActRIB (ALK4) is a type-I receptor that has been shown biochemically to mediate activin signals. Mice with mutations in both the activin bA and activin bB genes are born with craniofacial defects. In contrast, loss-of-function mutations in the ActRIB gene cause developmental arrest before gastrulation, with disorganization of both embryonic and extraembryonic ectoderm. Interestingly, chimeric embryos with ActRIB mutant and wild-type cells showed that ActRIB is required for primitive streak formation, but that other areas of the epiblast do not need functional ActRIB. Importantly, ActRIB−/− cells can contribute to mesoderm, but when ActRIB−/− cells form the extraembryonic tissues and wild-type cells form the epiblast, severe gastrulation defects are observed. As in the case of nodal, this suggests a dual function of ActRIB during early embryogenesis: ActRIB is required for primitive streak formation in the epiblast, but is also needed in the extraembryonic tissues to regulate epiblast gastrulation. One possible mechanism for this extraembryonic function is that signalling through ActRIB may regulate the expression or secretion of one or more signalling molecules that regulate gastrulation of the embryo. Since none of the activin ligands inactivated so far display a similar phenotype, other ligands might signal through ActRIB during gastrulation. The similarities between the ActRIB and nodal phenotypes raise the possibility that ActRIB might function as a nodal type-I receptor during gastrulation. The determination of the functional type-II receptor component of this presumed nodal/ActRIB receptor complex might be a more difficult task, since many of the ligands can bind several type-II receptors. Indeed, mice carrying an inactivating mutation in the ActRIIB locus exhibit no gastrulation defects, but die after birth with cardiac, skeletal, pulmonary, and splenic abnormalities (236). Therefore, ActRIIB might not be the main type-II receptor of a presumed nodal/ActRIB signalling pathway during gastrulation.

Smads in mouse gastrulation

Recently, the inactivation of several Smads by homologous recombination in ES cells has provided the first insights into the requirements for these intracellular mediators of TGFβ-like signals in mammalian development. Gene targeting of the Smad4 locus causes reduced cell proliferation and poor differentiation in the epiblast and the visceral endoderm prior to gastrulation (214, 215). Aggregation of mutant Smad4 ES cells with wild-type tetraploid morulae leads to the formation of embryos consisting of mutant epiblast and wild-type extraembryonic tissues. In these aggregation chimeras no defects in gastrulation are observed, indicating that the gastrulation defect in Smad4 mutant embryos is caused by impaired function of the visceral endoderm, and that epiblast cells can form mesoderm without functional Smad4. Later in development, however, these chimeric embryos display anterior defects similar to the head formation defects observed in chimeric nodal embryos, demonstrating that Smad4 activity in the epiblast is required for anterior development (214). It is somewhat surprising that Smad4, considered to be a 'common' Smad utilized by all TGFβ pathways, is not required during gastrulation in the epiblast, since the chimeric analysis of both the nodal and ActRIB knockouts indicated that these signalling components are imperative for primitive streak formation in the epiblast. One possible interpretation is that in some instances the receptor-regulated Smads can signal in the absence of a Smad4-like heterodimerization partner, as has been suggested by the observation of MEDEA-independent effects of DPP in *Drosophila* (108) and the TGFβ-dependent, yet Smad4-independent induction of fibronectin in mammalian cells (137).

Details of three independent knockouts of Smad2 have been published, and all result in gastrulation defects (210–212). However, while Weinstein *et al.* (212) and Nomura and Li (211) reported severe defects in the extraembryonic tissues and no mesoderm formation in Smad2 mutant embryos, Waldrip *et al.* (210) reported relatively normal extraembryonic tissues and lack of an anteroposterior axis in the mesoderm. The entire epiblast expresses Brachyury and appears to acquire a posterior, extraembryonic mesodermal fate, therefore suggesting that Smad2 plays a role in the development of anterior epiblast structures. The significant differences between the reported phenotypes might be related to the different targeting strategies employed, and the possibilities that hypomorphic, hypermorphic, or neomorphic alleles have been generated needs to be tested. Chimeric experiments by Waldrip *et al.* (1998) show that a wild-type epiblast cannot rescue the gastrulation defect, suggesting that Smad2 functions in extraembryonic tissues to control normal patterning of the embryo. The reverse experiments—generating chimeric embryos which consist of wild-type extra-embryonic tissues and Smad2 mutant epiblasts—have not yet been performed, so it is not clear if there is an embryonic, in addition to extraembryonic, requirement for Smad2 during gastrulation.

Two recent reports describe the complex phenotypes of Smad5 mutant mice (218, 219). These animals display no gastrulation defects, but die between days 9.5 and 11.5 of embryogenesis due to multiple embryonic and extraembryonic defects. Interestingly, lack of Smad5 leads to impaired organization and maintenance of blood vessels, resembling the phenotype of the TGFβ1 and TGFβRII knockouts. Moreover, the ALK1 receptor is also required for proper organization of blood vessels in humans (237). These results suggest that Smad5 might function as a component of a TGFβ signalling pathway during certain aspects of embryonic development.

Taken together, the results of knockout experiments in mice show that signalling through TGFβ pathways is required during gastrulation as well as at later stages of development. What has come as a surprise is that these pathways appear to play pivotal roles in the development and patterning of extraembryonic tissues, which in turn are essential for patterning of the embryo proper. Further chimeric analysis of components of the TGFβ signalling pathways will undoubtedly help to clarify the exact roles of these pathways in different cell layers during gastrulation.

Left–right axis

The internal organs of the vertebrate body display significant left–right (L–R) asymmetry. A second sign of asymmetry is manifested in the turning direction of the embryo. This asymmetry is believed to be initiated during gastrulation and is linked to the formation of dorsoventral and anteroposterior axes. In *Xenopus*, it has been shown that the TGFβ-related molecule Vg1 is both sufficient and necessary for the initiation of L–R patterning (238, 239). Vg1 induces expression of Xnr-1, a *Xenopus* nodal-related marker for left–right development that is only expressed in the left lateral plate mesoderm (238, 239).

In chick embryos, a signalling cascade involving several asymmetrically expressed molecules has been suggested to initialize and promote L–R patterning in the early embryo (reviewed in refs 240, 241). Activin B, expressed on the right side of the embryo at early stages (242, 243), is thought to induce local expression of the chick type-II receptor gene *cActRIIA* (242, 244), which in turn represses Sonic Hedgehog (Shh) expression in the right half of Hensen's node (243, 244). This yields left-sided expression of *Shh*, which is then thought to induce expression of nodal in the left lateral plate mesoderm (244). Misexpression of activin or Shh disrupts the normal expression pattern of nodal and randomizes heart-looping.

In the mouse, asymmetrical expression of nodal is first observed on the left side of the node and in the left lateral plate mesoderm (245, 246). In the mouse mutant *inv*, in which the L–R axis is reversed (situs invertus), nodal expression in the lateral plate mesoderm is restricted to the right side of the embryo and expression in the node is either symmetrical or enhanced on the right side (245, 246). In another mouse mutant, *iv*, in which the orientation of the L–R axis is randomized, nodal expression was bilaterally symmetrical or not expressed at all (246). These results indicate a strong relationship between sidedness of the embryo and nodal expression, and place nodal downstream of these mutations. While no L–R asymmetries were reported in a nodal knockout mouse (235), L–R defects were noticed in Nodal/Smad2 *trans*-heterozygous embryos (211), suggesting that Nodal as well as Smad2 function in the determination of L–R asymmetry. However, no L–R asymmetries were reported in a nodal knock-out mouse (235). Moreover, targeted mutagenesis of Shh, activinβB, follistatin, or ActRIIA does not affect L–R patterning (189, 190, 204, 247, 248). This might imply that the molecular events determining L–R patterning are not conserved between chick and mouse; however, it could also merely indicate functional redundancies due to the presence of related factors with similar activities. The latter might be closer to the truth, since mice carrying mutant alleles of ActRIIB, a molecule that is not asymmetrically expressed in the embryo, do exhibit defects in L–R asymmetries (236). Moreover, targeted mutagenesis of lefty-1, a TGFβ-related factor expressed on the left side of the ventral neural tube (249), causes L–R positional defects in visceral organs and bilateral expression of nodal and lefty-2, a factor closely related to lefty-1 (250). These observations suggest a role for lefty-1 upstream of lefty-2 and nodal to restrict their expression to the left side of the embryo.

Further evidence that the signals as well as the transcriptional programmes involved in L–R patterning might be similar in *Xenopus*, chick, and mouse comes from recent work on the Pitx2 homeobox transcription factor. Pitx2 is asymmetrically expressed in all three species from gastrulation on throughout organogenesis, and misexpression experiments in *Xenopus* and chick indicate that activin, Shh, nodal, and mouse lefty-1 can induce *Pitx2* expression. Moreover, misexpression of Pitx2 can alter L–R patterning. In the mouse, *Pitx2* expression is altered in *iv* and *inv* mice and closely resembles the altered patterns of *nodal* expression (251–254). Thus, the transcription factor Pitx2 might be a common downstream target of nodal or nodal-related signals in vertebrates that initiates the transcription of genes mediating L–R asymmetry.

A very recent paper sheds light on the mechanics of the initial establishment of a left–right symmetry in and around the node. The targeted mutagenesis of the KIF3B motor protein, a member of the kinesin superfamily of molecular motors, reveals a randomization of L–R asymmetry due to loss of cilia in the node, which, in normal embryos, causes a leftward flow of extraembryonic fluid around the node. This 'nodal flow' is lost in KIF3B mutant embryos (255). These authors propose a model in which the leftward flow of extraembryonic fluid might be able to transport a L–R asymmetry-initializing factor (a TGFβ-like molecule?), secreted by the node, preferentially to the left side of the node and adjacent tissues. This model could explain the initiation of asymmetry in a previously symmetrical embryo. Undoubtedly, further experiments will explore this novel finding and its implications in TGFβ signalling during L–R axis formation in the mouse embryo.

5.3   Somite patterning

The somites of the vertebrate embryo are derived from the paraxial mesoderm and develop lateral of the neural tube. The patterning of the somites leads to the formation of the ventrally located sclerotome (which will give rise to the axial bones and cartilage) and the dorsally located dermomyotome (which subsequently divides into dermatome to form dermis, and myotome to form skeletal muscle). The patterning and differentiation of the somites is thought to be controlled by signals from the adjacent tissues of neural tube, notochord, surface ectoderm, and lateral plate mesoderm. Recently, microsurgical manipulations and misexpression experiments in the chick have implicated BMPs as important signals that control myotome development and early myogenesis. Moreover, these studies showed that BMPs can act as morphogens by inducing different target genes and cell fates at different concentrations. A series of molecular events is thought to orchestrate the patterning and differentiation of the myotome. According to this model, BMP4 and BMP7 are expressed in the surface ectoderm overlaying the paraxial mesoderm and are able to induce *Pax-3* expression in early myotomal cells in the dermomyotome (256–258). Interestingly, low levels of BMP signalling maintain *Pax-3*-expressing, proliferating cells and delay differentiation, whereas high levels of BMP signalling prevent muscle development (257, 258). Maintenance of low levels of BMP activity in these tissues might be realized by the expression of the BMP inhibitors noggin and follistatin, in the dermomyotome (258–260; reviewed in ref. 261). Noggin expression in the most dorsomedial part of the dermomyotome, the medial lip, antagonizes the effects of BMP4 secreted from the dorsal neural tube (258–260). However, in the dorsal neural tube, BMP4 induces *Wnt-1* expression, a secreted factor which induces *Wnt-11* expression in *Pax-3*-positive dermomyotome cells (259). Induction of *Wnt-11* is necessary to allow muscle progenitor cells present in the dermomyotome to involute into the myotome and to further differentiate. Moreover, Wnt-1 can induce the expression of noggin in the medial lip (256, 258); thus, BMP4 activity induces its own antagonist to prevent detrimental effects on myotome development. High levels of BMP activity from dorsal neural tube, surface ectoderm, and lateral plate mesoderm are thought to specify the location of the dermomyotome by preventing an expansion of *Pax-3* expressing dermamyotome cells into medial or lateral regions (260; reviewed in ref. 261). A crucial role for BMP signalling during somitogenesis was also provided by the analysis of noggin knockout mice (227). In Noggin–/– embryos, the differentiation of somites is disturbed. Addition of BMP2 or BMP4 to paraxial mesoderm explants blocked Shh-mediated induction of *Pax-1*, a sclerotomal marker, whereas addition of noggin is sufficient to induce *Pax-1*. Noggin and Shh can induce *Pax-1* synergistically. These data confirm that inhibition of BMP signalling by somitically expressed noggin is important for normal somite patterning.

6.   TGFβ signalling and disease

The vertebrate TGFβ signalling pathways are involved in the control of multiple intracellular functions and the development and homeostasis of nearly all organs and tissues. Therefore, it comes as no surprise that alterations of normal function of components of these pathways can lead to a diverse array of disorders. Recent investigations in both mouse and human have suggested that the disregulation of TGFβ signalling plays an important role in several types of cancer as well as in three human hereditary disorders—persistent Müllerian duct syndrome (PMDS), Hunter–Thompson type acromesomelic chondrodysplasia, and hereditary haemorrhagic telangiectasia.

6.1 Disrupted TGFβ signalling in cancer

Since TGFβ signalling plays important roles in the negative control of cell proliferation, functional alterations of components of these pathways appear to be a logical mechanism through which cancer might develop or progress. Indeed, early studies showed that many epithelial cancer cell lines are resistant to TGFβ-mediated growth inhibition (reviewed in ref. 262). In colorectal carcinoma cell lines, increased resistance to TGFβ correlates with biological aggressiveness, suggesting that release from TGFβ growth inhibition is an important step in tumour progression. Recently, the molecular analysis of components of the TGFβ signalling pathways has led to a better understanding of TGFβ resistance and its role in tumour progression. The first molecular evidence that the loss of sensitivity to TGFβ in tumours could be a result of mutations in downstream signalling components came from the analysis of the receptors TβRI and TβRII. TβRII is frequently mutated in both sporadic and hereditary gastrointestinal cancers with microsatellite instability (262, 263). The primary defects in these tumours are mutations in DNA mismatch repair genes, causing addition or deletion of nucleotides in simple repetitive sequences (microsatellites) throughout the genome. The human *TβRII* gene contains a 10 base-pair polyadenine repeat at the junction of the extracellular and transmembrane domains of the receptor. Mutations in this repeat are found in most sporadic colon cancers with microsatellite instability, resulting in premature stop codons (263, 264). Moreover, mutations in this repeat region are also found in colorectal tumours of individuals with hereditary non-polyposis colon cancer (HNPCC) (265), whereas these mutations are rare in endometrial tumours in these patients. This suggests that mutations in the TβRII poly-adenine repeat are functionally important in the progression of colorectal tumours, and not just random consequences of the defects in mismatch repair. This conclusion has been validated with the recent discovery of germline mutations in the *TbRII* gene in HNPCC patients (266). Mutations in, or loss of expression of, TβRI or TβRII have also been identified in a large number of different tumours. Recently, the ligand TGFβ1 has been suggested to have a role as a tumour suppressor (267). Mice carrying homozygous mutations of the TGFβ1 gene die of a multifocal inflammatory disease soon after weaning (3, 268). However, when adult heterozygote animals were challenged with chemical carcinogens, enhanced tumorigenesis in liver and lung was observed (267). However, the wild-type allele of TGFβ1 was not lost in these tumours, leading to the suggestion that TGFβ1 is a new form of tumour suppressor with haploid insufficiency. The possibility of mutations in downstream components of the

TGFβ signalling pathway was not sufficiently addressed in this study, so it remains conceivable that these tumours acquired mutations in Smads or other components of the pathway, therefore compromising signalling in the presence of an intact TGFβ1 allele. α-Inhibin-deficient mice develop mixed granulosa/Sertoli cell tumours (191), indicating that α-inhibin is a potent negative regulator of gonadal stromal-cell proliferation. Mice deficient in both α-inhibin and MIS display accelerated tumour progression when compared to α-inhibin-deficient controls, suggesting that α-inhibin and MIS have synergistic effects in tumour development (269). Not only the ligands and receptors, but also the intracellular components of the TGFβ signalling pathway can function as tumour suppressors (49, 100, 270). Smad2 and Smad4/DPC4 map to chromosome 18q21, a region often deleted in pancreatic, lung, and colorectal tumours. Smad4/DPC4 is deleted or mutated in a significant portion of pancreatic carcinomas (112) and colorectal cancers (271), and has also been detected in lung cancers (272). Germline mutations in Smad4/DPC4 have been identified in familial juvenile polyposis (273), a form of gastrointestinal cancer. Mice carrying mutations in both the *Smad4/DPC4* and *Apc* tumour-suppressor genes show an increased malignant progression of colorectal tumours, compared to Apc single knockout mice (274). Whereas loss of heterozygosity (LOH) of Apc leads to the initiation of hyperplasia and dysplasia in normal epithelium to form polyp adenomas, the concomitant LOH of Smad4/Dpc4 allows the tumours to progress to more invasive adenocarcinomas. However, additional hits are required for further progression, since metastasis of these colorectal tumours was not observed in the double-mutant mice. Inactivating Smad2 mutations have been identified in colorectal carcinomas (120, 275), but appear to be rare in other types of cancers. Recently, targeted mutagenesis in the mouse addressed the role of Smad3 during development and disease (213, 216, 276). While Zhu *et al.* (213) report metastatic colorectal cancer in homozygous Smad3 mutant animals, Yang *et al.* (276) show a role of Smad3 in mucosal immunity and both Yang *et al.* (276) and Datto *et al.* (216) demonstrate that Smad3 modulates the response of T cells to TGFβ. Again, the possibility of hypomorphic or neomorphic alleles needs to be investigated to address these seemingly contradictory results. It is noteworthy, however, that Smad3 homozygous mutant mice in the study of Zhu *et al.* (213) do not only develop adenocarcinomas, but that these tumours can further progress and metastasize. Thus, loss of Smad3 includes a more severe tumorigenic phenotype than the loss of Smad4/Dpc4 (discussed above), suggesting that Smad3 functions in the prevention of metastasis and might be independent of Smad4/Dpc4 function.

6.2 Hereditary diseases

Persistent Müllerian duct syndrome (PMDS) is characterized by the presence of Müllerian duct derivatives (uterine and oviductal tissues) in males. Mutations in Müllerian inhibitory substance (MIS) or its receptor, important for the regression of the Müllerian duct, result in PMDS (277, 278). Therefore, it appears that both in mouse (see above) and in human, MIS and its receptor are highly specific for the development of the reproductive tract.

Mutations in the mouse *GDF-5* gene cause brachypodism, a shortening of the long bones in the limbs (207). It has recently been shown that the human homologue of GDF-5, called cartilage-derived morphogenetic protein-1 (CDMP-1), is frequently mutated in Hunter–Thompson type acromesomelic chondrodysplasia (279). This disorder is phenotypically very similar to brachypodism in the mouse, again suggesting that in mammals individual members of the TGFβ superfamily may have conserved functions.

Proximal Symphalangism (SYM1) and Multiple Synostoses Syndrome (SYNS1) are two human disorders that have multiple joint fusion as their principal feature. Recently, hereditary and sporadic mutations in noggin have been identified in these patients (280). Interestingly, noggin knockout mice which die from multiple defects also display joint fusion due to excess cartilage formation (227, 281). Thus, molecules like noggin, which antagonize TGFβ superfamily signals, also show functional conservation in mouse and man.

Hereditary haemorrhagic telangectasia (HHT) is a highly penetrant autosomal-dominant vascular dysplasia characterized by epistaxis, telangiectases, and arterio-venous malformations (282). Both Endoglin and ALK1 are highly expressed in vascular endothelial cells, and mutations in both receptors have been identified in HHT (237, 283), suggesting that both receptors mediate TGFβ-like signals that are important for vasculogenesis or the maintenance of blood vessels.

7. Conclusions and future perspectives

The recent advances in our understanding of molecular and biological aspects of TGFβ signalling provide an important framework for future studies. For example, the identification of intracellular mediators of TGFβ-like signals has resulted in the recent description of SARA, an important regulator of TGFβ signalling. These advances will undoubtedly aid in the further elucidation of intracellular regulatory and modulating mechanisms. Moreover, knowledge of these intracellular mediators now allows the analysis of cross-talk between different signalling pathways, a crucial first step in the examination of how cells integrate signals from different signalling pathways in order to initiate appropriate biological responses to complex environmental situations.

With the development of conditional mutagenesis strategies in mice, it is now possible to assess the biological function of TGFβ signalling molecules at specific stages of development. With double knockout as well as knockin strategies one can address the question of redundancy between different TGFβ pathways. The amalgamation of molecular and developmental studies will undoubtedly be crucial for an in-depth understanding of the biological importance of TGFβ signalling. Lastly, this in-depth knowledge will be important for the proper assessment of the events that interfere with TGFβ signalling in disease and, potentially, for the use of components of TGFβ signalling pathways as therapeutic agents.

Acknowledgements

Work in the laboratories of J.L.W. and L.A. is supported by grants from the National Cancer Institute of Canada and the Medical Research Council (MRC) of Canada. J.L.W. and L.A. are MRC Scholars, M.K. is a Hospital for Sick Children Restracom Fellow, and P.A.H. is an MRC Centennial Fellow.

References

1. Roberts, A. B. and Sporn, M. B. (1990). In *Peptide growth factors and their receptors* (ed. M. B. Sporn and A. B. Roberts), pp. 419. Springer-Verlag, Heidelberg.
2. Massagué, J. (1990). The transforming growth factor-β family. *Annu. Rev. Cell. Biol.*, **6**, 597.
3. Kulkarni, A. B., Huh, C.-G., Becker, D., Geiser, A., Lyght, M., Flanders, K. C., Roberts, A. B., Sporn, M. B., Ward, J. M., and Karlsson, S. (1993). Transforming growth factor β1 null mutation in mice causes excessive inflammatory response and early death. *Proc. Natl Acad. Sci. USA*, **90**, 770.
4. Rosen, V. and Thies, R. S. (1991). The BMP proteins in bone formation and repair. *Trends Genet.*, **8**, 97.
5. Hogan, B. L. (1996). Bone morphogenetic proteins in development. *Curr. Opin. Genet. Dev.*, **7**, 856.
6. Beddington, R. S. P. and Robertson, E. J. (1998). Anterior patterning in mouse. *Trends Genet.*, **14**, 277.
7. Padgett, R., Das, P., and Krishna, S. (1998). TGF-β signalling, Smads, and tumor suppressors. *BioEssays*, **20**, 382.
8. Ren, P., Lim, C.-S., Johnsen, R., Albert, P. S., Pilgrim, D., and Riddle, D. L. (1996). Control of *C. elegans* larval development by neuronal expressoin of a TGF-β homolog. *Science*, **274**, 1389.
9. Colavita, A., Krishna, S., Zheng, H., Padgett, R. W., and Culotti, J. G. (1998). Pioneer axon guidance by UNC-129, a *C. elegans* TGF-β. *Science*, **281**, 706.
10. Suzuki, Y., Yandell, M. D., Roy, P. J., Krishna, S., Savage-Dunn, C., Rosee, R. M., Padgett, R. W., and Wood, W. B. (1999). A BMP homolog acts as a dose-dependent regulator of body size and male tail patterning in *Caenorhabditis elegans*. *Development*, **126**, 241.
11. Morita, K., Chow, K. L., and Ueno, N. (1999). Regulation of body length and male tail ray pattern formation of *Caenorhabditis elegans* by a member of TGF-β family. *Development*, **126**, 1337.
12. Suzuki, A., Maeda, J., Kaneko, E., and Ueno, N. (1997). Mesoderm induction by BMP-4 and -7 heterodimers. *Biochem. Biophys. Res. Commun.*, **232**, 153.
13. Nishimatsu, S. and Thomsen, G. H. (1998). Ventral mesoderm induction and patterning by bone morphogenetic protein heterodimers in *Xenopus* embryos. *Mech. Dev.*, **74**, 75.
14. Constam, D. B. and Robertson, E. J. (1999). Regulation of bone morphogenetic protein activity by pro domains and proprotein convertases. *J. Cell Biol.*, **144**, 139.
15. Roebroek, A. J. M., Umans, L., Pauli, I. G. L., Robertson, E. J., van Leuven, F., van de Ven, W. J. M., and Constam, D. B. (1998). Failure of ventral closure and axial rotation in embryos lacking the proprotein convertase Furin. *Development*, **125**, 4863.
16. Thomsen, G. H. and Melton, D. A. (1993). Processed Vg1 protein is an axial mesoderm inducer in *Xenopus*. *Cell*, **74**, 433.
17. Dale, L., Matthews, G., and Colman, A. (1993). Secretion and mesoderm-inducing activity of the TGF-β-related domain of *Xenopus*-Vg1. *EMBO J.*, **12**, 4471.

18. Munger, J. S., Harpel, J. G., Gleizes, P.-E., Mazzieri, R., Nunes, I., and Rifkin, D. B. (1997). Latent transforming growth factor-β: structural feature and mechanisms of activation. *Kidney Int.*, **51**, 1376.

19. Crawford, S. E., Stellmach, V., Murphy-Ullrich, J. E., Ribeiro, S. M. F., Lawler, J., Hynes, R. O., Boivin, G. P., and Bouck, N. (1998). Thrombospondin-1 is a major activator of TGF-β1 *in vivo. Cell*, **93**, 1159.

20. Munger, J. S., Huang, X., Kawakatsu, H., Griffiths, M. J. D., Dalton, S. L., Wu, J., Pittet, J.-F., Kaminski, N., Garat, C., Matthay, M. A., Rifkin, D. B., and Sheppard, D. (1999). The integrin vb6 binds and activates latent TGFβ1: a mechanism for regulating pulmonary inflammation and fibrosis. *Cell*, **96**, 319.

21. Daopin, S., Piez, K. A., Ogawa, Y., and Davies, D. R. (1992). Crystal structure of transforming growth factor-β2: an unusual fold for the superfamily. *Science*, **257**, 369.

22. Schlunegger, M. P. and Grütter, M. (1992). An unusual feature revealed by the crystal structure at 2.2 Å resolution of human transforming growth factor-β2. *Nature*, **358**, 430.

23. Archer, S. J., Bax, A., Roberts, A. B., Sporn, M. B., Ogawa, Y., Plez, K. A., Weatherbee, J., Tsang, M., Lucas, R., Zheng, B., Wanker, J., and Torchia, D. A. (1993). Transforming growth factor β1: secondary structure as determined by heteronuclear magnetic resonance spectroscopy. *Biochemistry*, **32**, 1164.

24. McDonald, N. Q. and Hendrickson, W. A. (1993). A structural superfamily of growth factors containing a cystine knot motif. *Cell*, **73**, 421.

25. Griffith, D. L., Keck, P. C., Sampath, T. K., Rueger, D. C., and Carlson, W. D. (1996). Three-dimensional structure of recombinant human osteogenic protein 1: Structural paradigm for the transforming growth factor β superfamily. *Proc. Natl Acad. Sci. USA*, **93**, 878.

26. Qian, S. W., Burmester, J. K., Sun, P. D., Huang, A., Ohlsen, D. J., Suardet, L., Flanders, K. C., Davies, D., Roberts, A. B., and Sporn, M. B. (1994). Characterization of mutated transforming growth factor-beta's which possess unique biological properties. *Biochemistry*, **33**, 12298.

27. De Robertis, E. M. and Sasai, Y. (1996). A common plan for dorsoventral patterning in Bilateria. *Nature*, **380**, 37.

28. Harland, R. and Gerhart, J. (1997). Formation and function of Spemann's organizer. *Annu. Rev. Cell Dev. Biol.*, **13**, 611.

29. Holley, S. A., Jackson, P. D., Sasai, Y., Lu, B., De Robertis, E. M., Hoffmann, F. M., and Ferguson, E. L. (1995). A conserved system for dorsal-ventral patterning in insects and vertebrates involving sog and chordin. *Nature*, **376**, 249.

30. Schulte-Merker, S., Lee, K. J., McMahon, A. P., and Hammerschmidt, M. (1997). The zebrafish organizer requires chordino. *Nature*, **387**, 862.

31. Zimmerman, L. B., De Jesús-Escobar, J. M., and Harland, R. (1996). The Spemann organizer signal noggin binds and inactivates bone morphogenetic protein 4. *Cell*, **86**, 599.

32. Piccolo, S., Sasai, Y., Lu, B., and De Robertis, E. M. (1996). Dorsoventral patterning in *Xenopus*: inhibition of ventral signals by direct binding of chordin to BMP-4. *Cell*, **86**, 589.

33. Hsu, D. R., Economides, A. N., Wang, X., Eimon, P. M., and Harland, R. M. (1998). The *Xenopus* dorsalizing factor *Gremlin* identifies a new family of secreted proteins that antagonize BMP activities. *Mol. Cell*, **1**, 673.

34. Bouwmeester, T., Kim, S.-H., Sasai, Y., Lu, B., and De Robertis, E. M. (1996). Cerberus is a head-inducing secreted factor expressed in the anterior endoderm of Spemann's organizer. *Nature*, **382**, 595.

35. Shawlot, W., Deng, J. M., and Behringer, R. R. (1998). Expression of the mouse cerberus-related gene, Cerr1, suggests a role in anterior neural induction and somitogenesis. *Proc. Natl Acad. Sci. USA*, **95**, 6198.

36. Biben, C., Stanley, E., Fabri, L., Kotecha, S., Rhinn, M., Drinkwater, C., Lah, M., Wang, C.-C., Nash, C., Hilton, D., Ang, S.-L., Mohun, T., and Harvey, R. P. (1998). Murine Cerberus homologue mC er-1: a candidate anterior patterning molecule. *Dev. Biol.*, **194**, 135.
37. Belo, J. A., Bouwmeester, T., Leyns, L., Kertesz, N., Gallo, M., Follettie, M., and De Robertis, E. M. (1997). Cerberus-like is a secreted factor with neutralizing activity expressed in the anterior primitive endoderm of the mouse gastrula. *Mech. Dev.*, **68**, 45.
38. Piccolo, S., Agius, E., Leyns, L., Bhattacharyya, S., Grunz, H., Bouwmeester, T., and De Robertis, E. M. (1999). The head inducer Cerberus is a multifunctional antagonist of Nodal, BMP and Wnt signals. *Nature*, **397**, 707.
39. Marqués, G., Musacchio, M., Shimell, M. J., Wünnenberg-Stapleton, K., Cho, K. W. Y., and O'Connor, M. B. (1997). Production of a DPP activity gradient in the early *Drosophila* embryo through the opposing action of the SOG and TLD proteins. *Cell*, **91**, 417.
40. Piccolo, S., Agius, E., Lu, B., Goodman, S., Dale, L., and DeRobertis, E. M. (1997). Cleavage of Chordin by Xolloid metalloprotease suggests a role for proteolytic processing in the regulation of Spemann organizer activity. *Cell*, **91**, 407.
41. Hemmati-Brivanlou, A., Kelly, O. G., and Melton, D. A. (1994). Follistatin, an antagonist of activin is expressed in the Spemann organizer and displays direct neuralizing activity. *Cell*, **77**, 283.
42. Nakamura, T., Takio, K., Eto, Y., Shibai, H., Titani, K., and Sugino, H. (1990). Activin-binding protein from rat ovary is follistatin. *Science*, **247**, 836.
43. Fainsod, A., Deißler, K., Yelin, R., Marom, K., Epstein, M., Pillemer, G., Steinbeisser, H., and Blum, M. (1997). The dorsalizing and neural inducing gene *follistatin* is an antagonist of BMP-4. *Mech. Dev.*, **63**, 39.
44. Demetriou, M., Binker, C., Sukhu, B., Tenenbaum, H. C., and Dennis, J. W. (1996). Fetuin/α2-HS glycoprotein is a transforming growth factor-β type-II receptor mimic and cytokine antagonist. *J. Biol. Chem.*, **271**, 12755.
45. Zhu, Y., Oganesian, A., Douglas, R., Keene, D. R., and Sandell, L. J. (1999). Type IIA procollagen containing the cysteine-rich amino propeptide is deposited in the extracellular matrix of prechondrogenic tissue and binds to TGF-beta 1 and BMP-2. *J. Cell Biol.*, **144**, 1069.
46. Miyazono, K., Ichijo, H., and Heldin, C.-H. (1993). Transforming growth factor-β: latent forms, binding proteins and receptors. *Growth Factors*, **8**, 11.
47. Attisano, L. and Wrana, J. L. (1996). Signal transduction by members of the transforming growth factor-β superfamily. *Cytokine Growth Factor Rev.*, **7**, 327.
48. Kawabata, M., Imamura, T., and Miyazono, K. (1998). Signal transduction by bone morphogenetic proteins. *Cytokine Growth Factor Rev.*, **9**, 49.
49. Massagué, J. (1998). TGF-β signal transduction. *Annu. Rev. Biochem.*, **67**, 753.
50. Yamashita, H., ten Dijke, P., Franzén, P., Miyazono, K., and Heldin, C.-H. (1994). Formation of hetero-oligomeric complexes of type-I and type-II receptors for transforming growth factor-β. *J. Biol. Chem.*, **269**, 20172.
51. Chen, R.-H. and Derynck, R. (1994). Homomeric interactions between type-II transforming growth factor-β receptors. *J. Biol. Chem.*, **269**, 22868.
52. Henis, Y. I., Moustakas, A., Lin, H. Y., and Lodish, H. F. (1994). The type-II and III transforming growth factor-β receptors form homo-oligomers. *J. Cell Biol.*, **126**, 139.
53. Luo, K. and Lodish, H. F. (1997). Positive and negative regulation of type-II TGF-β receptor signal transduction by autophosphorylation on multiple serine residues. *EMBO J.*, **16**, 1970.
54. Wrana, J. L., Attisano, L., Wieser, R., Ventura, F., and Massague, J. (1994). Mechanism of activation of the TGF–β receptor. *Nature*, **370**, 341.

55. Wieser, R., Wrana, J. L., and Massague, J. (1995). GS domain mutations that constitutively activate TβR-I, the downstream signalling component in the TGF-β receptor complex. *EMBO J.*, **14**, 2199.

56. Souchelnytskyi, S., ten Dijke, P., Miyazono, K., and Heldin, C.-H. (1996). Phosphorylation of Ser165 in TGF-β type-I receptor modulates TGF-β1-induced cellular responses *EMBO J.*, **15**, 6231.

57. Chen, F. and Weinberg, R. A. (1995). Biochemical evidence for the autophosphorylation and transphosphorylation of transforming growth factor β receptor kinases. *Proc. Natl Acad. Sci. USA*, **92**, 1565.

58. Wieser, R., Attisano, L., Wrana, J. L., and Massagué, J. (1993). Signalling activity of TGF-β type-II receptors lacking specific domains in the cytoplasmic region. *Mol. Cell* Biol., **13**, 7239.

59. Cárcamo, J., Zentella, A., and Massagué, J. (1995). Disruption of transforming growth factor β signaling by a mutation that prevents transphosphorylation within the receptor complex. *Mol. Cell*. Biol., **15**, 1573.

60. Brummel, T., Abdollah, S., Haerry, T. E., Shimell, M. H., Merriam, J., Raftery, L., Wrana, J. L., and O'Connor, M. B. (1999). The *Drosophila* activin receptor Baboon signals through Dmad2 and controls cell proliferation but not patterning during larval development. *Genes Dev.*, **13**, 98.

61. Macìas-Silva, M., Hoodless, P. A., Tang, S. J., Buchwald, M., and Wrana, J. L. (1998). Specific activation of Smad1 signalling pathways by the BMP7 type I receptor, ALK 2. *J. Biol. Chem.*, **273**, 25628.

62. Georgi, L. L., Albert, P. S., and Riddle, D. L. (1990). *daf-1*, a *C. elegans* gene controling dauer larva development, encodes a novel receptor protein kinase. *Cell*, **61**, 635.

63. Krishna, S., Maduzia, L. L., and Padgett, R. W. (1999). Specificity of TGFβ signalling is conferred by distinct type-I receptors and their associated SMAD proteins in *Caenorhabditis elegans*. *Development*, **126**, 251.

64. Huse, M., Chen, Y.-G., Massagué, M., and Kuriyan, J. (1999). Crystal structure of the cytoplasmic domain of the type I TGFβ receptor in complex with FKBP12. *Cell*, **96**, 425.

65. Saitoh, M., Nishitoh, H., Amagasa, T., Miyazono, K., Takagi, M., and Ichijo, H. (1996). Identification of important regions in the cytoplasmic juxtamembrane domain of type-I receptor that separate signaling pathways of transforming growth factor-β. *J. Biol. Chem.*, **271**, 2769.

66. Feng, X.-H. and Derynck, R. (1997). A kinase subdomain of transforming growth factor-β (TGF-β) type 1 receptor determines the TGF-β intracellular signaling specificity. *EMBO J.*, **16**, 3912.

67. Wang, X.-F., Lin, H. Y., Ng-Eaton, E., Downward, J., Lodish, H. F., and Weinberg, R. A. (1991). Expression cloning and characterization of the TGF-β type III receptor. *Cell*, **67**, 797.

68. López-Casillas, F., Cheifetz, S., Doody, J., Andres, J. L., Lane, W. S., and Massagué, J. (1991). Structure and expression of the membrane proteoglycan betaglycan, a component of the TGF-β receptor system. *Cell*, **67**, 785.

69. López-Casillas, F., Wrana, J. L., and Massagué, J. (1993). Betaglycan presents ligand to the TGF-β signaling receptor. *Cell*, **73**, 1435.

70. Gougos, A. and Letarte, M. (1990). Primary structure of endoglin, an RGD-containing glycoprotein of human endothelial cells. *J. Biol. Chem.*, **265**, 8361.

71. Cheifetz, S., Bellón, T., Calés, C. S. V., Bernabeu, C., Massagué, J., and Letarte, M. (1992). Endoglin is a component of the TGF-β receptor system in human endothelial cells. *J. Biol. Chem.*, **267**, 19027.

72. Pece Barbara, N., Wrana, J. L., and Letarte, M. (1999). Endoglin is an accessory protein that interacts with the signalling receptor complex of multiple members of the transforming growth factor-β superfamily. *J. Biol. Chem.*, **274**, 584.

73. Letamendia, A., Lastres, P., Raab, U., Langa, C., Attisano, L., and Bernabeu, C. (1998). Role of Endoglin in cellular responses to transforming growth factor-β. *J. Biol. Chem.*, **273**, 33011.

74. Wang, T., Li, B.-Y., Danielson, P. D., Shah, P. C., Rockwell, S., Lechleider, R. J., Martin, J., Manganaro, T., and Donahoe, P. K. (1996). The immunophilin FKBP12 functions as a common inhibitor of the TGFβ family type I receptors. *Cell*, **86**, 435.

75. Charng, M.-J., Kinnunen, P., Hawker, J., Brandil, T., and Schneider, M. D. (1996). FKBP-12 recognition is dispensable for signal generation by type I TGFβ receptors. *J. Biol. Chem.*, **271**, 22941.

76. Okadome, T., Oeda, E., Saitoh, M., Ichijot, H., Moses, H. L., Miyazono, K., and Kawabata, M. (1996). Characterization of the interaction of FKBP-12 with the transforming growth factor-β type I receptor *in vivo*. *J. Biol. Chem.*, **271**, 21687.

77. Fruman, D. A., Burakoff, S. J., and Bierer, B. E. (1994). Immunophilins in protein folding and immunosuppression. *FASEB J.*, **8**, 391.

78. Chen, Y.-G., Liu, F., and Massagué, J. (1997). Mechanism of TGFβ receptor inhibition by FKBP12. *EMBO J.*, **16**, 3866.

79. Kawabata, M., Imamura, T., Miyazono, K., Engel, M. E., and Moses, H. L. (1995). Interaction of the transforming growth factor-β type I receptor with farnesyl-protein transferase-α. *J. Biol. Chem.*, **270**, 29628.

80. Wang, T., Danielson, P. D., Li, B., Shah, P. C., Kim, S. D., and Donahoe, P. K. (1996). The p21Ras farnesyltransferase-α subunit in TGFβ and activin signaling. *Science*, **271**, 1120.

81. Ventura, F., Liu, F., Doody, J., and Massague, J. (1996). Transforming growth factor-β receptor type-I interacts with farnesyl transferase-α in mammalian cells. *J. Biol. Chem.*, **271**, 13931.

82. Tamanoi, F. (1993). Inhibitors of Ras farnesyltransferases. *Trends Biochem. Sci.*, **18**, 349.

83. Kucich, U., Rosenbloom, J. C., Shen, G., Abrams, W. R., Blaskovich, M. A., Hamilton, A. D., Ohkanda, J., Sebti, S. M., and Rosenbloom, J. (1998). Requirement for geranylgeranyl transferase and acyltransferase in the TGF-β-stimulated pathway leading to elastin mRNA stabilization. *Biochem. Biophys. Res. Commun.*, **252**, 111.

84. Chen, R. H., Miettinen, P. J., Maruoka, E. M., Choy, L., and Derynck, R. (1995). A WD-domain protein that is associated with and phosphorylated by the type-II TGF-β receptor. *Nature*, **377**, 548.

85. Choy, L. and Derynck, R. (1998). The type II transforming growth factor (TGF)-β receptor-interacting protein TRIP-1 acts as a modulator of the TGF-β response. *J. Biol. Chem.*, **273**, 31455.

86. Griswold-Prenner, I., Kamibayashi, C., Maruoka, E. M., Mumby, M. C., and Derynck, R. (1998). Physical and functional interactions between type I transforming factor β receptors and Bα, a WD-40 repeat subunit of phosphatase 2A. *Mol. Cell. Biol.*, **18**, 6595.

87. Charng, M.-J., Zhang, D., Kinnunen, P., and Schneider, M. D. (1998). A novel protein distinguishes between quiescent and activated forms of type I transforming growth factor β receptor. *J. Biol. Chem.*, **273**, 9365.

88. Datta, P. K., Chytil, A., Gorska, A. E., and Moses, H. L. (1998). Identification of STRAP, a novel WD domain protein in transforming growth factor-β signalling. *J. Biol. Chem.*, **273**, 34671.

89. Yamaguchi, K., Shirakabe, K., Shibuya, H., Irie, K., Oishi, I., Ueno, N., Taniguchi, T.,

Nishida, E., and Matsumoto, K. (1995). Identification of a member of the MAPKKK family as a potential mediator of the TGF-β signal transduction. *Science*, **270**, 2008.

90. Shibuya, H., Yamaguchi, K., Shirakabe, K., Tonegawa, A., Gotoh, Y., Ueno, N., Irie, K., Nishida, E., and Matsumoto, K. (1996). TAB1: an activator of the TAK1 MAPKKK in TGF-β signal transduction. *Science*, **272**, 1179.

91. Yamaguchi, K., Nagai, S.-I., Ninomiya-Tsuji, J., Nishita, M., Tamai, K., Irie, K., Ueno, N., Nishida, E., Shibuya, H., and Matsumoto, K. (1999). XIAP, a cellular member of the inhibitor of apoptosis protein family, links the receptors to TAB1–TAK1 in the BMP signalling pathway. *EMBO J.*, **18**, 179.

92. Oeda, E., Oka, Y., Miyazono, K., and Kawabata, M. (1998). Interaction of *Drosophila* inhibitors of apoptosis with thick veins, a Type I serine/threonine kinase receptor for decapentaplegic. *J. Biol. Chem.*, **273**, 9353.

93. Wang, W., Zhou, G., Hu, M. C.-T., Yao, Z., and Tan, T.-H. (1997). Activation of the Hematopoietic Progenitor Kinase-1 (HPK1)-dependent, stress-activated c-Jun N-terminal kinase (JNK). Pathway by transforming growth factor β (TGF-β)-activated kinase (TAK1), a kinase mediator of TGFβ signal transduction. *J. Biol. Chem.*, **272**, 22771.

94. Shirakabe, K., Yamaguchi, K., Shibuya, H., Irie, K., Matsuda, S., Moriguchi, T., Gotoh, Y., Matsumoto, K., and Nishida, E. (1997). TAK1 mediates the ceramide signalling to stress-activated protein kinase/cJun N-terminal kinase. *J. Biol. Chem.*, **272**, 8141.

95. Sakurai, H., Shigemori, N., Hasegawa, K., and Sugita, T. (1998). TGF-β-activated kinase 1 stimulates NF-κB activation by an NF-κB-inducing kinase-independent mechanism. *Biochem. Biophys. Res. Commun.*, **243**, 545.

96. Attisano, L. and Wrana, J. L. (1998). Mads and Smads in TGFβ signalling. *Curr. Opin. Cell Biol.*, **10**, 188.

97. Derynck, R., Zhang, Y., and Feng, X.-H. (1998). Smads: transcriptional activators of TGF-β responses. *Cell*, **95**, 737.

98. Whitman, M. (1998). Smads and early developmental signalling by the TGF beta super-family. *Genes Dev.*, **12**, 2445.

99. Heldin, C.-H., Miyazono, K., and ten Dijke, P. (1997). TGF-β signalling from cell membrane to nucleus through SMAD proteins. *Nature.*, **390**, 465.

100. Hoodless, P. A. and Wrana, J. L. (1997). In *Current topics in microbiology and immunology*, Vol. 228 (ed. A. S. Pawson), pp. 235. Springer-Verlag, Berlin.

101. Raftery, L. A., Twombly, V., Wharton, K., and Gelbart, W. M. (1995). Genetic screens to identify elements of the decapentaplegic signaling pathway in *Drosophila*. *Genetics*, **139**, 241.

102. Sekelsky, J. J., Newfeld, S. J., Raftery, L. A., Chartoff, E. H., and Gelbart, W. M. (1995). Genetic characterization and cloning of Mothers against dpp, a gene required for decapentaplegic function in *Drosophila* melanogaster. *Genetics*, **139**, 1347.

103. Hoodless, P. A., Haerry, T., Abdollah, S., Stapleton, M., O'Connor, M. B., Attisano, L., and Wrana, J. L. (1996). MADR1, a MAD-related protein that functions in BMP2 signalling pathways. *Cell*, **85**, 489.

104. Wiersdorff, V., Lecuit, T., Cohen, S. M., and Mlodzik, M. (1996). Mad acts downstream of Dpp receptors, revealing a differential requirement for dpp signalling and propagation of morphogenesis in the *Drosophila* eye. *Development*, **122**, 2153.

105. Newfeld, S. J., Chartoff, E. H., Graff, J. M., Melton, D. A., and Gelbart, W. M. (1996). Mothers against dpp encodes a conserved cytoplasmic protein required in DPP/TGFβ responsive cells. *Development*, **122**, 2099.

106. Hudson, J. B., Podos, S. D., Keith, K., Simpson, S. L., and Ferguson, E. L. (1998). The

Drosophila Medea gene is required downstream of *dpp* and encodes a functional homolog of human Smad4. *Development*, **125**, 1407.

107. Das, P., Maduzia, L. L., Wang, H., Finelli, A. L., Cho, S.-H., Smith, M. M., and Padgett, R. (1998). The *Drosophila* gene *Medea* demonstrates the requirement for different classes of Smads in *dpp* signalling. *Development*, **125**, 1519.

108. Wisotzkey, R. G., Mehra, A., Sutherland, D. J., Dobens, L. L., Liu, X., Dohrmann, C., Attisano, L., and Raftery, L. (1998). *Medea* is a *Drosophila Smad 4* homolog that is differentially required to potentiate DPP responses. *Development*, **125**, 1433.

109. Tsuneizumi, K., Nakayama, T., Kamoshida, Y., Kornberg, T. B., Christian, J. L., and Tabata, T. (1997). Daughters against dpp modulates dpp organizing activity in *Drosophila* wing development. *Nature*, **389**, 627.

110. Savage, C., Das, P., Finelli, A., Townsend, S., Sun, C., Baird, S., and Padgett, R. (1996). The *C. elegans* **sma-2**, *sma-3* and *sma-4* genes define a novel conserved family of TGF-β pathway components. *Proc. Natl Acad. Sci. USA*, **93**, 790.

111. Patterson, G., Koweek, A., Wong, A., Liu, Y., and Ruvkun, G. (1997). The DAF-3 Smad protein antagonizes TGF-β-related receptor signalling in the C. elegans dauer pathway. *Genes and Dev.*, **11**, 2679.

112. Hahn, S. A., Schutte, M., Shamsul Hoque, A. T. M., Moskaluk, C. A., da Costa, L. T., Rozenblum, E., Weinstein, C. L., Fischer, A., Yeo, C. J., Hruban, R. H., and Kern, S. E. (1996). DPC4, a candidate tumor suppressor gene at human chromosome 18q21.1. *Science*, **271**, 350.

113. Baker, J. C. and Harland, R. (1996). A novel mesoderm inducer, Madr2, functions in the activin signal transduction pathway. *Genes Dev.*, **10**, 1880.

114. LeSueur, J. A. and Graff, J. A. (1999). Spemann organizer activity of Smad10. *Development*, **126**, 137.

115. Liu, F., Hata, A., Baker, J., Doody, J., Cárcamo, J., Harland, R., and Massagué, J. (1996). A human Mad protein acting as a BMP-regulated transcriptional activator. *Nature*, **381**, 620.

116. Chen, Y.-G. and Massagué, J. (1999). Smad1 recognition and activation by the ALK1 group of transforming growth factor-β family receptors. *J. Biol. Chem.*, **274**, 3672.

117. Newfeld, S. J., Mehra, A., Singer, M. A., Wrana, J. L., Attisano, L., and Gelbart, W. M. (1997). Mothers against dpp participates in a DPP/TGF-β responsive serine–threonine kinase signal transduction cascade. *Development*, **124**, 3167.

118. Suzuki, A., Chang, C., Yingling, J., Wang, X.-F., and Hemmati-Brivanlou, A. (1997). Smad5 induces ventral fates in *Xenopus* embryo. *Dev. Biol.*, **184**, 402.

119. Chen, Y., Bhushan, A., and Vale, W. (1997). Smad8 mediates the signalling of the receptor serine kinase. *Proc. Natl Acad. Sci. USA*, **94**, 12938.

120. Eppert, K., Scherer, S. W., Ozcelik, H., Pirone, R., Hoodless, P., Kim, H., Tsui, L.-C., Bapat, B., Gallinger, S., Andrulis, I., Thomsen, G., Wrana, J. L., and Attisano, L. (1996). MADR2 maps to 18q21 and encodes a TGFβ MAD-related protein that is functionally mutated in colorectal carcinoma. *Cell*, **86**, 543.

121. Nakao, A., Imamura, T., Souchelnytskyi, S., Kawabata, M., Ishisaki, A., Oeda, E., Tamaki, K., Hanai, J.-I., Heldin, C.-H., Miyazono, K., and ten Dijke, P. (1997). TGF-β receptor-mediated signalling through Smad2, Smad3 and Smad4. *EMBO J.*, **16**, 5353.

122. Liu, X., Sun, Y., Constantinescu, S. N., Karam, E., Weinberg, R. A., and Lodish, H. F. (1997). Transforming growth factor β-induced phosphorylation of Smad3 is required for growth inhibition and transcriptional induction in epithelial cells. *Proc. Natl Acad. Sci. USA*, **94**, 10669.

123. Hoodless, P. A., Tsukazaki, T., Nishimatsu, S.-I., Attisano, L., Wrana, J. L., and Thomsen,

G. H. (1999). Dominant-negative Smad2 mutants inhibit activin/Vg1 signalling and disrupt axis formation in *Xenopus*. *Dev. Biol.*, **207**, 364.

124. Zhang, Y., Feng, X.-H., Wu, R.-Y., and Derynck, R. (1996). Receptor-associated Mad homologues synergize as effectors of the TGF-β response. *Nature*, **383**, 168.

125. Lagna, G., Hata, A., Hemmati-Brivanlou, A., and Massagué, J. (1996). Partnership between DPC4 and SMAD proteins in TGF-β signalling pathways. *Nature*, **383**, 832.

126. Chen, Y., Lebrun, J.-J., and Vale, W. (1996). Regulation of transforming growth factor β- and activin-induced transcription by mammalian Mad proteins. *Proc. Natl Acad. Sci. USA*, **93**, 12992.

127. Tang, S. J., Hoodless, P. A., Lu, Z., Breitman, M. L., McInnes, R. R., Wrana, J. L., and Buchwald, M. (1998). Regulation of early mammalian development by a BMP/Tlx-2 signalling pathway. *Development*, **125**, 1877.

128. Macías-Silva, M., Abdollah, S., Hoodless, P. A., Pirone, R., Attisano, L., and Wrana, J. L. (1996). MADR2 is a substrate of the TGFβ receptor and its phosphorylation is required for nuclear accumulation and signalling. *Cell*, **87**, 1215.

129. Kretzschmar, M., Liu, F., Hata, A., Doody, J., and Massagué, J. (1997). The TGF-β family mediator Smad1 is phosphorylated directly and activated functionally by the BMP receptor kinase. *Genes Dev.*, **11**, 984.

130. Abdollah, S., Macías-Silva, M., Tsukazaki, T., Hayashi, H., Attisano, L., and Wrana, J. L. (1997). TβRI phosphorylation of Smad2 on Ser 465 and 467 is required for Smad2/Smad4 complex formation and signalling. *J. Biol. Chem.*, **272**, 27678.

131. Souchelnytskyi, S., Tamaki, K., Engström, U., Wernstedt, C., ten Dijke, P., and Heldin, C.-H. (1997). Phosphorylation of Ser465 and Ser467 in the C terminus of Smad2 mediates interaction with Smad4 and is required for transforming growth factor-β signaling. *J. Biol. Chem.*, **272**, 28107.

132. Kretzschmar, M., Doody, J., and Massagué, J. (1997). Opposing BMP and EGF signalling pathways converge on the TGF-β family mediator Smad1. *Nature*, **389**, 618.

133. de Caestecker, M. P., Parks, W. T., Frank, C. J., Castagnino, P., Bottaro, D. P., Roberts, A. B., and Lechleider, R. J. (1998). Smad2 transduces common signals from receptor serine–threonine and tyrosine kinases. *Genes Dev.*, **12**, 1587.

134. Chen, Y.-G., Hata, A., Lo, R. S., Wotton, D., Shi, Y., Pavletich, N., and Massagué, J. (1998). Determinants of specificity in TGF-β signal transduction. *Genes Dev.*, **12**, 2144.

135. Labbé, E., Silvestri, C., Hoodless, P. A., Wrana, J. L., and Attisano, L. (1998). Smad2 and Smad3 positively and negatively regulate TGFβ-dependent transcription through the forkhead DNA binding protein, FAST2. *Mol. Cell*, **2**, 109.

136. Liu, F., Pouponnot, C., and Massagué, J. (1997). Dual role of the Smad4/DPC4 tumor suppressor in TGFβ-inducible transcriptional complexes. *Genes Dev.*, **11**, 3157.

137. Hocevar, B. A., Brown, T. L., and Howe, P. H. (1999). TGF-β induces fibronectin synthesis through a c-Jun N-terminal kinase-dependent, Smad4-independent pathway. *EMBO J.*, **18**, 1345.

138. Riggins, G. J., Kinzler, K. W., Vogelstein, B., and Thiagalingam, S. (1997). Frequency of Smad gene mutations in human cancers. *Cancer Res.*, **57**, 2578.

139. Topper, J. N., Cai, J., Qiu, Y., Anderson, K. R., Xu, Y.-Y., Deeds, J. D., Feeley, R., Gimeno, C. J., Woolf, E. A., Tayber, O., Mays, G. G., Sampson, B. A., Schoen, F. J., Gimbrone, M. A. J., and Falb, D. (1997). Vascular MADs: two novel MAD-related genes selectively inducible by flow in human vascular endothelium. *Proc. Natl Acad. Sci. USA*, **94**, 9314.

140. Imamura, T., Takase, M., Nishihara, A., Oeda, E., Hanai, J.-I., Kawabata, M., and Miyazono, K. (1997). Smad6 inhibits signalling by the TGF-β superfamily. *Nature*, **389**, 622.

141. Hata, A., Lagna, G., Massagué, J., and Hemmati-Brivanlou, A. (1998). Smad6 inhibits BMP/Smad1 signalling by specifically competing with the Smad4 tumor suppressor. *Genes Dev.*, **12**, 186.

142. Hayashi, H., Abdollah, S., Qiu, Y., Cai, J., Xu, Y.-Y., Grinnell, B. W., Richardson, M. A., Topper, J. N., Gimbrone, Jr., M. A., Wrana, J. L., and Falb, D. (1997). The MAD-related protein Smad7 associates with the TGFβ receptor and functions as an antagonist of TGFβ signaling. *Cell*, **89**, 1165.

143. Nakao, A., Afrakhte, M., Morén, A., Nakayama, T., Christian, J. L., Heuchel, R., Itoh, S., Kawabata, M., Heldin, N.-E., Heldin, C.-H., and ten Dijke, P. (1997). Identification of Smad7, a TGFβ-inducible antagonist of TGF-β signalling. *Nature*, **389**, 631.

144. Souchelnytskyi, S., Nakayama, T., Nakao, A., Morén, A., Heldin, C.-H., Christian, J. L., and ten Dijke, P. (1998). Physical and functional interaction of murine and *Xenopus* Smad7 with bone morphogenetic protein receptors and transforming growth factor-β receptors. *J. Biol. Chem.*, **273**, 25364.

145. Nakayama, T., Snyder, M. A., Grewal, S. S., Tsuneizumi, K., Tabata, T., and Christian, J. L. (1998). *Xenopus* Smad8 acts downstream of BMP-4 to modulate its activity during vertebrate embryonic patterning. *Development*, **125**, 857.

146. Casellas, R. and Hemmati Brivanlou, A. (1998). *Xenopus* Smad7 inhibits both the activin and BMP pathways and acts as a neural inducer. *Dev. Biol.*, **198**, 1.

147. Bhushan, A., Chen, Y., and Vale, W. (1998). Smad7 inhibits mesoderm formation and promotes neural cell fate in *Xenopus* embryos. *Dev. Biol.*, **200**, 260.

148. Itoh, S., Landstrom, M., Hermansson, A., Itoh, F., Heldin, C.-H., Heldin, N.-E., and ten Dijke, P. (1998). Transforming growth factor β1 induces nuclear export of inhibitory Smad7. *J. Biol. Chem.*, **273**, 29195.

149. Ishiaki, A., Yamato, K., Nakao, A., Nonaka, K., Ohguchi, M., ten Dijke, P., and Nishihara, T. (1998). Smad7 is an activin-inducible inhibitor of activin-induced growth arrest and apoptosis in mouse B cells. *J. Biol. Chem.*, **273**, 24293.

150. Afrakhte, M., Morén, A., Jossan, S., Itoh, S., Sampath, K., Westermark, B., Heldin, C.-H., Heldin, N.-E., and ten Dijke, P. (1998). Induction of inhibitory Smad6 and Smad7 mRNA by TGF-β family members. *Biochem. Biophys. Res. Commun.*, **249**, 505.

151. Takase, M., Imamura, T., Sampath, T. K., Takeda, K., Ichijo, H., Miyazono, K., and Kawabata, M. (1998). Induction of Smad6 mRNA by bone morphogenetic proteins. *Biochem. Biophys. Res. Commun.*, **244**, 26.

152. Ulloa, L., Doody, J., and Massagué, J. (1999). Inhibition of transforming factor-β/SMAD signalling by the interferon-v/STAT pathway. *Nature*, **397**, 710.

153. Kim, J., Johnson, K., Chen, H. J., Carroll, S., and Laughon, A. (1997). *Drosophila* Mad binds to DNA and directly mediates activation of *vestigial* by decapentaplegic. *Nature*, **388**, 304.

154. Thatcher, J. D., Haun, C., and Okkema, P. G. (1999). The DAF-3 Smad binds DNA and represses gene expression in the *Caenorhabditis elegans* pharynx. *Development*, **126**, 97.

155. Yagi, K., Goto, D., Hamamoto, T., Takenoshita, S., Kato, M., and Miyazono, K. (1999). Alternatively spliced variant of Smad2 lacking exon3. *J. Biol. Chem.*, **274**, 703.

156. Zawel, L., Dai, J. L., Buckhaults, P., Zhou, S., Kinzler, K. W., Vogelstein, B., and Kern, S. E. (1998). Human Smad3 and Smad4 are sequence-specific transcription activators. *Mol. Cell*, **1**, 611.

157. Yingling, J. M., Datto, M. B., Wong, C., Frederick, J. P., Liberati, N. T., and Wang, X.-F. (1997). Tumour suppressor Smad4 is a transforming growth factor β-inducible DNA binding protein. *Mol. Cell. Biol.*, **17**, 7019.

158. Dennler, S., Itoh, S., Vivien, D., ten Dijke, P., Huet, S., and Gauthier, J.-M. (1998). Direct

binding of Smad3 and Smad4 to critical TGFβ-inducible elements in the promoter of human plasminogen activator inhibitor-type 1 gene. *EMBO J.*, **17**, 3091.

159. Jonk, L. J. C., Itoh, S., Heldin, C.-H., ten Dijke, P., and Kruijer, W. (1998). Identification and functional characterization of a Smad binding element (SBE) in the *JunB* promoter that acts as a transforming growth factor-β, activin, and bone morphogenetic protein-inducible enhancer. *J. Biol. Chem.*, **273**, 21145.

160. Song, C.-Z., Siok, T. E., and Gelehrter, T. D. (1998). Smad4/DPC4 and Smad3 mediate transforming growth factor-β (TGF-β). Signalling through direct binding to a novel TGF-β-responsive element in the human plasminogen activator inhibitor-1 promoter. *J. Biol. Chem.*, **273**, 29287.

161. Vindevoghel, L., Lechleider, R. J., Kon, A., de Caestecker, M. P., Uitoo, J., Roberts, A. B., and Mauviel, A. (1998). SMAD3/4-dependent transcriptional activation of the human type VII collagen gene (COL7A1) promoter by transforming growth factor β. *Proc. Natl Acad. Sci. USA*, **95**, 14769.

162. Hua, X., Lin, X., Ansari, D. O., and Lodish, H. F. (1998). Synergistic cooperation of TFE3 and Smad proteins in TGF-β-induced transcription of the plasminogen activator inhibitor-1 gene. *Genes Dev.*, **12**, 3084.

163. Chen, X., Weisberg, E., Fridmacher, V., Watanabe, M., Naco, G., and Whitman, M. (1997). Smad4 and FAST-1 in the assembly of activin-responsive factor. *Nature*, **389**, 85.

164. Zhou, S., Zawel, L., Lengauer, C., Kinzler, K. W., and Vogelstein, B. (1998). Characterization of human FAST-1, a TGFβ and activin signal transducer. *Mol. Cell*, **2**, 121.

165. Zhang, Y., Feng, X.-H., and Derynck, R. (1998). Smad3 and Smad4 cooperate with c-Jun/c-Fos to mediate TGFβ-induced transcription. *Nature*, **394**, 909.

166. Kurokawa, M., Mitani, K., Irie, K., Matsuyama, T., Takahashi, T., Chiba, S., Yazaki, Y., Matsumoto, K., and Hirai, H. (1998). The oncoprotein Evi-1 represses TGF-β signalling by inhibiting Smad3. *Nature*, **394**, 92.

167. Liu, F., Massagué, J., and Altaba, A. R.I. (1999). Carboxy-terminally truncated Gli3 proteins associate with Smads. *Nature Genet.*, **20**, 325.

168. Yanagisawa, J., Yanagi, Y., Masuhiro, Y., Suzawa, M., Watanabe, M., Kashiwagi, K., Toriyabe, T., Kawabata, M., Miyazono, K., and Kato, S. (1999). Convergence of transforming growth factor-β and vitamin D signalling pathways on SMAD transcriptional coactivators. *Science*, **283**, 1317.

169. Chen, X., Rubock, M. J., and Whitman, M. (1996). A transcriptional partner for MAD proteins in TGF-β signalling. *Nature*, **383**, 691.

170. Liu, B., Dou, C.-L., Prabhu, L., and Lai, E. (1999). FAST-2 is a mammalian winged-helix protein which mediates transforming growth factor β signals. *Mol. Cell. Biol.*, **19**, 424.

171. Weisberg, E., Winnier, G. E., Chen, X., Farnsworth, C. L., Hogan, B. L. H., and Whitman, M. (1998). A mouse homologue of FAST-1 transduces TGFβ superfamily signals and is expressed during early embryogenesis. *Mech. Dev.*, **79**, 17.

172. Janknecht, R., Wells, N. J., and Hunter, T. (1998). TGF-β-stimulated cooperation of Smad proteins with the coactivators CBP/p300. *Genes Dev.*, **12**, 2114.

173. Topper, J. N., DiChiara, M. R., Brown, J. D., Williams, A. J., Falb, D., Collins, T., and Gimbrone Jr., M. A. (1998). CREB binding protein is a required coactivator for Smad-dependent, transforming growth factor β transcriptional responses in endothelial cells. *Proc. Natl Acad. Sci. USA*, **95**, 9506.

174. Pouponnot, C., Jayaraman, L., and Massagué, J. (1998). Physical and functional interaction of SMADs and p300/CBP. *J. Biol. Chem.*, **273**, 22865.

175. Shen, X., Hu, P. P.-C., Liberati, N. T., Datto, M. B., Frederick, J. P., and Wang, X.-F. (1998).

TGF-beta-induced phosphorylation of Smad3 regulates its interaction with coactivator p300/CREB-binding protein. *Mol. Biol. Cell*, **9**, 3309.

176. Feng, X.-H., Zhang, Y., Wu, R.-Y., and Derynck, R. (1998). The tumor suppressor Smad4/DPC4 and transcriptional adaptor CBP/p300 are coactivators for Smad3 in TGF-β-induced transcriptional activation. *Genes Dev.*, **12**, 2153.

177. Waltzer, L. and Bienz, M. (1999). A function of CBP as a transcriptional co-activator during Dpp signalling. *EMBO J.*, **18**, 1630.

178. Shioda, T., Lechleider, R. J., Dunwoodie, S. L., Li, H., Yahata, T., de Caestecker, M. P., Fenner, M. H., Roberts, A. B., and Isselbacher, K. J. (1998). Transcriptional activating activity of Smad4: Roles of SMAD hetero-oligomerization and enhancement by an associating transactivator. *Proc. Natl Acad. Sci. USA*, **95**, 9785.

179. Wotton, D., Lo, R. S., Lee, S., and Massagué, J. (1999). A Smad transcriptional corepressor. *Cell*, **97**, 29.

180. Hata, A., Lo, R. S., Wotton, D., Lagna, G., and Massagué, J. (1997). Mutations increasing autoinhibition inactivate tumour suppressors Smad2 and Smad4. *Nature*, **388**, 82.

181. Shi, Y., Hata, A., Lo, R. S., Massagué, J., and Pavletich, N. P. (1997). A structural basis for mutational inactivation of the tumour suppressor Smad4. *Nature*, **388**, 87.

182. Kawabata, M., Inoue, H., Hanyu, A., Imamura, T., and Miyazono, K. (1998). Smad proteins exist as monomers *in vivo* and undergo homo- and hetero-oligomerization upon activation by serine/threonine kinase receptors. *EMBO J.*, **17**, 4056.

183. Lo, R. S., Chen, Y.-G., Shi, Y., Pavletich, N. P., and Massagué, J. (1998). The L3 loop: a structural motif determining specific interactions between SMAD proteins and TGF-β receptors. *EMBO J.*, **17**, 996.

184. Shi, Y., Wang, Y.-F., Jayaraman, L., Yang, H., Massagué, J., and Pavletich, N. P. (1998). Crystal structure of a Smad MH1 domain bound to DNA: insights on DNA binding in TGF-β signalling. *Cell*, **94**, 585.

185. Tsukazaki, T., Chiang, T. A., Davison, A. F., Attisano, L., and Wrana, J. L. (1998). SARA, a FYVE domain protein that recruits Smad2 to the TGF-β receptor. *Cell*, **95**, 779.

186. Manova, K., Paynton, B. V., and Bachvarova, R. F. (1992). Expression of activins and TGFβ1 and β2 RNAs in early postimplantation mouse embryos and uterine decidua. *Mech. Dev.*, **36**, 141.

187. Albano, R. M., Groome, N., and Smith, J. C. (1993). Activins are expressed in pre-implantation mouse embryos and in ES and EC cells and are regulated on their differentiation. *Development*, **117**, 711.

188. Albano, R. M., Arkell, R., Beddington, R. S. P., and Smith, J. C. (1994). Expression of inhibin subunits and follistatin during postimplantation mouse development: decidual expression of activin and expression of follistatin in primitive streak, somites and hindbrain. *Development*, **120**, 803.

189. Matzuk, M. M., Kumar, T. R., Vassalli, A., Bickenbach, J. R., Roop, D. R., Jaenisch, R., and Bradley, A. (1995). Functional analysis of activins during mammalian development. *Nature*, **374**, 354.

190. Vassali, A., Matzuk, M. M., Gardner, H. A. R., Lee, K.-F., and Jaenish, R. (1994). Activin/inhibin βB subunit gene disruption leades to defects in eyelid development and female reproduction. *Genes Dev.*, **8**, 414.

191. Matzuk, M. M., Finegold, M. J., Su, J.-G. J., Hsueh, A. J. W., and Bradley, A. (1992). α-Inhibin is a tumour-suppressor gene with gonadal specificity in mice. *Nature*, **360**, 313.

192. Conlon, F. L., Lyons, K. M., Takaesu, N., Barth, K. S., Kispert, A., Herrmann, B., and Robertson, E. J. (1994). A primary requirement for nodal in the formation and maintenance of the primitive streak in the mouse. *Development*, **120**, 1919.

193. Zhou, X., Sasaki, H., Lowe, L., Hogan, B. L. M., and Kuehn, M. R. (1993). Nodal is a novel TGF-β-like gene expressed in the mouse node during gastrulation. *Nature*, **361**, 543.

194. Lyons, K. M., Hogan, B. L. M., and Robertson, E. J. (1995). Colocalization of BMP 7 and BMP 2 RNAs suggests that these factors cooperatively mediate tissue interactions during murine development. *Mech. Dev.*, **50**, 71.

195. Winnier, G., Blessing, M., Labosky, P. A., and Hogan, B. L. M. (1995). Bone morphogenetic protein-4 is required for mesoderm formation and patterning in the mouse. *Genes Dev.*, **9**, 2105.

196. Zhang, H. and Bradley, A. (1996). Mice deficient for BMP2 are nonviable and have defects in amnion/chorion and cardiac development. *Development*, **122**, 2977.

197. Kingsley, D. M., Bland, A. E., Grubber, J. M., Marker, P. C., Russell, L. B., Copeland, N. G., and Jenkins, N. A. (1992). The mouse short ear skeletal morphogenesis locus is associated with defects in a bone morphogenetic member of the TGF-β superfamily. *Cell*, **71**, 399.

198. King, J. A., Marker, P. C., Seung, K. J., and Kingsley, D. M. (1994). BMP5 and the molecular, skeletal, and soft-tissue alterations in short ear mice. *Dev. Biol.*, **166**, 112.

199. Dudley, A. T., Lyons, K. M., and Robertson, E. J. (1995). A requirement for bone morphogenetic protein-7 during development of the mammalian kidney and eye. *Genes Dev.*, **9**, 2795.

200. Luo, G., Hofmann, C., Bronckers, A. L. J. J., Sohocki, M., Bradley, A., and Karsenty, G. (1995). BMP-7 is an inducer of nephrogenesis, and is also required for eye development and skeletal patterning. *Genes Dev.*, **9**, 2808.

201. Zhao, G. Q., Deng, K., Labosky, P. A., Liaw, L., and Hogan, B. L. M. (1996). The gene encoding bone morphogenic protein 8B is required for the initiation and maintenance of spermatogenesis in the mouse. *Genes Dev.*, **10**, 1657.

202. Mishina, Y., Suzuki, A., Ueno, N., and Behringer, R. R. (1995). Bmpr encodes a type-I bone morphogenetic protein receptor that is essential for gastrulation during mouse embryogenesis. *Genes Dev.*, **9**, 3027.

203. Manova, K., DeLeon, V., Angeles, M., Kalantry, S., Giarre, M., Attisano, L., Wrana, J. L., and Bachvarova, R. F. (1994). mRNAs for activin receptors II and IIB are expressed in mouse oocytes and in the epiblast of pregastrula and gastrula stage mouse embryos. *Mech. Dev.*, **49**, 3.

204. Matzuk, M. M., Lu, N., Vogel, H., Sellheyer, K., Roop, D. R., and Bradley, A. (1995). Multiple defects and perinatal death in mice deficient in follistatin. *Nature*, **347**, 360.

205. Behringer, R. R., Finegold, M. J., and Cate, R. L. (1994). Müllerian-inhibiting substance function during mammalian sexual development. *Cell*, **79**, 415.

206. Mishina, Y., Rey, R., Finegold, M. J., Matzuk, M. M., Josso, N., Cate, R. L., and Behringer, R. R. (1996). Genetic analysis of the Müllerian-inhibiting substance signal transduction pathway in mammalian sexual differentiation. *Genes Dev.*, **10**, 2577.

207. Storm, E. E., Huynh, T. V., Copeland, N. G., Jenkins, N. A., Kingsley, D. M., and Lee, S.-J. (1994). Limb alterations in brachypodism mice due to mutations in a new member of the TGFβ-superfamily. *Nature*, **368**, 639.

208. Dong, J., Albertini, D. F., Nishimori, K., Kumar, T. R., Lu, N., and Matzuk, M. M. (1996). Growth differentiation factor-9 is required during early ovarian folliculogenesis. *Nature*, **383**, 531.

209. Gu, Z., Nomura, M., Simpson, B. B., Lei, H., Feijen, A., van den Eijnden-van Raaij, J., Donahoe, P. K., and Li, E. (1998). The type-I activin receptor ActRIB is required for egg cylinder organization and gastrulation in the mouse. *Genes Dev.*, **12**, 844.

210. Waldrip, W. R., Bikoff, E. K., Hoodless, P. A., Wrana, J. L., and Robertson, E. J. (1998).

Smad2 signalling in extraembryonic tissues determines anterior-posterior polarity of the early mouse embryo. *Cell*, **92**, 797.

211. Nomura, M. and Li, E. (1998). Smad2 role in mesoderm formation, left-right patterning and craniofacial development. *Nature*, **393**, 786.

212. Weinstein, M., Yang, X., Li, C., Xu, X., Goday, J., and Deng, C.-X. (1998). Failure of egg cylinder elongation and mesoderm induction in mouse embryos lacking the tumor suppressor Smad2. *Proc. Natl Acad. Sci. USA*, **95**, 9378.

213. Zhu, Y., Richardson, J. A., Parada, L. F., and Graff, J. M. (1998). Smad3 mutant mice develop metastatic colorectal cancer. *Cell*, **94**, 703.

214. Sirard, C., de la Pompa, J. L., Elia, A., Itie, A., Mirtsos, C., Cheung, A., Hahn, S., Wakeham, A., Schwartz, L., Kern, S. E., Rossant, J., and Mak, T. W. (1997). The tumor suppressor gene *Smad4/Dpc4* is required for gastrulation and later for anterior development of the mouse embryo. *Genes Dev..*, **12**, 107.

215. Yang, X., Li, C., Xu, X., and Deng, C. (1998). The tumor suppressor SMAD4/DPC4 is essential for epiblast proliferation and mesoderm induction in mice. *Proc. Natl Acad. Sci. USA*, **95**, 3667.

216. Datto, M. B., Frederick, J. P., Pan, L., Borton, A. J., Zhuang, Y., and Wang, X.-F. (1999). Targeted disruption of Smad3 reveals an essential role in transforming growth factor beta-mediated signal transduction. *Mol. Cell. Biol.*, **19**, 2495.

217. Sanford, L. P., Ormsby, I., Gittenberger-de Groot, A. C., Sariola, H., Friedman, R., Boivin, G. P., Cardell, E. L., and Doetschman, T. (1997). TGFbeta2 knockout mice have multiple developmental defects that are non-overlapping with other TGFbeta knockout phenotypes. *Development*, **124**, 2659.

218. Chang, H., Huylebroeck, D., Verschueren, K., Guo, Q., Matzuk, M. M., and Zwijsen, A. (1999). Smad5 knockout mice die at mid-gestation due to multiple embryonic and extraembryonic defects. *Development*, **126**, 1631.

219. Yang, X., Castilla, L. H., Xu, X., Li, C., Gotay, J., Weinstein, M., Liu, P. P., and Deng, C.-X. (1999). Angiogenesis defects and mesenchymal apoptosis in mice lacking SMAD5. *Development*, **126**, 1571.

220. Katagiri, T., Boorla, S., Frendo, J. L., Hogan, B. L., and Karsenty, G. (1998). Skeletal abnormalities in double heterozygous Bmp4 and Bmp7 mice. *Dev. Genet.*, **22**, 340.

221. McPherron, A. C., Lawler, A. M., and Lee, S. J. (1997). Regulation of skeletal muscle mass in mice by a new TGF-beta superfamily member. *Nature*, **387**, 83.

222. Solloway, M. J., Dudley, A. T., Bikoff, E. K., Lyons, K. M., Hogan, B. L., and Robertson, E. J. (1998). Mice lacking Bmp6 function. *Dev. Genet.*, **22**, 321.

223. Kaartinen, V., Voncken, J. W., Shuler, C., Warburton, D., Bu, D., Heisterkamp, N., and Groffen, J. (1995). Abnormal lung development and cleft palate in mice lacking TGF-beta3 indicates defect of epethilial-mesenchymal interactions. *Nature Genet.*, **11**, 415.

224. Proetzel, G., Pawlowski, S. A., Wiles, M. V., Yin, M., Boivin, G. P., Howles, P. N., Ding, J., Ferguson, M. V., and Doetschman, T. (1995). Transforming growth factor beta 3 is required for secondary palate fusion. *Nature Genet.*, **11**, 409.

225. Thomsen, G. H. (1997). Antagonism within and around the Spemann Organizer: BMP inhibitors in vertebrate body patterning. *Trends Genet.*, **13**, 209.

226. Zhang, J., Houston, D. W., King, M. L., Payne, C., Wylie, C., and Heasman, J. (1998). The role of maternal VegT in establishing the primary germ layers in *Xenopus* embryos. *Cell*, **94**, 515.

227. McMahon, J. A., Takada, T., Zimmerman, L. B., Fan, C.-M., Harland, R. M., and McMahon, A. P. (1998). Noggin-mediated antagonism of BMP signalling is required for growth and patterning of the neural tube and somite. *Genes Dev.*, **12**, 1438.

228. Graff, J. M., Bansal, A., and Melton, D. A. (1996). *Xenopus* Mad proteins transduce distinct subsets of signals for the TGFβ superfamily. *Cell*, **85**, 479.

229. Wilson, P. A., Lagna, G., Suzuki, A., and Hemmati-Brivanlou, A. (1997). Concentration-dependent patterning of the *Xenopus* ectoderm by BMP4 and its signal transducer Smad1. *Development*, **124**, 3177.

230. Zhang, Y., Musci, T., and Derynck, R. (1997). The tumor suppressor Smad4/DPC4 as a central mediator of Smad function. *Curr. Biol.*, **7**, 270.

231. Candia, A. F., Watabe, T., Hawley, S. B., Onichtchouk, D., Zhang, Y., Derynck, R., Niehrs, C., and Cho, K. Y. (1997). Cellular interpretation of multiple TGF-β signals: intracellular antagonism between activin/BVg1 and BMP-2/4 signalling mediated by Smads. *Development*, **124**, 4467.

232. Shibuya, H., Iwata, H., Masuyama, N., Gotoh, Y., Yamaguchi, K., Irie, K., Matsumoto, K., Nishida, E., and Ueno, N. (1998). Role of TAK1 and TAB1 in BMP signalling in early *Xenopus* development. *EMBO J.*, **17**, 1019.

233. Hogan, B. L. M., Constantini, F., and Lacy, E. (1986). *Manipulating the mouse embryo: a laboratory manual*. Cold Spring Harbor Laboratory Press, Cold Spring Harbor, New York.

234. Coucouvanis, E. and Martin, G. R. (1999). BMP signaling plays a role in visceral endoderm differentiation and cavitation in the early mouse embryo. *Development*, **126**, 535.

235. Varlet, I., Collingnon, J., and Robertson, E. J. (1997). *nodal* expression in the primitive endoderm is required for specification of the anterior axis during mouse gastrulation. *Development*, **124**, 1033.

236. Oh, S. P. and Li, E. (1997). The signaling pathway mediated by the type IIB activin receptor controls axial patterning and lateral assymetry in the mouse. *Genes Dev.*, **11**, 1812.

237. Johnson, D. W., Berg, J. N., Baldwin, M. A., Gallione, C., Marondel, I., Stenzil, T., Speer, M., Pericak-Vance, M., Qumsiyeh, W. A., Schwartz, C., Diamond, A., Guttmacher, A. E., Jackson, C. E., Attisano, L., Kucherlapati, R., Porteous, M. E. M., and Marchuk, D. A. (1996). Mutations in the activin receptor-like kinase 1 gene in hereditary hemorrhagic telangiectasia type 2. *Nature Genet.*, **13**, 189.

238. Hyatt, B. A., Lohr, J. L., and Yost, H. J. (1996). Initiation of vertebrate left–right axis formation by maternal Vg1. *Nature*, **384**, 62.

239. Hyatt, B. A. and Yost, H. J. (1998). The left–right coordinator: the role of Vg1 in organizing left–right axis formation. *Cell*, **93**, 37.

240. King, T. and Brown, N. A. (1997). Embryonic asymmetry: left TGFβ at the right time? *Curr. Biol.*, **7**, 212.

241. Beddington, R. S. P. and Robertson, E. J. (1999). Axis development and early asymmetry in mammals. *Cell*, **96**, 195.

242. Stern, C. D., Yu, R. T., Kakizuka, A., Kintner, C. R., Mathews, L. S., Vale, W. W., Evans, R. M., and Umesono, K. (1995). Activin and its receptors during gastrulation and the later phases of mesoderm development in the chick embryo. *Dev. Biol.*, **172**, 192.

243. Levin, M., Pagan, S., Roberts, D., Cooke, J., Kuehn, M., and Tabin, C. (1997). Left/right patterning signals and the independent regulation of different aspects of situs in the chick embryo. *Dev. Biol.*, **189**, 57.

244. Levin, M., Johnson, R. L., Stern, C. D., Kuehn, M., and Tabin, C. (1995). A molecular pathway determining left–right asymmetry in chick embryogenesis. *Cell*, **82**, 803.

245. Collignon, J., Varlet, I., and Robertson, E. J. (1996). Relationship between asymmetric nodal expression and the direction of embryonic turning. *Nature*, **381**, 155.

246. Lowe, L. A., Supp, D. M., Sampath, K., Yokoyama, T., Wright, C. V. E., Potter, S. S., Over-

beek, P., and Kuehn, M. R. (1996). Conserved left–right asymmetry of nodal expression and alterations in murine situs inversus. *Nature*, **381**, 158.

247. Matzuk, M. M., Kumar, T. R., and Bradley, A. (1995). Different phenotypes for mice deficient in either activins or activin receptor type-II. *Nature*, **374**, 356.

248. Chiang, C., Litingtung, Y., Lee, E., Young, K. E., Corden, J. L., Westphal, H., and Beachy, P. A. (1996). Cyclopia and defective axial patterning in mice lacking Sonic Hedgehog gene function. *Nature*, **383**, 407.

249. Meno, C., Ito, Y., Saijoh, Y., Matsuda, Y., Tashiro, K., Kuhara, S., and Hamada, H. (1997). Two closely related left–right asymmetrically expressed genes, lefty-1 and lefty-2: their distinct expression domains, chromosomal linkage and direct neuralizing activity in *Xenopus* embryos. *Genes Cells*, **8**, 513.

250. Meno, C., Shimono, A., Saijoh, Y., Yashiro, K., Mochida, K., Ohishi, S., Noji, S., Kondoh, H., and Hamada, H. (1998). lefty-1 is required for left–right determination as a regulator of lefty-2 and nodal. *Cell*, **94**, 287.

251. Logan, M., Pagan-Westphal, S. M., Smith, D. M., Paganessi, L., and Tabin, C. J. (1998). The transcription factor Pitx2 mediates situs-specific morphogenesis in response to left–right asymmetric signals. *Cell*, **94**, 307.

252. Piedra, M. E., Icardo, J. M., Albajar, M., Rodriguez-Rey, J. C., and Ros, M. A. (1998). Pitx2 participates in the late phase of the pathway controlling left–right asymmetry. *Cell*, **94**, 319.

253. Ryan, A. K., Blumberg, B., Rodriguez-Esteban, C., Yonei-Tamura, S., Tamura, K., Tsukui, T., de la Pena, J., Sabbagh, W., Greenwald, J., Choe, S., Norris, D. P., Robertson, E. J., Evans, R. M., Rosenfeld, M. G., and Belmonte, J. C. I. (1998). Pitx2 determines left–right asymmetry of internal organs in vertebrates. *Nature*, **394**, 545.

254. Yoshioka, H., Meno, C., Koshiba, K., Sugihara, M., Itoh, H., Ishimaru, Y., Inoue, T., Ohuchi, H., Semina, E. V., Murray, J. C., Hamada, H., and Noji, S. (1998). Pitx2, a bicoid-type homeobox gene, is involved in a lefty-signaling pathway in determination of left–right asymmetry. *Cell*, **94**, 299.

255. Nonaka, S., Tanaka, Y., Okada, Y., Takeka, S., Harada, A., Kanai, Y., Kido, M., and Hirokawa, N. (1998). Randomization of left–right asymmetry due to loss of nodal cilia generating leftward flow of extraembryonic fluid in mice lacking KIF3B motor protein. *Cell*, **95**, 829.

256. Amthor, H., Christ, B., and Patel, K. (1999). A molecular mechanism enabling continuous embryonic muscle growth—a balance between proliferation and differentiation. *Development*, **126**, 1041.

257. Amthor, H., Christ, B., Weil, M., and Patel, K. (1998). The importance of timing differentiation during limb muscle development. *Curr. Biol.*, **8**, 642.

258. Reshef, R., Maroto, M., and Lassar, A. B. (1998). Regulation of dorsal somitic cell fates: BMPs and Noggin control the timing and pattern of myogenic regulator expression. *Genes Dev.*., **12**, 209.

259. Marcelle, C., Stark, M. R., and Bronner-Fraser, M. (1997). Coordinate actions of BMPs, Wnts, Shh and Noggin mediate patterning of the dorsal somite. *Development*, **124**, 3955.

260. Tonegawa, A. and Takahashi, Y. (1998). Somitgenesis controlled by Noggin. *Dev. Biol.*, **202**, 172.

261. Miller, J. B., Schaefer, L., and Dominov, J. A. (1999). Seeking muscle stem cells. *Curr. Top. Dev. Biol.*, **43**, 191.

262. Markowitz, S. D. and Roberts, A. B. (1996). Tumor suppressor activity of the TGF-β pathway in human cancers. *Cytokine Growth Factor Rev.*, **7**, 93.

263. Markowitz, S., Wang, J., Myeroff, L., Parsons, R., Sun, L. Z., Lutterbaugh, J., Fan, R. S., Zborowska, E., Kinzler, K. W., Vogelstein, B., Brattain, M., and Willson, J. K. V. (1995). Inactivation of the Type II TGF-β receptor in colon cancer cells with microsatellite instability. *Science*, **268**, 1336.

264. Myeroff, L. L., Parsons, R., Kim, S.-J., Hedrick, L., Cho, K. R., Orth, K., Mathis, M., Kinzler, K. W., Lutterbaugh, J., Park, K., Bang, Y.-J., Lee, H. Y., Park, J.-G., Lynch, H. T., Roberts, A. B., Vogelstein, B., and Markowitz, S. D. (1995). A transforming growth factor β receptor type-II gene mutation common in colon and gastric but rare in endometrial cancers with microsatellite instability. *Cancer Res.*, **55**, 5545.

265. Lu, S. L., Zhang, W. C., Akiyama, Y., Nomizu, T., and Yuasa, Y. (1996). Genomic structure of the transforming growth factor beta type-II receptor gene and its mutations in hereditary nonpolyposis colorectal cancers. *Cancer Res.*, **56**, 4595.

266. Lu, S. L., Kawabata, M., Imamura, T., Akiyama, Y., Nomizu, T., Miyazono, K., and Yuasa, Y. (1998). HNPCC associated with germline mutations in the TGF-beta type-II receptor gene. *Nature Genet.*, **19**, 17.

267. Tang, B., Bottinger, E. P., Jakowlew, S. B., Bagnall, K. M., Mariano, J., Anver, M. R., Letterio, J. J., and Wakefield, L. M. (1998). Transforming growth factor-beta 1 is a new form of tumor suppressor with true haploid insuffiency. *Nature Med.*, **4**, 802.

268. Shull, M. M., Ormsby, I., Kier, A. B., Pawlowski, S., Diebold, R. J., Yin, M., Allen, R., Sidman, C., Proetzel, G., Calvin, D., Annunziata, N., and Doetschman, T. (1992). Targeted disruption of the mouse transforming growth factor-β1 gene results in multifocal inflammatory disease. *Nature*, **359**, 693.

269. Matzuk, M. M., Finegold, M. J., Mishina, Y., Bradley, A., and Behringer, R. R. (1995). Synergistic effects of inhibins and müllerian-inhibiting substance on testicular tumorigenesis. *Mol. Endocrinol.*, **9**, 1337.

270. Hata, A., Shi, Y., and Massague, J. (1998). TGF-signaling and cancer: structural and functional consequences of mutations in Smads. *Mol. Med. Today*, **4**, 257.

271. Takagi, Y., Kohmura, H., Futamura, M., Kida, H., Tanemura, H., Shimokawa, K., and Saji, S. (1996). Somatic alterations of the DPC4 gene in human colorectal cancers *in vivo*. *Gastroenterology*, **111**, 1369.

272. Nagatake, M., Takagi, Y., Osada, H., Uchida, K., Mitsudomi, T., Saji, S., Shimokata, K., and Takahashi, T. (1996). Somatic *in vivo* alteration of the DPC4 gene at 18q21 in human lung cancers. *Cancer Res.*, **56**, 2718.

273. Howe, J. R., Roth, S., Ringold, J. C., Summers, R. W., Jarvinen, H. J., Sistonen, P., Tomlinson, I. P. M., Houlston, R. S., Bevan, S., Mitros, F. A., Stone, E. M., and Aaltonen, L. A. (1998). Mutations in the SMAD4/DPC4 gene in juvenile polyposis. *Science*, **280**, 1086.

274. Takaku, K., Oshima, M., Miyoshi, H., Matsui, M., Seldin, M. F., and Taketo, M. M. (1998). Intestinal tumorigenesis in compound mutant mice of both DPC4 (Smad4) and Apc genes. *Cell*, **92**, 645.

275. Riggins, G. J., Thiagalingam, S., Rozenblum, E., Weinstein, C. L., Kern, S. E., Hamilton, S. R., Willson, J. K.V., Markowitz, S. D., Kinzler, K. W., and Vogelstein, B. (1996). MAD-related genes in the human. *Nature Genet.*, **13**, 347.

276. Yang, X., Letterio, J. J., Lechleider, R. J., Chen, L., Hayman, R., Gu, H., Roberts, A. B., and Deng, C. (1999). Targeted disruption of Smad3 results in impaired mucosal immunity and diminished T cell responsiveness to TGF beta. *EMBO J.*, **18**, 1280.

277. Imbeaud, S., Carré-Eusebe, D., Rey, R., Belville, C., Josso, N., and Picard, J. Y. (1994). Molecular genetics of the persistent Müllerian duct syndrome: a study of 19 families. *Hum. Mol. Genet.*, **3**, 125.

278. Imbeaud, S., Faure, E., Lamarre, I., Mattei, M. G., di Clemente, N., Tizard, R., *et al.* (1995). Insensitivity to anti-Müllerian hormone due to a mutation in the human anti-Müllerian hormone receptor. *Nature Genet.*, **11**, 382.

279. Thomas, J. T., Lin, K., Nandedkar, M., Camargo, M., Cervanka, J., and Luyten, F. P. (1996). A human chondrodysplasia due to a mutation in a TGF-β superfamily member. *Nature Genet.*, **12**, 315.

280. Gong, Y., Krakow, D., Marcelino, J., Wilkin, D., Chitayat, D., Babul-Hirji, R., Hudgins, L., Cremers, C. W., Cremers, F. P. M., Brunner, H. G., Reinker, K., Rimoin, D., Cohn, H., Goodman, F. R., Reardon, W., Patton, M., Francomano, C. A., and Warman, M. L. (1999). Heterozygous mutations in the gene encoding noggin affect human joint morphogenesis. *Nature Genetics*, **21**, 302.

281. Brunet, L. J., McMahon, J. A., McMahon, A. P., and Harland, R. M. (1999). Noggin, cartilage morphogenesis, and joint formation in the mammalian skeleton. *Science*, **280**, 1455.

282. Guttmacher, A. E., Marchuk, D. A., and White, R. I., Jr. (1995). Hereditary hemorrhagic telangiectasia. *N. Engl. J. Med.*, **333**, 918.

283. McAllister, K. A., Grogg, K. M., Johnson, D. W., Gallione, C. J., Baldwin, M. A., Jackson, C. E., Helmbold, E. A., Markel, D. S., McKinnon, W. C., Murell, J., McCormick, M. D., Pericak-Vance, M. A., Heutink, P., Oostra, B. A., Haitjema, T., Westerman, C. J., Porteous, M. E., Guttmacher, A. E., Letarte, M., and Marchuk, D. A. (1994). Endoglin, a TGF-β binding protein of endothelial cells is the gene for hereditary haemorrhagic telangiectasia type-I. *Nature Genet.*, **8**, 345.

284. Zhu, Y., Richardson, J. A., Parada, L. F., and Graff, J. M. (1998). Smad3 mutant mice develop metastatic colorectal cancer. *Cell* **94**, 703.

8 | ATM and DNA-PK: the sensing, signalling, and repair of DNA damage

ROBERT BRISTOW, HANS BLUYSSEN, ANNELIES DE KLEIN, and DIK VAN GENT

1. Overview: DNA damage responses

A significant and ongoing problem for every living cell is the occurrence of DNA damage caused by exogenous and endogenous agents. For example, DNA base damage and DNA strand breaks are induced as consequences of the normal cellular processes of DNA replication and oxidative stress, or after exposure to exogenous DNA-damaging agents including UV-light or ionizing irradiation. As a result, cells have developed a sophisticated approach to the integration of DNA damage-sensing and subsequent DNA repair as a means of preserving genomic stability. This chapter will mainly focus on the cellular activity and biochemistry of two members of the PI3K-related protein kinase family named ATM (for 'mutated in ataxia telangiectasia'), and DNA-PK (DNA-dependent protein kinase), which are thought to play important roles in the sensing, signalling, and repair of DNA strand breaks.

1.1 DNA repair

Cells have evolved multiple DNA repair systems to remove or bypass potentially mutagenic lesions (1–3). UV-induced lesions, for example, can be repaired by nucleotide excision repair. DNA double-strand breaks (dsb) can be repaired by the homologous recombination or DNA end-joining pathways (4). This latter type of damage, especially, is very deleterious to the cell, since genetic information can easily be lost if a dsb is left unrepaired at mitosis. Indeed, cell survival following ionizing radiation correlates best with the number of unrepaired dsb (5). Homologous recombination requires extensive homology between the DNA ends at the site of breakage and the sister chromatid or homologous chromosome from which repair is directed. In contrast, end-joining requires little or no homology on the ends of the strands being joined.

In yeast, the *RAD52* epistasis group (*RAD50–57, -59, MRE11*, and *XRS2*) is involved in homologous recombination, which is thought to be the predominant dsb repair pathway in this organism. The end-joining pathway of dsb repair is thought to predominate in mammalian cells, although the homologous recombination pathway is conserved from yeast to humans, and is also active in mammals. Two main discrete repair-protein complexes are implicated in the end-joining pathway: (1) the DNA-PK complex and (2) the ligase IV/XRCC4 complex. The importance of these two pathways of dsb repair is underscored by the fact that mutations in genes involved in either pathway can cause profound recombination defects and increased cellular sensitivity to dsb-inducing agents (4).

1.2 Cell-cycle checkpoints and genomic stability

Mammalian cells respond to ionizing radiation by delaying their progression through the G_1, S, and G_2 phases of the cell cycle. These delays are mediated by checkpoint control proteins and are thought to ensure an appropriate period for cells to repair DNA damage before continuing on to DNA replication or mitosis. For example, the molecular basis for the G_1 cell-cycle arrest following ionizing radiation damage in mammalian cells appears to involve the TP53 and retinoblastoma (pRb) tumour suppressor proteins. In normal cells, activation of the TP53 protein following DNA strand breaks leads to the activation of a number of downstream genes, including the *p21*$^{WAF1/CIP1}$ gene, and subsequent inhibition of the phosphorylation activity of G_1 cyclin-dependent kinases. As a result, the pRb protein remains hypophosphorylated and the cells arrest in the G_1 phase. Other molecular mechanisms are responsible for the S- and G_2-phase checkpoint control; the reader is directed to recent reviews of this topic (6).

The TP53-mediated G_1 checkpoint is reduced in cells derived from patients with ataxia telangiectasia who harbour a mutation in the *ATM* gene. Cells that are cell-cycle checkpoint-defective due to mutations in checkpoint-related genes are usually more sensitive to DNA-damaging agents, and they may lack certain aspects of the cellular sensing and signalling cascade which occurs after DNA damage (6, 7). As we will see in subsequent sections, the early detection and sensing of DNA breaks by ATM is an important signal for the signal transduction following DNA damage.

1.3 Evolutionary conservation and biochemical function within the PI3K-related kinase family

In this chapter, we will briefly describe members of the PI3K-related kinase family involved in DNA damage responses, focusing mainly on the role of the DNA-PK and ATM proteins in linking DNA repair and cell-cycle checkpoint control. The kinases discussed in this chapter are mostly serine/threonine protein kinases without lipid kinase activity. They are characterized by a well-conserved C-terminal kinase domain (see Fig. 1) and have extensive homology throughout the evolutionary tree. However,

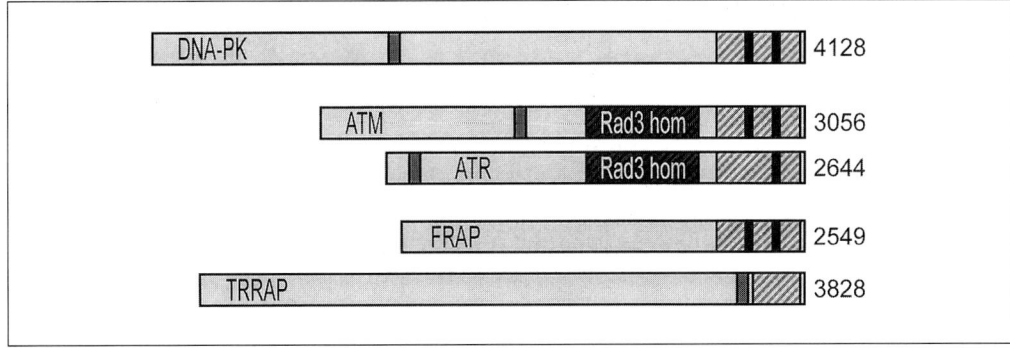

Fig. 1 Schematic alignment of PI3K-like genes in mammals. DNA-PK, ATM, ATR, FRAP, and TRAPP are aligned at the C-terminal kinase domain. Two highly conserved motifs within the kinase domain are represented by black bars. Note that TRAPP lacks both these elements, and is probably devoid of protein kinase activity. Potential leucine zipper motifs are represented by dark grey bars, and the Rad3 homology domain shared between ATM and ATR is also highlighted. The number of amino acids in each human polypeptide is given to the right.

there are some exceptions, e.g. DNA-PK$_{cs}$ does not have a clear homologue in *Saccharomyces cerevisiae* or *Caenorhabditis elegans*.

Other PI3K family members which will be briefly described include: ATR (ATM and Rad3-related protein, see Section 4.1); mTOR (and its *Saccharomyces cerevisiae* homologues, Tor1p and Tor2p; see Section 4.2), which controls mRNA translation in response to growth and nutrient stimuli; and TRRAP (Tra1p in *S. cerevisiae*, see Section 4.2), which associates with histone acetyltransferase complexes, affecting transcriptional control. A summary of the PI3K-related kinase superfamily, including the ATM and DNA-PK proteins, is provided in Table 1.

Table 1 Protein kinases in the PI3-kinase family

Member	Synonyms	*Homologues	Function	Chromosomal localization	Molecular weight (kDa)
DNA-PK$_{cs}$	SCID XRCC7 Prkdc	None	Protein kinase-catalytic subunit	8q11	470
ATM	–	TEL1 (Sc)	Protein kinase	11q22–23	350
ATR	FRP MCCS1 MRK1	Rad3(Sp) MEC1/ESR1 (Sc) Mei 41 (Dm)	Protein kinase	3q23–24	302
TRRAP	PAF400	TRA1p (Sc)	Cofactor	7q21–22	434
mTOR	RAPT mTOR RAFT SEP	TOR1 (Sc) TOR2 (Sc)	Kinase	1p36	290

*Sc, *Saccharomyces cerevisiae*; Sp, *Schizosaccharomyces pombe*; Dm, *Drosophila melanogaster*.

2. DNA-dependent protein kinase

The DNA-dependent protein kinase (DNA-PK) is the largest of the class of PI3K-related protein kinases. It consists of a catalytic subunit (DNA-PK$_{cs}$), and the Ku heterodimer (8). Originally, DNA-PK$_{cs}$ was called p350, on the basis of its molecular weight as estimated from SDS-PAGE analysis. However, subsequent sequence analysis of the cDNA encoding this protein revealed that this was an underestimate and that the calculated molecular weight is approximately 470 kDa. The catalytic subunit shows homology to the PI3K superfamily at its C-terminus, which contains the actual protein kinase domain. Outside its kinase domain, DNA-PK$_{cs}$ shows little homology to other proteins. The only obvious sequence feature in the rest of the protein is a leucine zipper between amino acids 1500 and 1540, suggesting that this region is involved in protein–protein interactions.

2.1 Biochemistry of DNA binding and phosphorylation by DNA-PK

DNA-PK has an important role in the end-joining pathway of DNA double-strand break repair. Mutations in either DNA-PK$_{cs}$ or in one of the Ku genes results in ionizing radiation sensitivity, probably caused by a reduced ability to repair the radiation-induced DNA double-strand breaks (4). Furthermore, mice in which one of these genes is inactivated show a severe combined immunodeficiency (SCID) disorder, caused by an inability to complete the V(D)J recombination reaction required for immunoglobulin and T-cell receptor gene assembly (see Fig. 2 for discussion) (9–12). More specifically, DNA-PK is involved in the processing and coupling of the DNA ends formed by the RAG proteins at antigen-receptor loci in pre-B or pre-T cells.

The Ku70 and Ku80 genes have been conserved from yeast to humans. The Ku70 and Ku80 gene products form the Ku heterodimer, which has a high affinity for DNA ends and for single-strand to double-strand transitions in the DNA (13). A clear DNA-PK$_{cs}$ homologue has not been found in yeast, suggesting the specific DNA-dependent kinase functions developed later in evolution. DNA-PK$_{cs}$ does not form a complex with Ku unless the Ku protein is first bound to a DNA end. When the trimeric DNA-PK$_{cs}$/Ku complex is formed on DNA, the catalytic subunit acquires the ability to phosphorylate a variety of protein substrates. In the absence of Ku, DNA-PK$_{cs}$ has an approximately 100-fold lower affinity for DNA ends, resulting in a low DNA-dependent protein kinase activity *in vitro*, and probably a complete lack of activity *in vivo* (14).

Obviously, the high affinity of Ku for DNA ends indicates that this protein may have a very early role in the repair of dsbs. One model suggests that after binding, the Ku heterodimer can protect the DNA ends against exonucleases, which might otherwise cause deletions at the site of the break (15). It is not completely clear what the exact sequence of events is which follows Ku protein binding to a DNA end, but it appears logical that the complex first attracts the catalytic subunit, which can then phosphorylate target proteins.

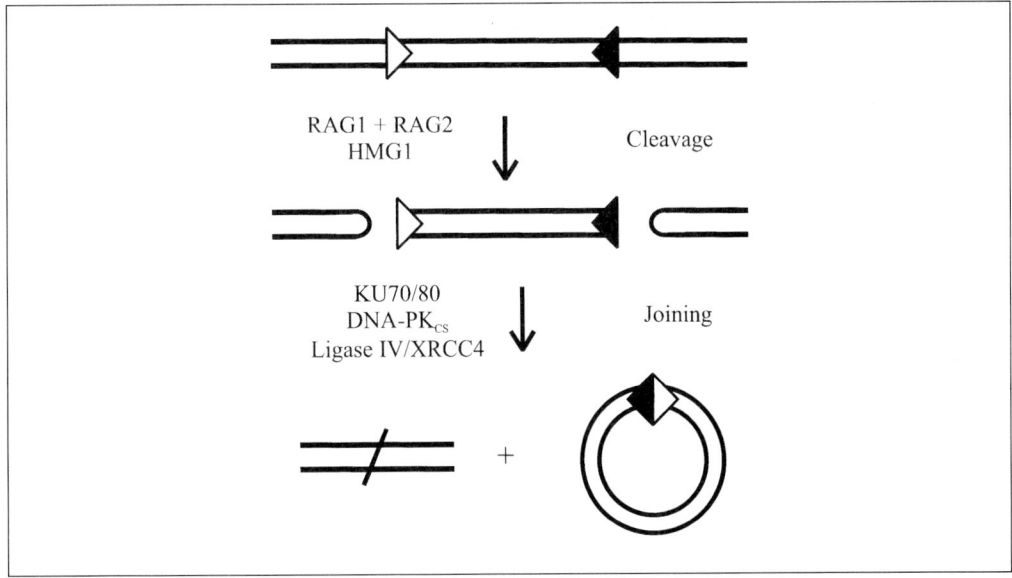

Fig. 2 Outline of the V(D)J recombination reaction. Recombination takes place between two recombination signal sequences (RSSs), depicted as triangles. RSSs contain a conserved heptamer and nonamer sequences, which are separated by a spacer of 12 or 23 base pairs, depicted as open and closed triangles, respectively. Recombination always takes place between one RSS with a 12-bp spacer, and one with a 23-bp spacer. The RAG1 and RAG2 proteins are required and sufficient to cleave an RSS, forming a blunt, 5′-phosphorylated signal end, and a coding end in which the top strand is covalently coupled to the bottom strand in a hairpin structure (110). This reaction can be stimulated *in vitro* by DNA-bending proteins, such as HMG1. After this cleavage reaction, the breaks are most probably repaired by the non-homologous end-joining machinery, which also repair dsbs introduced by, for example, ionizing radiation (81–83). Cells or animals in which the Ku70, Ku80, DNA-PK$_{cs}$, ligase IV, or XRCC4 genes have been inactivated do not support efficient V(D)J recombination (4, 111). The Ku heterodimer has high affinity for DNA ends, suggesting an early role in the joining reaction, whereas the ligase IV/XRCC4 complex is involved in the final ligation reaction. At this point it is unclear how many other proteins are involved in this step, but exonuclease and DNA polymerase functions are expected to be required for completion of the reaction. In addition, terminal deoxynucleotidyl transferase can add untemplated nucleotides on to coding ends before ligation, although this step is not obligatory for a complete V(D)J recombination reaction.

Many phosphorylation targets have been identified, both *in vitro* and *in vivo*, although it is unclear which phosphorylation events are functionally significant (8). Two of the most interesting proteins in the long list of *in vitro* phosphorylation targets are the single-stranded DNA binding protein RPA and the tumour suppressor protein TP53. DNA-PK can also phosphorylate its own subunits (autophosphorylation) (16). Interestingly, autophosphorylation of the catalytic subunit causes inhibition of kinase activity due to a decrease in its affinity for Ku protein. This may have important consequences for repair (e.g. the catalytic subunit may dissociate from the repair complex to make room for other repair factors) or for possible signalling functions (the signalling could be attenuated after the appropriate repair complex has assembled on the DNA ends).

2.2 Protein–protein interactions

The sheer size of DNA-PK$_{cs}$ and other PI3K family members is another fascinating feature. They are all in the order of several thousand amino acids in length: DNA-PK$_{cs}$ itself contains over 4000 amino acids in a single polypeptide chain, while the kinase domain is confined to approximately 400 amino acids at its C-terminus. Why are these proteins so large? An attractive hypothesis is that the DNA-PK and ATM proteins have multiple domains to interact with various partner proteins. These interactions are probably required to communicate with many different cellular factors in order to integrate upstream signals following DNA damage, and to subsequently transduce them to several downstream pathways that affect cell-cycle regulation, DNA repair, or apoptosis.

We propose the following categories of possible DNA-PK$_{cs}$ interacting cellular proteins, in which most of the known interactors can be placed: (1) factors directly involved in repair; (2) positive and negative regulators of kinase activity; (3) phosphorylation substrates; and (4) factors that may remodel protein–protein or protein–DNA complexes. Some candidate interactors are described for each category below.

1. Its large size makes DNA-PK$_{cs}$ an excellent candidate to serve as a 'landing plat-form' or 'scaffold' for DNA repair proteins. For example, it could serve as a central organizer of dsb repair complexes. In addition to its stimulatory interaction with the Ku heterodimer, DNA-PK$_{cs}$ may also interact with repair factors that act later in the end-joining reaction. The XRCC4 protein may be one such factor: it is a good substrate for DNA-PK phosphorylation, and may even have a direct physical interaction with DNA-PK (17, 18). XRCC4 forms a stable complex with DNA ligase IV, and could potentially link the initial lesion detected by DNA-PK to the actual ligation reaction carried out by ligase IV (4).

2. If we assume that PI3Ks are central regulators of cell-cycle responses, cellular metabolism, and/or DNA repair reactions, it is to be expected that they integrate signals from several upstream signalling factors. As described above, DNA-PK$_{cs}$ receives signals from the Ku protein bound to DNA, activating its protein kinase activity. On top of that, one can speculate that it may also respond to protein kinases or phosphatases which may modulate its association with Ku/DNA complexes, thereby influencing its activity. The Lyn-tyrosine kinase, which was recently described as a DNA-PK$_{cs}$ interactor, would fit such a negative regulatory function, although it is not completely clear what the functional significance of this par-ticular regulation would be (19).

3. A long list of DNA-PK phosphorylation targets has been described (8). Most of these substrates were identified in kinase reactions *in vitro*, and it remains to be deter-mined whether they are also phosphorylated by DNA-PK *in vivo*. Nevertheless, it is possible that at least some of the cellular phenotypes of mammalian DNA-PK mutants can be attributed to a failure to interact with phosphorylation substrates that do not directly participate in the end-joining process (but, for example, in cell-cycle regulation). The radiation-induced interaction with and phosphorylation of

the c-Abl tyrosine kinase could fall into this category (20) Especially interesting is the ability of DNA-PK to phosphorylate serine 15 of human TP53 and the homologous serine 18 of murine TP53, suggesting a link between these two proteins in a DNA-damage cascade. Although recent studies have shown that serine 18 was also phosphorylated in irradiated DNA-PK$_{cs}$−/− mouse cells (21) and that the absence of DNA-PK activity does not result in failure of the G_1 cell-cycle checkpoint (22, 23), it is possible that DNA-PK is redundant with other PI3K-related kinases in this respect (e.g. with the ATM protein, see below).

4. The problem of all biochemical interactions is that they have to be reversible in order to react to changing conditions. The more stable a complex is, the more difficult it is to disassemble such a biochemical complex. The DNA-PK trimer binds with high affinity to DNA ends (through its Ku component), but probably this complex must be disassembled, or at least remodelled, before later steps in the repair reaction can occur. Specific chaperone proteins or proteases could be involved in remodelling complexes containing DNA-PK bound to DNA. It is not yet clear whether such factors exist, but the observation that DNA-PK$_{cs}$ can be degraded by ICE (interleukin-1 converting enzyme)-like protease upon induction of apoptosis may point to such an event (24).

2.3   Other functions of DNA-PK

The discovery that the Ku-dependent pathway of non-homologous end-joining exists in yeast has been of great help in elucidating other possible functions of these proteins. Yeast mutants defective in either the Ku70 or Ku80 homologue not only show defective end-joining, but also have defects in their ability to maintain telomere length (2, 3). At the telomeres, the Ku proteins may be associated with proteins that are important for maintaining a closed chromatin structure, such as the SIR2, SIR3, and SIR4 proteins. Like SIR mutants, yeast Ku mutants have defects in transcriptional silencing of telomeric loci, suggesting that Ku proteins have a structural role in yeast chromatin.

In addition to a defect in telomeric length maintenance, there is also an altered organization of telomere position within the nucleus of Ku yeast mutants (25). Wild-type cells have telomeres located at the nuclear periphery, whereas Ku mutants have a more random distribution of telomeres throughout the nucleus. The majority of Ku proteins are concentrated directly on the telomeres, but this nuclear distribution changes dramatically upon the induction of dsbs in the DNA. For example, after inducing DNA breaks with DNA restriction enzymes, the telomeric Ku protein relocalizes (together with the SIR proteins) to those DNA breaks (26, 27). This relocalization depends on the action of the cell-cycle checkpoint proteins Rad9 and Mec1 (the yeast homologue of the ATR protein described in Section 3.3).

It is not yet clear whether similar functions can be attributed to the Ku heterodimer in mammalian cells. However, some recent observations might be explained by a function for Ku/DNA-PK in telomeric functioning. For example, mouse embryonic fibroblasts defective in Ku70 or Ku80 suffer premature senescence, which could indicate defective telomere metabolism (9, 11). Recently, Ku proteins have been found at telomeric sequences and in the nuclear periphery in mammalian cells (28, 29). Interestingly, cells derived from the SCID mouse have been reported to have elongated telomeres, suggesting that Ku and DNA-PK$_{cs}$ may have opposite effects on telomeric length. More careful studies on the localization of these proteins and the effects of various mutations on the structure of chromosome ends in mammalian cells will probably elucidate this issue in the near future.

3. The cell-cycle checkpoint gene *ATM*

The mammalian PI3K family members ATM and ATR are required for arrest of the cell cycle when DNA damage or unreplicated DNA is present. In addition to the PI3K domain, they share another region of homology known as the Rad3 homologous region (30, 31).

The *ATM* gene, mutated in the human genetic disorder ataxia telangiectasia (AT), was first identified in 1995 (32–34). Mutations in this gene lead to a number of clinical symptoms in the homozygous state, including increased cellular radiosensitivity (but not UV sensitivity), cerebellar neurodegeneration, and an increased incidence of some forms of cancer (see Section 5.2). The underlying defect in AT cells involves altered G_1-, S-, and G_2-phase checkpoint functions, probably (in part) due to a reduced or delayed induction of TP53 protein levels following irradiation. This may also be the cause of the increased number of spontaneous chromosomal aberrations observed in these cells.

3.1 Basic biochemistry and phosphorylation substrates of ATM

The approx. 350-kDa ATM phosphoprotein is ubiquitously expressed in various cell types and tissues. Consistent with its role in DNA damage signal transduction, most of the ATM protein is located in the nucleus—although the protein lacks a classic nuclear localization signal (35)—with a minor fraction being located in cytoplasmic vesicles and/or microsomal fractions (36).

Purified ATM protein has been shown to possess protein kinase activity, which is stimulated by DNA, although this stimulation is much less robust than for DNA-PK (37). The majority of mutations in the ATM protein characterized thus far result in a C-terminal truncation and loss of the serine/threonine kinase function (35). Small, in-frame deletions, or missense mutations in the conserved PI3K region, have also been reported, suggesting that, in particular, the loss of this kinase function is essential for the development of the AT phenotype.

The importance of the C-terminal kinase domain was confirmed by a series of studies demonstrating the effect of transfecting wild-type or mutated forms of the *ATM* gene into AT and normal cells (38–40). Several cellular defects of AT cells, such as radiosensitivity, radioresistant DNA synthesis, and increased numbers of radiation-induced chromosomal aberrations, could be corrected by transfection of a wild-type sequence, or by a truncated ATM protein containing the active C-terminal kinase domain. However, a kinase-dead mutant form of the *ATM* gene did not correct the AT phenotype. Furthermore, overexpression of an ATM protein truncated in the C-terminus, but still containing the leucine-zipper motif (see Fig. 1) caused an AT-like cellular phenotype with increased radiosensitivity, increased chromosomal aberrations following irradiation, and defects in the G_1 cell-cycle arrest following DNA damage.

Two motifs have been implicated in ATM protein–protein interactions: a proline-rich region and a leucine-zipper motif. Similar to DNA-PK, the serine/threonine kinase activity of ATM resides in the carboxy-terminal portion of the protein. ATM has been observed to be involved in both the autophosphorylation and phoshorylation of other proteins, including c-Abl, TP53, and BRCA1, following irradiation of cells (41). ATM can interact with, and phosphorylate, BRCA1 in a region that contains clusters of serine/glutamine residues. Such a phosphorylation of BRCA1 by ATM may be critical for proper responses to DNA double-strand breaks and may provide a molecular explanation for the role of ATM in breast cancer.

3.2 ATM in DNA repair and cell-cycle checkpoint control

One of the first observations pointing to a defect in cell-cycle control was the observation that AT cells, unlike normal cells, continue DNA synthesis following exposure to ionizing radiation. This phenomenon of 'radioresistant DNA synthesis' has been interpreted as a failure to arrest the cell cycle in response to DNA damage. The failure of AT cells to arrest in the G_1, S, and G_2 phases of the cell cycle following irradiation may explain the radiosensitivity and genomic instability of these cells.

Subsequent experiments revealed that the induction of TP53 by ionizing radiation is reduced or delayed in AT cells, consistent with a loss of tight G_1-checkpoint control (42). In response to γ-irradiation, TP53 is phosphorylated on serine 15 by the ATM protein resulting in activation of TP53 (43–45).

In addition to a checkpoint defect in G_1, a G_2-checkpoint defect has also been reported for AT cells. Two possible explanations for the mechanism underlying this defect have been proposed. First, dephosphorylation of serine 376 of TP53 in response to ionizing radiation requires the presence of an active ATM protein. This dephosphorylated form can associate with a 14–3–3 protein, which leads to cell-cycle arrest in G_2 (46). Another explanation may be found in the observation that the G_2-related checkpoint kinases Chk1 and Chk2 both have been found to have decreased kinase activity in AT cells. In these studies, transfection of Chk1 complemented both the G_2-checkpoint defect and the radiosensitivity of AT cells, but had little effect on the G_1- or S-phase checkpoints (47, 48)

ATM protein levels are not increased in response to DNA damage by ionizing radiation (49, 50), but the activity of the ATM kinase is increased following exposure to ionizing radiation (43, 44, 51). Substrates of the ATM protein include the radiation-induced proteins, c-Abl and I-κBα (52–54). The ATM protein binds to the SH3 domain of c-Abl through a proline-rich region and phosphorylates it in response to ionizing radiation. Other experiments have observed that a radiation-induced complex of the Rad51 and Rad52 proteins (involved in the homologous recombination DNA-dsb repair pathway) requires both ATM and c-Abl (47).

4. Other PI3K family members

4.1 ATR

The mammalian *ATR* gene was identified through its sequence similarity with the *Schizosaccharomyces pombe Rad3* and the *S. cerevisiae Mec1* genes (55, 56). ATR is constitutively expressed and the highest levels of expression are found in testis (57). ATR, like ATM and DNA-PK, is a serine/threonine kinase (57) and one of its substrates is also TP53 (also phosphorylation *in vitro* on serine 15 (58)). Human ATR can complement the UV sensitivity of an *S. cerevisiae* Mec1 mutant, indicating that ATR is a functional homologue of this protein kinase in yeast (56). *S. pombe* Rad3 and *S. cerevisiae* Mec1 play a central role in signalling in response to many types of DNA damage and unreplicated DNA (31, 59, 60), suggesting that the *ATR* gene may serve a similar function in mammalian cells. Overexpression of a kinase-inactive form of ATR causes sensitivity to UV and ionizing radiation and further defects in cell-cycle control (61, 62).

4.2 TRRAP and mTOR

TRRAP (transformation/transcription-domain-associated protein) is a 434-kDa protein which is highly conserved in evolution (homologues are present in, for example, *S. cerevisiae, S. pombe, Caenorhabditis elegans, and A. rabidopis*). This protein has been linked to cellular transformation mediated by the oncogene c-myc and E2F transcription factors. TRRAP protein binding to C-Myc has been directly correlated with C-Myc oncogenic transforming capability (63). In addition, TRRAP is involved in general transcriptional regulation as a component of the large, multi-protein transcriptional complex, SAGA (i.e. complex of Spt, Ada and Gcn5 proteins), which includes histone acetyltransferases that affect chromatin structure and the regulation of gene expression (64). As such, it is now thought that TRRAP may provide a link between C-Myc and a histone acetylase-containing transcriptional complex.

The C-terminal portion of TRRAP exhibits significant homology to the PI3K domain of the ATM protein family (63). Curiously, the critical amino acids that map to the catalytic site of the PI3K family proteins are not found in any of the TRRAP homologues (although the flanking residues are conserved, implying that the kinase domain of TRRAP is not catalytically active).

mTOR (also known as FRAP: 'FKBP12 and rapamycin-binding protein kinase') is the 290 kDa mammalian homologue of the yeast TOR proteins, with which it shares about 50% identity. The yeast *TOR1* and *TOR2* genes were identified by genetic studies exploring how rapamycin exerts its antifungal effects. Mutations in the *TOR1* and *TOR2* genes confer resistance to rapamycin (65–68). Rapamycin inhibits TOR function by binding with high affinity to FKBP, after which this complex binds to TOR1 and TOR2. Both these proteins function to regulate protein synthesis and turnover in yeast, but TOR2 has an additional function in regulating cytoskeletal organization.

In mammals, mTOR is involved in the integration of mitogenic and nutritional signalling in the cell through translational regulation. mTOR signals to two translational control factors: 4E-BP1, which is a repressor of the protein synthesis initiation factor eIF-4E; and the 40S ribosomal protein p70^{s6k}, which functions to increase translation mRNAs that encode ribosomal proteins. Yeast TOR genes may also function to regulate the transcription of rRNA and tRNA genes, although further work is required to understand whether the mammalian mTOR protein also functions in a similar manner (69).

5. DNA-PK- and ATM-deficient mice as models for human disease

In the previous sections the biochemical and cell biological aspects of defects in the DNA-PK and ATM proteins have been described. In this section we will concentrate on the implications of mutations in components of DNA-PK or ATM at the level of a complete animal or in human patients. The characteristics of animals deficient in DNA-PK or ATM are summarized in Table 2.

5.1 Mutations in DNA-PK

At present, unlike ataxia telangiectasia, there are no known human syndromes resulting from defects in DNA-PK protein function. In human tissues, DNA-PK$_{cs}$ and

Table 2 Comparison of phenotypes between murine knockout models for AT and SCID syndromes

Phenotype	ATM	Scid/DNA-PK$_{cs}$	Ku70/80
Radiosensitivity	Increased	Increased	Increased
Heterozygous phenotype	Possible	No	No
DNA-dsb repair	Defective?	Defective	Defective
Spontaneous apoptosis	Increased	Increased	Unknown
Increased radiation-induced apoptosis	Yes	?	No
V(D)J recombination	Normal	Few T or B cells Reduced signal joints Very few coding joints	Few B-cells Decreased T cells Reduced coding and signal joints
Class-switch recombination	Normal	No	No
Oncological	Thymic lymphomas	Thymic lymphomas	T-cell lymphoma (Ku70)
Animal growth	Retarded	Normal	Retarded
Germline	Aspermatogonia	Normal	Normal
Cell cycle	Defective G$_1$ and G$_2$ checkpoints Radioresistant DNA synthesis	Normal	Normal

Ku80 protein expression are usually parallel, but can vary widely among different tissues. Interestingly, the mRNA's of the components of DNA-PK do not vary among tissues, suggesting that the variability observed among protein levels in a variety of tissues is due to a post-transcriptional mechanism (70).

Much information regarding the cellular activity of the DNA-PK complex stems from research utilizing the SCID mouse model. The SCID mouse is immunodeficient due to a complete lack of mature T and B cells. Lymphocytes derived from SCID mice show defective V(D)J recombination, caused by an inability to process the broken DNA molecules produced by the RAG1 and RAG2 proteins at the immunoglobulin or T-cell receptor loci. More specifically, SCID cells were observed to accumulate coding ends with a DNA hairpin structure (71) (see Fig. 2). Coupling of the signal ends was relatively normal, although the efficiency and accuracy of this reaction is also somewhat reduced. The DNA-PK$_{cs}$ protein in the SCID mouse is 83 amino acids shorter than the wild-type protein, resulting in an unstable protein without DNA-dependent protein kinase activity (72). Recently, mice have been generated in which the DNA-PK$_{cs}$ gene has been completely disrupted (73–75). Very similar results have been documented for DNA-PK$_{cs}$−/− mice, suggesting that the original SCID mouse harbours a complete loss of function mutation.

In addition to disruption of the catalytic subunit, mice with inactivated *Ku70* or *Ku80* genes have been generated (9, 11, 12) (see Table 2). Like the SCID mouse, mice lacking either a functional *Ku70* or *Ku80* gene show severely impaired V(D)J recombination. Both mouse models are characterized by an almost complete absence of mature B cells. However, there are some differences in T-cell development: the lack of Ku80 protein is incompatible with mature T-cell development, whereas Ku70-deficient animals show development of a low number of T cells with rearranged *TCRα* and *TCRβ* genes. This finding suggests that V(D)J recombination at *TCR* loci is not completely dependent on Ku70. In addition to this immunological phenotype, the Ku70- and Ku80-deficient mice have a lower birth weight than their wild-type littermates and show signs of early senescence, suggesting a role for Ku which is independent of the catalytic subunit (76). Furthermore, deletion of *Ku70*, but not *Ku80*, leads to the development of T-cell lymphomas, suggesting a role as a tumour suppressor gene, as well (77).

The SCID, Ku70, and Ku80 knockout mice are not only defective in V(D)J recombination, but also show a defect in another DNA rearrangement process at immunoglobulin loci, called class-switching (78–80). After assembly of a functional V(D)J exon at the immunoglobulin heavy chain locus, this site-specific recombination event leads to the production of other subclasses of immunoglobulins. Mice deficient in one of the DNA-PK components, that harbour already rearranged V(D)J exons at their immunoglobulin heavy and light chain loci, are unable to complete such a class switch. Apparently, this recombination reaction requires the action of the DNA-PK complex, suggesting that also in this case the end-joining machinery is used for the coupling of DNA ends.

In addition to these immunological phenotypes, DNA-PK-deficient mice are also hypersensitive to ionizing radiation, indicating that a similar end-joining mechanism, utilizing the same enzymes, is required for the repair of the random dsbs introduced into DNA by ionizing radiation, as well as site-specific dsbs at antigen-receptor loci (81–83). Indeed, fibroblasts derived from DNA-PK$_{cs}$-, Ku70-, or Ku80-deficient animals have defects in the kinetics and overall amount of dsb-rejoining following ionizing radiation.

5.2 Mutations in ATM

Ataxia telangiectasia is a human autosomal-recessive disorder with a variety of clinical symptoms. These include a progressive cerebellar degeneration leading to an abnormality of limb movement (ataxia), eye movement, and speech. This neuro-degeneration has been linked to a loss of Purkinje cells within the cerebellum. Furthermore, there is hypogammaglobulinaemia resulting in immunodeficiency, abnormal vascularity within the conjunctiva of the eye (telangiectasia), features of premature ageing, and increased clinical and cellular sensitivity to ionizing radiation (35). The cellular and molecular defects pertaining to cell-cycle control have been previously described in Section 4.

AT patients have a 100–250-fold excess risk of cancer due to an increased genomic instability. There is both an increased spontaneous and radiation-induced level of chromosomal aberrations in cells derived from affected individuals. The incidence of leukaemias and lymphomas is markedly elevated in AT patients, with as many as 30–40% developing tumours. These include T-cell tumours (i.e. T-ALL, acute lymphocytic leukaemia; T-PLL, prolymphocytic leukaemia), and, less frequently, B-cell tumours. Recently, somatic mutation of the *ATM* gene due to point mutations or complex rearrangements has been implicated in the development of sporadic T-PLL, suggesting that *ATM* acts as a tumour suppressor gene (84).

The development of murine models in which the *ATM* gene is disrupted, has been of great help in the investigation of this gene's function (85, 86). In general, ATM-deficient mice show severely impaired growth when compared to normal littermates, and fibroblasts derived from these animals show premature senescence. ATM–/– animals usually die within 4 months due to the presence of thymic lymphomas. Interestingly, these lymphomas are dependent on the action of the RAG proteins, suggesting that ATM may specifically function as a tumour suppressor gene for tumorigenesis initiated by V(D)J recombination (87).

Gonadal atrophy is also present in ATM–/– animals, suggesting a functional ATM defect in meiosis. For example, the spermatocytes of ATM-deficient mice are arrested in early meiotic prophase and their chromosomes are fragmented (88, 89). The neuronal phenotype observed in AT patients has not been reproduced in ATM-deficient mice, probably due to the very early development of lymphomas. The longer life span of ATM–/–RAG1–/– mice may help investigations into the neuronal degenerative aspects of the AT phenotype.

ATM and ATR interact directly with meiotic chromosomes, thus suggesting a role for these proteins in meiotic recombination (57, 88, 89). Such a function could explain the developmental arrest and chromosomal breaks observed in spermatocytes in the early phases of the meiotic division in ATM-deficient mice (see above).

Similar to the observations for its yeast homologue Tel1, ATM is involved in maintaining the length of chromosomal telomeres (90). Loss of ATM, as seen in AT patients or ATM-deficient mice results in an accelerated telomere shortening, premature senescence, and end-to-end fusion of chromosomes (91–96).

5.3 ATM and DNA-PK protein function in cancer prognosis and therapy

Currently there is interest in clinical oncology as to whether the activity and/or expression of the DNA-PK and ATM protein kinase activity can provide prognostic information for the response of normal and tumour tissues following cancer treatment, or provide avenues for novel treatment approaches. Penetrance studies initially suggested that intermediate levels of radiosensitivity and risk of cancer exist for AT heterozygotes, although this may depend on the exact nature of the ATM mutations (41, 97). However, in prospective clinical studies, AT heterozygotes were not proven to be radiosensitive (98–103). Similarly, there is no correlation between overall ATM protein expression and cellular radiosensitivity and/or dsb repair in tumour cells derived from patient biopsies. Other studies have attempted to correlate the expression and activity of DNA-PK with the response of tumour and normal cells to ionizing radiation (104–107). Again, no simple relationship exists between cellular radiosensitivity, dsb repair, and DNA-PK expression or activity.

Novel cancer-treatment strategies have been hypothesized, based on the potential alteration of DNA-PK or ATM protein function in tumour cells as a means of sensitizing tumours to DNA-damaging agents (108, 109). Such a therapeutic ratio will have to be reached by sensitizing tumour cells more than the surrounding normal cells. At the moment there is no clear indication that such differences exist, but further study will be needed to investigate these issues more thoroughly.

6. Conclusions and outstanding questions

In conclusion, the relative sensitivity to DNA-damaging agents does not solely depend on the activity of PI3K family members, but is probably the result of many factors that can influence DNA repair, susceptibility to apoptosis, and cell-cycle checkpoint control.

However, the study of the PI3K family has led to an incredible wealth of information regarding the biochemistry, cell biology, and potential clinical ramifications of the expression and activity of these kinases in normal and tumour tissues. Future investigations may shed further light on their roles in dsb repair and carcinogenesis, and their impact on the chemosensitivity and radiosensitivity of tissues. We consider the following questions to be of prime importance in lifting our current knowledge to the next level of understanding:

1. How is the interplay between DNA damage recognition, repair, and cell-cycle checkpoints regulated? How does the cell decide when to arrest the cell cycle in order to start repair, or to die by apoptosis or mitotic catastrophe?

2. Are ATM and Ku70 really safeguards against translocations (as suggested by some recent papers), and how does that work?

3. What is the exact location of these proteins in the cell, and how is that regulated in response to DNA damage? Are proteins relocalized in response to ionizing radiation, and how would this influence nuclear architecture? Is there, for example, a pool of telomere-bound proteins that can be recruited, as suggested by experiments in yeast? For this purpose proteins tagged with the Green Fluorescent Protein® may be very useful tools.

4. What are the true targets of DNA-PK and ATM? The development of more specific mouse models, such as kinase-dead versions of DNA-PK and ATM, will help in answering this question.

No doubt further information gained from these types of studies will further our knowledge of the relative roles of the PI3K family members in cell-cycle checkpoint control and DNA repair.

References

1. Friedberg, E., Walker, G., and Siede, W. (1995). *DNA repair and mutagenesis*. American Society for Microbiology Press, Washington D.C.
2. Tsukamoto, Y., Kato, J., and Ikeda, H. (1997). Silencing factors participate in DNA repair and recombination in *Saccharomyces cerevisiae*. *Nature*, **388**, 900.
3. Boulton, S. and Jackson, S. (1998). Components of the Ku-dependent non-homologous end-joining pathway are involved in telomeric length maintenance and telomeric silencing. *EMBO J.*, **17**, 1819.
4. Kanaar, R., Hoeijmakers, J., and Van Gent, D. (1998). Molecular mechanisms of DNA double strand break repair. *Trends Cell Biol.*, **8**, 483.
5. Bristow, R. and Hill, R. (1998). Molecular and cellular basis of radiotherapy. In *The basic science of oncology* (3rd edn) (ed. I. Tannock and R. Hill), p. 295. McGraw-Hill, New York.
6. Weinert, T. and Lydall, D. (1993). Cell cycle checkpoints, genetic instability and cancer. *Semin. Cancer Biol.*, **4**, 129.
7. Elledge, S. J. (1996). Cell cycle checkpoints: preventing an identity crisis. *Science*, **274**, 1664.
8. Anderson, C. W. and Carter, T. H. (1996). The DNA-activated protein kinase—DNA-PK. *Curr. Top. Microbiol. Immunol.*, **217**, 91.
9. Gu, Y., Seidl, K., Rathbun, G., Zhu, C., Manis, J., van der Stoep, N., Davidson, L., Cheng, H., Sekiguchi, J., Frank, K., Stanhope-Baker, P., Schlissel, M., Roth, D., and Alt, F. (1997). Growth retardation and leaky SCID phenotype of Ku70-deficient mice. *Immunity*, **7**, 653.
10. Ouyang, H., Nussenzweig, A., Kurimasa, A., Soares, V. C., Li, X., Cordon-Cardo, C., Li, W., Cheong, N., Nussenzweig, M., Iliakis, G., Chen, D. J., and Li, G. C. (1997). Ku70 is required for DNA repair but not for T cell antigen receptor gene recombination *in vivo*. *J. Exp. Med.*, **186**, 921.

11. Nussenzweig, A., Chen, C., da Costa Soares, V., Sanchez, M., Sokol, K., Nussenzweig, M. C., and Li, G. C. (1996). Requirement for Ku80 in growth and immunoglobulin V(D)J recombination. *Nature*, **382**, 551.

12. Zhu, C., Bogue, M. A., Lim, D. S., Hasty, P., and Roth, D. B. (1996). Ku86-deficient mice exhibit severe combined immunodeficiency and defective processing of V(D)J recombination intermediates. *Cell*, **86**, 379.

13. Dynan, W. S. and Yoo, S. (1998). Interaction of Ku protein and DNA-dependent protein kinase catalytic subunit with nucleic acids. *Nucleic Acids Res.*, **26**, 1551.

14. West, R., Yaneva, M., and Lieber, M. (1998). Productive and nonproductive complexes of Ku and DNA-dependent protein kinase at DNA termini. *Mol. Cell Biol.*, **18**, 5908.

15. Liang, F. and Jasin, M. (1996). Ku80-deficient cells exhibit excess degradation of extra-chromosomal DNA. *J. Biol. Chem.*, **271**, 14405.

16. Chan, D. W. and Lees-Miller, S. P. (1996). The DNA-dependent protein kinase is inactivated by autophosphorylation of the catalytic subunit. *J. Biol. Chem.*, **271**, 8936.

17. Critchlow, S., Bowater, R., and Jackson, S. (1997). Mammalian DNA double-strand break repair protein XRCC4 interacts with DNA ligase IV. *Curr. Biol.*, **7**, 588.

18. Leber, R., Wise, T. W., Mizuta, R., and Meek, K. (1998). The XRCC4 gene product is a target for and interacts with the DNA-dependent protein kinase. *J. Biol. Chem.*, **273**, 1794.

19. Kumar, S., Pandey, P., Bharti, A., Jin, S., Weichselbaum, R., Weaver, D., Kufe, D., and Kharbanda, S. (1998). Regulation of DNA-dependent protein kinase by the Lyn tyrosine kinase. *J. Biol. Chem.*, **273**, 25654.

20. Kharbanda, S., Pandey, P., Jin, S., Inoue, S., Bharti, A., Yuan, Z., Weichselbaum, R., Weaver, D., and Kufe, D. (1997). Functional interaction between DNA-PK and c-Abl in response to DNA damage. *Nature*, **386**, 732.

21. Burma, S., Kurimasa, A., Xie, G., Taya, Y., Araki, R., Abe, M., Crissman, H., Ouyang, H., Li, G., and Chen, D. (1999). DNA-dependent protein kinase-independent activation of p53 in response to DNA damage. *J. Biol. Chem.*, **74**, 17139.

22. Rathmell, W. K., Kaufmann, W. K., Hurt, J. C., Byrd, L. L., and Chu, G. (1997). DNA-dependent protein kinase is not required for accumulation of p53 or cell cycle arrest after DNA damage. *Cancer Res.*, **57**, 68.

23. Jimenez, G., Bryntesson, F., Torres-Arzayus, M., Priestly, A., Beeche, M., Saito, S., Sakaguchi, K., Appella, E., Jeggo, P., Taccioli, G., Wahl, G., and Hubank, M. (1999). DNA-dependent protein kinase is not required for the p53-dependent response to DNA damage. *Nature*, **400**, 81.

24. Song, Q., Lees-Miller, S., Kumar, S., Zhang, Z., Chan, D., Smith, G., Jackson, S., Alnemri, E., Litwack, G., Khanna, K., and Lavin, M. (1996). DNA-dependent protein kinase catalytic subunit: a target for an ICE-like protease in apoptosis. *EMBO J.*, **15**, 3238.

25. Laroche, T., Martin, S., Gotta, M., Gorham, H., Pryde, F., Louis, E., and Gasser, S. (1998). Mutation of yeast Ku genes disrupts the subnuclear organization of telomeres. *Curr. Biol.*, **8**, 653.

26. Martin, S., Laroche, T., Suka, N., Grunstein, M., and Gasser, S. (1999). Relocalization of telomeric Ku and SIR proteins in response to DNA strand breaks in yeast. *Cell*, **97**, 621.

27. Mills, K., Sinclair, D., and Guarente, L. (1999). MEC1-dependent redistribution of the Sir3 silencing protein from telomeres to DNA double-strand breaks. *Cell*, **97**, 609.

28. Higashiura, M., Shimizu, Y., Tanimoto, M., Morita, T., and Yagura, T. (1992). Immuno-localization of Ku-proteins (p80/p70): localization of p70 to nucleoli and periphery of both interphase nuclei and metaphase chromosomes. *Exp. Cell Res.*, **201**, 444.

29. Hsu, H., Gilley, D., Blackburn, E., and Chen, D. (1999). Ku is associated with the telomere in mammals. *Proc. Natl Acad. Sci. USA*, **96**, 12454.

30. Zakian, V. A. (1995). ATM-related genes: what do they tell us about functions of the human gene? *Cell*, **82**, 685.

31. Hoekstra, M. F. (1997). Responses to DNA damage and regulation of cell cycle checkpoints by the ATM protein kinas family. *Curr. Opin. Genet. Dev.*, **7**, 170.

32. Savitsky, K., Bar-Shira, A., Gilad, S., Rotman, G., Ziv, Y., Vanagaite, L., Tagle, D. A., Smith, S., Uziel, T., Sfez, S., *et al.* (1995). A single ataxia telangiectasia gene with a product similar to PI-3 kinase. *Science*, **268**, 1749.

33. Savitsky, K., Sfez, S., Tagle, D., Zif, Y., Sariel, A., Collins, F., Shiloh, Y., and Rotman, G. (1995). The complete sequence of the coding region of the ATM gene reveals similarity to cell cycle regulators in different species. *Hum. Mol. Genet.*, **4**, 2025.

34. Rasio, D., Negrini, M., and Croce, C. M. (1995). Genomic organization of the ATM locus involved in ataxia-telangiectasia. *Cancer Res.*, **55**, 6053.

35. Shiloh, Y. (1997). Ataxia-telangiectasia and the Nijmegen breakage syndrome: related disorders but genes apart. *Annu. Rev. Genet.*, **31**, 635.

36. Lim, D., Kirsch, D., Canman, C., Ahn, J., Ziv, Y., Newman, L., Darnell, R., Shiloh, Y., and Kastan, M. (1998). ATM binds to beta-adaptin in cytoplasmic vesicles. *Proc. Natl Acad. Sci. USA*, **95**, 10146.

37. Smith, G., di Fagagna, F., Lakin, N., and Jackson, S. (1999). Cleavage and inactivation of ATM during apoptosis. *Mol. Cell. Biol.*, **19**, 6076.

38. Morgan, S. E., Lovly, C., Pandita, T. K., Shiloh, Y., and Kastan, M. B. (1997). Fragments of ATM which have dominant-negative or complementing activity. *Mol. Cell. Biol.*, **17**, 2020.

39. Ziv, Y., Bar-Shira, A., Pecker, I., Russell, P., Jorgensen, T. J., Tsarfati, I., and Shiloh, Y. (1997). Recombinant ATM protein complements the cellular A-T phenotype. *Oncogene*, **15**, 159.

40. Zhang, N., Chen, P., Khanna, K. K., Scott, S., Gatei, M., Kozlov, S., Watters, D., Spring, K., Yen, T., and Lavin, M. F. (1997). Isolation of full-length ATM cDNA and correction of the ataxia-telangiectasis cellular phenotype. *Proc. Natl Acad. Sci. USA*, **94**, 8021.

41. Lakin, N. D., Weber, P., Stankovic, T., Rottinghaus, S. T., Taylor, A. M., and Jackson, S. P. (1996). Analysis of the ATM protein in wild-type and ataxia telangiectasia cells. *Oncogene*, **13**, 2707.

42. Kastan, M. B., Zhan, Q., el-Deiry, W. S., Carrier, F., Jacks, T., Walsh, W. V., Plunkett, B. S., Vogelstein, B., and Fornace, A. J., Jr. (1992). A mammalian cell cycle checkpoint pathway utilizing p53 and GADD45 is defective in ataxia-telangiectasia. *Cell*, **71**, 587.

43. Banin, S., Moyal, L., Shieh, S., Taya, Y., Anderson, C., Chessa, L., Smorodinsky, N., Prives, C., Reiss, Y., Shiloh, Y., and Ziv, Y. (1998). Enhanced phosphorylation of p53 by ATM in response to DNA damage. *Science*, **281**, 1674.

44. Canman, C. E., Lim, D. S., Cimprich, K. A., Taya, Y., Tamai, K., Sakaguchi, K., Appella, E., Kastan, M. B., and Siliciano, J. D. (1998). Activation of the ATM kinase by ionizing radiation and phosphorylation of p53. *Science*, **281**, 1677.

45. Khanna, K. K., Keating, K. E., Kozlov, S., Scott, S., Gatei, M., Hobson, K., Taya, Y., Gabrielli, B., Chan, D., Lees-Miller, S. P., and Lavin, M. F. (1998). ATM associates with and phosphorylates p53: mapping the region of interaction. *Nature Genet.*, **20**, 398.

46. Waterman, M. J., Stavridi, E. S., Waterman, J. L., and Halazonetis, T. D. (1998). ATM-dependent activation of p53 involves dephosphorylation and association with 14–3–3 proteins. *Nature Genet.*, **19**, 175.

47. Chen, P., Gatei, M., O'Connell, M., Khanna, K., Bugg, S., Hogg, A., Scott, S., Hobson, K., and Lavin, M. (1999). Chk1 complements the G_2/M checkpoint defect and radiosensitivity of ataxia-telangiectasia cells. *Oncogene*, **18**, 249.

48. Tominaga, K., Morisaki, H., Kaneko, Y., Fujimoto, A., Tanaka, T., Ohtsubo, M., Hirai, M., Okayama, H., Ikeda, K., and Nakanishi, M. (1999). Role of human Cds1 (Chk2) kinase in DNA damage checkpoint and its regulation by p53. *J. Biol. Chem.*, **274**, 31463.

49. Brown, K. D., Ziv, Y., Sadanandan, S. N., Chessa, L., Collins, F. S., Shiloh, Y., and Tagle, D. A. (1997). The ataxia-telangiectasia gene product, a constitutively expressed nuclear protein that is not up-regulated following genome damage. *Proc. Natl Acad. Sci. USA*, **94**, 1840.

50. Watters, D., Khanna, K. K., Beamish, H., Birrell, G., Spring, K., Kedar, P., Gatei, M., Stenzel, D., Hobson, K., Kozlov, S., Zhang, N., Farrell, A., Ramsay, J., Gatti, R., and Lavin, M. (1997). Cellular localisation of the ataxia-telangiectasia (ATM) gene product and discrimination between mutated and normal forms. *Oncogene*, **14**, 1911.

51. Canman, C. and Lim, D. (1998). The role of ATM in DNA damage responses and cancer. *Oncogene*, **17**, 3301.

52. Baskaran, R., Wood, L. D., Whitaker, L. L., Canman, C. E., Morgan, S. E., Xu, Y., Barlow, C., Baltimore, D., Wynshaw-Boris, A., Kastan, M. B., and Wang, J. Y. (1997). Ataxia telangiectasia mutant protein activates c-Abl tyrosine kinase in response to ionizing radiation. *Nature*, **387**, 516.

53. Shafman, T., Khanna, K. K., Kedar, P., Spring, K., Kozlov, S., Yen, T., Hobson, K., Gatei, M., Zhang, N., Watters, D., Egerton, M., Shiloh, Y., Kharbanda, S., Kufe, D., and Lavin, M. F. (1997). Interaction between ATM protein and c-Abl in response to DNA damage. *Nature*, **387**, 520.

54. Jung, M., Kondratyev, A., Lee, S. A., Dimtchev, A., and Dritschilo, A. (1997). ATM gene product phosphorylates I kappa B-alpha. *Cancer Res.*, **57**, 24.

55. Cimprich, K. A., Shin, T. B., Keith, C. T., and Schreiber, S. L. (1996). cDNA cloning and gene mapping of a candidate human cell cycle checkpoint protein. *Proc. Natl Acad. Sci.*, **93**, 2850.

56. Bentley, N. J., Holtzman, D. A., Flaggs, G., Keegan, K. S., DeMaggio, A., Ford, J. C., Hoekstra, M., and Carr, A. M. (1996). The *Schizosaccharomyces pombe* rad3 checkpoint gene. *EMBO J.*, **15**, 6641.

57. Keegan, K. S., Holtzman, D. A., Plug, A. W., Christenson, E. R., Brainerd, E. E., Flaggs, G., Bentley, N. J., Taylor, E. M., Meyn, M. S., Moss, S. B., Carr, A. M., Ashley, T., and Hoekstra, M. F. (1996). The Atr and Atm protein kinases associate with different sites along meiotically pairing chromosomes. *Genes Dev.*, **10**, 2423.

58. Tibbetts, R., Brumbaugh, K., Williams, J., Sarkaria, J., Cliby, W., Shieh, S., Taya, Y., Prives, C., and Abraham, R. (1999). A role for ATR in the DNA damage-induced phosphorylation of p53. *Genes Dev.*, **13**, 152.

59. Longhese, M. P., Foiani, M., Muzi-Falconi, M., Lucchini, G., and Plevani, P. (1998). DNA damage checkpoint in budding yeast. *EMBO J.*, **17**, 5525.

60. Carr, A. (1997). Control of cell cycle arrest by the Mec1sc/Rad3sp DNA structure checkpoint pathway. *Curr. Opin. Genet. Dev.*, **7**, 93.

61. Cliby, W. A., Roberts, C. J., Cimprich, K. A., Stringer, C. M., Lamb, J. R., Schreiber, S. L., and Friend, S. H. (1998). Overexpression of a kinase-inactive ATR protein causes sensitivity to DNA-damaging agents and defects in cell cycle checkpoints. *EMBO J.*, **17**, 159.

62. Wright, J. A., Keegan, K. S., Herendeen, D. R., Bentley, N. J., Carr, A. M., Hoekstra, M. F., and Concannon, P. (1998). Protein kinase mutants of human ATR increase sensitivity to UV and ionizing radiation and abrogate cell cycle checkpoint control. *Proc. Natl Acad. Sci. USA*, **95**, 7445.

63. McMahon, S., Van Buskirk, H., Dugan, K., Copeland, T., and Cole, M. (1998). The novel

ATM-related protein TRRAP is an essential cofactor for the c-Myc and E2F oncoproteins. *Cell*, **94**, 363.

64. Sakamuro, D. and Prendergast, G. (1999). New Myc-interacting proteins: a second Myc network emerges. *Oncogene*, **18**, 2942.

65. Kunz, J. and Hall, M. (1993). Cyclosporin A, FK506 and rapamycin: more than just immunosuppression. *Trends Biochem. Sci.*, **18**, 334.

66. Brown, W., DeWald, D., Emr, S., Plutner, H., and Balch, W. (1995). Role for phosphatidyl-inositol 3-kinase in the sorting and transport of newly synthesized lysosomal enzymes in mammalian cells. *J. Cell Biol.*, **130**, 781.

67. Sabatini, D., Erdjument-Bromage, H., Lui, M., Tempst, P., and Snyder, S. (1994). RAFT1: a mammalian protein that binds to FKBP12 in a rapamycin-dependent fashion and is homologous to yeast TORs. *Cell*, **78**, 35.

68. Sabers, C. J., Martin, M. M., Brunn, G. J., Williams, J. M., Dumont, F. J., Wiederrecht, G., and Abraham, R. T. (1995). Isolation of a protein target of the FKBP12-rapamycin complex in mammalian cells. *J. Biol. Chem.*, **270**, 815.

69. Dennis, P., Fumagalli, S., and Thomas, G. (1999). Target of rapamycin (TOR): balancing the opposing forces of protein synthesis and degradation. *Curr. Opin. Genet. Dev.*, **9**, 49.

70. Moll, U., Lau, R., Sypes, M., Gupta, M., and Anderson, C. (1999). DNA-PK, the DNA-activated protein kinase, is differentially expressed in normal and malignant human tissues. *Oncogene*, **18**, 3114.

71. Roth, D., Menetski, J., Nakajima, P., Bosma, M., and Gellert, M. (1992). V(D)J recombination: broken DNA molecules with covalently sealed (hairpin) coding ends in scid mouse thymocytes. *Cell*, **70**, 983.

72. Danska, J., Holland, D., Mariathasan, S., Williams, K., and Guidos, C. (1996). Biochemical and genetic defects in the DNA-dependent protein kinase in murine SCID lymphocytes. *Mol. Cell. Biol.*, **16**, 5507.

73. Jhappan, C., Morse, H., Fleischmann, R., Gottesman, M., and Merlino, G. (1997). DNA PKcs: A T cell tumour suppressor encoded at the mouse SCID locus. *Nature Genet.*, **17**, 483.

74. Gao, Y., Chaudhuri, J., Zhu, C., Davidson, L., Weaver, D., and Alt, F. (1998). A targeted DNA-PKcs-null mutation reveals DNA-PK-independent functions for Ku in V(D)J recombination. *Immunity*, **9**, 367.

75. Taccioli, G., Amatucci, A., Beamish, H., Gell, D., Xiang, X., Arzayus, M., Priestley, A., Jackson, S., Rothstein, A., Jeggo, P., and Herrera, V. (1998). Targeted disruption of the catalytic subunit of the DNA-PK gene in mice confers severe combined immunodeficiency and radiosensitivity. *Immunity*, **9**, 355.

76. Vogel, H., Lim, D., Karsenty, G., Finegold, M., and Hasty, P. (1999). Deletion of Ku86 causes early onset of senescence in mice. *Proc. Natl Acad. Sci. USA*, **96**, 10770.

77. Li, G., Ouyang, H., Li, X., Nagasawa, H., Little, J., Chen, D., Ling, C., Fuks, Z., and Cordon-Cardo, C. (1998). Ku70: a candidate tumor suppressor gene for murine T cell lymphoma. *Mol. Cell*, **2**, 1.

78. Rolink, A., Melchers, F., and Andersson, J. (1996). The SCID but not the RAG-2 gene product is required for S mu-S epsilon heavy chain class switching. *Immunity*, **5**, 319.

79. Manis, J., Gu, Y., Lansford, R., Sonoda, E., Ferrini, R., Davidson, L., Rajewsky, K., and Alt, F. (1998). Ku70 is required for late B cell development and immunoglobulin heavy chain class switching. *J. Exp. Med.*, **187**, 2081.

80. Casellas, R., Nussenzweig, A., Wuerffel, R., Pelanda, R., Reichlin, A., Suh, H., Qin, X., Besmer, E., Kenter, A., Rajewsky, K., and Nussenzweig, M. (1998). Ku80 is required for immunoglobulin isotype switching. *EMBO J.*, **17**, 2404.

81. Fulop, G. and Phillips, R. (1990). The SCID mutation in mice causes a general defect in DNA repair. *Nature*, **347**, 479.

82. Biedermann, K., Sun, J., Giaccia, A., Tosto, L., and Brown, J. (1991). SCID mutation in mice confers hypersensitivity to ionizing radiation and a deficiency in DNA double-strand break repair. *Proc. Natl Acad. Sci. USA*, **88**, 1394.

83. Hendrickson, E., Qin, X., Bump, E., Schatz, D., Oettinger, M., and Weaver, D. (1991). A link between double-strand break related repair and V(D)J recombination in the SCID mutation. *Proc. Natl Acad. Sci. USA*, **88**, 4061.

84. Vorechovsky, I., Luo, L., Dyer, M., Catovsky, D., Amlot, P., Yaxley, J., Foroni, L., Hammarstrom, L., Webster, A., and Yuille, M. (1997). Clustering of missense mutations in the ataxia-telangiectasia gene in a sporadic T-cell leukaemia. *Nature Genet.*, **17**, 96.

85. Xu, Y., Ashley, T., Brainerd, E., Bronson, R., Meyn, M., and Baltimore, D. (1996). Targeted disruption of ATM leads to growth retardation, chromosomal fragmentation during meiosis, immune defects, and thymic lymphoma. *Genes Dev.*, **10**, 2411.

86. Barlow, C., Hirotsune, S., Paylor, R., Liyanage, M., Eckhaus, M., Collins, F., Shiloh, Y., Crawley, J., Ried, T., Tagle, D., and Wynshaw-Boris, A. (1996). Atm-deficient mice: a paradigm of ataxia telangiectasia. *Cell*, **86**, 159.

87. Liao, M. and Van Dyke, T. (1999). Critical role for Atm in suppressing V(D)J recombination-driven thymic lymphoma. *Genes Dev.*, **13**, 1246.

88. Moens, P., Tarsounas, M., Morita, T., Habu, T., Rottinghaus, S., Freire, R., Jackson, S., Barlow, C., and Wynshaw-Boris, A. (1999). The association of ATR protein with mouse meiotic chromosome cores. *Chromosoma*, **108**, 95.

89. Plug, A. W., Peters, A. H., Xu, Y., Keegan, K. S., Hoekstra, M. F., Baltimore, D., de Boer, P., and Ashley, T. (1997). ATM and RPA in meiotic chromosome synapsis and recombination. *Nature Genet.*, **17**, 457.

90. Dahlen, M., Olsson, T., Kanter-Smoler, G., Ramne, A., and Sunnerhagen, P. (1998). Regulation of telomere length by checkpoint genes in *Schizosaccharomyces pombe*. *Mol. Biol. Cell.*, **9**, 611.

91. Metcalfe, J. A., Parkhill, J., Campbell, L., Stacey, M., Biggs, P., Byrd, P. J., and Taylor, A. M. (1996). Accelerated telomere shortening in ataxia telangiectasia. *Nature Genet.*, **13**, 350.

92. Pandita, T. K., Pathak, S., and Geard, C. R. (1995). Chromosome end associations, telomeres and telomerase activity in ataxia telangiectasia cells. *Cytogenet. Cell. Genet.*, **71**, 86.

93. Vaziri, H., West, M., Allsopp, R., Davison, T., Wu, Y., Arrowsmith, C., Poirier, G., and Benchimol, S. (1997). A TM-dependent telomere loss in aging human diploid fibroblasts and DNA damage lead to the post-translational activation of p53 protein involving poly (ADP-ribose) polymerase. *EMBO J.*, **16**, 6018.

94. Greenwell, P. W., Kronmal, S. L., Porter, S. E., Gassenhuber, J., Obermaier, B., and Petes, T. D. (1995). TEL1, a gene involved in controlling telomere length in *S. cerevisiae*, is homologous to the human ataxia telangiectasia gene. *Cell*, **82**, 823.

95. Morrow, D. M., Tagle, D. A., Shiloh, Y., Collins, F. S., and Hieter, P. (1995). TEL1, an *S. cerevisiae* homolog of the human gene mutated in ataxia telangiectasia, is functionally related to the yeast checkpoint gene MEC1. *Cell*, **82**, 831.

96. Naito, T., Matsuura, A., and Ishikawa, F. (1998). Circular chromosome formation in a fission yeast mutant defective in two ATM homologues. *Nature Genet.*, **20**, 203.

97. Shigeta, T., Takagi, M., Delia, D., Chessa, L., Iwata, S., Kanke, Y., Asada, M., Eguchi, M., and Mizutani, S. (1999). Defective control of apoptosis and mitotic spindle checkpoint in heterozygous carriers of ATM mutations. *Cancer Res.*, **59**, 2602.

98. Appleby, J., Barber, J., Levine, E., Varley, J., Taylor, A., Stankovic, T., Heighway, J., Warren, C., and Scott, D. (1997). Absence of mutations in the ATM gene in breast cancer patients with severe responses to radiotherapy. *Br. J. Cancer*, **76**, 1546.

99. Clarke, R., Goozee, G., Birrell, G., Fang, Z., Hasnain, H., Lavin, M., and Kearsley, J. (1998). Absence of ATM truncations in patients with severe acute radiation reactions. *Int. J. Radiat. Oncol. Biol. Phys.*, **41**, 1021.

100. Hall, E., Schiff, P., Hanks, G., Brenner, D., Russo, J., Chen, J., Sawant, S., and Pandita, T. (1998). A preliminary report: frequency of A-T heterozygotes among prostate cancer patients with severe late responses to radiation therapy. *Cancer J. Sci. Am.*, **4**, 385.

101. Oppitz, U., Bernthaler, U., Schindler, D., Sobeck, A., Hoehn, N., Platzer, M., Rosenthal, A., and Flentje, M. (1999). Sequence analysis of the ATM gene in 20 patients with RTOG grade 3 or 4 acute and/or late tissue radiation side effects. *Int. J. Radiat. Oncol. Biol. Phys.*, **44**, 981.

102. Proud, C. G. (1996). p70 S6 kinase: an enigma with variations. *Trends Biochem. Sci.*, **21**, 181.

103. Shayeghi, M., Seal, S., Regan, J., Collins, N., Barfoot, R., Rahman, N., Ashton, A. M. M., Wooster, R., Owen, R., Bliss, J., Stratton, M., and Yarnold, J. (1998). Heterozygosity for mutations in the ataxia telangiectasia gene is not a major cause of radiotherapy complications in breast cancer patients. *Br. J. Cancer*, **78**, 922.

104. Sirzen, F., Nilsson, A., Zhivotovsky, B., and Lewensohn, R. (1999). DNA-dependent protein kinase content and activity in lung carcinoma cell lines: correlation with intrinsic radiosensitivity. *Eur. J. Cancer*, **35**, 111.

105. Polischouk, A., Cedervall, B., Ljungquist, S., Flygare, J., Hellgren, D., Grenman, R., and Levensohn, R. (1999). DNA double-strand break repair, DNA-PK, and DNA ligases in two human squamous carcinoma cell lines with different radiosensitivity. *Int. J. Radiat. Oncol. Biol. Phys.*, **43**, 191.

106. Kasten, U., Plottner, N., Johansen, J., Overgaard, J., and Dikomey, E. (1999). Ku70/80 gene expression and DNA-dependent protein kinase (DNA-PK) activity do not correlate with double-strand break (dsb) repair capacity and cellular radiosensitivity in normal human fibroblasts. *Br. J. Cancer*, **79**, 1037.

107. Kasten, U., Borgmann, K., Burgmann, P., Li, G., and Dikomey, E. (1999). Overexpression of human Ku70/Ku80 in rat cells resulting in reduced DSB repair capacity with appropriate increase in cell radiosensitivity but with no effect on cell recovery. *Radiat. Res.*, **151**, 532.

108. Uhrhammer, N., Fritz, E., Boyden, L., and Meyn, M. (1999). Human fibroblasts transfected with an ATM antisense vector respond abnormally to ionizing radiation. *Int. J. Mol. Med.*, **4**, 43.

109. Kastan, M. (1999). Molecular determinants of sensitivity to antitumor agents. *Biochim. Biophys. Acta*, 1**424**, R37.

110. McBlane, J., van Gent, D., Ramsden, D., Romeo, C., Cuomo, C., Gellert, M., and Oettinger, M. (1995). Cleavage at a V(D)J recombination signal requires only RAG1 and RAG2 proteins and occurs in two steps. *Cell*, **83**, 387.

111. van Gent, D., Hiom, K., Paull, T., and Gellert, M. (1997). Stimulation of V(D)J cleavage by high mobility group proteins. *EMBO J.*, **16**, 2665.

Index

14-3-3 proteins 60, 181, 284, 365
4E-BP 29, 51, 367
Abl tyrosine kinase 122, 223, 226, 255, 292, 363, 365, 366
ACTH 266
actin binding protein 99
activin 119, 304, 307–309, 315, 317, 326, 330
acute phase reactive factor 209
AFX 23
Akt, see PKB
ANF, see atrial natriuretic factor
angiotensin 95
anisomycin 66, 79, 99
anoikis 25, 111
anthrax lethal factor 54, 55
AP-1 65, 71, 72, 74, 77, 127, 160, 319, 320
APC/C 287, 289, 292
apoptosis
apoptosis receptors 104
role of phosphatidylinositol 3' kinase 24–26
role of SAPKs 125–127
apoptosis-signalling kinase 81, 83, 84, 87, 88, 108, 117–119
Armadillo 132
ASK1, see apoptosis-signalling kinase
ataxia telangiectasia mutated 122, 123, 357, 358, 363–366, 369–371
ataxia telangiectasia related 358, 363, 366, 369–371
ATF 73
ATM, see ataxia telangiectasia mutated
ATR, see ataxia telangiectasia related
atrial natriuretic factor 127
Avian sarcoma virus 23
Axin 132, 181

Bad 24
Basket, see Drosophila SAPK/JNK
Bax 263
Bcl-2 24, 263, 331
Bcl-X_L 219
Bcr-Abl 224, 225, 231
betaglycan 310

BH domain 4,
BMK1, see ERK5
BMP, see bone morphogenic protein
bone morphogenic protein 119, 120, 131, 304–306, 310, 312, 314–318, 324–326, 330–332
BRCA1 365
bride of sevenless 159, 161
Bruton's tyrosine kinase 96, 102

C3G 62
CAAX domain 8, 57, 59, 113
CAK see Cdk7
calcineurin 59, 71, 127
cardiac hypertrophy 127, 128
caspases 111, 112
β-catenin 132, 133
CD44 204, 205, 218
CD69 205,
Cdc2 278–285, 291, 293
Cdc6 293
Cdc25 284, 285, 290–293
CDC28 278 -280, 284, 285
CDC37 61
Cdc42 26, 60, 90, 110, 121, 122, 125, 128
Cdk
Cdk1, see cdc2
Cdk2 277, 279–281, 287, 289–291, 293
Cdk3 279, 280
Cdk4 277, 279, 280, 282–284, 291, 293
Cdk5 279, 280
Cdk6 279, 280, 291
Cdk7 279, 280, 283, 289
Cdk8 280
nomenclature 280
substrates 292
C/EBP 30
Cerberus 306
CHOP 71
chronic myelogenous leukaemia 224, 225
cIAP 120
Cis 247–251, 253, 260–262, 264, 266

cis-platinum 122
CKI 286
CLN 281–283
CNTF 198, 200, 266
Coffin-Lowry syndrome 47
concanavalin A 203, 209
corkscrew 160, 164
CRD 58
CREB 48, 50, 71, 73,
CRIB domain 87, 92, 93, 100, 110, 121, 122
Crk 62, 103, 104, 111–113, 125
CRM1 290
CSAID, see SB203580
Csk 289, 290
Csw, see corkscrew
Cullin 256
cyclic AMP-dependent protein kinase 45, 61, 62
in Dictyostelium discoideum 169–171
cyclins 56, 218, 280
cyclin A 280–282, 289, 290
cyclin B 280, 281, 284
cyclin D 281–283, 286, 289, 291, 293
cyclin E 281, 288
nomenclature 280
cycloheximide 65
cytochrome P450 216, 217

DAF-16 23
Daxx 87, 108, 117, 118
Dbl 91, 255
decapentaplegic 130, 304, 306, 307, 313, 316, 318, 325, 333
delayed type hypersensitivity 214
diacylglycerol 1, 19, 96
DH domain 90
Dictyostelium discoideum signalling 169–174
DIF 170, 173, 174, 222
Dishevelled 97, 131, 132
DLK, see mixed lineage kinases
DNA damaging agents 122
DNA-dependent protein kinase 357–364, 366–371

docking sites 63, 64, 75–77
dorsal closure 130, 131, 135
daughter of sevenless 160, 163
double strand breaks 357, 358, 363, 370
DPC4, *see* Smad4
dpp, see decapentaplegic
drk 160, 162
Drosophila SAPK/JNK 130–133

E2F 293, 366
ECB 282
EGF, *see* epidermal growth factor
eIF2B 29, 32
eIF4-E 29, 32, 51, 367
eIF-4F 51
Elk 52, 64, 75, 77, 78
Elongin 256–261, 266
Endoglin 310, 339
endothelin 95, 128
EPAC 62
epidermal growth factor 21, 52, 66, 103, 104, 312, 315, 318
Epstein–Barr virus 105
ERK, *see* mitogen-activated protein kinase
ERK5 67, 68, 80
  ERK5 substrates 68, 72
erythropoetin 200, 203, 206, 215, 217, 219, 229, 231, 246, 250, 260, 261, 266
Evi-1 319, 320

F-box 257, 258, 288
FAK 25
Far1 286
farnesyl transferase 311
farnesylation 8
Fas 24, 104, 105, 108, 117, 118, 127
FAST-1 319, 321
FGF 208
FK506 59, 114, 310
FK1012 59, 114, 115
Fos 52, 229, 252, 292, 319
FRAP 358, 366
Frizzled 131
FLT3 receptor 252, 254
FYVE 27, 322

G-CSF 199, 202, 215, 247
GADD153 71
GADD genes 123

gag 23
gamma activated sequence 195, 197, 206, 219, 265
GCKR 101–104, 114–116
germinal centre kinase 42, 90, 92, 97, 100–104, 108, 110, 113–116, 124, 131, 134
Gli3 320
GLK 101, 103
GLUT4 28, 31
glycogen synthase 45
glycogen synthase kinase -3 30, 31, 45, 71, 72, 77, 78, 132, 171–174
GM-CSF 199–202, 215, 261, 263
Golgi 6, 290, 292
G protein coupled receptors 9, 11–13, 95–97
Grb-2 103, 104, 110, 124, 125, 255
growth hormone 216, 217, 266, 267
GS domain 307, 309–311, 317
GSK-3, *see* glycogen synthase kinase -3
GTPase 4; *see also* Ras
GTPase activating factor 10
guanine nucleotide exchange factor 10, 90

heat shock proteins 60, 61, 68, 126
HECT domain 259
hemipterous, see MKK7
hepatocyte growth factor 315
heterotrimeric G proteins 11–13, 95–97
HGK 104
HIF-1 259
HNPCC 337
hopscotch 220, 222, 223, 227, 230
HPK1 97, 101, 102–104, 110, 113–116, 124, 125, 313
HTLV-1 223

ICAM 200, 207
IGF, *see* insulin-like growth factor
ILK, *see* integrin-linked kinase
INK, *see* CKI
insulin
  in metabolism 29, 30
  activation of Ras 57
insulin-like growth factor 24, 30, 264
integrin-linked kinase 17
interferon 195, 199, 202, 206–208, 211, 214, 231, 250, 263, 264, 265, 318
interleukin-1 104, 108, 210

interleukin-2 199, 203–204, 206, 209, 210, 212, 215, 218, 223, 224, 227, 247, 250, 265
interleukin-3 199, 202, 204, 212, 215, 246, 247, 250
interleukin-4 203, 212, 213, 227, 231, 263
interleukin-5 199, 202, 206, 215
interleukin-6 199, 202, 209, 210, 247, 251, 252, 260
interleukin-7 199, 203, 206, 209, 215
interleukin-10 199, 206, 208–211, 265
interleukin-11 200, 266
interleukin-12 211, 212, 214, 231
interleukin-13 213
interleukin-15 199, 203, 215
IRAK 108
IRF1 201, 206–208, 252, 264
ischaemia 67
ISGF 195, 196, 208, 265

JAB, *see* suppressor of cytokine signalling
Jaks
  gene targeting of 198
  in *Caenorhabditis elegans* 221
  in *Dictyostelium discoideum* 221, 222
  in *Drosophila* 220, 211
  Jak1 195–200, 223, 265
  Jak2 195, 200–203, 225, 251–253, 264, 266
  Jak3 199, 202, 223, 227–229, 265
  Tel-Jak2 225–227, 229, 232
  Tel-Jak3 225
  Tyk2 196, 206, 265
JIP1 97
JNK, *see* Jun N-terminal kinase
Jun N-terminal kinase 44, 66,
  agonists of 66
  binding to Jun 75–77
  nomenclature 44
  phosphorylation of NFAT 72
  phosphorylation of Jun 72–74

kainate seizure 127
Kin28 280
Kit receptor 251–253
KHS1, *see* GCKR
Kinase suppressor of ras 160, 162, 166
KIP, *see* CKI

K-Rev1, *see* Rap1
Ksr, *see* Kinase suppressor of ras
Ku 359, 362–364, 368, 371

LAP 305
Leptin 266
leukaemia inhibitory factor 198, 200,
  252, 253, 255, 260, 264, 266
lipopolysaccharide 199, 203, 208,
  211
LMP1 105, 107
LY294002 7–9, 18
lysophosphatidic acid 21, 99

MAD, *see* Smad
major urinary protein 216, 217
mammary gland factor 215
MAPK, *see* mitogen -activated protein
  kinase
MAPKAP kinases 44–46, 68, 69
Max 72
Mec1 366
MEDEA 313, 317
MEF2C 74, 78
MEK
  binding to MAPK 64
  binding to MP1 63
  in *Caenorhabditis elegans* 166, 167
  in *Drosophila* 160, 163
  MEK5 80
  nomenclature 53
  phosphorylation by Raf 55, 64
  proteins 41
  structural features 53, 54
MEKK
  downstream of Rho GTPases 110
  binding to GCK 114–116
  binding to TRAFs 114–116
  MEKK1 activation of SAPK/JNK
    85, 86
  MEKK1 -/- ES cells 112
  MEKK2/3/4 (MTK1) 86, 87, 94
  nomenclature 84
  phosphorylation of MEK 55
  structure 81
MEK partner 1 42, 53, 63,
MH domain 314, 317, 320, 321, 330
Mik1 285
misshapen 92, 104, 116, 131
mitogen-activated protein kinase
  nuclear translocation 63, 64
  in *Dictyostelium discoideum* 172
  in *Drosophila* 161

in yeast 176–180
  nomenclature 44
  regulation by MEK phosphorylation
    52–55
  substrates 44–52
mixed lineage kinases
  binding to other proteins 124, 125,
    134
  dimerization 125
  nomenclature 84, 89
  scaffold association 97, 98
  structure 81
  substrates 89
MKK3 80, 81
MKK4, *see* SEK1
MKK6 80, 81
MKK7
  cloning 78
  differential splice products 79
  hemipterous 79
  phosphorylation by MEKK1 86
  regulatory phosphorylation sites
    81
  scaffold binding 97, 98, 100
MLK, *see* mixed lineage kinases
MNK 51, 71
Mom-4 133
Mos 56
MPF 56, 278–280
MSG-1 320
MSK 50, 70, 71
MTK1, *see* MEKK
mTOR 358, 366, 367
myristoylation 8
Myt1 285, 291

natural killer cells 199, 204, 211,
  218
NckIK 101, 102, 104, 113–115, 131
NEDD4 258, 259
Nemo-like kinase 133
nerve growth factor 24, 56, 62, 126,
  200, 305
NFAT 71, 72, 127, 128
NGF, *see* nerve growth factor
Nim1 285
nitric oxide 207, 210
notch
  in *Caenorhabditis elegans* 166, 168

ob/ob 266
oncostatin M 198
osmotic shock 66, 68, 179

p21 WAF1 208, 358
p38 MAPK
  agonists of 66
  in cell cycle regulation 128
  isoforms 44, 66–68
  phosphorylation of ATFs 73, 74
  phosphorylation of PRAK 69, 70
  substrates 68–75
p53 123, 292, 358, 361, 363–366
p70 S6 kinase
  activation by phosphatidylinositol
    3′ kinase 18
  regulation by mTOR 367
Pangolin 132
PAK 60, 92–94, 100, 134
PAK interacting exchange factors 93,
  104
PD98059 50, 55, 56, 135
PDK, *see* phosphatidylinositol 3,4,5,
  trisphosphate-dependent kinase
PDZ domain 132
PEST domain 101, 114
PH domain 3, 16, 87, 90
Pho 85 280
phorbol myristate acetate 199, 203,
  204, 209, 218, 315
phosphatase 2A 311
phosphatidylinositol 3′ kinase
  activation of p70 S6 kinase 18–20
  activation of PDKs 16–18
  in apoptosis 24–26
  in *Caenorhabditis elegans* 22, 23
  in *Drosophila* 22, 160
  inhibitors of 7–9; *see also*
    wortmannin, LY294002
  lipid substrates 1–3
  serine kinase activity of 6, 7
  structural features of 3–5
phosphatidylinositol 3,4,5,
  trisphosphate-dependent kinase
  PDK1 16, 48, 49
  PDK2 17
  phosphorylation of p70 S6 kinase
    18
phosphatidylinositol 4′ kinase
  3
phospholipase C 96, 255
PIX, *see* PAK interacting exchange
  factors
PKA, *see* cyclic AMP-dependent
  protein kinase
PKB, *see* protein kinase B
platelet-derived growth factor
  activation of phosphatidylinositol
    3′ kinase 10

activation of PDK1   16, 17
  as a survival signal   24
  in actin rearrangement   26
  in mitogenesis   21
POLO   284
POMC   266
POP-1   133, 134
POSH   93, 94
PRAK   69, 70
protein kinase B
  activation of   15–17
  structure of   15, 16
  phosphorylation of GSK-3   30
protein kinase C   1, 18–20, 28
protein phosphatase 1   45
PSTAIRE   290
PTB domain   160
PTEN
  mutations in cancer   14
  3′ lipid phosphatase activity   14
Pyk2   96, 104

Rab, see vesicle fusion
Rac
in regulation of cytoskeleton   26, 90,
    91
  in heterotrimeric G protein
      signalling   96
  in HPK regulation   104, 124, 125
RAD   52   357, 366
Raf
  activation by Ras   56–58
  B-Raf   61
  coumermycin-Raf   59
  in Caenorhabditis elegans   166–168
  in Drosophila   160–162
  FKBP-Raf   59
  phosphorylation of MEK   55
  Raf-CAAX   59
  Ras binding domain   58
RANK   104, 108
Rap1   61, 62
rapamycin   312, 366
Ras
  activation of phosphatidylinositol
      3′ kinase   10–12
  activation of Raf   56–58
  in Caenorhabditis elegans   166–168
  in Drosophila   160–164
RBD, see Ras binding domain
Receptor interacting protein   105,
    107, 108, 117
Restriction-enzyme mediated
    integration   174

Retinoblastoma   293, 358
RGS protein   95
Rho GTPases   26, 60, 90, 93, 94, 110
RING finger   107, 108, 114–116
RIP, see receptor interacting protein
Rsk, see MAPKAP kinase
Rum1   286

SAGA   366
SAPK, see stress-activated protein
    kinase
SARA   322, 323, 339
Saxophone   309
SB203580   67, 135
scaffolding proteins   42, 43, 53, 63, 97,
    98
SCF   257–259, 288, 289
SCID   205, 227, 228, 359, 364, 368
SEK1
  activation of SAPK/JNK   78–80
  differential splicing   79
  hemipterous   79, 130, 131
  knockout mice   130
  null ES cells   79, 130
  phosphorylation by MEKK1   86
  regulatory phosphorylation sites
      81
  SEK1 as a scaffold   64, 100
serum response element   52, 74
serum response factor   52, 74
sevenless   57, 61, 158–164
SH2 domain   4, 103, 160, 163, 195, 230,
    246, 252, 254, 256
SH2 domain-containing inositol
    5′ phosphatase   13
SH3 domain   4, 94, 103, 104, 110, 124,
    130, 163, 255, 365
SHIP, see SH2 domain-containing
    inositol   5′ phosphatase
SIR mutants   363
Smads
  Smad1   315, 317, 318, 321
  Smad2   314, 315, 316, 318–322, 329,
      333, 338
  Smad3   315, 316, 318–320, 322, 329,
      338
  Smad4   314, 316, 317, 319–321, 329,
      330, 333, 338
  Smad5   314, 315, 329
  Smad6   317, 318, 330
  Smad7   317, 318, 330
  Smad 8   330
  Smad 10   318, 330
SNAREs, see vesicle fusion

SOCS, see suppressor of cytokine
    signalling
son of sevenless   160, 161
sonic hedgehog   334, 335
sorbitol, see osmotic shock
Spemann organizer   306, 325, 330
spindle pole bodies   291
Src   2, 60, 159, 223, 292
SRE, see serum response element
SRF, see serum response factor
SSI, see suppressor of cytokine
    signalling
START   278, 286
Stats
  domain structure   195
  gene targeting of   198
  Stat1   195, 206–208, 210, 217, 224,
      225, 250, 264
  Stat2   195, 208–209
  Stat3   209–211, 217, 225, 247
  Stat4   211, 212, 214
  Stat5   203, 210, 215, 223, 224, 225,
      230, 250, 260, 261, 266
  Stat6   212, 213, 214, 223, 225, 227
Ste5   43, 97
Ste20   43, 90, 176–178
Steel   254
stem cell factor   202
STRAP   312
stress-activated protein kinase
  agonists of   66
  binding to Jun   75–77
  differential splicing   44, 75, 77
  Drosophila SAPK/JNK   130–133
  in T cell function   129–131
  nomenclature   44
  phosphorylation by SEK1/MKK4
      78–80
  phosphorylation of NFAT   72
  phosphorylation of Jun   72–74
Suc1, see Csk
suppressor of cytokine signalling
    202, 229, 246–268
  expression   249, 250
  genomic structure   248
  homologues   249
  knockout mouse   262, 263
  nomenclature   248
Swe1   285
Swi4/6   283

TAB1   120, 121, 133, 310, 312, 313, 331
TAK1   87, 119–121, 133, 310, 312, 313,
    331

TAO kinase 81–84, 89, 90
TEC 255
telomere 363
TFIIH 283
TGFβ, see transforming growth factor β
thioredoxin 87, 118
thrombopoetin 202, 217
TNF, see tumour necrosis factor
torso 57, 162
Tpl-2 81, 83, 84, 88
TRADD 105, 120
TRAFs 88, 102, 104–108, 114–119, 131, 134
transforming growth factor β 87, 119, 130, 131, 284, 303–312, 315, 316–318, 322–326, 330–339
transforming growth factor β receptor
  type I 307, 309, 311, 312, 316, 322, 337
  type II 307, 308, 311, 312, 322, 337
TRAP-1 312

TRAPP 358, 366
β-TrCP 249
TRIP1 310, 311
Trk 62, 226
tumour necrosis factor 66, 67, 79, 80, 88, 97, 99, 101, 104–106, 116–118, 208, 210
tumour necrosis factor receptors 104, 105
TUNEL stain 219

Ubiquitin ligase 256–259, 287

Vav 21, 255, 256, 263
vesicle fusion
  Rab 27, 28
  SNAREs 27
VCB 256, 257
V(D)J recombination 360, 361, 367–369

Vesicular stomatitis virus 207, 264
von Hippel Lindau 256, 257, 259, 260
Vps34 6, 27
vulval precursor cells 164–166

WD40 domain 249
Wee1 285
whey acidic protein 215
Wingless 132, 221, 306
Wnt 131–133, 306, 336
wortmannin 7–9, 18, 19, 29
WW domain 259

XIAP 120, 312, 331
XRCC4 358, 361, 362

Zeste-white3 132
zone of polarising activity 249